Lecture Notes in Computer Science

Lecture Notes in Artificial Intelligence 16197

Founding Editor

Jörg Siekmann

The series Lecture Notes in Artificial Intelligence (LNAI) was established in 1988 as a topical subseries of LNCS devoted to artificial intelligence.

The series publishes state-of-the-art research results at a high level. As with the LNCS mother series, the mission of the series is to serve the international R & D community by providing an invaluable service, mainly focused on the publication of conference and workshop proceedings and postproceedings.

Masatoshi Yoshikawa · Xiaofeng Meng ·
Yang Cao · Chuan Xiao · Weitong Chen ·
Yanda Wang
Editors

Advanced Data Mining and Applications

21st International Conference, ADMA 2025
Kyoto, Japan, October 22–24, 2025
Proceedings, Part I

Springer

Editors
Masatoshi Yoshikawa
Osaka Seikei University
Osaka, Japan

Yang Cao
Institute of Science Tokyo
Tokyo, Japan

Weitong Chen
University of Adelaide
Adelaide, SA, Australia

Xiaofeng Meng
Renmin University of China
Beijing, China

Chuan Xiao
Osaka University
Osaka, Japan

Yanda Wang
Nanjing Normal University
Nanjing, China

ISSN 0302-9743 ISSN 1611-3349 (electronic)
Lecture Notes in Artificial Intelligence
ISBN 978-981-95-3452-4 ISBN 978-981-95-3453-1 (eBook)
https://doi.org/10.1007/978-981-95-3453-1

LNCS Sublibrary: SL7 – Artificial Intelligence

This Springer imprint is published by the registered company Springer Nature Singapore Pte Ltd.
The registered company address is: 152 Beach Road, #21-01/04 Gateway East, Singapore 189721, Singapore

Preface

The 21st International Conference on Advanced Data Mining and Applications (ADMA 2025) was held in Kyoto, Japan, during October 22-24, 2025. Researchers and practitioners from around the world came together at this leading international forum to share innovative ideas, original research findings, case study results, and experienced insights into advanced data mining and its applications. With the ever-growing importance of appropriate methods in these data-rich times, ADMA has become a flagship conference in this field.

ADMA 2025 received a total of 327 valid submissions to the following tracks: Research Paper, Industry Paper, Special Sessions, Posters/Encore Presentations. After a rigorous double-blind review process by 259 reviewers, 147 regular papers were accepted to be published in the proceedings, corresponding to an acceptance rate of 44.9%. 3 papers were accepted as posters.

The Program Committee (PC), composed of international experts in relevant fields, did a thorough and professional job of reviewing the papers submitted to ADMA 2025, and each paper was double-blindly reviewed by at least three PC members. With the growing importance of data in this digital age, papers accepted by ADMA 2025 covered a wide range of research topics in the field of data mining, including machine learning, graph mining, time series analysis, information retrieval, and security and trust.

We thank the PC and Senior PC members for completing the review process and providing valuable comments within tight schedules. The high-quality program would not have been possible without the expertise and dedication of our PC and Senior members. Moreover, we would like to take this valuable opportunity to thank all authors who submitted technical papers and contributed to the tradition of excellence at ADMA. We are confident that the papers presented in these proceedings will be both exciting and beneficial for colleagues seeking to advance their research. We extend our gratitude to Microsoft for providing the complimentary CMT system for conference organization and to Springer for their enduring support. Additionally, we appreciate the generous contributions from Cygames Inc., Corporation, and Recruit Co., Ltd.

We are grateful for the guidance of the steering committee members, Xue Li, Michael Sheng, Osmar R. Zaiane, Chenqi Zhang, Jianlin Li, Guodong Long, and Weitong Chen, whose leadership and support helped the conference to run smoothly. We also would like to acknowledge the support of the other members of the organizing committee. All of them helped to make ADMA 2025 a success. Finally, we would like to thank all the session chairs and volunteers who contributed to the conference organization.

October 2025

Masatoshi Yoshikawa
Xiaofeng Meng
Yang Cao
Chuan Xiao
Weitong Chen
Yanda Wang

Organization

General Chairs

Masatoshi Yoshikawa	Osaka Seikei University, Japan
Xiaofeng Meng	Renmin University of China, China

Program Chairs

Yang Cao	Institute of Science Tokyo, Japan
Chuan Xiao	Osaka University, Japan

Industry Track Chairs

Yuya Sasaki	Osaka University, Japan
Masafumi Oyamada	NEC Corporation, Japan

Special Session Track Chairs

Yasuhiko Morimoto	Hiroshima University, Japan
Ye Zhu	Deakin University, Australia
Chen Li	Osaka University, Japan

Short Paper Track Chairs

Daichi Amagata	Osaka University, Japan
Kento Sugiura	Nagoya University, Japan

Local Chairs

Qiang Ma	Kyoto Institute of Technology, Japan
Shinsuke Nakajima	Kyoto Sangyo University, Japan
Kenta Oku	Ryukoku University, Japan

Finance Chair

Hisashi Miyamori | Kyoto Sangyo University, Japan

Sponsorship Chairs

Masato Oguchi | Ochanomizu University, Japan
Kazutaka Sakurai | Recruit Co., Ltd., Japan

Panel Discussion Chair

Xuyun Zhang | Macquarie University, Australia

Proceeding Chairs

Weitong Chen | University of Adelaide, Australia
Yanda Wang | Nanjing Normal University, China

Tutorial Chair

Makoto Onizuka | Osaka University, Japan

Publicity Chair

Yuanyuan Wang | Yamaguchi University, Japan

Web Chair

Shuyuan Zheng | Osaka University, Japan

Registration Chair

Pengpeng Qiao | Institute of Science Tokyo, Japan

Steering Committee

Xue Li	University of Queensland, Australia
Michael Sheng	Macquarie University, Australia
Osmar R. Zaiane	University of Alberta, Canada
Chengqi Zhang	University of Technology Sydney, Australia
Jianxin Li	Deakin University, Australia
Guodong Long	University of Technology Sydney, Australia
Weitong Chen	University of Adelaide, Australia

Program Committee

Adita Kulkarni	SUNY Brockport, USA
Adittya Soukarjya Saha	North Carolina State University, USA
Aiting Yao	Pengcheng Laboraroty, China
Akiyoshi Matono	AIST, Japan
Alex Delis	National and Kapodistrian University of Athens, Greece
Alexander Zhou	Hong Kong Polytechnic University, China
An Liu	Soochow University, China
Anan Du	Nanjing Vocational University of Industry Technology, China
Baobao Song	University of Technology Sydney, Australia
Baoliang Chen	South China Normal University, China
Binbin Zhou	Hangzhou City University, China
Chaoran Huang	University of New South Wales, Australia
Chen Wang	Shaoxing University, China
Chen Li	University of Osaka, Japan
Chen Li	University of Technology Sydney, Australia
Chen Li	Northeastern University, China
Chen Shen	Megagon Labs, USA
Chen Chen	University of Wollongong, Australia
Cheng Cheng	Liaoning Normal Univeristy, China
Chengcheng Yang	East China Normal University, China
Chengcheng Mai	Nanjing Normal University, China
Chenhao Zhang	University of Queensland, Australia
Chenxu Wang	Xi'an Jiaotong University, China
Chuan Xiao	Osaka University, Japan
Chuan Ma	Chongqing University, China
Daichi Amagata	University of Osaka, Japan
Dan Li	Sun Yat-sen University, China

Dan Zhang	Deakin University, Australia
Denis Ponomaryov	Ershov Institute of Informatics Systems, Russia
Derong Shen	Northeastern University, China
Dhaval Patel	IBM Research, USA
Di Wu	Nanjing University of Science and Technology, China
Dong Li	Liaoning University, China
Dong Wen	University of New South Wales, Australia
Dongjin Yu	Hangzhou Dianzi University, China
Dunlei Rong	Harbin Institude of Technology, China
Eiji Uchino	Yamaguchi University, Japan
Essam Rashed	University of Hyogo, Japan
Faming Li	Northeastern University, China
Bibo Fu	Ocean University of China, China
Peiliang Gong	Nanjing University of Aeronautics and Astronautics, China
Guangyou Zhou	Central China Normal University, China
GuanNan Dong	Nanjing Tech University, China
Guixian Zhang	China University of Mining and Technology, China
Guojing Zhou	South China Normal University, China
Haïfa Nakouri	ISG Tunis, Tunisia
Hailong Liu	Northwestern Polytechnical University, China
Hairong Chen	Beijing Jiaotong University, China
Haiyang Xia	Australian National University, Australia
Hanbing Zhang	Fudan University, China
Hang Yu	Shanghai University, China
Hang Zhang	Nanjing University, China
Hang Zhang	Changchun University of Science and Technology, China
Hang Yuan	South China Normal University, China
Hao Long	Huazhong Agricultural University, China
Hao Du	Hokkaido University, Japan
Hao Tian	Nanjing University, China
Hao Xin	Hong Kong University of Science and Technology, China
Haobing Liu	Ocean University of China, China
Haochen Zhang	NEC Corporation, Japan
Haochen Yuan	Harbin Institute of Technology, China
Haoyang Li	Hong Kong Polytechnic University, China
Heng Liang	University of Hong Kong, China
Hieu Hanh Le	Ochanomizu University, Japan

Hiroaki Ohshima	University of Hyogo, Japan
Hiroaki Shiokawa	University of Tsukuba, Japan
Hongsheng Hu	University of Newcastle, Australia
Hongzhi Yin	University of Queensland, Australia
Hu Tianze	Beijing Institute of Technology, China
Huadong Mo	University of New South Wales, Australia
Huaming Chen	University of Sydney, Australia
Huidong Tang	Hiroshima University, Japan
Ian Holmes	North Carolina State University, USA
Indika Weerasingha Dewage	Tilburg University, Netherlands
Jiajun Li	Beijing Institute of Technology, China
Jiali Mao	East China Normal University, China
Jian Li	Northeastern University, China
Jianqiu Xu	Nanjing University of Aeronautics and Astronautics, China
Jianye Yang	Guangzhou University, China
Jiaxiang Wang	Anhui University, China
Jie Zhang	Xi'an University of Technology, China
Jieyu Zhan	South China Normal University, China
Jing Zhao	Jianghan University, China
Jinwei Zhu	Huawei Technologies Co., Ltd., China
Jun Zhou	RIKEN, Japan
Junchang Xin	Northeastern University, China
Junhua Zhang	University of New South Wales, Australia
Junjie Yao	East China Normal University, China
Junya Arai	Nippon Telegraph and Telephone Corporation, Japan
Kai Peng	Huaqiao University, China
Kai Zhong	Beijing Institution of Technology, China
Kangzheng Liu	Huazhong University of Science and Technology, China
Kento Sugiura	Nagoya University, Japan
Kesheng Wu	Lawrence Berkeley National Laboratory, USA
Lei Duan	Sichuan University, China
Lei Li	Hong Kong University of Science and Technology (Guangzhou), China
Leixia Wang	Northeastern University, China
Leong Hou U	University of Macau, China
Li Li	Southwest University, China
Liang Zhang	Lanzhou University, China
Liangwei Zheng	University of Adelaide, Australia
Liangze Ma	Huangzhong Agricultural University, China

Lihua Cai	South China Normal University, China
Ling Chen	National Yang Ming Chiao Tung University, Taiwan
Lingwei Chen	Rochester Institute of Technology, USA
Linlin Ding	Liaoning University, China
Lishan Yang	University of Adelaide, Australia
Longchao Da	Arizona State University, USA
Longkun Guo	Fuzhou University, China
Lu Chen	Swinburne University of Technology, Australia
Lyuyang Tong	Wuhan University, China
Man Wu	Keio University, Japan
Manan Buddhadev	IEEE, USA
Mariusz Bajger	Flinders University, Australia
Marwan Hassani	TU Eindhoven, Netherlands
Meghjeet Vartak	TransUnion, USA
Mehmet Ali Kaygusuz	Middle East Technical University, Turkey
Mengxuan Zhang	Australian National University, Australia
Michele Melchiori	University of Brescia, Italy
Ming Zhong	Wuhan University, China
Minghao Zhao	East China Normal University, China
Minghe Yu	Northeastern University, China
Mo Li	Liaoning University, China
Mourad Ellouze	Lille Catholic University, France
Muhammad Haris	TIB – Leibniz-Informationszentrum Technik und Naturwissenschaften und Universitätsbibliothek, Germany
Nicolas Travers	De Vinci Higher Education, France
Ning Liu	Shandong University, China
Ningning Cui	Chang'an University, China
Ozan Kahramanogullari	Free University of Bozen-Bolzano, Italy
Peiquan Jin	University of Science and Technology of China, China
Peng Peng	Hunan University, China
Peng Cheng	Tongji University, China
Pengpeng Zhao	Soochow University, China
Pengpeng Qiao	Institute of Science Tokyo, Japan
Ping Lu	Beihang University, China
Qi Song	University of Science and Technology of China, China
Qian Chen	Hefei University, China
Qiang Qu	Shenzhen Institutes of Advanced Technology, Chinese Academy of Sciences, China

Qing Liu	Data61, CSIRO, Australia
Qingzhi Ma	Soochow University, China
Qiong Chang	Tokyo Institute of Technology, Japan
Qiyu Liu	Southwest University, China
Rajesh Debnath	North Carolina State University, USA
Renjie Sun	East China Normal University, China
Riccardo Cantini	University of Calabria, Italy
Rong Zhu	Alibaba Group, China
Rong Gao	Hubei University of Technology, China
Ruiqi Xu	Beijing Institute of Technology (Zhuhai), China
Savong Bou	University of Tsukuba, Japan
Shang Liu	China University of Mining and Technology, China
Shatrughna Upadhyay	Intuit Inc, USA
Sheng Li	Institute of Science Tokyo, Japan
Sheng Hu	Hokkaido University, Japan
Shinya Kitajima	Fujitsu Limited, Japan
Shi-ting Wen	NingboTech University, China
Shuai Xu	Nanjing University of Aeronautics and Astronautics, China
Shuang Wu	South China University of Technology, China
Shuchao Pang	Nanjing University of Science and Technology, China
Shuhao Zhang	Huazhong University of Science and Technology, China
Shuiqiao Yang	University of New South Wales, Australia
Shun Mao	South China Normal University, China
Shunmei Meng	Nanjing University of Science and Technology, China
Shuting Yang	Huazhong Agricultural University, China
Shuxiang Lin	Central South University, China
Shuyuan Zheng	University of Osaka, Japan
Sijie Mai	South China Normal University, China
Silvestro Roberto Poccia	University of Turin, Italy
Siyuan Chen	Guangzhou University, China
Siyuan Wu	Nanjing University, China
Takeshi Yamamuro	NTT, Japan
Tangpeng Dan	Renmin University of China, China
Tanmaykumar Shah	JPMorgan Chase, USA
Tao He	SuperScaling Technologies, LLC, USA
Tao Qiu	Shenyang Aerospace University, China
Tarique Anwar	University of York, UK

Taro Yano	NEC Corporation, Japan
Tetsushi Ohki	Shizuoka University, Japan
Thanh Tam Nguyen	Griffith University, Australia
Tianyu Liu	University of Hong Kong, China
Tianzi Zang	Nanjing University of Aeronautics and Astronautics, China
Tieke He	Nanjing University, China
Tiezheng Nie	Northeastern Univeristy, China
Wei Li	Harbin Engineering University, China
Wei Chen	Huazhong University of Science and Technology, China
Weijian Ma	Fudan University, China
Weixin Zeng	National University of Defense Technology, China
Wenda Tang	China Telecom Cloud Computing Research Institute, China
Wenhao Liang	University of Adelaide, Australia
Wenjun Ma	South China Normal University, China
Wenpeng Lu	Qilu University of Technology (Shandong Academy of Sciences), China
Wenqi Sun	Renmin University of China, China
Wentao Li	University of Leicester, UK
Wenwen Gong	China Agricultural University, China
Xiangwen Yang	University of Sydney, Australia
Xiangyu Song	Chang'an University, China
Xiao Liu	China University of Mining and Technology, China
Xiao Pan	Shijiazhuang Tiedao University, China
Xiaojuan Cheng	Deakin University, Australia
Xiaolin Xiao	South China Normal University, China
Xiaoxiao Chi	Macquarie University, Australia
Xiaoxiao Yang	Intretech, China
Xiaoyan Wang	Institute of Information Engineering, China
Xiaoyang Li	Adelaide University, Australia
Xin Han	Deakin University, Australia
Xinbai Li	Nara Institute of Science and Technology, Japan
Xiongxiao Xu	Illinois Institute of Technology, USA
Xiu Fang	Donghua University, China
Xiuhua Li	Chongqing University, China
Xiujing Guo	Osaka University, Japan
Xudong Yuan	Anhui University of Science and Technology, China

Xuefeng Zhu	Kunming University of Science and Technology, China
Yalun Wu	Beijing Jiaotong University, China
Yanfeng Zhang	Northeastern University, China
Yang Xu	Nanjing University, China
Yang Cao	Great Bay University/Tsinghua University, China
Yang Li	Beijing Institute of Technology, China
Yang-Sae Moon	Kangwon National University, South Korea
Yanhao Wang	East China Normal University, China
Yanmei Hu	Chengdu University of Technology, China
Yanmin Zhu	Shanghai Jiao Tong University, China
Yibo Sun	University of Adelaide, Australia
Yihong Yang	China University of Geosciences (Beijing), China
Yijun Su	JD.com, China
Yikai Mao	Keio University, Japan
Yike Li	Beijing Jiaotong University, China
Yinuo Ren	Stanford University, USA
Yishuo Zhang	Deakin University, Australia
Yixiang Fang	Chinese University of Hong Kong, Shenzhen, China
Yixin Zhang	Kyoto University, Japan
Yong Zhang	Tsinghua University, China
Yonghong Yu	Nanjing University of Posts and Telecommunications, China
Yongzhe Jia	Nanjing University, China
Yu Sun	Nankai University, China
Yu Yang	Education University of Hong Kong, China
Yu Liu	Huazhong University of Science and Technology, China
Yuanfang Zhang	Beijing Institute of Technology, China
Yuankai Fan	Institute of Artificial Intelligence (TeleAI), China
Yuchen Ji	Osaka University, Japan
Yue Kou	Northeastern University, China
Yuliang Ma	Northeastern University, China
Yunjun Gao	Zhejiang University, China
Yunzhe Tian	Beijing Jiaotong University, China
Yurong Cheng	Beijing Institute of Technology, China
Yutong Qu	University of Adelaide, Australia
Yuwei Peng	Wuhan University, China
Yuxiang Zeng	Hong Kong University of Science and Technology, China
Yu-Xuan Qiu	Beijing Institute of Technology, China

Contents – Part I

Machine Learning for Data Mining

Advanced Time Series Analysis

Exploring Causal Relationships Across Shale Gas Wells: Granger Causality-Based Temporal Production Prediction

Run Yang, Jiajie Zhu, Pengfei Ding, and Yan Wang(✉)

School of Computing, Macquarie University, Sydney, NSW 2109, Australia
run.yang@hdr.mq.edu.au, {jiajie.zhu,pengfei.ding,yan.wang}@mq.edu.au

Abstract. The shale gas temporal production prediction (SGTPP) aims to estimate the capacity of gas resources. However, existing SGTPP methods suffer from data sparsity and neglect the correlations across different wells, resulting in performance degradation. To overcome these limitations, in this paper, we propose a novel Granger Causality-based Temporal Production Prediction model, named as GCTPP. In GCTPP, we firstly propose a Granger causality detection module, which utilizes the historical production data of shale gas wells to identify the Granger causal well pairs. Secondly, we propose a temporal attention mechanism to selectively extract the most important temporal features from the Granger causal well pairs across different timestamps. Finally, we leverage the extracted temporal features of the Granger reason well to help predict the future production of the Granger result well. The results of extensive experiments on real-world datasets demonstrate that our GCTPP model significantly outperforms the state-of-the-art methods in terms of prediction accuracy.

Keywords: Granger causality · Shale gas production prediction

1 Introduction

As the cleanest unconventional energy, shale gas has become an important resource in the natural gas industry. Predicting shale gas production is crucial for estimating the capacity of resources in reservoirs. In recent years, there has been increasing attention on leveraging historical production data of shale gas wells to capture production trends and patterns for better prediction, namely, shale gas temporal production prediction (SGTPP). Traditional SGTPP methods typically adopt decline curve analysis (DCA) [4,10,15,22] based on domain knowledge. In contrast, some advanced SGTPP methods adopt deep learning techniques, such as recurrent neural networks (RNNs) [23] and artificial neural networks (ANNs) [16] to improve the prediction accuracy.

However, existing SGTPP methods heavily rely on high-quality historical production data and focus on individual wells only, leading to two limitations as

M. Yoshikawa et al. (Eds.): ADMA 2025, LNAI 16197, pp. 3–19, 2026.
https://doi.org/10.1007/978-981-95-3453-1_1

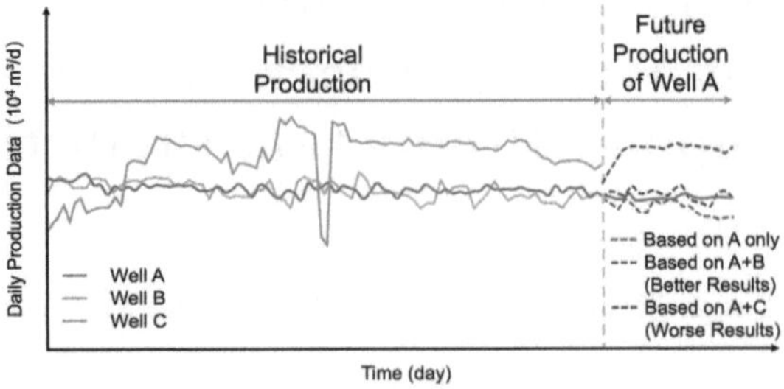

Fig. 1. Schematic diagram of the Granger causal well pair.

follows. Firstly, in the industry, due to data privacy, many wells have limited historical production data [25]. Consequently, these methods [4,10,15,16,18,22,23] struggle to predict future production of wells with such poor-quality historical production data. Secondly, existing methods only use the historical production data of an individual well to predict its future production, overlooking the valuable information available from the production of its neighboring wells. The above limitations impact the prediction performance of existing methods and restrict their applications in real-world scenarios.

In fact, two neighboring wells may exhibit correlations in historical production trends and patterns, which may arise from the interfering well pressure, similar reservoir properties, and interconnected fracture networks between these wells [1]. This observation suggests that a Granger causal relationship may exist between two neighboring wells [20]. To examine this hypothesis, we conducted a preliminary study using real-world data from adjacent wells (A, B and C) from the Changning block[1]. We designed three scenarios to predict the future production for the target well A based on historical production data from (1) well A only, (2) wells A and B, and (3) wells A and C, with a representative and state-of-the-art prediction model [26]. As shown in Fig. 1, incorporating data from well B improves the prediction accuracy for well A compared to using well A alone. In contrast, including well C's data does not provide similar benefits and even degraded the performance. This result indicates that well B has a Granger causal influence on well A, while well C does not.

These observations highlight the importance of identifying well pairs with relevant production patterns. According to Granger causality theory [6], such a pair can be termed a *Granger causal well pair*, where one well's (denoted as "Granger reason well") historical data can improve the prediction accuracy for the other well's (denoted as "Granger result well") future production. Building on this insight, we posit that by utilizing relevant production patterns between wells in the Granger causal well pair, we can not only overcome data sparsity, but also capture correlations between the Granger reason well and the Granger result well, thereby enhancing the overall accuracy of temporal production prediction.

[1] Changning block is one of blocks of the Southwest shale gas fields in China.

The above discussion reveals a novel problem of *cross-well temporal production prediction*, which aims to use historical production data from the Granger reason well to help predict the future production of the corresponding Granger result well. To tackle this novel problem, we identify two key challenges:
CH1: *How to identify the Granger causal well pairs based on historical production data?* Existing SGTPP methods [4,10,15,16,18,22,23] focus primarily on statistical correlations across timestamps, while often neglecting Granger causality between shale gas wells. In fact, exploring such relationships is challenging, because we need to consider the production trends and patterns of different wells. Hence, identifying Granger causal well pairs requires new methods to distinguish Granger causality from potential correlations in historical production data.
CH2: *How to selectively capture temporal features from a Granger causal well pair to enhance production prediction?* The production of shale gas wells exhibits different temporal features. Existing SGTPP methods fail to fully capture temporal dependencies across timestamps [23], and often overlook temporal features between causally related wells for production prediction. This limitation hinders accurate cross-well modeling.

To address the above two challenges, in this paper, we propose a novel **G**ranger **C**ausality-based **T**emporal **P**roduction **P**rediction model, called GCTPP. To the best of our knowledge, this is the first work in the literature that proposes a novel problem of *cross-well temporal production prediction* and provides a solution for it. Our primary contributions are summarized as follows.

1) Targeting **CH1**, we propose a Granger causality detection module to identify Granger causal well pairs. Specifically, we perform Augmented Dickey-Fuller (ADF) tests and Granger causality tests to identify the Granger causal well pairs based on historical production data.
2) Targeting **CH2**, we propose a novel temporal attention mechanism to selectively capture the most important temporal features across different timestamps. We combine feature representations of the Granger reason well and the Granger result well to predict the future production of the Granger result well.
3) We conduct extensive experiments on real-world datasets. The results demonstrate that our GCTPP model outperforms existing state-of-the-art methods with better accuracy.

2 Related Work

2.1 Prediction of Temporal Production for Shale Gas

Existing methods for SGTPP aim to predict future production based on the historical data. These methods can be divided into two categories.

(1) DCA Methods. Traditional DCA methods include the Arps decline model [15], the power-law exponential (PLE) model [10], the stretched exponential production decline (SEPD) model [22], and the Duong model [4]. These methods rely on the initial temporal production data to make production predictions.

However, in real-world industry, it is common for the initial production data of shale gas wells to be unavailable or of low quality. In such cases, DCA methods cannot predict production accurately, thereby limiting their applicability.

(2) Deep Learning-Based Methods. In recent years, deep learning-based methods have been increasingly applied to SGTPP. These methods typically utilize architectures such as GRU [18], ANNs [16], RNNs [23], and LSTM networks [26], which are adept at handling temporal data. Additionally, the improved DCA method [29] applies ensemble learning to enhance production prediction accuracy. However, existing methods often overlook correlations between neighboring wells, leading to downgraded prediction performance.

2.2 Cross-Well Analysis

Cross-well analysis allows better characterization of the correlations between wells. For instance, LogRegX [14] aims to generate missing geophysical logs in cross-well analysis by aligning source and target data while addressing domain discrepancies. The SpatialâĂŞTemporal Graph Convolutional Network [24] incorporates the graph structure to model well interference to predict production. However, current studies do not consider causal relationships, particularly temporal causality, which is critical to the production correlations over time. This lack of causal analysis limits their applicability to production prediction.

2.3 Granger Causality

Granger causality is used to determine whether one time series is helpful in predicting another [6]. For example, the method proposed in [19] applied Granger causality to verify the relationship between exchange rates and stock prices. Granger-based temporal causal modeling [21] was used to explore the causality between public opinion and the death penalty. The TC-GATN model [13] was proposed to capture inherent dependencies in industrial multivariate time series. However, most existing methods have not incorporated Granger causality into prediction systems. Although the TC-GATN model has utilized Granger causality in the relationship analysis among variables and the prediction system, it did not discover Granger causal relationships among multiple univariate time series. So far, no studies have been reported that consider the Granger causality-based prediction system among multiple univariate time series. Due to the fact that univariate time series contain only information about individual variables and lack the interaction data between variables, it is more challenging to conduct a prediction based on Granger causality detection among them.

3 Problem Definition

The cross-well prediction of temporal production problem is defined as follows:

Table 1. Important notations.

Symbol	Definition
$A = \{a_1, a_2, \ldots, a_N\}$	set of wells on the same platform
$(a_c, a_e) \in \Omega_G$	Granger well pair
a_i, $i \in \{1, 2, \cdots, N\}$	arbitrary well of A
$(a_p, a_q) \in \Omega$	arbitrary well pair
$\mathbf{h}_c^t$	temporal feature representation of the Granger reason well
$\mathbf{h}_e^t$	temporal feature representation of the Granger result well
$X_i = \{x_i^1, x_i^2, \ldots, x_i^{m_i}\}$	set of the historical daily production data
$\hat{Y}_e$	predicted future production
$\mathbf{z}_c$	context vector of the Granger reason well
$\mathbf{z}_e$	context vector of the Granger result well

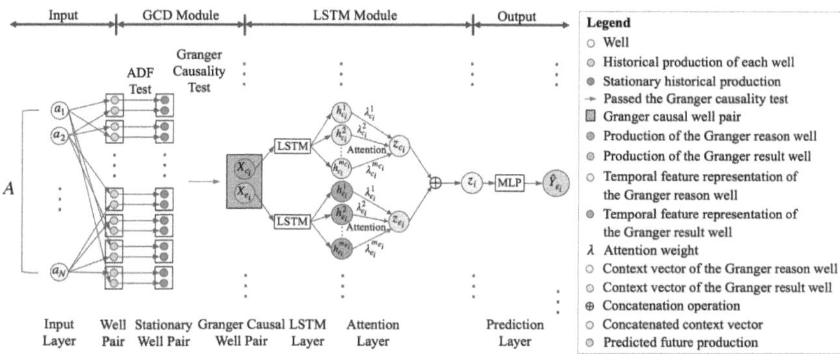

Fig. 2. The structure of our GCTPP framework.

Let $A = \{a_1, a_2, \ldots, a_N\}$ represent the set of N shale gas wells on the same platform. For a well $a_i \in A$, its historical daily production data is denoted as $X_i = \{x_i^1, x_i^2, \ldots, x_i^{m_i}\}$, where m_i represents the length of the time series (i.e., the number of daily production data) for well a_i, and x_i^t represents the daily production data of well a_i at time t, where $t \in \{1, 2, \ldots, m_i\}$.

For each obtained Granger causal well pair (a_c, a_e), as well as temporal production data X_c and X_e, respectively, the goal is to predict the future production $\hat{Y}_e$ of the Granger result well a_e, which can be expressed as:

$$\hat{Y}_e = \{\hat{y}_e^{m_i+1}, \hat{y}_e^{m_i+2}, \ldots, \hat{y}_e^{m_i+l}\} = F(X_c, X_e) \quad \forall (a_c, a_e) \in \Omega_G, \tag{1}$$

where $F(\cdot)$ is the predictive model that uses historical temporal production data of Granger reason well a_c and Granger result well a_e to make predictions, and l is the length of future timestamps. Important notations are listed in Table 1.

4 Methodology

4.1 Overview of GCTPP

Algorithm 1. GCTPP Algorithm

Input: A set of shale gas wells A on the same platform, and the historical daily production data X_i for each shale gas well a_i.
Output: The predicted future production $\hat{Y}_e$ of the Granger result well a_e.

construct well pairs (a_p, a_q) for each platform,
extract the overlapping timestamps and corresponding production values and update (X_p, X_q) for each well pair (a_p, a_q),
for X_i of each $(a_p, a_q) \in \Omega$ **do**
  update X_i with difference function using Eq. (8) until X_i passes the ADF test,
end for
for (X_p, X_q) of each $(a_p, a_q) \in \Omega$ that passed the ADF test **do**
  if (X_p, X_q) meets Granger causality test **then**
    form a Granger causal well pair (a_c, a_e) where $a_c = a_p$ and $a_e = a_q$,
  end if
end for
for (X_p, X_q) of each $(a_c, a_e) \in \Omega_G$ **do**
  generate embedding $\mathbf{h}_c^t$ and $\mathbf{h}_e^t$ through LSTM encoder,
  generate attention weights λ_c^t and λ_e^t using Eq. (19),
  calculate the context vectors $\mathbf{z}_c$ and $\mathbf{z}_e$ using Eq. (20),
  calculate the concatenated context vector $\mathbf{z}$ using Eq. (21),
  predict the future production $\hat{Y}_e$ using Eq. (22),
  return $\hat{Y}_e$.
end for

To address the cross-well prediction of the temporal production problem, we propose a Granger Causality-based Temporal Production Prediction model, called GCTPP. As illustrated in Fig. 2, the GCTPP consists of two key modules: the Granger Causality Detection (GCD) module, which identifies the Granger causal well pairs, and the LSTM module, which predicts the future production of the Granger result well. The algorithm for the GCTPP is presented in Algorithm 1.

4.2 GCD Module

In the GCD module, we identify Granger causal well pairs by following steps.

Well Pairing. We pair the wells on the same platform, considering the arrangement order. Consequently, for a platform containing n shale gas wells, a set Ω of $n*(n-1)$ well pairs (a_p, a_q) could be generated. Since the Granger causality test requires that the two input time series have the same length and timestamps, we then update the values of X_p and X_q to reflect the overlapping timestamps extracted from the well pair (a_p, a_q).

ADF Test. Since Granger causality tests assume that the input data series are stationary, it is crucial to ensure that the production data for each shale gas well meet this condition. Without stationarity, the Granger test results may be unreliable, potentially leading to incorrect conclusions about causal relationships. To assess stationarity, we employ the ADF test [2] in the production data of each shale gas well. We will introduce the detailed explanation of each step.

(1) Formulate Hypotheses. The ADF test starts with the assumption that the time series is non-stationary ($H_{0,\text{ADF}}$), which is called the null hypothesis. Conversely, the alternative hypothesis is that the time series is stationary ($H_{1,\text{ADF}}$). Our goal of the ADF test is to identify whether the time series is stationary by determining if the $H_{0,\text{ADF}}$ can be rejected.

(2) Construct the ADF Regression Equation. To model the time series, we use the following ADF regression equation:

$$\Delta x_i^t = \alpha + \tau t + \gamma x_i^{t-1} + \sum_{r=1}^{s} \delta_r \Delta x_i^{t-r} + \epsilon_t, \tag{2}$$

where x_i^t is the production data for well a_i at time t, $\Delta x_i^t = x_i^t - x_i^{t-1}$ represents the first-order difference of the time series, and s is the number of lags. α is the constant term, τt is the linear trend term, and γ is the autoregressive coefficient tested to determine the stationarity of the time series. Besides, δ_r are the coefficients of the lagged difference terms, and ϵ_t is the white noise error term. We aim to estimate the unknown parameters α, τ, γ, and δ_r using the known x_i^t.

(3) Estimate the Autoregressive Coefficient γ. Based on Eq. (2), we estimate the autoregressive coefficient γ using the ordinary least squares (OLS) method [3]. In this step, we denote the response vector based on the known Δx_i^t values:

$$\mathbf{R} = \begin{bmatrix} \Delta x_i^1 \\ \Delta x_i^2 \\ \vdots \\ \Delta x_i^{m_i} \end{bmatrix}. \tag{3}$$

Then, we denote the regression matrix as follows:

$$\mathbf{M} = \begin{bmatrix} 1 & 1 & x_i^0 & \Delta x_i^0 & \dots & \Delta x_i^{-s} \\ 1 & 2 & x_i^1 & \Delta x_i^1 & \dots & \Delta x_i^{1-s} \\ \vdots & \vdots & \vdots & \vdots & \ddots & \vdots \\ 1 & m_i & x_i^{m_i-1} & \Delta x_i^{m_i-1} & \dots & \Delta x_i^{m_i-s} \end{bmatrix}. \tag{4}$$

The parameter estimated values can be calculated using the OLS method:

$$[\hat{\alpha}, \hat{\tau}, \hat{\gamma}, \hat{\delta_1}, \hat{\delta_2}, ..., \hat{\delta_s}]^T = (\mathbf{M}^T\mathbf{M})^{-1}\mathbf{M}^T\mathbf{R}. \tag{5}$$

(4) Calculate the ADF Statistic. Then, we calculate the standard error of $\hat{\gamma}$:

$$\text{SE}(\hat{\gamma}) = \sqrt{\frac{1}{m_i - k} \sum_{t=1}^{m_i} \epsilon_t^2 \cdot [\text{Var}(\hat{\gamma})]}, \tag{6}$$

where k is the number of model parameters, and $\mathrm{Var}(\hat{\gamma})$ represents the variance of $\hat{\gamma}$. Next, we can obtain the ADF statistic by the following formula:

$$\text{ADF statistic} = \frac{\hat{\gamma}}{\mathrm{SE}(\hat{\gamma})}. \tag{7}$$

(5) Determine the p-Value. After obtaining the ADF statistic, we can determine the p-value for the ADF test (p_{ADF}), which represents the probability of observing the current or more extreme ADF statistic value, under the condition that the $H_{0,\mathrm{ADF}}$ is true. To determine the p_{ADF}, we refer to the critical values from the Dickey-Fuller distribution table [12].

(6) Make a decision. The decision rule for the ADF test is as follows:

- If the p_{ADF} is lower than the significance level, we then reject the $H_{0,\mathrm{ADF}}$, indicating that the time series is stationary.
- If the p_{ADF} is higher than the significance level, we fail to reject the $H_{0,\mathrm{ADF}}$, suggesting that the time series is non-stationary.

As aforementioned, stationary sequences are required to be the input of Granger causality tests. Therefore, for non-stationary sequences, we apply differencing operations to make them stationary:

$$\Delta x_i^t = x_i^t - x_i^{t-1}. \tag{8}$$

We repeatedly apply differencing operations on the time series until it passes the ADF test, ensuring the reliability of subsequent Granger causality analysis.
Granger Causality Test. To determine if incorporating one time series can lead to more accurate predictions of another time series, we conduct the Granger causality test on each stationary well pair (a_p, a_q) with following steps.
(1) Formulate Hypotheses. Given the historical production data (X_p, X_q) of each well pair (a_p, a_q), the null hypothesis ($H_{0,\mathrm{GC}}$) of the Granger causality test is that X_p is not the Granger reason of X_q, while the alternative hypothesis ($H_{1,\mathrm{GC}}$) is that X_p is the Granger reason of X_q. Our goal of the Granger causality test is to determine whether well a_p is the Granger reason of well a_q in each well pair (a_p, a_q) by determining if the null hypothesis can be rejected.

(2) Construct the Granger Regression Equation. The Granger regression equation can be defined as follows:

$$x_q^t = \mu + \sum_{j=1}^{d} \phi_j x_q^{t-j} + \sum_{j=1}^{d} \psi_j x_p^{t-j} + \nu_t, \tag{9}$$

where d is the number of lags, μ is the intercept term, ϕ_j and ψ_j are the regression coefficients, and ν_t is the residual term.

(3) Estimate the Parameters. Based on Eq. (9), we use the OLS method to estimate the unknown parameters μ, ϕ_j and ψ_j using the known historical production data x_p^t and x_q^t, respectively. The goal of OLS is to find the parameter values that minimize the sum of squared residuals:

$$\min_{\mu,\phi_j,\psi_j} \sum_{t=1}^{n} (\nu_t)^2 = \min_{\mu,\phi_j,\psi_j} \sum_{t=1}^{n} (x_q^t - \mu - \sum_{j=1}^{d} \phi_j x_q^{t-j} - \sum_{j=1}^{d} \psi_j x_p^{t-j})^2. \tag{10}$$

Thus, we can obtain the estimated values for μ, ϕ_j and ψ_j based on Eq. (10).
(4) Calculate the F-Statistic. Then, we can calculate the predicted production value $\hat{x}_q^t$ by substituting the obtained estimated μ, ϕ_j, and ψ_j, as well as the known historical data x_q^{t-j} and x_p^{t-j} into Eq. (9). Next, we calculate the residual sum of squares of the model:

$$\text{RSS}_R = \sum_{t=1}^{n} (\nu_t)^2 = \sum_{t=1}^{n} (x_q^t - \hat{x}_q^t)^2. \tag{11}$$

To evaluate the effect of x_p^{t-j} terms in Eq. (9), we set all ψ_j terms to zero, thereby constructing a model that does not include any x_p^t terms. We calculate the residual sum of squares of the model without any x_p^t terms and denote it as RSS_U. To determine if x_p^t terms are significant, we calculate the F-statistic:

$$\mathcal{F} = \frac{(\text{RSS}_R - \text{RSS}_U)/d}{\text{RSS}_U/(m_p - 2d - 1)}, \tag{12}$$

where m_p is the overlapped length of the production time series for well a_p and a_q. Since m_p is equal to m_q, we use m_p in Eq. (12).

(5) Determine the p-Value. After obtaining $\mathcal{F}$, we can determine the p-value for the Granger causality test (p_{GC}) by referring to the F-distribution table [5].

(6) Make a Decision. If the p_{GC} is lower than the significance level, we reject the null hypothesis, suggesting that well a_p is the Granger reason of well a_q. Therefore, the well pair (a_p, a_q) is referred to as a Granger causal well pair (a_c, a_e), where $a_c = a_p$ and $a_e = a_q$. Otherwise, we fail to reject the null hypothesis, suggesting that well a_p is not the Granger reason of well a_q. In GCTPP, well pairs without Granger causality are excluded from further analysis, as they do not exhibit the necessary causal relationships that are required for the cross-well prediction of temporal production problem.

4.3 LSTM Module

Based on the GCD module, we obtain the set of Granger causal well pairs. For a Granger causal well pair (a_c, a_e), we can utilize the historical production data of the Granger reason well a_c and the Granger result well a_e to predict the future production data of well a_e. We then employ the LSTM module to jointly extract the temporal feature representation of the two wells. The LSTM module includes two parts: (1) the LSTM encoder and (2) the temporal attention mechanism.

(1) LSTM Encoder. Given a Granger causal well pair (a_c, a_e), we first extract the temporal feature representation of the Granger reason well a_c, based on the historical production data X_c. The LSTM encoder processes this data to obtain the temporal feature representation $\mathbf{h}_c^t$. The LSTM can effectively capture long-term dependencies and patterns in time series data, making it particularly advantageous for modeling the complex and sequential nature of production data. The input at timestamp t $(t \in 1, 2, \ldots, m_c)$ is the production data x_c^t. Then, the LSTM processes the input sequence step-by-step. At each timestamp t, it updates its hidden state $\mathbf{h}_c^t$ and cell state C_c^t using the following equations:

$$f_c^t = \sigma(\mathbf{W}_f[\mathbf{h}_c^{t-1}, x_c^t] + b_f), \tag{13}$$

$$I_c^t = \sigma(\mathbf{W}_I[\mathbf{h}_c^{t-1}, x_c^t] + b_I), \tag{14}$$

$$\widetilde{C}_c^t = \tanh(\mathbf{W}_C[\mathbf{h}_c^{t-1}, x_c^t] + b_C), \tag{15}$$

$$C_c^t = f_c^t \odot C_c^{t-1} + I_c^t \odot \widetilde{C}_c^t, \tag{16}$$

$$o_c^t = \sigma(\mathbf{W}_o[\mathbf{h}_c^{t-1}, x_c^t] + b_o), \tag{17}$$

$$\mathbf{h}_c^t = o_c^t \odot \tanh(C_c^t), \tag{18}$$

where f_c^t, I_c^t, and o_c^t are the forget gate, input gate, and output gate values at timestamp t for well a_c, respectively. σ denotes the sigmoid activation function, while $\mathbf{W}_f$, $\mathbf{W}_I$, and $\mathbf{W}_C$ represent the weight matrices. $\widetilde{C}_c^t$ denotes the candidate cell state, C_c^t represents the current cell state, and $\mathbf{h}_c^t$ is the current hidden state. b_f, b_I, b_c, and b_o are the bias terms associated with the forget gate, input gate, candidate cell state, and output gate, respectively.

For the Granger result well a_e in the Granger causal well pair (a_c, a_e), the LSTM encoder follows a similar process to extract the temporal feature representation $\mathbf{h}_e^t$. Similarly, we can obtain $\mathbf{h}_e^t$ by inputting x_e^t and $\mathbf{h}_e^{t-1}$.

(2) Temporal Attention Mechanism. Timestamps are not equally important in prediction. Thus, we incorporate the temporal attention mechanism to capture varying temporal features and assign different weights to features accordingly.

First, we compute attention weights of each timestamp using the hidden states. For the Granger reason well a_c:

$$\lambda_c^t = \frac{\exp(\mathbf{h}_c^t \cdot \mathbf{v}_c)}{\sum_{k=1}^{m_c} \exp(\mathbf{h}_c^k \cdot \mathbf{v}_c)}, \tag{19}$$

where $\mathbf{v}_c$ is the trainable parameter. Then, we can obtain the context vector of the Granger reason well by taking the weighted sum of the hidden states:

$$\mathbf{z}_c = \sum_{t=1}^{m_c} \lambda_c^t \mathbf{h}_c^t. \tag{20}$$

Similarly, we can obtain the context vector $\mathbf{z}_e$ of the Granger result well a_e. Then, the concatenated context vector $\mathbf{z}$ for the Granger causal well pair (a_c, a_e) is obtained by concatenating the context vectors $\mathbf{z}_c$ and $\mathbf{z}_e$:

$$\mathbf{z} = [\mathbf{z}_c \oplus \mathbf{z}_e]. \tag{21}$$

Using the combined feature representation $\mathbf{z}$, we predict the future production of the effect well a_e:

$$\hat{Y_e} = \sigma(F(\mathbf{z})), \tag{22}$$

where $F(\cdot)$ denotes the predictive model of multilayer perceptron (MLP).

The loss function used to train the model is the mean squared error (MSE) between the predicted production $\hat{Y_e}$ and the actual production Y_e:

$$\mathcal{L} = \frac{1}{l} \sum_{t=m_e+1}^{m_e+l} (\hat{y}_e^t - y_e^t)^2, \tag{23}$$

where $\hat{y}_e^t$ represents the predicted production at time t, and y_e^t denotes the actual production at time t.

4.4 Time Complexity Analysis

Our GCTPP model primarily consists of the GCD module and the LSTM module. In the GCD module, firstly, the well pairing step generates $N \times (N-1)$ well pairs, with a complexity of $O(N^2)$. For each well pair, timestamp alignment is performed, which depends on the time series length (denoted as m). In the worst case, this involves iterating over all timestamps for each well pair, leading to a complexity of approximately $O(N^2 \times m)$. Then, each well pair's time series undergoes the ADF test. With a complexity of $O(m)$ per test, this step has a total complexity of $O(N^2 \times m)$. Finally, the Granger causality test is conducted for each well pair, and involves building a regression model with a complexity of around $O(m \times d)$. Therefore, this step has a total complexity of $O(N^2 \times m \times d)$. Overall, the GCD module's total time complexity is $O(N^2 \times m \times d)$.

Next, the complexity for encoding each well pair in the LSTM is approximately $O(m \times D^2)$, where D represents the hidden dimensions. Thus, a total complexity of $O(n \times m \times D^2 \times L)$, where n denotes the number of the Granger causal well pairs of the platform, and L denotes the number of layers. Furthermore, assuming each context vector generated at each timestamp iterates over all

Table 2. Datasets statistics.

Block names [1]	# Platforms [2]	# Wells [3]	# Well pairs
Changning	87	517	2774
Weiyuan	89	469	2424
Luzhou	27	114	452
Yuxi	7	34	96
Zhaotong	61	267	1068
Total	271	1401	6814

[1,2,3] A "block" refers to a large geographical area that contains multiple "platforms", and each platform, in turn, hosts several individual "wells".

LSTM hidden layer dimensions, the complexity for the temporal attention mechanism is $O(D)$, resulting in a total complexity of $O(n \times m \times D \times L)$. Therefore, the total time complexity for the LSTM module is approximately $O(n \times m \times D^2 \times L)$.

Combining the complexities of the two modules, the total time complexity of the GCTPP model is $O(N^2 \times m \times d + n \times m \times D^2 \times L)$.

5 Experiments and Analysis

We conduct extensive experiments to answer the following research questions:

- **RQ1:** How does our GCTPP model perform compared to baseline models?
- **RQ2:** How does each proposed module in GCTPP affect the performance?

5.1 Experimental Settings

(1) Experimental Environment. The CPU is a Core i5-10500, with a capacity of 32GB. The graphics card is NVIDIA GeForce RTX 3080 with 10GB of VRAM. The operating system is Windows 10, the programming language is Python 3.8, and the deep learning frameworks used are TensorFlow 2.5 and Keras 2.4.3.

(2) Datasets. The datasets used in this study contain historical production data from shale gas wells located in the Changning, Weiyuan, Luzhou, Yuxi and Zhaotong blocks of the Southwest shale gas fields in China, from the time of exploitation to December 2, 2022, as shown in Table 2[2]. These blocks exhibit differences in geological properties, such as formation depth, fracture density, and reservoir pressure, which lead to variability in well production profiles.

(3) Datasets Division. To assess the performance of our GCTPP model and the baseline models across different temporal levels, we follow the dataset division method for single-well-based production prediction by Yang et al. [26]. For the dataset division of well pairs, as described in Sect. 4.2, the production data of each well pair obtained after preprocessing have equal length time series and the same timestamps, and thus can be divided by the same method. For short-term prediction, the last 10% length of the time range is used for testing, while the rest 90% of the data samples are used for training. Similarly, for medium- and long-term prediction, the ratio is 8:2 and 7:3, respectively. 80% of the training data samples are used for real training, while the rest 20% are validation sets.

(4) Parameter Setting. In the GCD module, we set both the significance level of the ADF test and the Granger causality test to 0.05, both s and d to 3, and k to 6. To achieve the best prediction performance, we use Optuna[3] to automatically tune the hyperparameters of the LSTM module in the model. We search for the optimal hyperparameters over the following candidate sets: LSTM hidden dimensions {64, 96, 128}, attention hidden dimensions {32, 64, 96, 128},

[2] Due to the confidential nature of the dataset, it is not publicly available.
[3] https://optuna.org/.

learning rate {0.001, 0.005, 0.01}, batch size {16, 32, 64, 128}, and LSTM layers {1, 2, 3}. For a fair comparison, this optimization method is used in both our GCTPP model and the comparison methods introduced in following sections.

(5) Comparison Methods. The GCTPP model is designed to predict multiple univariate time series, utilizing Granger causality relationships between wells. Unlike TC-GATN [13], PCA-DLSTM [11], and T-GCN [27], which are tailored for multivariate time series, our model deals with predicting multiple independent time series. Due to the different nature of our problem, we do not compare GCTPP with methods designed for multivariate time series. As mentioned in the related work, so far, no studies have been reported that proposed a model for Granger causality-based prediction among multiple univariate time series. Therefore, we select MLP, GRU, and LSTM as baseline models for comparison, as they are well-established approaches for time series prediction.

- **MLP**: A multilayer perceptron model [8], which is a type of feedforward neural network with multiple hidden layers to learn non-linear relationships.
- **GRU**: A gated recurrent unit model [28], which is a type of RNN that use gating mechanisms to capture temporal features in sequential data.
- **LSTM**: A long short-term memory model [26], which is a specialized type of RNN that can learn long-term dependencies in data more effectively.

The hyperparameter tuning for the baseline models is conducted independently, allowing each model to be optimized using the full range. This ensures that the baseline models are configured with their respective optimal hyperparameter settings, enabling a fair comparison against our GCTPP model. In MLP, we set the range of the number of hidden layers to 1, 2, and 3. In GRU and LSTM, we set the range of the number of recurrent layers to 1, 2, and 3.

(6) Evaluation Metrics. To quantify the accuracy of the model, the mean absolute error (MAE), the mean square error (MSE), and the root mean square error (RMSE) are widely used evaluation metrics in temporal prediction [7,9].

5.2 Results and Analysis

Performance Comparison with Baselines (for RQ1). Table 3 shows the MSE, RMSE and MAE values of our GCTPP model and baseline models for short-, medium- and long-term prediction tasks. Among five datasets, GCTPP outperforms the best-performing baseline by an average of 38.51%, 19.86%, and 26.91% in terms of MSE, RMSE and MAE, respectively. Compared to baseline models, our GCTPP model considers cross-well information and utilizes the temporal attention mechanism, both contributing to its superior performance.

Specifically, from Table 3, we can further observe that compared to the best-performing baseline, the average improvements between MSE, RMSE and MAE of GCTPP are 26.96%, 23.78%, and 33.87% in short-, medium- and long-term prediction, respectively. Thus, GCTPP performs much better than the best-performing baseline in the long-term prediction task. For long-term predictions,

Table 3. Prediction results of the baseline models and our GCTPP model.

Datasets	Tasks	Baselines (*10^−2)									Our model (*10^−2)			Improvements (%)		
		MLP			GRU			LSTM			GCTPP			GCTPP vs.best baselines		
		MSE	RMSE	MAE	MSE	RMSE	MAE	MSE	RMSE	MAE	MSE	RMSE	MAE	MSE	RMSE	MAE
Changning	S	12.39	43.20	38.72	7.28	27.69	24.11	4.57*	18.92*	15.10*	**2.68**[1]	**15.70**	**11.81**	41.35	17.01	21.78
	M	17.27	51.94	44.45	11.91	40.13	36.32	5.03*	23.88*	17.02*	**3.56**	**19.78**	**13.05**	29.22	17.16	23.32
	L	23.62	56.90	50.00	17.12	41.10	34.46	9.39*	30.46*	26.29*	**4.62**	**23.44**	**17.83**	50.79	23.04	32.17
Weiyuan	S	14.04	46.43	40.89	8.46	30.33	27.78	4.88*	20.12*	16.42*	**2.91**	**16.41**	**12.02**	40.36	18.43	26.79
	M	18.12	55.68	48.07	11.85	40.48	36.19	5.72*	25.08*	19.21*	**4.47**	**20.69**	**14.21**	21.85	17.50	26.02
	L	26.67	62.80	51.83	21.60	45.35	39.95	10.61*	32.62*	25.86*	**5.88**	**24.96**	**18.53**	44.58	23.48	28.34
Luzhou	S	13.48	44.61	39.53	7.87	28.36	25.10	4.17*	18.58*	15.26*	**2.53**	**16.07**	**11.88**	39.32	13.50	22.14
	M	18.15	55.72	47.86	11.17	39.41	35.84	5.45*	24.76*	18.83*	**3.90**	**20.33**	**13.67**	28.44	17.89	27.40
	L	24.89	56.09	50.22	19.52	43.98	38.12	9.91*	31.90*	25.10*	**5.37**	**24.69**	**18.10**	45.81	22.60	27.88
Yuxi	S	13.67	45.58	39.02	8.46	29.15	27.01	4.66*	20.03*	15.69*	**2.77**	**15.91**	**11.96**	40.55	20.56	23.77
	M	17.99	54.53	47.62	11.59	39.91	35.72	5.84*	24.97*	19.44*	**4.34**	**20.42**	**13.91**	25.68	18.22	28.44
	L	22.97	60.18	55.08	20.87	45.42	39.48	11.19*	31.92*	25.62*	**5.78**	**24.73**	**18.28**	51.42	22.52	28.64
Zhaotong	S	12.74	43.65	38.98	7.60	27.88	24.53	4.04*	18.86*	15.93*	**2.54**	**15.82**	**11.84**	37.12	16.11	25.67
	M	17.62	52.03	45.37	10.59	38.65	34.33	5.70*	24.81*	17.55*	**3.88**	**19.93**	**13.33**	31.92	19.66	24.04
	L	24.66	58.88	55.33	19.28	44.17	37.32	9.79*	29.86*	28.59*	**4.97**	**23.81**	**17.92**	49.23	20.26	37.32

[1] Best results are in bold. * Indicates the results of the best-performing baseline. S: Short-term. M: Medium-term. L: Long-term.

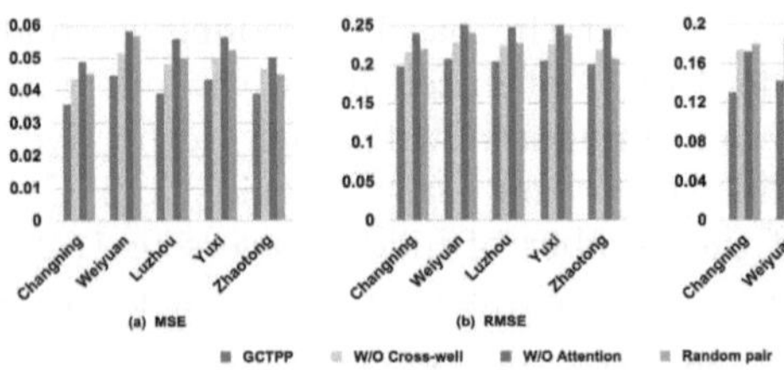

Fig. 3. Medium-term prediction performance of GCTPP and its three variants.

the model needs to learn trends and patterns from more training data. The temporal attention mechanism of the LSTM module enables GCTPP to adaptively emphasize the most informative temporal features as the prediction horizon increases, leading to superior predicting performance compared to baseline models that lack such a temporal feature weighting mechanism. As long-term shale gas production prediction task is more important for practical applications [17], our proposed GCTPP model provides a valuable reference for the industry.

Ablation Study (for RQ2). To explore the impact of the core modules in our GCTPP model, we create three variants: (1) to explore the effectiveness of the temporal attention mechanism in the LSTM module, we construct a variant of GCTPP, called **W/O Attention**. We combine the hidden states by averaging them, replacing the temporal attention mechanism. (2) To determine if the cross-well information captured by the GCD module of GCTPP leads to improved prediction accuracy, we construct a variant of GCTPP, **W/O Cross-well**, which only includes the single-well information. (3) To investigate the impact of the well pairs selection within the GCD module, we develop a GCTPP variant termed **Random pair**. This variant forms pairs by randomly selecting two wells on the same platform to conduct cross-well production prediction.

Our experiments show similar results across short-, medium-, and long-term prediction tasks[4]. As shown in Fig. 3, our findings are as follows: (1) using only single-well information decreases 16.59% of the prediction accuracy, proving that the GCD module of GCTPP can capture the Granger causal dependencies among wells, and exploit the rich cross-well features to enhance the prediction accuracy. (2) We can find that removing the temporal attention mechanism from the LSTM module of GCTPP degrades 22.18% of the prediction accuracy, verifying that selective capture of temporal features can emphasize the most informative features and disregard the less relevant ones, thus realize the more accurate prediction. (3) Randomly pairing two arbitrary wells reduces 18.06% of the prediction accuracy. This exhibits that the Granger causality test of the GCTPP is capable of selecting relevant well pairs to improve prediction accuracy.

6 Conclusion and Future Work

In this paper, we propose a novel Granger Causality-based Temporal Production Prediction model, called GCTPP, which leverages the Granger causal relationship to improve the performance of cross-well temporal production prediction. Our Granger causality detection module can effectively identify the Granger causal well pairs based on historical production data. We also use the temporal attention mechanism to selectively capture the most important temporal features, which helps predict future production. Extensive experiments demonstrate that our GCTPP model outperforms the existing state-of-the-art methods.

In our future work, we plan to discover the multi-well production prediction and address potential confounding factors (e.g., operational changes, environmental conditions), which enables us to expand the application of the model to the spatio-temporal production prediction of shale gas.

References

1. Al-Shami, T.M., et al.: A comprehensive review of interwell interference in shale reservoirs. Earth-Sci. Rev. **237** (2023). https://doi.org/10.1016/j.earscirev.2023.104327
2. Cheung, Y.W., Lai, K.S.: Lag order and critical values of the augmented dickey-fuller test. J. Bus. Econ. Stat. **13**(3), 277–280 (1995)
3. Dismuke, C., Lindrooth, R.: Ordinary least squares. Meth. Des. Outcomes Res. **93**(1), 93–104 (2006)

[4] Due to space limitations, we analyze medium-term prediction results as an example. Similar trends can be observed for the results on the other omitted tasks.

4. Duong, A.N.: An unconventional rate decline approach for tight and fracture-dominated gas wells. In: SPE Canada Unconventional Resources Conference. SPE (2010)
5. Geisser, S., Greenhouse, S.W.: An extension of box's results on the use of the f distribution in multivariate analysis. Ann. Math. Stat. **29**(3), 885–891 (1958)
6. Granger, C.W.: Testing for causality: a personal viewpoint. J. Econ. Dyn. Control **2**, 329–352 (1980)
7. Hodson, T.O.: Root mean square error (RMSE) or mean absolute error (MAE): when to use them or not. Geosci. Model Dev. Discuss. **2022**, 1–10 (2022)
8. Hornik, K., Stinchcombe, M., White, H.: Multilayer feedforward networks are universal approximators. Neural Netw. **2**(5), 359–366 (1989)
9. Hyndman, R.J., et al.: Another look at forecast-accuracy metrics for intermittent demand. Foresight Int. J. Appl. Forecast. **4**(4), 43–46 (2006)
10. Ilk, D., Rushing, J.A., Perego, A.D., Blasingame, T.A.: Exponential vs. hyperbolic decline in tight gas sands—understanding the origin and implications for reserve estimates using ARPs' decline curves. In: SPE Annual Technical Conference and Exhibition. SPE (2008)
11. Kim, G.B., Hwang, C.I., Choi, M.R.: PCA-based multivariate LSTM model for predicting natural groundwater level variations in a time-series record affected by anthropogenic factors. Environ. Earth Sci. **80**(18), 657 (2021)
12. Leybourne, S.J.: Testing for unit roots using forward and reverse dickey-fuller regressions. Oxford B. Econ. Stat. **57**(4) (1995)
13. Li, J., Shi, Y., Li, H., Yang, B.: TC-GATN: temporal causal graph attention networks with nonlinear paradigm for multivariate time-series forecasting in industrial processes. IEEE Trans. Ind. Inform. **19**(6), 7592–7601 (2023)
14. Lv, W., Yuan, C., Wang, J., Zhu, J., Kang, Y., Chang, J.: Logregx: an explainable regression network for cross-well geophysical logs generation. IEEE Trans. Instrum. Meas. **72**, 1–11 (2023)
15. Ma, X., Liu, Z.: Predicting the oil production using the novel multivariate nonlinear model based on ARPs decline model and kernel method. Neural Comput. Appl. **29**, 579–591 (2018)
16. Manda, P., Nkazi, D.: Accuracy assessment of single and hybrid models for predicting shale gas production. Energy Fuels **35**(7), 6068–6080 (2021)
17. Nguyen-Le, V., Shin, H., Little, E.: Development of shale gas prediction models for long-term production and economics based on early production data in Barnett reservoir. Energies **13**(2, 424) (2020)
18. Qin, X., Hu, X., Liu, H., Shi, W., Cui, J.: A combined gated recurrent unit and multi-layer perception neural network model for predicting shale gas production. Processes **11**(3, 806) (2023)
19. Ruan, J.: Fuzzy correlation measurement algorithms for big data and application to exchange rates and stock prices. IEEE Trans. Ind. Inform. **16**(2), 1296–1309 (2020)
20. Shojaie, A., Fox, E.B.: Granger causality: a review and recent advances. Annu. Rev. Stat. Appl. **9**(1), 289–319 (2022)
21. Tatro, K., Oliver, J.R.: Exploring causality between public opinion and the death penalty using granger testing. Adv. Appl. Sociol. **13**(6), 441–456 (2023)
22. Valkó, P.P., Lee, W.J.: A better way to forecast production from unconventional gas wells. In: SPE Annual Technical Conference and Exhibition. SPE (2010)
23. Xu, Z., Leung, J.Y.: A novel formulation of RNN-based neural network with real-time updating-an application for dynamic hydraulic fractured shale gas production forecasting. Geoenergy Sci. Eng. **233**, 212491 (2024)

24. Xu, Z., Leung, J.Y.: Shale gas production forecasting with well interference based on spatial-temporal graph convolutional network. SPE J. **29**(10), 5120–5131 (2024)
25. Yang, R., Hu, Z., Gu, Z., Liu, X., Duan, X.: Estimation of porosity in gas shale using 1H NMR measurement: implications for enhanced shale gas recovery. Energy Fuels **35**(6), 4943–4953 (2021)
26. Yang, R., Liu, X., Yu, R., Hu, Z., Duan, X.: Long short-term memory suggests a model for predicting shale GAS production. Appl. Energy **322**, 119415 (2022)
27. Zhao, L., et al.: T-GCN: a temporal graph convolutional network for traffic prediction. IEEE Trans. Intell. Transp. Syst. **21**(9), 3848–3858 (2019)
28. Zheng, W., Chen, G.: An accurate GRU-based power time-series prediction approach with selective state updating and stochastic optimization. IEEE Trans. Cybern. **52**(12), 13902–13914 (2021)
29. Zhou, Y., Gu, Z., He, C., Yang, J., Xiong, J.: An improved decline curve analysis method via ensemble learning for shale gas reservoirs. Energies **17**(23), 5910 (2024)

Continuous Blood Pressure Dataset Featuring Arrhythmia and Diverse Baselines for Blood Pressure Estimation

Shuangdu Li[1], Ziyou Li[1], Yixuan Li[1], Xiaomao Fan[1(✉)], Yumeng Liu[1], Wenjun Ma[2], Bowen Zhang[1], Jianhua Ye[1], and Ye Li[3]

[1] Shenzhen Technology University, Shenzhen, China
2410263053@mails.szu.edu.cn, astrofan2008@gmail.com, liuyumeng@sztu.edu.cn

[2] South China Normal University, Guangzhou, China
mawenjun@scnu.edu.cn

[3] Shenzhen Institute of Advanced Technology, Chinese Academy of Sciences, Shenzhen, China
ye.li@siat.ac.cn

Abstract. Continuous blood pressure (BP) monitoring holds significant clinical value for the early diagnosis and prevention of cardiovascular conditions. Although arterial invasive lines remain the gold standard for BP assessment, they can cause discomfort and increase infection risk. Deep learning-based methods are widely employed for non-invasive continuous BP monitoring. However, their accuracy may be affected by arrhythmic conditions in physiological signals. This study aims to investigate the impact of rhythm changes on the predictive outcomes of various methods by constructing a novel continuous blood pressure dataset featuring arrhythmia called BP-ARR. Specifically, we constructed the BP-ARR dataset to include arrhythmic conditions and assessed the performance of established baseline methods under both normal and arrhythmic scenarios. Our comprehensive evaluation on the BP-ARR dataset reveals that the established baseline methods exhibit a decline in performance in BP estimation under arrhythmic scenarios compared to normal conditions, with an average increase in SBP MAE of 1.6 mmHg. Notably, the most significant increases were observed in ResUNet + Self-Attention (SBP MAE increased by 3.65 mmHg, 35.2%) and U-Net (SBP MAE increased by 2.21 mmHg, 40.8%).

Keywords: Deep Neural Network · Feature Fusion · Blood Pressure Estimation

1 Introduction

Hypertension is a prevalent chronic condition in clinical practice and a major risk factor for acute cardiovascular events, including stroke, heart failure, and kidney disease [7,33,37]. Blood pressure exhibits significant variability, influenced

M. Yoshikawa et al. (Eds.): ADMA 2025, LNAI 16197, pp. 20–34, 2026.
https://doi.org/10.1007/978-981-95-3453-1_2

by physiological cycles, emotions, environmental factors, and circadian rhythms. Research has shown that this variability strongly predicts the incidence and mortality of cardiovascular diseases, contributing to events such as intracerebral hemorrhage, myocardial infarction, and sudden cardiac death [32,35]. Consequently, continuous blood pressure monitoring is essential for cardiovascular disease prevention. While invasive blood pressure measurement is the gold standard for accuracy [3], its risks, including bleeding and infection, have prompted research into non-invasive, continuous measurement methods (Fig. 1).

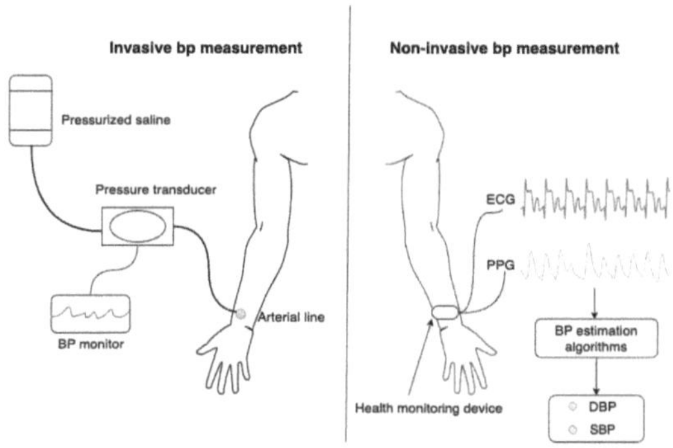

Fig. 1. Comparison of invasive and non-invasive blood pressure measurement methods. Left: Invasive measurement using an arterial line and BP monitor. Right: Non-invasive estimation using ECG and PPG signals from a health monitoring device processed by bp estimation algorithms to output diastolic and systolic blood pressure.

Current non-invasive, continuous blood pressure estimation methods include parametric [6,31] and end-to-end approaches [14]. Parametric methods often rely on physiological parameters such as pulse transit time (PTT) [11,30] and pulse wave velocity (PWV). PTT is defined as the time interval for the arterial pulse pressure wave to travel from the heart to a peripheral site, such as the wrist or finger. It is inversely correlated with blood pressure because higher blood pressure increases arterial stiffness, which accelerates the propagation of the pulse wave and reduces PTT. Conversely, lower blood pressure results in more compliant arteries, slowing the pulse wave and increasing PTT. Common methodologies for PTT estimation utilize synchronously recorded electrocardiogram (ECG) and photoplethysmogram (PPG) signals. The calculation involves identifying specific temporal markers on the ECG, the PPG, and often the first derivative of the PPG waveform. However, these methods are sensitive to arrhythmias, as demonstrated in Fig. 2. Arrhythmias alter the waveform morphology and rhythmic characteristics of physiological signals [8,13], which can disrupt the accurate measurement of PTT. Consequently, variations in PTT and PWV parameters degrade the performance of blood pressure estimation models, posing significant challenges for achieving reliable measurements.

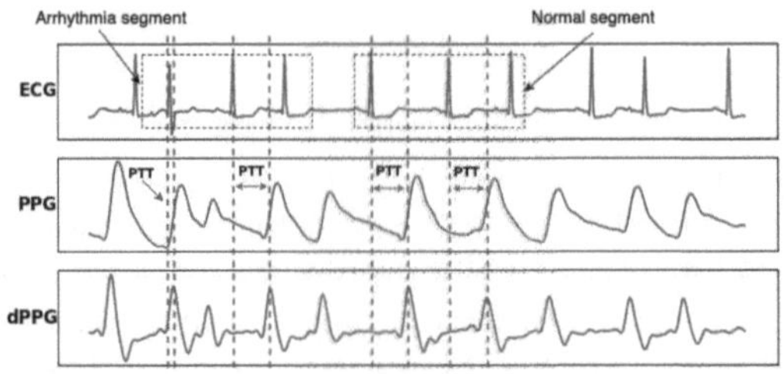

Fig. 2. The distance between the two red dashed lines represents PTT. PTT exhibits a more significant change in the arrhythmia segment compared to the normal segment. (Color figure online)

The end-to-end approach employs deep learning models to automatically extract features and map raw physiological signals to blood pressure values [36]. Specifically, these models typically take synchronized segments of ECG and PPG waveforms as direct input. Architectures like Convolutional Neural Networks (CNNs) are often utilized to capture morphological characteristics and local patterns within each signal [37]. Recurrent Neural Networks (RNNs), such as LSTMs or GRUs, are frequently incorporated, either sequentially or in parallel, to model the temporal dependencies and relationships between the ECG and PPG signals over time, effectively learning how changes in one signal relate to changes in the other preceding the blood pressure pulse [5]. Hybrid models combining CNNs and RNNs are also common, leveraging the strengths of both for feature extraction and sequence modeling [4,24]. The deep learning network processes these complex spatio-temporal features learned directly from the data to establish a mapping function that predicts systolic and diastolic blood pressure. Unlike parametric methods, it learns direct relationships without explicitly constructing physiological models, enhancing robustness through deep learning's ability to capture complex data patterns.

This study investigates how waveform and rhythm changes, particularly due to arrhythmias, impact blood pressure estimation by introducing a novel continuous blood pressure dataset that includes arrhythmic conditions. We evaluate established baseline methods to assess their effectiveness in estimating blood pressure under normal and arrhythmic scenarios. Our findings indicate a performance decline in these methods under arrhythmic conditions. This unique dataset, incorporating arrhythmias, provides a foundation for developing future models to improve blood pressure estimation amidst variations in physiological signal waveforms and rhythms. The main contributions of this study can be summarized as follows:

- We construct a continuous blood pressure dataset featuring arrhythmia, providing a foundation for future models to enhance blood pressure estimation accuracy.
- We evaluate established baseline methods to assess their blood pressure estimation performance under normal and arrhythmic conditions.

- Our comprehensive evaluation on the BP-ARR dataset reveals that the established baseline methods exhibit a decline in BP estimation performance under arrhythmic conditions compared to normal conditions.

2 Related Work

Early datasets for blood pressure research, such as UQVS [23], provided a high-resolution, multi-parameter vital sign data repository focused on anesthetized patients during surgery. Unlike databases predominantly sourced from intensive care unit (ICU) patients, UQVS specifically collected monitoring data from patients under anesthesia. Most cases included electrocardiogram (ECG) and non-invasive arterial blood pressure (NIBP) data. However, UQVS is constrained by its limited sample size of only 32 cases, which restricts patient diversity and lacks detailed clinical information. Similarly, the 2015 CHARIS dataset [20] targeted traumatic brain injury (TBI) patients, offering multi-channel recordings from 29 individuals in a surgical intensive care unit (SICU). This dataset includes continuous arterial blood pressure (ABP) and ECG recordings but is hindered by its small patient cohort and insufficient clinical context.

In recent years, several large-scale, publicly accessible databases containing relevant physiological signals have become available. The MIMIC-III [16] database serves as a comprehensive single-center repository, capturing data from patients in a large tertiary hospital's ICU. It includes vital signs, medications, laboratory measurements, caregiver observations, fluid balance, procedure codes, diagnostic codes, imaging reports, length of stay, survival data, and physiological waveforms such as ECG and BP. Nevertheless, for direct application in training deep learning models for BP estimation, MIMIC-III's primary limitation lies in the raw nature of its physiological signals, which frequently exhibit significant noise, artifacts, and irregularities inherent to real-world clinical recordings.

HEMOBP [22] is another extensive, publicly available dataset designed to support BP prediction in chronic hemodialysis (HD) management. It comprises data from 1,075 HD patients, totaling 4,366,298 records, though it predominantly provides only BP information. The Autonomic Aging dataset records resting ECG and continuous NIBP signals from 1,121 healthy volunteers, accompanied by basic demographic information. However, a key drawback of Autonomic Aging is its exclusion of individuals with diseases that might affect cardiovascular function, resulting in a lack of comprehensive data.

VitalDB [21] offers high-resolution, multi-parameter data from 6,388 cases, encompassing 196 intraoperative monitoring parameters with 486,451 waveform and numerical data trajectories, 73 perioperative clinical parameters, and 34 time-series laboratory result parameters. MIMIC-IV, the latest iteration in the MIMIC series [16,17,29], provides detailed clinical data from over 40,000 ICU patients, enriched with extensive clinical context. While both MIMIC-IV and VitalDB supply vast physiological signal data and clinical information, neither is directly suitable for model training without preprocessing.

PulseDB [34] addresses this gap by providing a high-quality physiological signal dataset constructed from the MIMIC-III Waveform Database Matched Subset and VitalDB. It features rigorously cleaned data and standardized training and testing sets, making it directly applicable for model training. Although PulseDB leverages MIMIC-III's waveform data and includes basic demographic information, it does not incorporate the rich clinical information available in MIMIC-III.

The BIDMC PPG and Respiration Dataset [27], derived from MIMIC-II, provides PPG signals from critically ill patients with detailed manual breath annotations. Initially used to benchmark PPG-based respiratory rate algorithms, it's valued for high-quality annotations and clinical relevance in non-invasive respiratory monitoring research.

The UCI Cuffless Blood Pressure Estimation Dataset [18] offers PPG/ECG signals and BP values for research on cuffless, continuous BP monitoring, supporting wearable health tech. Key limitations are its small sample size and demographic homogeneity, potentially affecting model generalizability, and signal noise requiring preprocessing.

To address these limitations, this study enhances PulseDB by integrating de-identified patient information from MIMIC-III, thereby creating a comprehensive BP dataset tailored specifically for use under arrhythmia conditions.

3 Dataset Construction

The dataset construction involves three key processes: patients selection, physiological segment selection, and signal preprocessing, detailed in the subsequent subsections.

3.1 Patient Selection

The BP-ARR dataset leverages the publicly available PulseDB-V2.0 dataset, which compiles continuous blood pressure measurements from the MIMIC-III Waveform Matched Subset and VitalDB. These sources organize data by unique patient identifiers, enabling the identification of individuals with arrhythmia labels. Due to the absence of clinical information in VitalDB, we focus exclusively on the MIMIC-III Waveform Matched Subset. To integrate comprehensive clinical and demographic information, we align 10,282 patients from this subset with corresponding records in the MIMIC-III Clinical Database. This alignment process uses patient IDs as the unique identifier to match each patient's waveform data with their clinical profiles, including diagnostic codes, age, gender, and other relevant attributes. Specifically, for each patient ID in the Waveform Matched Subset, we retrieve associated records from the Clinical Database, ensuring accurate correspondence between physiological signals and clinical annotations. This enables the extraction of arrhythmia-related diagnostic information, such as ICD-9 codes, to determine arrhythmic status. After filtering for data completeness and clinical relevance, we classify patients into two groups: arrhythmic (928

Table 1. The dataset description of BP-ARR

Item	Characteristics
Number of patients	2341
Number of normal patients	1413 (60%)
Number of arrhythmia patients	928 (40%)
Age range	<80
Gender	1357 male, 984 female
Arrhythmia types	14

Table 2. The statistics of physiological segments.

Item	Normal	Arrhythmia
Number of ECGs	2,460,201 (64%)	1,330,721 (36%)
Number of PPGs	2,460,201	1,330,721
Number of ABPs	2,460,201	1,330,721
SBP (mmHg, mean ± SD)	124.2 ± 22.6	122.58 ± 23.60
DBP (mmHg, mean ± SD)	62.4 ± 13.2	58.73 ± 13.62

patients, 40%) and non-arrhythmic (1,413 patients, 60%), totaling 2,341 participants. All are under 80 years old, with 1,357 males and 984 females. The dataset includes 14 types of arrhythmias, enabling comprehensive studies of blood pressure variability.

3.2 Physiological Segment Selection

According to the patient ID, physiological signal segments of ECG, PPG, and ABP are extracted from PulseDB for both arrhythmic and non-arrhythmic patients. Table 2 summarizes the statistics of these segments, highlighting differences between non-arrhythmic and arrhythmic patients. For each patient, multiple synchronized signal triplets, each consisting of ECG, PPG, and ABP segments, are extracted. Each triplet serves as the basic unit, where all three signals span 10 s with a sampling rate 125 Hz, ensuring consistent temporal resolution. In total, 2,460,201 triplets are extracted for non-arrhythmic patients (64%) compared to 1,330,721 for arrhythmic patients (36%), with equal counts of ECG, PPG, and ABP segments within each triplet. Across all triplets, the systolic blood pressure averages 124.2 ± 22.6 mmHg for non-arrhythmic patients and 122.58 ± 23.60 mmHg for arrhythmic patients, while the diastolic blood pressure averages 62.4 ± 13.2 mmHg and 58.73 ± 13.61 mmHg, respectively, calculated over the entire 10-second duration of each triplet. Figure 3 shows the overall distribution of SBP and DBP under normal and arrhythmic conditions. To ensure data quality, we retain only triplets with synchronized ECG, PPG, and ABP signals, excluding those with significant noise or missing data through automated preprocessing. The inclusion of 14 arrhythmia types, as detailed in Table 1 and identified during patient selection, enables BP-ARR to support comprehensive evaluation of non-invasive blood pressure estimation models, particularly in scenarios where waveform morphology exhibits significant variations due to arrhythmias.

3.3 Signal Preprocessing

The PulseDB paper outlines multiple strategies for preprocessing raw MIMIC-III data to obtain clean and high-quality signals. Given the substantial size of the dataset, additional steps are necessary to eliminate noisy segments while preserving physiological fidelity. We employ NeuroKit [25] to assess the quality of ECG and PPG signals within each 10-second triplet, discarding any triplet where either signal exhibits low quality, such as excessive noise, baseline wander, or missing data. To maintain arrhythmic characteristics critical for BP-ARR, we avoid excessive filtering and outlier removal for ECG and PPG data, ensuring that waveform variations due to arrhythmias are retained. For ABP signals, we apply minimal preprocessing to preserve continuous blood pressure dynamics, verifying that each 10-second segment aligns temporally with its corresponding ECG and PPG signals at 125 Hz sampling rate (Table 3).

Table 3. Arrhythmia types and their corresponding ICD-9 codes

Arrhythmia types	ICD-9 CODE
Parox atrial tachycardia	4270
Parox atrial tachycardia	4271
Parox ventric tachycard	4272
Atrial fibrillation	42731
Atrial flutter	42732
Ventricular fibrillation	42741
Ventricular flutter	42742
Cardiac arrest	4275
Premature beats NOS	42760
Atrial premature beats	42761
Premature beats NEC	42769
Sinoatrial node dysfunct	42781
Cardiac dysrhythmias NEC	42789
Cardiac dysrhythmia NOS	4279

In addition, we extract patient demographic information, including age and gender, and clinical diagnostic data, such as ICD-9 codes for arrhythmia types, from the MIMIC-III Clinical Database. These auxiliary data enhance the dataset's utility, enabling comprehensive analyses of blood pressure variability across diverse patient profiles. Specifically, the inclusion of 14 arrhythmia types, as identified during patient selection, supports detailed investigations into their impact on blood pressure estimation. This enriched dataset is particularly valuable for training advanced models, such as attention-based architectures, which can leverage demographic and diagnostic information to improve performance in non-invasive blood pressure estimation under arrhythmic conditions.

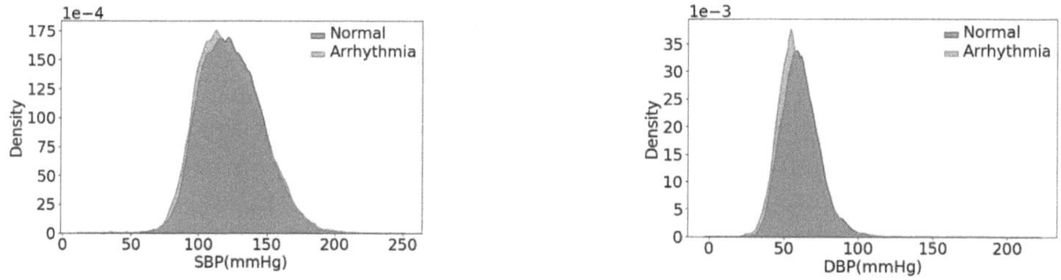

Fig. 3. The data distributions of SBP and DBP.

4 Deep Learning-Based Methods

Deep learning models have demonstrated remarkable efficacy across various domains, with architectures like U-Net and ResNet setting benchmarks through innovative designs. These foundational models serve as baselines for evaluation on our BP-ARR dataset. These architectures essentially cover the overall structure of mainstream deep learning models.

U-Net [28]: it is a convolutional neural network (CNN) specifically designed for biomedical image segmentation. Its distinctive U-shaped architecture combines a contracting path for context and an expanding path for localization, utilizing skip connections to preserve spatial details. It performs exceptionally well on small datasets for pixel-level predictions in medical imaging tasks.

ResNet [10]: it, *i.e.* Residucal Network, addresses the vanishing gradient problem in deep networks through its innovative residual learning mechanism. By using skip connections within residual blocks. it allows gradients to flow more effectively, enabling the training of extremely deep architectures. ResNet achieved benchmark performance in image classification and serves as a versatile backbone for many computer vision applications.

V-Net [26]: it is a 3D convolutional neural network specifically developed for volumetric medical image segmentation, processing inputs like MRI or CT scans. It adapts the U-Net architecture with 3D convolutions and integrates residual connections within its contracting and expanding paths to enhance gradient flow. V-Net shows strong results in segmenting complex 3D structures such as organs and lesions.

MLP-BP [12]: it is an end-to-end framework adapted from MLP-Mixer for blood pressure estimation using ECG and PPG signals. It features a novel MFMC preprocessing technique avoiding manual feature extraction and has variants utilizing gMLP or LSTM components. Evaluated on the MIMIC II dataset, MLP-BP showed competitive performance meeting AAMI and BHS standards.

Conv-LSTM [19]: it is a blood pressure prediction models using Conv1D and hybrid Conv1D-LSTM architectures, trained on the MIMIC-III dataset. The hybrid Conv1D-LSTM model demonstrated superior performance, especially for DBP estimation, compared to a stacked Conv1D model, while requiring fewer parameters. Despite its strengths, SBP predictions fell short of the BHS A-grade standard, and dataset limitations potentially impacted full AAMI compliance.

ResUNet [15]: it is enhanced with a self-attention mechanism for blood pressure estimation from ECG and PPG signals. This integrates ResUNet's feature extraction with self-attention for contextual understanding, outperforming prior work in calibration-based DBP estimation on the PulseDB dataset. Its performance diminished significantly in calibration-free tests, suggesting reliance on individual-specific calibration.

TransfoRhythm [1]: it is a Transformer-based deep neural network for cuffless blood pressure estimation relying solely on PPG signals. Utilizing a multi-head attention mechanism, it captures dependencies within the PPG signal and achieved superior accuracy on the MIMIC-IV dataset. However, its performance depends heavily on PPG signal quality, requiring careful handling of morphology, motion artifacts, and noise.

5 Experiments

5.1 Experimental Setup

Training and evaluation were conducted on our custom BP-ARR dataset. To ensure patient diversity, we uniformly sampled approximately equal signal segments from each patient, constructing training and test sets. Experiments showed that 500,000 segments optimized performance. The dataset was split into training, validation, and test sets in a 7:2:1 ratio. Each model input synchronized 10-second ECG and PPG signals, extracted directly from the dataset. SBP and DBP outputs were computed as the mean SBP and DBP from corresponding 10-second ABP signals, representing blood pressure over the interval.

Multiple models were trained using the Adam optimizer with Mean Absolute Error (MAE) as the loss function to minimize the average absolute difference between predicted and ground truth blood pressure values. Performance was evaluated using MAE (average absolute error magnitude for accuracy), Mean Error (ME, average signed error for bias) $\pm$ Standard Deviation (SD, error variability) to quantify accuracy and consistency in SBP and DBP predictions. For these evaluation metrics, lower MAE indicates better accuracy, an ME closer to zero (smaller absolute value) signifies lower bias, and a smaller SD reflects less variability in the prediction errors, with lower values generally being preferable for all three.

$$\mathrm{MAE} = \frac{1}{n}\sum_{i=1}^{n}|\hat{y}_i - y_i| \tag{1}$$

$$\mathrm{ME} = \frac{1}{n}\sum_{i=1}^{n}(\hat{y}_i - y_i) \tag{2}$$

$$\mathrm{SD} = \sqrt{\frac{1}{n}\sum_{i=1}^{n}(\hat{y}_i - y_i - \mathrm{ME})^2} \tag{3}$$

Additionally, clinical acceptability was evaluated against standards set by the Association for the Advancement of Medical Instrumentation (AAMI) and the British Hypertension Society (BHS). The AAMI standard requires a mean difference of ≤ 5 mmHg and a standard deviation of ≤ 8 mmHg compared to a reference. The BHS protocol assigns grades (A, B, C) based on the percentage of measurements within 5, 10, and 15 mmHg, with grade A requiring at least 60%, 85%, and 95%, respectively, as detailed in Table 4.

Table 4. BHS Standard Grades and Corresponding Percentages

Grades	$\leq$ 5 mmHg	$\leq$ 10 mmHg	$\leq$ 15 mmHg
Grade A	60%	85%	95%
Grade B	50%	75%	90%
Grade C	40%	65%	85%

5.2 Results and Analysis

Tables 5 and 6, respectively, present the results of aforementioned models on the normal and arrhythmia subsets of our proposed BP-ARR dataset. Notably, our study utilized 500,000 signal segments uniformly sampled from BP-ARR for training and testing, a volume exceeding that used in the existing evaluations for the majority of these baseline models. This substantial difference in data scale is a key factor to consider when comparing performance across studies. A notable observation is that the majority of models fail to comply with AAMI and BHS standards on our dataset. For instance, TransfoRhythm [1] achieved an SBP MAE of 1.37 mmHg on MIMIC-IV but recorded an SBP MAE exceeding 17 mmHg on both BP-ARR subsets. These discrepancies likely arise not only from the increased data volume, which may introduce greater signal variability and noise challenges, but also from variations in other dataset properties, such as signal quality distribution, patient demographics, and the breadth of physiological conditions captured.

Table 5. Model performance on the normal subset of the BP-ARR dataset

Model	SBP			DBP			AAMI	BHS
	MAE	ME	SD	MAE	ME	SD		
U-Net [2]	5.42	−4.50	7.62	2.44	−0.41	4.30	pass	B
ResNet [9]	12.44	0.07	14.63	7.26	−0.98	9.66	fail	C
V-Net [9]	25.16	−25.30	12.84	7.14	4.42	8.10	fail	C
ConvLSTM [19]	8.58	0.20	11.52	5.61	−0.43	7.96	fail	C
MLP-BP [12]	7.54	5.13	7.56	3.86	1.29	5.66	pass	C
ResUNet+SA [15]	10.36	−2.71	13.60	6.42	−1.35	8.85	fail	B
TransfoRhythm [1]	17.27	−0.41	21.77	9.57	−0.80	12.56	fail	C

Table 6. Model performance on the arrhythmia subset of the BP-ARR dataset

Model	SBP			DBP			AAMI	BHS
	MAE	ME	SD	MAE	ME	SD		
U-Net [2]	7.63	−2.08	7.68	2.68	0.25	4.10	pass	B
ResNet [9]	12.47	0.40	16.06	7.37	−0.54	10.35	fail	B
V-Net [9]	26.23	−26.12	12.82	8.16	2.89	9.44	fail	C
ConvLSTM [19]	9.31	0.56	12.41	6.36	0.03	8.78	fail	C
MLP-BP [12]	8.69	7.02	8.35	5.01	3.76	5.42	fail	C
ResUNet+SA [15]	14.01	−0.01	17.94	8.25	−0.55	11.13	fail	C
TransfoRhythm [1]	18.18	0.98	22.62	10.10	0.78	13.43	fail	C

Focusing on the normal subset of the BP-ARR dataset, as shown in Table 6, which includes signal segments without arrhythmias, the models displayed a range of performance levels. U-Net stood out as the top performer, achieving the lowest MAE for both SBP (5.42 mmHg) and DBP (2.44 mmHg). It satisfied the AAMI standard and earned a BHS Grade B. MLP-BP also met the AAMI requirements, with an SBP MAE of 7.54 mmHg and a DBP MAE of 3.86 mmHg, though it received a BHS Grade C and showed a tendency to overestimate SBP (ME of 5.13 mmHg). Other models, including ResNet, V-Net, ConvLSTM, ResUNet+SA, and TransfoRhythm, did not meet the AAMI standard, as they produced higher errors or greater variability (SD). In particular, V-Net and TransfoRhythm performed poorly, with V-Net underestimating SBP significantly (ME of -25.30 mmHg), while TransfoRhythm recorded the highest SBP MAE (17.27 mmHg) and SD (21.77 mmHg). Most of these models received a Grade C from BHS, while ResUNet+SA [1] achieved a Grade B despite failing the AAMI standard. These results indicate considerable inaccuracy and inconsistency, even under normal conditions in our challenging dataset.

Analysis of the arrhythmia subset (Table 6) showed a consistent drop in performance for all models compared to the normal subset, emphasizing the negative effect of arrhythmias on blood pressure estimation. U-Net remained the best performer, retaining its AAMI 'pass' status and BHS Grade B, but its accuracy declined, with the SBP MAE rising by 2.21 mmHg (a 40.8% increase) to 7.63 mmHg. Strikingly, all other models failed to meet the AAMI standard on this subset. Even MLP-BP, which succeeded on the normal subset, could not maintain its performance on arrhythmic signals. On average, the SBP MAE across all models increased by 1.6 mmHg. Some models experienced larger declines; for instance, ResUNet+SA saw its SBP MAE rise by 3.65 mmHg (a 35.2% increase). Additionally, the variability of errors (SD) typically increased with arrhythmias, suggesting less reliable predictions. These findings reveal that current deep learning methods for estimating blood pressure are highly sensitive to arrhythmic conditions, possibly because irregular signal patterns interfere with the features the models have learned. This underscores the need to develop models that can handle arrhythmias effectively, and the BP-ARR dataset offers a useful tool for this purpose.

6 Conclusion

This study investigates the impact of arrhythmic conditions on the accuracy of deep learning-based non-invasive BP estimation models on the proposed BP-ARR dataset. The BP-ARR dataset, a large-scale resource designed to address challenges in cardiovascular monitoring, includes 500,000 10-second physiological signal segments from individuals under normal cardiac rhythms and an equivalent number from those with arrhythmic conditions. By analyzing established baseline models under both scenarios, our findings reveal a notable decline in BP estimation performance during arrhythmic episodes compared to normal conditions. This performance gap underscores the inherent sensitivity of current deep learning frameworks to irregular cardiac patterns, which may compromise their reliability in real-world clinical settings where arrhythmias are prevalent.

Our findings emphasize the critical need for improving the robustness of BP estimation algorithms to account for cardiac rhythm variability. The BP-ARR dataset emerges as a pivotal tool in this endeavor, offering a structured platform to evaluate and refine models against arrhythmia-induced signal distortions. By enabling the development of more resilient algorithms, this resource has the potential to advance continuous, non-invasive BP monitoring technologies, particularly for patients with cardiovascular diseases who are at higher risk of arrhythmias.

These insights not only highlight a key limitation in existing methodologies but also validate the BP-ARR dataset's utility in driving innovation for patient-specific BP management solutions. Future work leveraging this dataset could lead to enhanced model generalizability, improved clinical decision-making, and better outcomes in cardiovascular care, ultimately bridging the gap between theoretical model performance and practical healthcare applications.

Acknowledgement. Shuangdu Li and Ziyou Li contribute equally to this work. This work is partially supported by the Guangdong Basic and Applied Basic Research Foundation (No. 2025A1515011614), the National Natural Science Foundation of China (No. 62473267, No. U2241210), and the Natural Science Foundation of Top Talent of SZTU (No. GDRC202318).

References

1. Arjomand, A., Boudesh, A., Bayatmakou, F., Kent, K.B., Mohammadi, A.: Transforhythm: a transformer architecture conductive to blood pressure estimation via solo PPG signal capturing. arXiv preprint arXiv:2404.15352 (2024)
2. Athaya, T., Choi, S.: An estimation method of continuous non-invasive arterial blood pressure waveform using photoplethysmography: a u-net architecture-based approach. Sensors **21**(5), 1867 (2021)
3. Cheroni, C., Caporale, N., Testa, G.: Autism spectrum disorder at the crossroad between genes and environment: contributions, convergences, and interactions in ASD developmental pathophysiology. Mol. Autism **11**(1), 69 (2020)
4. El Hajj, C., Kyriacou, P.A.: Cuffless and continuous blood pressure estimation from PPG signals using recurrent neural networks. In: 2020 42nd Annual International Conference of the IEEE Engineering in Medicine & Biology Society (EMBC), pp. 4269–4272. IEEE (2020)
5. El-Hajj, C., Kyriacou, P.A.: Cuffless blood pressure estimation from PPG signals and its derivatives using deep learning models. Biomed. Signal Process. Control **70**, 102984 (2021)
6. Esmaili, A., Kachuee, M., Shabany, M.: Nonlinear cuffless blood pressure estimation of healthy subjects using pulse transit time and arrival time. IEEE Trans. Instrum. Meas. **66**(12), 3299–3308 (2017)
7. Ettehad, D., et al.: Blood pressure lowering for prevention of cardiovascular disease and death: a systematic review and meta-analysis. Lancet **387**(10022), 957–967 (2016)
8. Fiori, G., Fuiano, F., Scorza, A., Conforto, S., Sciuto, S.A.: Non-invasive methods for PWV measurement in blood vessel stiffness assessment. IEEE Rev. Biomed. Eng. **15**, 169–183 (2021)
9. González, S., Hsieh, W.T., Chen, T.P.C.: A benchmark for machine-learning based non-invasive blood pressure estimation using photoplethysmogram. Sci. Data **10**(1), 149 (2023)
10. He, K., Zhang, X., Ren, S., Sun, J.: Deep residual learning for image recognition. In: Proceedings of the IEEE Conference on Computer Vision and Pattern Recognition, pp. 770–778 (2016)
11. Heydari, F., Ebrahim, M.P., Redoute, J.M., Joe, K., Walker, K., Yuce, M.R.: A chest-based continuous cuffless blood pressure method: Estimation and evaluation using multiple body sensors. Inf. Fusion **54**, 119–127 (2020)
12. Huang, B., Chen, W., Lin, C.L., Juang, C.F., Wang, J.: MLP-BP: a novel framework for cuffless blood pressure measurement with PPG and ECG signals based on MLP-mixer neural networks. Biomed. Signal Process. Control **73**, 103404 (2022)
13. Huang, X., Lu, Y., Guo, M., Du, S., Han, N.: Recent strategies for Nano-based PTT combined with immunotherapy: from a biomaterial point of view. Theranostics **11**(15), 7546 (2021)
14. Ismail, S.N.A., Nayan, N.A., Jaafar, R., May, Z.: Recent advances in non-invasive blood pressure monitoring and prediction using a machine learning approach. Sensors **22**(16), 6195 (2022)

15. Jamil, Z., Lui, L.T., Chan, R.H.: Blood pressure estimation using self-attention mechanism built-in ResuNet on PulseDB: demographic fairness, and generalization. IEEE Sens. J. (2024)
16. Johnson, A., et al.: Mimic-III, a freely accessible critical care database. Sci. Data **3**(1), 1–9 (2016)
17. Johnson, A.E., et al.: Mimic-IV, a freely accessible electronic health record dataset. Sci. Data **10**(1), 1 (2023)
18. Kachuee, M., Kiani, M., Mohammadzade, H., Shabany, M.: Cuff-Less Blood Pressure Estimation. UCI Machine Learning Repository (2015)
19. Kamanditya, B., Fuadah, Y.N., Mahardika T, N.Q., Lim, K.M.: Continuous blood pressure prediction system using conv-LSTM network on hybrid latent features of photoplethysmogram (PPG) and electrocardiogram (ECG) signals. Sci. Rep. **14**(1), 16450 (2024)
20. Kim, N., et al.: Trending autoregulatory indices during treatment for traumatic brain injury. J. Clin. Monit. Comput. **30**, 821–831 (2016)
21. Lee, H.C., Park, Y., Yoon, S.B., Yang, S.M., Park, D., Jung, C.W.: Vitaldb, a high-fidelity multi-parameter vital signs database in surgical patients. Scientific Data **9**(1), 279 (2022)
22. Lin, C.J., Chen, Y.Y., Pan, C.F., Wu, V., Wu, C.J.: Dataset supporting blood pressure prediction for the management of chronic hemodialysis. Sci. Data **6**(1), 313 (2019)
23. Liu, D., Görges, M., Jenkins, S.A.: University of queensland vital signs dataset: development of an accessible repository of anesthesia patient monitoring data for research. Anesthesia Analgesia **114**(3), 584–589 (2012)
24. Mahmud, S., et al.: NABNet: a nested attention-guided BiConvLSTM network for a robust prediction of blood pressure components from reconstructed arterial blood pressure waveforms using PPG and ECG signals. Biomed. Signal Process. Control **79**, 104247 (2023)
25. Makowski, D., et al.: Neurokit2: a python toolbox for neurophysiological signal processing. Behavior research methods, pp. 1–8 (2021)
26. Milletari, F., Navab, N., Ahmadi, S.A.: V-net: fully convolutional neural networks for volumetric medical image segmentation. In: 2016 Fourth International Conference on 3D Vision (3DV), pp. 565–571. IEEE (2016)
27. Pimentel, M.A.F., et al.: Towards a robust estimation of respiratory rate from pulse oximeters. IEEE Trans. Biomed. Eng. **64**(8), 1914–1923 (2016)
28. Ronneberger, O., Fischer, P., Brox, T.: U-net: convolutional networks for biomedical image segmentation. In: Navab, N., Hornegger, J., Wells, W.M., Frangi, A.F. (eds.) MICCAI 2015, Part III. LNCS, vol. 9351, pp. 234–241. Springer, Cham (2015). https://doi.org/10.1007/978-3-319-24574-4_28
29. Saeed, M., Lieu, C., Raber, G., Mark, R.G.: Mimic ii: a massive temporal ICU patient database to support research in intelligent patient monitoring. In: Computers in Cardiology, pp. 641–644. IEEE (2002)
30. Seeberg, T.M., et al.: A novel method for continuous, noninvasive, cuff-less measurement of blood pressure: evaluation in patients with nonalcoholic fatty liver disease. IEEE Trans. Biomed. Eng. **64**(7), 1469–1478 (2016)
31. Shao, J., Shi, P., Hu, S., Liu, Y., Yu, H.: An optimization study of estimating blood pressure models based on pulse arrival time for continuous monitoring. J. Healthcare Eng. **2020**(1), 1078251 (2020)
32. Umemura, S., et al.: The Japanese society of hypertension guidelines for the management of hypertension (JSH 2019). Hypertens. Res. **42**(9), 1235–1481 (2019)

33. Visseren, F.L., et al.: 2021 esc guidelines on cardiovascular disease prevention in clinical practice: developed by the task force for cardiovascular disease prevention in clinical practice with representatives of the European society of cardiology and 12 medical societies with the special contribution of the european association of preventive cardiology (eapc). Eur. Heart J. **42**(34), 3227–3337 (2021)
34. Wang, W., Mohseni, P., Kilgore, K.L., Najafizadeh, L.: PulseDB: a large, cleaned dataset based on mimic-iii and VitalDB for benchmarking cuff-less blood pressure estimation methods. Frontiers in Digital Health **4**, 1090854 (2023)
35. Wang, Z., et al.: Status of hypertension in china: results from the China hypertension survey, 2012–2015. Circulation **137**(22), 2344–2356 (2018)
36. Yang, Z., et al.: CVAN: a novel sleep staging method via cross-view alignment network. IEEE J. Biomed. Health Inform. (2024)
37. Zabihi, S., Rahimian, E., Marefat, F., Asif, A., Mohseni, P., Mohammadi, A.: BP-net: cuff-less and non-invasive blood pressure estimation via a generic deep convolutional architecture. Biomed. Signal Process. Control **78**, 103850 (2022)

Fast Fourier Transform Diffusion Model for ECG Denoising

Yitong Li(✉)

KU Leuven, Oude Markt 13, 3000 Leuven, Belgium
yitong.li1997@outlook.com

Abstract. Biomedical signals such as electrocardiograms (ECG) are distorted by diverse types of noise such as baseline wander, undermining diagnostic accuracy. Current denoising methods struggle to balance noise reduction and preservation of critical waveforms such as P/QRS/T waves. For this purpose, this study proposes a Fast Fourier Transform Diffusion (FFT Diffusion) framework. FFT Diffusion implements Fourier transform to decompose ECG into multiscale frequency subbands, enabling adaptive noise reduction while retaining frequency-specific biological features. A U-shaped architecture based on Transformers captures local waveform details and models long-term dependencies. This study also uses a composite loss function to train the U-shaped model. By combining adaptive frequency decomposition and transformer architecture, FFT Diffusion outperforms state-of-the-art methods and provides a robust solution for ECG denoising and effective preservation of diagnostic information from complex noise.

Keywords: ECG denoising · Fast Fourier transform · Diffusion model · Transformer

1 Introduction

Electrocardiogram (ECG) signals are essential for cardiac assessment. However, their clinical features are frequently compromised by various types of noise, which diminishes diagnostic accuracy [1]. Research [6] illustrates that high noise levels negatively affect heartbeat detection and further obscure the identification of cardiac diseases. Several studies have indicated that traditional denoising methods, such as wavelet transform, suffer from boundary effects, a lack of translation invariance, and aliasing [8]. Deep learning-based models are state-of-the-art in biomedical data analysis tasks ([10,23,24] and [9]). The study [29] shows that they can substantially enhance denoising performance but they are also limited by incomplete retention of information.

The cold diffusion strategy may offer a solution. However, its application to ECG is impeded by two unresolved challenges: (1) the lack of frequency-specific feature preservation for P/QRS/T waves, and (2) the inability to model

M. Yoshikawa et al. (Eds.): ADMA 2025, LNAI 16197, pp. 35–50, 2026.
https://doi.org/10.1007/978-981-95-3453-1_3

long-range temporal dependencies in cardiac cycles, which are essential for distinguishing repetitive noise from pathological rhythms.

In light of these challenges, this study proposes a Fast Fourier Transform Diffusion (FFT Diffusion) model for effective noise removal. The primary contribution of this work lies in decomposing ECG into multiscale frequency subbands using Fourier transform and integrating a U-shaped Transformer network for modeling temporal dependencies. Experimental results from two standard databases demonstrate that FFT Diffusion outperforms state-of-the-art methods in mixed noise scenarios and across varying levels of signal-to-noise ratio. This work not only advances ECG denoising techniques but also establishes a generalizable framework for applying cold diffusion to other biomedical signals with distinct spectral-temporal characteristics.

The rest of this paper is organized as follows: Sect. 2 reviews related work on ECG denoising techniques. Section 3 details the proposed methodology. Section 4 presents experimental results, comparing the proposed method with baseline approaches and analyzing the impact of key architectural components. Finally, Sect. 5 concludes the study and outlines future research directions.

2 Related Works

2.1 Traditional Denoising Methods

Traditional methods such as wavelet transform [15] and empirical mode decomposition [25] have been widely used for noise reduction. These methods decompose the signal into different frequency components and eliminate noise-related artifacts. Adaptive filtering techniques [4] iteratively update filter coefficients based on reference noise signals to achieve effective denoising. However, limitations such as aliasing reduce denoising accuracy.

2.2 Deep Learning-Based Denoising Methods

Convolutional neural network (CNN) and recurrent neural network (RNN) have demonstrated more effective performance in ECG signal denoising compared to traditional methods. The fully convolutional network (FCN) [5] not only effectively improves the signal-to-noise ratio of ECG signals but also facilitates signal compression. The deep recurrent neural network (DRNN) [2] significantly outperformed traditional denoising methods, particularly for significantly corrupted signals. This study also showed that networks pretrained on synthetic data exhibited superior performance compared to those trained exclusively on real data. The dual recurrent structure [26] allows the exploitation of both temporal waveform characteristics and spatial image details, achieving more efficient noise reduction. Furthermore, the researches ([20] and [18]) demonstrate that generative adversarial network (GAN) model improves denoising performance and reconstructs high-quality ECG signals.

2.3 Diffusion Based Denoising Methods

Diffusion models have become a notable approach, particularly in tasks involving image restoration and generation. Initially, these models are based on adding Gaussian noise to the images, followed by training a denoising network to reconstruct clean images [7]. However, they often underperform when applied to non-Gaussian noise.

The cold diffusion framework [3] replaces stochastic Gaussian noise with deterministic degradation operators to model image transformations. The work found that diffusion-based reconstruction remains effective without Gaussian noise when forward/reverse processes align with degradation-restoration logic, expanding the framework to tasks like denoising and inpainting via non-stochastic operations. Subsequent work [27] used cold diffusion for real-world image denoising by aligning the forward process with actual noise distributions. This framework incorporated a noise-aware training strategy with real noisy-clean pairs and featured an adaptive noise scheduler. The reverse process employed a score-matching network optimized for non-Gaussian noise to enhance robustness against complex corruptions. In domain-specific applications, the CDiffSD approach [21] customized the degradation operators to seismic noise patterns, improving the reconstruction of earthquake signals from noisy data. The study on speech enhancement [28] applied deterministic degradations in conjunction with unfolded training, enabling effective noise reduction under diverse acoustic conditions.

The cold diffusion framework demonstrates significant potential for ECG denoising due to its ability to manage non-Gaussian noise distributions frequently encountered in biomedical signals. The methodology, proven effective in processing one-dimensional signals such as speech and seismic data, supports its adaptability to ECG denoising. However, unlike other one-dimensional signals, ECG signals possess distinct characteristics, such as the presence of P/QRS/T waves, which require modification to the cold diffusion framework.

3 Methodology

The cold diffusion framework comprises two main components: the degradation process and the reverse process. The forward process involves the iterative addition of noise to the clean signal, while the reverse process reconstructs the clean signal from the noisy version in the backward direction.

3.1 Degradation and Sampling Process

The degradation process is defined by a linear interpolation between a clean signal x_0 and noisy signal x_T [27]. The intermediate noisy signal x_t at time step t is controlled by the parameter $\alpha = \frac{t}{T}$, where T denotes the total number of diffusion steps:

$$x_t = D_{x_T}(x_0, t) = (1 - \alpha)x_0 + \alpha x_T \quad \text{for} \quad t = 0, 1, \ldots, T. \tag{1}$$

The sampling process follows the improved sampling algorithm based on [27]. Given the degraded signal x_t, the reconstructed signal is obtained by the restoration model R_θ, which is specifically designed to estimate the desired output x_0. The improved sampling algorithm can be written as the following formula:

$$x_{t-1} = x_t - \tilde{x}_t + \tilde{x}_{t-1} = x_t - 1/T \cdot [x_T - R_\theta(x_t, t)] \tag{2}$$

3.2 Model Architecture

Main Architecture. The proposed model is founded on a U-shaped neural network [30]. It consists of an encoder-decoder structure with skip connections (see Fig. 1). The input signal is transformed into the frequency domain using a fast Fourier transform (FFT), where it is decomposed into multiple frequency bands. The encoder and decoder are structured as transformer blocks, following upsampling and downsampling modules. The sampling module also integrates temporal information through sinusoidal time embeddings. Finally, an inverse FFT is applied to recover the denoised signal in the time domain.

Multi-scale Frequency Embedding via Fast Fourier Transform. Given an input signal $x_t \in \mathbb{R}^T$, the discrete Fourier transform (DFT) converts a time-domain signal into its frequency-domain representation, which can be expressed as:

$$X_k = \sum_{n=0}^{N-1} x_n \cdot e^{-j2\pi kn/N}, \quad k = 0, 1, ..., N-1 \tag{3}$$

where x_n denotes the input time-domain signal of length N, and X_k is the corresponding complex frequency-domain coefficient at frequency bin k. The inverse process, known as the Inverse Discrete Fourier Transform (IDFT), reconstructs the original time-domain signal from its frequency components:

$$x_n = \frac{1}{N} \sum_{k=0}^{N-1} X_k \cdot e^{j2\pi kn/N}, \quad n = 0, 1, ..., N-1 \tag{4}$$

In this study, fast Fourier transform (FFT) is applied to decrease time complexity of DFT. The MultiScaleFFTEmbedding module first computes the FFT of the input signal $x \in \mathbb{R}^{B \times C \times T}$, where B is the batch size, C is the number of input channels and T is the temporal length. The FFT result $X \in \mathbb{C}^{B \times C \times T}$ is then divided into S non-overlapping frequency bands:

$$X^{(i)} = M^{(i)} \odot X, \quad i = 1, 2, ..., S \tag{5}$$

where $M^{(i)} \in \{0, 1\}^{B \times C \times T}$ is a binary mask selecting the frequency subband, and $\odot$ denotes element-wise multiplication. These masked subband components $X^{(i)}$ are concatenated along the channel dimension to form the multiscale embedding:

$$X_{\text{embed}} = \text{Concat}\left(X^{(1)}, X^{(2)}, ..., X^{(S)}\right) \tag{6}$$

The MultiScaleFFTReconstruction module processes each frequency subband independently using separate convolutional layers for the real and imaginary components:

$$Y^{(i)} = \text{Conv}_{\text{real}}^{(i)}\left(\Re\left(X^{(i)}\right)\right) + j \cdot \text{Conv}_{\text{imag}}^{(i)}\left(\Im\left(X^{(i)}\right)\right) \tag{7}$$

The processed frequency subbands are concatenated and fused via an additional convolution:

$$Y_{\text{fused}} = \text{FusionConv}_{\text{real}}\left(\Re(\text{Concat}(Y^{(i)}))\right) + j \cdot \text{FusionConv}_{\text{imag}}\left(\Im(\text{Concat}(Y^{(i)}))\right) \tag{8}$$

Finally, the fused frequency-domain signal Y_{fused} is transformed back to the time domain using the Inverse FFT:

$$\tilde{x} = \Re\left(\text{IFFT}(Y_{\text{fused}})\right) \tag{9}$$

Frequency information is crucial in time series analysis, especially for electrocardiogram (ECG) signals. Different ECG segments such as P/QRS/T waves have unique spectral patterns, and their accurate decomposition aids in detecting arrhythmias and other pathologies. Traditional methods such as convolutional layers with varying kernel sizes [30] aim to extract multiscale features by modeling local patterns, but often lack sensitivity to actual frequency content and struggle with generalization between signals of different periodic structures. In contrast, the FFT directly analyzes signals in the frequency domain, providing an interpretable and robust representation of periodic patterns. It enables adaptive decomposition that matches the true spectral behavior of signals, which is particularly useful for ECG where different physiological components occupy distinct frequency ranges.

Encoder/Decoder Blocks. The network's encoder-decoder architecture is built around transformer-based modules, which are designed to model both local waveform details and long-range temporal correlations in frequency-domain signals. The encoder integrates two transformer blocks, each paired with a patch merging layer that uses convolutional operations to upsample features (reducing sequence length while increasing channel depth). Conversely, the decoder employs symmetric transformer blocks followed by patch separating layers, which use transposed convolutions to downsample features, restoring sequence length and reducing channel dimensions. Each transformer block consists of a standardized pipeline: layer normalization, multi-head self-attention (MHSA) and a feed-forward network (FFN) for feature transformation, repeated with residual connections. This design allows the model to capture clinically short-duration patterns through local feature extraction. It also uses self-attention to track long-range dependencies which are essential for distinguishing periodic cardiac signals from noise that exhibits similar temporal structures. The hybrid architecture improves noise discrimination in ECG signals [30].

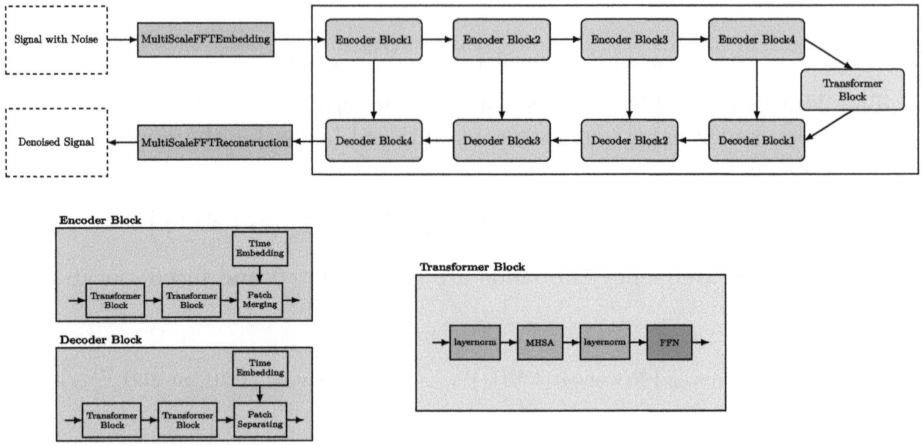

Fig. 1. Architecture of the proposed model. The top part shows the main flow of the model, including the multi-scale patch embedding, encoder-decoder structure, and bottleneck block. The bottom left part illustrates the structure of the encoder and decoder blocks, while the bottom right part shows the internal components of a transformer block.

3.3 Loss Function

To train the denoising model within the proposed diffusion framework, this study applies a composite loss function that captures spatial, temporal, and spectral differences between the predicted clean signal $\tilde{x}_0$ and the ground truth clean signal x_0. The loss function used to optimize the model is defined as:

$$\mathcal{L} = \lambda_1 \cdot \mathcal{L}_{\text{MSE}} + \lambda_2 \cdot \mathcal{L}_{\text{DTW}} + \lambda_3 \cdot \mathcal{L}_{\text{FreDF}}, \tag{10}$$

where:

- $\mathcal{L}_{\text{MSE}} = \|\tilde{x}_0 - x_0\|_2^2$ is the mean squared error (MSE) which calculates point-wise signal differences.
- $\mathcal{L}_{\text{DTW}}$ is the Soft Dynamic Time Warping (Soft-DTW) loss ([13] and [14]), which is a differentiable version of the DTW distance. It captures the temporal alignment between the predicted and clean signals, allowing for non-linear warping of the time axis to minimize the distance between them.
- $\mathcal{L}_{\text{FreDF}} = \|\mathcal{F}(\tilde{x}_0) - \mathcal{F}(x_0)\|_1$ is the mean absolute error between the real-valued frequency spectra of the signals using the one-dimensional real FFT $\mathcal{F}$ [22].

3.4 Evaluation Metrics

The performance of the proposed model is evaluated using the following metrics.

Signal-to-Noise Ratio (SNR). Signal to Noise Ratio (SNR) is a measure used to compare the level of a signal with the level of background noise. A higher SNR indicates that the signal stands out clearly from the background noise.

$$\text{SNR (dB)} = 10 \cdot \log_{10}\left(\frac{\sum_{i=1}^{N} x_i^2}{\sum_{i=1}^{N}(x_i - \tilde{x}_i)^2}\right) \tag{11}$$

Root Mean Square Error (RMSE). Root Mean Square Error (RMSE) quantifies the average difference between the predicted signal and the clean signal. A lower RMSE indicates that the predicted signal closely follows the true waveform morphology.

$$\text{RMSE} = \sqrt{\frac{1}{N}\sum_{i=1}^{N}(x_i - \tilde{x}_i)^2} \tag{12}$$

Percentage Root Mean Square Difference (PRD). Percentage Root Mean Square Difference (PRD) is a normalized version of RMSE, which expresses the RMSE as a percentage of the total energy of the clean signal. A lower PRD means more accurate signal reconstruction. It is useful to preserve peak magnitudes such as those in the QRS complex.

$$\text{PRD}(\%) = \frac{\sqrt{\sum_{i=1}^{N}(x_i - \tilde{x}_i)^2}}{\sqrt{\sum_{i=1}^{N} x_i^2}} \times 100 \tag{13}$$

Correlation Coefficient (CC). Correlation Coefficient (CC) measures the strength and direction of the linear relationship between the predicted signal and the clean signal. CC evaluates the overall similarity between the original and reconstructed signals.

$$\text{CC} = \frac{\sum_{i=1}^{N}(x_i - \bar{x})(\tilde{x}_i - \bar{\tilde{x}})}{\sqrt{\sum_{i=1}^{N}(x_i - \bar{x})^2} \cdot \sqrt{\sum_{i=1}^{N}(\tilde{x}_i - \bar{\tilde{x}})^2}} \tag{14}$$

where $\bar{x}$ and $\bar{\tilde{x}}$ are the means of x and $\tilde{x}$, respectively.

4 Experiments and Results

4.1 Model Setting and Training

Clean ECG signals sourced from the MIT-BIH Arrhythmia database [16] and the QT database [11] were groundtruth references for evaluating the proposed denoising framework. Each ECG recording was segmented into fixed-length intervals of 512 samples, corresponding to individual heartbeats. To construct the training dataset, noise segments were randomly sampled from the MIT-BIH Noise Stress Test database [17] and superimposed onto clean heartbeat segments to simulate noisy inputs. The intensity of the added noise was modulated by a scaling factor to achieve specific signal-to-noise ratio (SNR) levels of −5 dB, 0 dB, and 5 dB. The noise types incorporated in the experiments included seven distinct categories: three representing isolated artifact sources, including baseline wander (BA), muscle artifact (MA) and electrode motion (EM), and four comprising various combinations of these noise types (BA+EM, BA+MA, MA+EM, and BA+EM+MA).

The constructed dataset was divided into training, validation and testing subsets following an 8:1:1 ratio. The model was trained over 100 epochs with a batch size of 512. An initial learning rate of 0.001 was adopted and subsequently decreased after 50 epochs to promote convergence. Optimization was performed using the Adam algorithm, minimizing a composite loss function as defined in Eq. 10. This loss function incorporated multiple terms, each weighted by a corresponding hyperparameter to balance their influence during training. Specifically, the weights were set as $\lambda_1 = 0.3$, $\lambda_2 = 0.3$, and $\lambda_3 = 0.4$. The number of diffusion steps, a key parameter in the denoising framework, was fixed at $T = 100$ to guide the reverse process of signal reconstruction.

4.2 Results

In the following we present our results and discuss the effectiveness of our model by adopting quantitative metrics. To compare the performance of the proposed diffusion algorithm, we selected the following denoising methods: wavelet transform with Savitzky-Golay filter (WT [19]), fully connected convolutional neural network (FCN [5]), deep recurrent neural network (DRNN [2]) and Deep Score-Based Diffusion Model (DeScod [12]). The performance of these methods was evaluated using the same dataset and noise conditions as the FFT Diffusion. The results are presented in the following sections.

Evaluation Results

Table 1 presents a detailed evaluation of denoising performance on the MIT-BIH Arrhythmia database under various input SNR levels (−5, 0, and 5 dB). Across all metrics, FFT Diffusion demonstrates consistent superiority over baseline models. For SNR, FFT Diffusion achieves leading values in the BW noise scenario: 28.338, 30.878, and 34.386 at −5, 0, and 5 dB respectively, significantly outperforming FCN and DRNN, for instance. Even in complex mixed-noise conditions such as BW+EM+MA, FFT Diffusion maintains the highest SNR values. In terms of RMSE and PRD, FFT Diffusion achieves the lowest values in all settings. For example, under BW noise at −5 dB, its RMSE is just 0.027 compared to WT's 1.178, and PRD drops to 0.050 versus WT's 2.205. It indicates excellent signal fidelity across all types and levels of noise. The CC metric (closer to 1 denotes higher waveform similarity) further validates the effectiveness of FFT diffusion. It consistently exceeds 0.99 in most conditions, reaching up to

0.997 under BW and BW+MA noise at 5 dB and outperforming other models. In summary, FFT Diffusion not only offers superior denoising performance, but also preserves in preserving ECG waveform features across varying noise types and levels.

The evaluation results of the QT database are presented in Table 2. It is evident that the FFT Diffusion model outperforms baseline models across metrics. For example, at −5 dB, its SNR (30.412 in BW) surpasses FCN (17.412) and DRNN (21.33). At 0 dB (32.398 in BW) and 5 dB (34.81 in BW), FFT diffusion shows the strongest denoising efficiency. For RMSE, at −5 dB, the value of 0.023 for BW+EM+MA is lower than those for DRNN (0.051) and FCN (0.066). At 0 dB and 5 dB, FFT Diffusion exhibits lower values than other methods, indicating minimal distortion and effective waveform preservation. Similarly, the PRD values of FFT Diffusion are the smallest among all baselines across different conditions. Regarding CC, at −5 dB, the value of BW (0.992) exceeds those for other models. FFT Diffusion continues to demonstrate high waveform similarity at 0 dB (0.995) and 5 dB (0.997). Similar results are observed under other noise conditions. The higher SNR, lower RMSE and PRD, along with higher CC of FFT Diffusion confirm its effectiveness in denoising and waveform preservation across all SNR levels and noise types.

Figure 2 and Fig. 3 display average evaluation metrics of denoised ECG signals for different methods for the MIT-BIH Arrhythmia and QT databases. In terms of SNR (subplot a), FFT Diffusion demonstrates superior denoising performance at −5 dB, 0 dB, and 5 dB. For RMSE (subplot b), its low values (notably at −5 dB and 0 dB) indicate minimal distortion and effective waveform preservation. Regarding CC (subplot c), FFT Diffusion achieves the highest values, reflecting strong waveform similarity to the original signal. In PRD (subplot d), lower values highlight its efficient noise removal and waveform preservation compared to other models. In summary, FFT Diffusion surpasses other baseline models due to high SNR, low RMSE and PRD, and high CC.

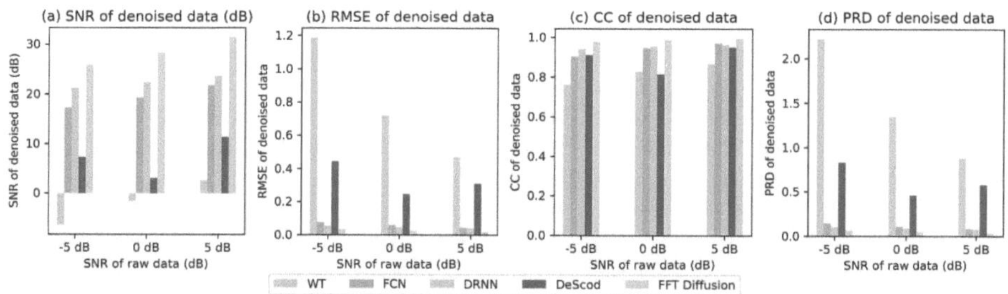

Fig. 2. Average evaluation metrics of the denoised ECG signals using different methods on the MIT-BIH Arrhythmia database.

Table 1. Model comparisons of different denoising methods on the MIT-BIH Arrhythmia database, presenting SNR, RMSE, PRD, and CC for each noise type and input SNR level.

Metrics	Input SNR (dB)	model	BW	EM	MA	BW+EM	BW+MA	EM+MA	BW+EM+MA
SNR	−5	WT	−6.388	−6.563	−6.299	−6.419	−6.361	−6.380	−6.387
		FCN	18.365	15.740	17.610	17.031	17.176	17.135	17.338
		DRNN	21.853	19.477	21.808	20.521	22.402	20.964	21.130
		DeScod	−3.115	7.636	0.288	14.729	20.241	4.286	6.846
		FFT Diffusion	28.338	23.075	26.332	24.915	27.273	25.190	26.101
	0	WT	−1.594	−1.772	−1.547	−1.637	−1.561	−1.605	−1.590
		FCN	19.545	17.971	19.561	19.098	19.859	19.520	19.570
		DRNN	23.102	20.806	23.134	21.909	23.379	21.962	22.611
		DeScod	−36.991	−4.243	19.577	−3.559	21.762	17.427	7.561
		FFT Diffusion	30.878	24.91	29.046	27.406	30.447	27.429	28.669
	5	WT	2.613	2.389	2.594	2.613	2.636	2.585	2.626
		FCN	22.934	20.554	22.415	21.17	22.082	22.013	21.597
		DRNN	24.242	22.122	23.903	23.099	24.749	23.415	23.973
		DeScod	23.852	0.081	15.992	2.612	19.487	15.338	2.314
		FFT Diffusion	34.386	27.506	31.934	30.334	33.771	30.671	32.092
RMSE	−5	WT	1.178	1.225	1.174	1.188	1.173	1.182	1.178
		FCN	0.067	0.090	0.072	0.079	0.074	0.076	0.076
		DRNN	0.050	0.065	0.050	0.057	0.047	0.054	0.053
		DeScod	1.171	0.217	0.692	0.102	0.055	0.579	0.288
		FFT Diffusion	0.027	0.047	0.032	0.036	0.028	0.035	0.031
	0	WT	0.717	0.743	0.714	0.723	0.713	0.719	0.717
		FCN	0.057	0.069	0.057	0.061	0.055	0.058	0.056
		DRNN	0.044	0.055	0.043	0.048	0.043	0.048	0.045
		DeScod	0.141	0.106	0.061	0.990	0.045	0.077	0.291
		FFT Diffusion	0.020	0.036	0.023	0.026	0.019	0.026	0.022
	5	WT	0.470	0.481	0.471	0.469	0.469	0.471	0.469
		FCN	0.039	0.051	0.041	0.046	0.043	0.042	0.045
		DRNN	0.039	0.047	0.041	0.043	0.037	0.042	0.040
		DeScod	0.037	0.665	0.100	0.563	0.064	0.111	0.616
		FFT Diffusion	0.014	0.026	0.016	0.018	0.013	0.017	0.014
PRD	−5	WT	2.205	2.293	2.197	2.224	2.195	2.212	2.206
		FCN	0.126	0.169	0.136	0.148	0.139	0.143	0.141
		DRNN	0.094	0.121	0.094	0.106	0.088	0.101	0.099
		DeScod	2.192	0.406	1.296	0.190	0.102	1.084	0.540
		FFT Diffusion	0.050	0.088	0.060	0.068	0.053	0.066	0.058
	0	WT	1.342	1.391	1.336	1.353	1.335	1.345	1.342
		FCN	0.107	0.129	0.108	0.114	0.104	0.109	0.105
		DRNN	0.083	0.103	0.081	0.091	0.080	0.090	0.085
		DeScod	0.265	0.198	0.114	1.854	0.085	0.144	0.545
		FFT Diffusion	0.038	0.067	0.044	0.049	0.036	0.048	0.041
	5	WT	0.881	0.901	0.882	0.878	0.878	0.882	0.878
		FCN	0.073	0.096	0.076	0.087	0.080	0.079	0.084
		DRNN	0.074	0.088	0.076	0.081	0.070	0.078	0.074
		DeScod	0.070	1.246	0.188	1.054	0.121	0.207	1.152
		FFT Diffusion	0.026	0.049	0.030	0.035	0.024	0.032	0.027

(continued)

Table 1. (*continued*)

Metrics	Input SNR (dB)	model	BW	EM	MA	BW+EM	BW+MA	EM+MA	BW+EM+MA
CC	−5	WT	0.803	0.670	0.780	0.745	0.794	0.759	0.777
		FCN	0.930	0.851	0.916	0.896	0.916	0.907	0.910
		DRNN	0.947	0.915	0.948	0.935	0.954	0.940	0.941
		DeScod	0.932	0.851	0.920	0.911	0.971	0.855	0.931
		FFT Diffusion	0.987	0.960	0.981	0.976	0.985	0.977	0.982
	0	WT	0.848	0.777	0.839	0.823	0.845	0.830	0.839
		FCN	0.953	0.918	0.953	0.941	0.959	0.950	0.953
		DRNN	0.959	0.938	0.960	0.951	0.961	0.951	0.956
		DeScod	0.210	0.713	0.981	0.882	0.983	0.961	0.972
		FFT Diffusion	0.993	0.976	0.990	0.987	0.993	0.987	0.991
	5	WT	0.876	0.844	0.872	0.866	0.874	0.869	0.872
		FCN	0.980	0.955	0.980	0.965	0.975	0.975	0.973
		DRNN	0.966	0.953	0.964	0.960	0.969	0.962	0.966
		DeScod	0.992	0.751	0.991	0.971	0.994	0.982	0.970
		FFT Diffusion	0.997	0.987	0.995	0.993	0.997	0.994	0.996

Ablation Study

An ablation study was conducted to evaluate the contributions of specific elements to the model's performance. Focusing on the Transformer-based U-shaped architecture and FFT related modules, it replaced the former with Resnet CNN-based U-Net and removed the latter to evaluate their impacts. Table 3 provides the results of ablation study on the MIT-BIH Arrhythmia database, comparing FFT Diffusion and UNet model. For SNR, at −5 dB, FFT Diffusion achieves 28.338 (BW) compared to UNet's 25.895; at 0 dB, it reaches 30.878 versus UNet's 29.332; and at 5 dB, 34.386 compared to 33.427. At −5 dB (BW), the RMSE of FFT Diffusion is 0.027, lower than UNet's 0.032, a trend that continues across all scenarios. For PRD, at −5 dB (BW), FFT Diffusion records 0.05 compared to UNet's 0.061. For CC, at −5 dB (BW), FFT Diffusion achieves 0.987 compared to UNet's 0.979. This pattern repeats across all input SNR levels and noise types. Overall, FFT Diffusion has a better performance than UNet, which indicates that the transformer U-shaped structure and the FFT related modules are crucial for effective denoising and waveform preservation. The results of the ablation study confirm that the proposed design significantly enhances its performance in ECG denoising.

5 Conclusion

High-quality denoising of ECG signals meets two main challenges in biomedical signal processing. One challenge is non-Gaussian noise, which includes low-frequency baseline wander and high-frequency muscle artifacts that change over time. The other challenge is the clinical sensitivity of ECG waveform features, where small changes in the amplitude or timing of P/QRS/T waves can

Table 2. Model comparisons of different denoising methods on the QT database, presenting SNR, RMSE, PRD, and CC for each noise type and input SNR level.

Metrics	Input SNR (dB)	model	BW	EM	MA	BW+EM	BW+MA	EM+MA	BW+EM+MA
SNR	−5	WT	−6.798	−6.996	−6.783	−6.826	−6.777	−6.782	−6.807
		FCN	17.412	14.755	17.237	16.428	17.591	16.408	17.068
		DRNN	21.330	18.99	21.148	19.772	20.970	20.152	20.604
		DeScod	−0.611	−6.465	8.026	9.061	12.722	13.393	9.017
		FFT Diffusion	30.412	25.147	27.335	26.799	28.614	26.787	27.181
	0	WT	−2.510	−2.706	−2.434	−2.561	−2.499	−2.54	−2.509
		FCN	20.394	17.375	19.233	18.144	18.652	18.810	19.370
		DRNN	22.058	20.115	21.829	20.876	22.275	21.011	21.404
		DeScod	10.382	−69.865	0.336	11.852	11.977	−1.700	11.530
		FFT Diffusion	32.398	26.992	29.883	28.771	30.965	27.835	29.309
	5	WT	1.031	0.818	1.014	0.983	1.055	1.008	1.038
		FCN	20.070	19.561	19.919	19.046	20.893	21.179	19.437
		DRNN	22.512	21.208	22.822	21.786	22.832	21.886	22.222
		DeScod	2.182	8.959	19.903	−16.555	5.842	15.053	17.377
		FFT Diffusion	34.810	28.868	32.401	30.709	33.773	30.857	31.713
RMSE	−5	WT	1.197	1.257	1.199	1.215	1.195	1.205	1.201
		FCN	0.065	0.086	0.065	0.072	0.063	0.071	0.066
		DRNN	0.049	0.062	0.05	0.056	0.050	0.054	0.051
		DeScod	0.812	2.480	0.238	0.199	0.124	0.118	0.204
		FFT Diffusion	0.018	0.030	0.024	0.025	0.020	0.024	0.023
	0	WT	0.793	0.818	0.790	0.800	0.793	0.801	0.795
		FCN	0.047	0.065	0.051	0.060	0.053	0.054	0.050
		DRNN	0.046	0.054	0.046	0.051	0.044	0.050	0.048
		DeScod	0.203	0.685	0.971	0.164	0.177	1.732	0.149
		FFT Diffusion	0.014	0.024	0.018	0.019	0.016	0.021	0.018
	5	WT	0.588	0.601	0.589	0.592	0.587	0.588	0.588
		FCN	0.047	0.051	0.047	0.054	0.043	0.041	0.051
		DRNN	0.044	0.049	0.042	0.046	0.042	0.046	0.044
		DeScod	0.388	0.228	0.055	0.218	0.495	0.114	0.077
		FFT Diffusion	0.011	0.019	0.013	0.015	0.011	0.015	0.013
PRD	−5	WT	2.418	2.538	2.421	2.453	2.413	2.433	2.425
		FCN	0.131	0.174	0.131	0.145	0.128	0.144	0.134
		DRNN	0.099	0.124	0.100	0.113	0.101	0.110	0.103
		DeScod	1.639	5.008	0.481	0.401	0.250	0.238	0.411
		FFT Diffusion	0.036	0.061	0.048	0.050	0.041	0.049	0.047
	0	WT	1.601	1.652	1.596	1.615	1.600	1.617	1.605
		FCN	0.094	0.132	0.103	0.121	0.106	0.110	0.102
		DRNN	0.092	0.11	0.093	0.103	0.089	0.100	0.097
		DeScod	0.410	0.138	1.961	0.330	0.356	3.496	0.300
		FFT Diffusion	0.028	0.049	0.036	0.039	0.031	0.043	0.036
	5	WT	1.187	1.215	1.190	1.195	1.185	1.188	1.187
		FCN	0.094	0.102	0.094	0.108	0.086	0.083	0.103
		DRNN	0.089	0.098	0.085	0.094	0.085	0.092	0.089
		DeScod	0.783	0.461	0.112	0.441	0.999	0.230	0.156
		FFT Diffusion	0.022	0.039	0.027	0.031	0.023	0.030	0.027

(continued)

Table 2. (*continued*)

Metrics	Input SNR (dB)	model	BW	EM	MA	BW+EM	BW+MA	EM+MA	BW+EM+MA
CC	−5	WT	0.770	0.634	0.749	0.707	0.761	0.722	0.740
		FCN	0.910	0.831	0.907	0.878	0.917	0.882	0.900
		DRNN	0.941	0.904	0.939	0.919	0.937	0.925	0.934
		DeScod	0.948	0.756	0.956	0.908	0.971	0.927	0.932
		FFT Diffusion	0.992	0.978	0.986	0.985	0.990	0.985	0.987
	0	WT	0.823	0.743	0.814	0.790	0.820	0.799	0.81
		FCN	0.955	0.901	0.948	0.924	0.948	0.934	0.945
		DRNN	0.949	0.926	0.947	0.935	0.952	0.938	0.943
		DeScod	0.982	0.157	0.946	0.939	0.985	0.888	0.966
		FFT Diffusion	0.995	0.986	0.992	0.991	0.994	0.989	0.992
	5	WT	0.854	0.814	0.85	0.839	0.853	0.845	0.849
		FCN	0.957	0.943	0.956	0.940	0.962	0.964	0.947
		DRNN	0.952	0.941	0.957	0.946	0.957	0.949	0.952
		DeScod	0.962	0.912	0.991	0.637	0.993	0.979	0.987
		FFT Diffusion	0.997	0.991	0.995	0.994	0.996	0.994	0.995

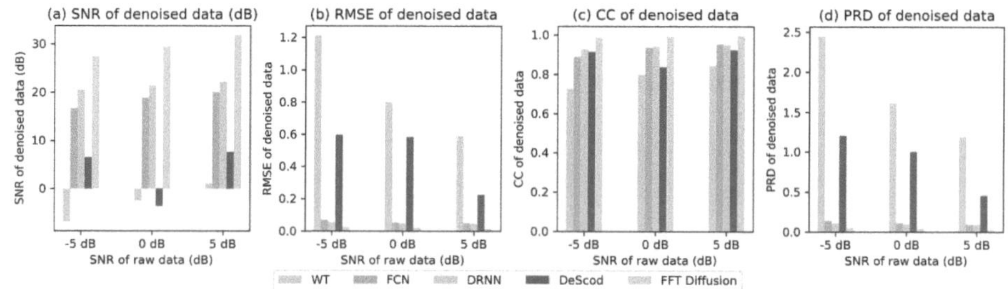

Fig. 3. Average evaluation metrics of the denoised ECG signals using different methods on the QT database.

affect diagnostic accuracy. Conventional methods, which rely on fixed frequency-domain assumptions, struggle to handle non-stationary noise effectively. There are also limitations of deep learning models, as they do not explicitly account for ECG-specific spectral features and often distort waveforms in real clinical settings. To solve these issues, this study introduces the FFT Diffusion framework, which uses hybrid frequency-temporal modeling. This framework decomposes the signal using the fast Fourier transform, creating multi-scale frequency subbands that match various frequency ranges. It applies Transformer-based architectures to capture long-range time dependencies. The proposed framework uses an integrated multiple loss optimization strategy, including Soft-DTW for time alignment, frequency-domain regularization for spectral fidelity, and MSE to minimize point-wise differences. This approach significantly improves waveform reconstruction quality. However, the proposed framework still struggles with extreme noise and real-time use. Future work will focus on developing an

Table 3. Results of ablation study on the MIT-BIH Arrhythmia database, presenting SNR, RMSE, PRD, and CC for each noise type and input SNR level.

	Input SNR (dB)	model	BW	EM	MA	BW+EM	BW+MA	EM+MA	BW+EM+MA
SNR	−5	FFT Diffusion	28.338	23.075	26.332	24.915	27.273	25.190	26.101
		UNet	25.895	19.699	24.500	22.240	25.563	22.448	23.731
	0	FFT Diffusion	30.878	24.910	29.046	27.406	30.447	27.429	28.669
		UNet	29.332	22.346	27.574	25.190	29.135	25.489	26.970
	5	FFT Diffusion	34.386	27.506	31.934	30.334	33.771	30.671	32.092
		UNet	33.427	25.181	31.140	28.372	33.065	29.152	30.737
RMSE	−5	FFT Diffusion	0.027	0.047	0.032	0.036	0.028	0.035	0.031
		UNet	0.032	0.060	0.037	0.045	0.032	0.043	0.037
	0	FFT Diffusion	0.020	0.036	0.023	0.026	0.019	0.026	0.022
		UNet	0.023	0.044	0.026	0.031	0.021	0.030	0.025
	5	FFT Diffusion	0.014	0.026	0.016	0.018	0.013	0.017	0.014
		UNet	0.016	0.031	0.017	0.022	0.014	0.020	0.017
PRD	−5	FFT Diffusion	0.050	0.088	0.060	0.068	0.053	0.066	0.058
		UNet	0.061	0.113	0.068	0.084	0.060	0.081	0.069
	0	FFT Diffusion	0.038	0.067	0.044	0.049	0.036	0.048	0.041
		UNet	0.043	0.082	0.049	0.059	0.040	0.056	0.047
	5	FFT Diffusion	0.026	0.049	0.030	0.035	0.024	0.032	0.027
		UNet	0.029	0.059	0.032	0.041	0.026	0.037	0.031
CC	−5	FFT Diffusion	0.987	0.960	0.981	0.976	0.985	0.977	0.982
		UNet	0.979	0.927	0.973	0.960	0.980	0.962	0.972
	0	FFT Diffusion	0.993	0.976	0.990	0.987	0.993	0.987	0.991
		UNet	0.989	0.961	0.986	0.980	0.991	0.981	0.987
	5	FFT Diffusion	0.997	0.987	0.995	0.993	0.997	0.994	0.996
		UNet	0.995	0.980	0.994	0.990	0.996	0.992	0.994

adaptive noise scheduling module that can operate in dynamic environments, adjusting diffusion parameters based on real-time noise monitoring. The loss function will also be refined by incorporating prior medical knowledge, including an attention-based local feature loss to preserve critical diagnostic features and improve the model's clinical reliability.

References

1. Alickovic, E., Subasi, A.: Effect of multiscale PCA De-noising in ECG beat classification for diagnosis of cardiovascular diseases. Circ. Syst. Signal Process. **34**(2), 513–533 (2015). https://doi.org/10.1007/s00034-014-9864-8
2. Antczak, K.: Deep recurrent neural networks for ECG signal denoising. arXiv:abs/1807.11551 (2018). https://doi.org/10.48550/arXiv.1807.11551
3. Bansal, A., et al.: Cold diffusion: inverting arbitrary image transforms without noise (2022). https://arxiv.org/abs/2208.09392
4. Chandrakar, C., Kowar, M.: Denoising ECG signals using adaptive filter algorithm. Int. J. Soft Comput. Eng. (IJSCE) **2**(1), 120–123 (2012)

5. Chiang, H.T., Hsieh, Y.Y., Fu, S.W., Hung, K.H., Tsao, Y., Chien, S.Y.: Noise reduction in ECG signals using fully convolutional denoising autoencoders. IEEE Access **7**, 60806–60813 (2019). https://doi.org/10.1109/ACCESS.2019.2912036
6. Fariha, Z., Ikeura, R., Hayakawa, S., Tsutsumi, S.: An analysis of the effects of noisy electrocardiogram signal on heartbeat detection performance. Bioengineering **7**, 53 (2020). https://doi.org/10.3390/bioengineering7020053
7. Ho, J., Jain, A., Abbeel, P.: Denoising diffusion probabilistic models. CoRR abs/2006.11239 (2020). https://arxiv.org/abs/2006.11239
8. Jia, Y., et al.: Preprocessing and denoising techniques for electrocardiography and magnetocardiography: a review. Bioengineering **11**(11), 1109 (2024). https://doi.org/10.3390/bioengineering11111109, https://www.mdpi.com/2306-5354/11/11/1109, number: 11 Publisher: Multidisciplinary Digital Publishing Institute
9. Jia, Z., Liang, H., Liu, Y., Wang, H., Jiang, T.: Distillsleepnet: heterogeneous multi-level knowledge distillation via teacher assistant for sleep staging. IEEE Trans. Big Data **11**(3), 1273–1284 (2025). https://doi.org/10.1109/TBDATA.2024.3453763
10. Jia, Z., et al.: Multi-view spatial-temporal graph convolutional networks with domain generalization for sleep stage classification. IEEE Trans Neural Syst. Rehabil. Eng. **29**, 1977–1986 (2021). https://doi.org/10.1109/TNSRE.2021.3110665
11. Laguna, P., Mark, R., Goldberg, A., Moody, G.: A database for evaluation of algorithms for measurement of QT and other waveform intervals in the ECG. In: Computers in Cardiology 1997, pp. 673–676 (1997). https://doi.org/10.1109/CIC.1997.648140
12. Li, H., Ditzler, G., Roveda, J., Li, A.: Descod-ECG: deep score-based diffusion model for ECG baseline wander and noise removal. IEEE J. Biomed. Health Inform. **28**(9), 5081–5091 (2024). https://doi.org/10.1109/JBHI.2023.3237712
13. Maghoumi, M.: Deep recurrent networks for gesture recognition and synthesis. Electronic Theses and Dissertations, 2020–2023 (2020). https://stars.library.ucf.edu/etd2020/379
14. Maghoumi, M., Taranta, E.M., LaViola, J.: DeepNAG: deep non-adversarial gesture generation. In: 26th International Conference on Intelligent User Interfaces, pp. 213–223. IUI '21, ACM (2021). https://doi.org/10.1145/3397481.3450675
15. Mikhled, A.: ECG signal denoising by wavelet transform thresholding. Am. J. Appl. Sci. **5** (2008). https://doi.org/10.3844/ajassp.2008.276.281
16. Moody, G., Mark, R.: The impact of the MIT-BIH arrhythmia database. IEEE Eng. Med. Biol. Mag. **20**(3), 45–50 (2001). https://doi.org/10.1109/51.932724
17. Moody, G.B., Muldrow, W.E., Mark, R.G.: A noise stress test for arrhythmia detectors. In: Computers in Cardiology. vol. 11, pp. 381–384. IEEE Computer Society Press (1984)
18. Mvuh, F.L., Ebode Ko'a, C.O.V., Bodo, B.: Multichannel high noise level ECG denoising based on adversarial deep learning. Sci. Rep. **14**(1), 801 (2024). https://doi.org/10.1038/s41598-023-50334-7
19. Samann, F., Schanze, T.: An efficient ECG denoising method using discrete wavelet with savitzky-golay filter. Current Directi. Biomed. Eng. **5**(1), 385–387 (2019). https://doi.org/10.1515/cdbme-2019-0097
20. Singh, P., Pradhan, G.: A new ECG denoising framework using generative adversarial network **18**(2), 759–764. https://doi.org/10.1109/TCBB.2020.2976981
21. Trappolini, D., et al.: Cold diffusion model for seismic denoising. J. Geophys. Res. Mach. Learn. Comput. **1**(2), e2024JH000179 (2024). https://doi.org/10.1029/2024JH000179, e2024JH000179 2024JH000179
22. Wang, H., et al.: FreDF: learning to forecast in the frequency domain (2024). https://doi.org/10.48550/arXiv.2402.02399

23. Wang, J., Feng, Z., Ning, X., Lin, Y., Chen, B., Jia, Z.: Two-stream dynamic heterogeneous graph recurrent neural network for multi-label multi-modal emotion recognition. IEEE Trans. Affect. Comput. 1–14 (2025). https://doi.org/10.1109/TAFFC.2025.3561439
24. Wang, J., Wang, X., Ning, X., Lin, Y., Phan, H., Jia, Z.: Subject-adaptation salient wave detection network for multimodal sleep stage classification. IEEE J. Biomed. Health Inform. **29**(3), 2172–2184 (2025). https://doi.org/10.1109/JBHI.2024.3512584
25. Weng, B., Blanco-Velasco, M., Barner, K.E.: Ecg denoising based on the empirical mode decomposition. In: 2006 International Conference of the IEEE Engineering in Medicine and Biology Society, pp. 1–4 (2006). https://doi.org/10.1109/IEMBS.2006.259340
26. xiao, S., Yang, W., Cao, B., Wu, J.: ECGDeDRDNet: a deep learning-based method for electrocardiogram noise removal using a double recurrent dense network (2025). https://arxiv.org/abs/2505.05477
27. Yang, C., Liang, L., Su, Z.: Real-world denoising via diffusion model (2023). https://arxiv.org/abs/2305.04457
28. Yen, H., Germain, F.G., Wichern, G., Roux, J.L.: Cold diffusion for speech enhancement (2023). https://arxiv.org/abs/2211.02527
29. Zhang, D., Yuan, M., Li, F., Zhang, L., Sun, Y., Ling, Y.: Attention-based residual dense shrinkage network for ECG denoising. Comput. Model. Eng. Sci. **138**(3), 2809–2824 (2023). https://doi.org/10.32604/cmes.2023.029181, https://www.techscience.com/CMES/v138n3/54919, publisher: Tech Science Press
30. Zhu, D., Chhabra, V.K., Khalili, M.M.: ECG signal denoising using multi-scale patch embedding and transformers (2024). https://arxiv.org/abs/2407.11065

MAVI: MLLM-Enhanced Anomaly Validator and Interpreter for Astronomical Time Series

Xinli Hao[1], Chaohong Ma[2], Wei Li[1], Yihan Tao[3,4], Bingbing Xu[1], and Xiaofeng Meng[1(✉)]

[1] Renmin University of China, Beijing 100872, China
{xinli_hao,leeway,xvbingbing,xfmeng}@ruc.edu.cn
[2] Hebei Normal University, Shijiazhuang 050024, China
chaohma@hebtu.edu.cn
[3] National Astronomical Data Center, Beijing 100101, China
y.tao@nao.cas.cn
[4] National Astronomical Observatories, Chinese Academy of Sciences, Beijing, China

Abstract. Time series anomaly detection (TSAD) is a critical task in scientific domains like astronomy, where identifying rare and scientifically meaningful events is important. However, existing TSAD methods face limitations when applied to astronomical settings. Generic methods often exhibit high false positive rates due to their neglect of domain-specific semantics. Moreover, the lack of interpretability undermines scientific trust, and single-modality approaches fail to leverage the diverse sources of knowledge. To address these challenges, we propose **MAVI**, a **M**ultimodal large language model-enhanced **A**nomaly **V**alidator and **I**nterpreter tailored for astronomical TSAD. Rather than serving as a primary detector, MAVI operates as a lightweight post-processing framework that filters and explains anomaly candidates produced by base TSAD methods. MAVI introduces two novel innovations: first, a multimodal in-context learning strategy that retrieves numerical, visual, and textual anomaly templates to integrate multi-source knowledge; second, a domain-guided chain-of-thought prompting mechanism that emulates astronomers' analytical reasoning to enhance both accuracy and interpretability. Experiments on six astronomical datasets show that MAVI substantially reduces false positives while maintaining high recall, and provides expert-aligned, interpretable rationales for anomaly decisions.

Keywords: Time series anomaly detection · Multimodal Large language model · AI for science

1 Introduction

Time series anomaly detection (TSAD) is a fundamental task in data mining, with broad applicability across critical domains such as scientific discovery, financial fraud prevention, and IT system reliability assessment. In astronomical discovery, TSAD plays a crucial role in identifying unusual and valuable phenomena

M. Yoshikawa et al. (Eds.): ADMA 2025, LNAI 16197, pp. 51–67, 2026.
https://doi.org/10.1007/978-981-95-3453-1_4

in celestial observations. However, astronomical time series present unique challenges, limiting the effectiveness of existing TSAD methods.

1) Domain-agnostic methods cause high false positives. Most TSAD methods are designed as general-purpose detectors, overlooking domain-specific semantics and expert knowledge. For example, SPOT [1] relies on intrinsic statistical features; transformer-based methods (e.g., TranAD [2]) capture temporal dependencies; and GNN-based methods (e.g., GDN [3]) focus on inter-variable correlations. However, the definition of an anomaly varies across domains. In astronomy, anomalies are typically rare and scientifically significant phenomena that occur in isolation [4]—unlike the anomalies in IT systems, which are often caused by equipment factors and tend to appear in groups. Without adaptation to the astronomical domain, general models tend to misclassify routine or noisy patterns as anomalies, leading to a high false positive rate.

2) Poor interpretability undermines scientific trust and practical utility. Current TSAD models typically assign anomaly scores and binary labels without providing explanations for their decisions [5–7]. This lack of interpretability is particularly problematic in astronomy, where scientists follow evidence-driven practices and require scientifically grounded reasoning to trust model outputs. Without interpretability, the practical utility of these models within scientific workflows is greatly limited.

3) Single-modality approaches constrain the potential to leverage diverse knowledge. In real-world scenarios, domain experts typically determine anomalies in time series data by integrating multiple sources of information, including visual patterns and contextual knowledge, rather than relying solely on numerical time series [8,9]. This highlights the potential of multimodal methods, which can incorporate various forms of information to improve the reliability and interpretability of anomaly detection results.

Recent advances in Multimodal Large Language Models (MLLMs) show strong capabilities in processing visual, textual, and numerical inputs for contextual reasoning, offering promising opportunities for astronomical time series analysis. By integrating diverse modalities and mimicking expert reasoning, MLLMs can address key limitations of traditional TSAD methods.

However, applying MLLMs to large-scale astronomical data [4] poses significant challenges due to high computational demands or associated costs. Open-source models require substantial computational resources, while commercial models are priced based on usage. These factors limit the scalability of using MLLMs as standalone detectors for continuous or exploratory use in astronomy.

Methodology. To harness the power of MLLMs while addressing these challenges, this paper introduces **MAVI**, an **M**LLM-Enhanced **A**nomaly **V**alidator and **I**nterpreter tailored for astronomical TSAD. Rather than replacing existing detectors, MAVI functions as a lightweight post-processing framework that filters and explains anomaly candidates generated by the base TSAD methods, thereby enhancing both accuracy and interpretability while avoiding the high costs associated with using MLLMs as primary detectors. More specifically, MAVI consists of two key designs. First, a multimodal in-context learning (MICL) strategy retrieves domain-relevant anomaly templates, including numerical, visual, and textual modalities, from a curated template pool. These templates provide contextual guidance for MLLM-based reasoning without the need for fine-tuning. Second, MAVI designs a domain-guided chain-of-thought prompting strategy (DCoTP), formulating the validation process into interpretable reasoning steps aligned with astronomers' analytical workflows. Through its flexible design, MAVI effectively reduces false positives by improving validation accuracy and enhancing interpretability through domain-guided reasoning.

The **contributions** of our work are summarized as follows:

1) We propose MAVI, an anomaly detection-validation–interpretation framework that augments existing TSAD methods with MLLMs, improving both accuracy and interpretability in astronomical anomaly detection.
2) We introduce MICL mechanism that retrieves and integrates numerical, visual, and textual anomaly templates to guide the reasoning of MLLMs without additional fine-tuning.
3) We design DCoTP, a domain-adapted chain-of-thought prompting strategy that enables step-by-step, interpretable validation by injecting domain knowledge and aligning with astronomers' analytical workflows.
4) We conduct extensive experiments on six astronomical datasets, demonstrating that MAVI reduces false positives while maintaining high recall and providing expert-aligned interpretability.

2 Related Work

2.1 Time Series Anomaly Detection (TSAD)

TSAD methods have evolved from traditional statistical models to deep learning techniques. Early approaches rely on statistical assumptions, such as Extreme Value Theory (e.g., SPOT [1]), which are often limited in capturing complex temporal dependencies. With the rise of deep learning, Variational Autoencoders [10] became widely adopted to capture latent temporal and inter-variable dependencies, such as OmniAnomaly [11].

Recent studies also explore Graph Neural Networks (GNNs) [12] to explicitly model complex variable interactions. For instance, GDN [3] models static inter-variate dependencies, while ESG [13] learns dynamic graph structures to capture temporal variations. Additionally, Transformer-based architectures have been developed to enhance anomaly detection capabilities, such as Anomaly Transformer [14] and TranAD [2]. Furthermore, AERO [4] combines GNNs and Transformer architecture to address anomaly detection tailored to astronomical time series.

Despite these advances, most existing methods operate solely on numerical signals and lack mechanisms for integrating semantic knowledge or domain-specific constraints, limiting their generalization and interpretability.

2.2 Large Language Models for Time Series

Recent work introduces Large Language Models (LLMs) to time series tasks, bridging numerical time series and semantic knowledge. For example, LLM-TIME [15] directly converts time series into textual prompts. However, LLMs are not inherently designed for continuous temporal signals, so this approach suffers from modality mismatch issues. To mitigate this, AutoTimes [16] introduces intermediate representations, while TimeLLM [17] proposes a fine-tuning scheme to align temporal patterns with the semantic space of LLMs. These alignment strategies aim to preserve temporal dependencies while harnessing the reasoning power of LLMs. Although most existing work focuses on forecasting, recent efforts have extended LLMs to anomaly detection. For example, LLMAD [18] applies LLMs to few-shot anomaly detection by leveraging both positive and negative reference segments.

However, deploying LLMs as standalone analytic engines remains constrained by high computational and financial costs, particularly for large-scale or high-frequency scenarios, highlighting the need for more efficient hybrid approaches.

2.3 Time Series as Image

Recent studies have explored transforming time series into image representations to enable the use of computer vision models for pattern recognition. ViTST [19] converts irregular time series into line graph images and applies vision transformers for classification. GAF/MTF [20] encodes time series using Gramian Angular Fields and Markov Transition Fields to support classification. TimesNet [21], a foundation model for time series analysis, further incorporates computer vision techniques by reshaping time series into two-dimensional tensors.

However, image-based approaches for anomaly detection remain underexplored. Leveraging multimodal large language models for time series anomaly detection emerges as a promising and valuable research direction [22].

3 Preliminary

In this section, we first introduce the three key concepts, followed by a formal definition of the problem studied here.

Time Series. Let $\{x_t\}_{t=1}^{T}$ denote time series over T timestamps, where each $x_t \in R^N$ is an observation at timestamp t consisting of N variates. When $N = 1$, this is a univariate time series.

Time Series Anomaly Detection. TSAD aims to assign each data point a real-valued anomaly score $s_t^n \in [0, 1]$. Under thresholding, we get a binary flag $y_t^n \in \{0, 1\}$, indicating whether the point is anomalous.

Anomaly Canditate. An anomaly candidate C_i is a continuous data subsequence $\{x_t\}_{t=start}^{t=end}$ with relatively high anomaly scores $S_i^C = \{s_t\}_{t=start}^{t=end}$ in astronomical time series, where *start* and *end* denote the time boundaries of the anomaly candidate. These candidates are typically generated by a base TSAD method and have not yet been confirmed as true anomalous via validation.

Problem Definition. Given a time series anomaly candidate C_i and its anomaly score S_i^C, our objective is not only to validate whether it is a true anomaly but also to interpret the reason for the decision-making. Formally, we aim to design a framework F that acts as a validator and interpreter:

$$(\mathcal{Y}_i, \mathcal{A}_i, \mathcal{S}_i, \mathcal{E}_i) = F(C_i, S_i^C) \tag{1}$$

where $\mathcal{Y}_i$ is the anomaly flag, indicating whether the identified anomaly candidate is a true or false positive(anomaly). Anomaly types $\mathcal{A}_i$, anomaly scores $\mathcal{S}_i$, and text explanations $\mathcal{E}_i$ make up interpretations. Interpreting time series anomalies can provide explanations that humans can understand, facilitating informed decision-making.

4 Proposed Method

In this section, we first introduce an overview of MAVI and then describe the key designs in the subsequent sections.

4.1 MAVI Overview

As illustrated in Fig. 1, our proposed MAVI framework follows a detection–validation–interpretation pipeline. Firstly, given the input time series, a base time series anomaly detection method will generate a set of anomaly candidates, which may include false positives (Fig. 1a), and each is assigned an anomaly

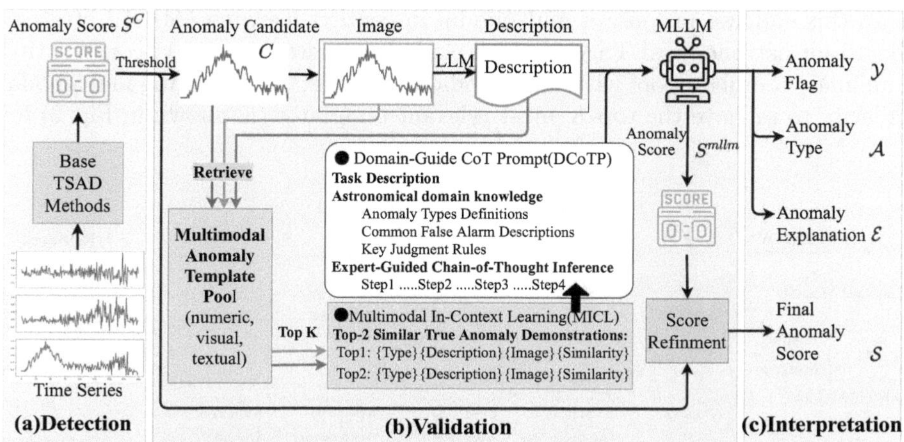

Fig. 1. An overview of MAVI. (a) Process of preliminary anomaly detection by base TSAD Methods. (b) Validation process for candidates, including ❶ MICL: Multimodal In-Context Learning, and ❷ DCoTP: Domain-guide Chain-of-Thought Prompt. (c) The interpretable output results.

score. Secondly, MAVI validates these candidates by integrating domain knowledge through multimodal in-context learning (MICL) (Fig. 1b-❶) and domain-guide chain-of-thought prompting (DCoTP) (Fig. 1b-❷) enhanced by MLLMs. Specifically, in the MICL process, MAVI retrieves historical anomaly templates that are similar to current candidates (detailed in Sect. 4.2). Then, in the DCoTP module, MAVI incorporates knowledge of anomaly types and domain-specific rules into a prompt for candidate validation (detailed in Sect. 4.3). These two strategies enable the MLLM to understand the context and domain knowledge without fine-tuning. Finally, MAVI generates comprehensive interpretations (Fig. 1c), including a binary anomaly flag, anomaly type classification, anomaly score, and natural language explanations (detailed in Sect. 4.4).

4.2 Multimodal In-Context Learning (MICL)

Accurate anomaly validation in astronomy often involves comparing candidates to representative historical cases. Besides, astronomers often seek to classify anomalies into fine-grained event types (e.g., flare, microlensing [23]). However, high-quality labeled data is often limited and expensive to obtain. Therefore, the validation process relies not only on the original numerical values of time series, but also on visual cues and textual summaries that contain type information.

In-context learning (ICL) enables effective use of limited labeled data by constructing analogical labeled demonstrations at inference time [24], without requiring model fine-tuning [25]. To adapt ICL for astronomical TSAD, it is essential to construct demonstrations using domain-specific anomaly templates. Furthermore, to align with scientific validation practices, ICL should be extended into a multimodal mechanism that integrates visual and textual modalities with the original numerical time series for robust decision-making.

To this end, we propose a multimodal in-context learning (MICL) strategy tailored for astronomical TSAD, which consists of three stages: 1) constructing an anomaly template pool with multimodal templates; 2) computing multimodal similarity to retrieve the top-K most relevant templates (as shown in Fig. 2) for

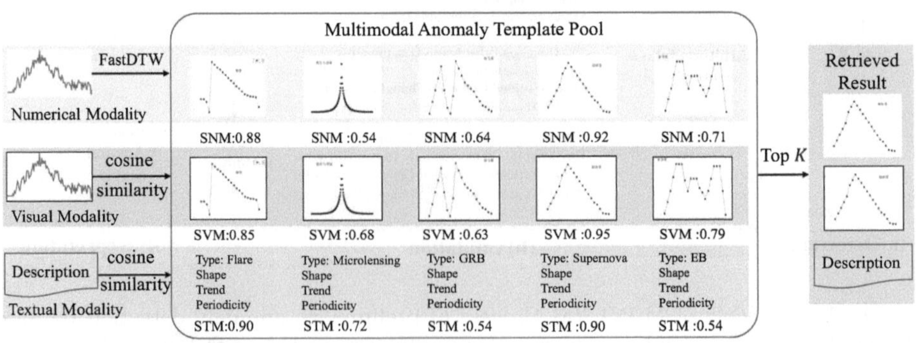

Fig. 2. An illustration of multimodal anomaly template pool and retrieval process.

each candidate; and 3) enhancing in-context learning by the retrieved templates as demonstrations for domain-aligned reasoning. We next detail each stage.

Stage 1: Construction of Anomaly Template Pool. To ensure that retrieved templates are both informative and aligned with the domain, we construct a high-quality multimodal anomaly template pool based on labeled data from the kaggle[1]. This pool includes 14 types of rare and transient astronomical events. For each template, we retain the original numerical time series and generate corresponding images and textual descriptions. The images are derived directly by plotting to highlight morphological patterns, while textual descriptions are manually crafted by domain experts based on anomaly type, shape, trend, and duration. The multimodal representation allows templates to comprehensively characterize each anomaly type, supporting accurate retrieval.

Stage 2: Multimodal Similarity Computation. With the anomaly template pool constructed, MAVI retrieves the K templates most similar to a given time series anomaly candidate C_i in the three modalities:

1) Numerical Modality. To measure the similarity in the numerical modality, denoted as SNM_i^k, between candidate C_i^{num} and a template TM_k^{num}, we use Fast Dynamic Time Warping (FastDTW) [26], a linear-time approximation for aligning time series. The similarity is computed as:

$$SNM_i^k = \frac{1}{1 + \text{FastDTW}(C_i^{num}, TM_k^{num})} \tag{2}$$

2) Visual Modality. Images capture structural features such as symmetry and shape irregularities that may be missed in numerical data. We embed each image using a pre-trained image encoder E_v, and compute the cosine similarity between candidate C_i^{img} and template TM_k^{img}, denoted as SVM_i^k:

$$SVM_i^k = cosine(E_v(C_i^{img}), E_v(TM_k^{img})) \tag{3}$$

3) Textual Modality. The textual modality captures high-level semantics and domain-specific concepts (e.g., supernova-like light curve, periodic with drift). We encode each description using a pre-trained language encoder E_{text}, and compute the cosine similarity between candidate C_i^{text} and template TM_k^{text}, denoted as STM_i^k. Formally:

$$STM_i^k = cosine(E_{text}(C_i^{text}), E_{text}(TM_k^{text})) \tag{4}$$

By integrating these complementary similarity signals, we get Sim_i^k. Then we select the top-K most relevant anomaly templates with candidate C_i as set $\mathcal{K}_i$.

$$\begin{aligned} Sim_i^k &= SNM_i^k + SVM_i^k + STM_i^k \\ \mathcal{K}_i &= \{k \in \arg \text{topK}\, \text{min}_\text{k} \text{Sim}_i^k\} \end{aligned} \tag{5}$$

[1] https://www.kaggle.com/competitions/PLAsTiCC-2018/data.

Stage 3: Enhancing In-Context Learning with Retrieved Templates. Based on the retrieved top-K anomaly templates, we construct the input prompt for the MLLM by inserting these templates as in-context demonstrations (highlighted in blue in Fig. 3). Each template provides a cosine representation of numerical, visual, and textual characteristics, enabling the model to align its reasoning with domain-relevant patterns. This design supports effective in-context learning and strengthens the model's ability to perform fine-grained anomaly discrimination.

4.3 Domain-Guided Chain-of-Thought Prompting (DCoTP)

Reliable anomaly validation in astronomy not only requires accurate classification but also demands interpretable and logical reasoning that aligns with scientific workflows. However, standard prompting strategies in MLLMs, such as standard chain-of-thought (CoT) prompting [27], focus primarily on general reasoning coherence. As a result, they often lack the scientific rigor and domain-specific reasoning necessary for astronomical TSAD.

To bridge this gap, we propose domain-guided chain-of-thought prompting (DCoTP), which enhances anomaly validation by integrating astronomical knowledge into the reasoning process. As shown in Fig. 3, DCoTP involves two core components: astronomical domain knowledge injection, and expert-guided chain-of-thought inference. Next, we detail each operation.

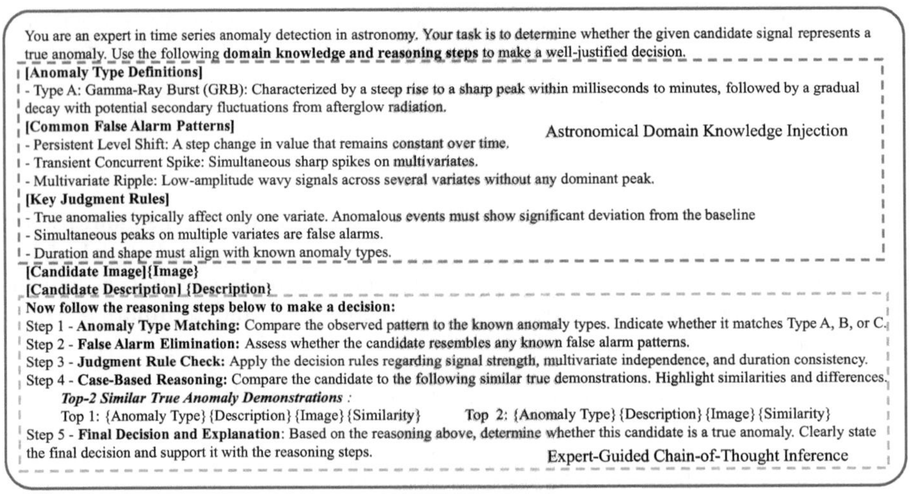

Fig. 3. The overall structure of domain-guided chain-of-thought prompt (DCoTP).

Astronomical Domain Knowledge Injection. To facilitate the step-by-step reasoning, we incorporate three complementary forms of domain knowledge within the prompt, each guiding a specific inference stage: 1) Anomaly Type Definitions, which provide formal semantic characteristics of astrophysical anomaly categories; 2) Common False Alarm Descriptions, that summarize typical features of non-anomalous variations such as noise occurring on multivariates simultaneously; and 3) Key Judgment Rules, which define decision heuristics based on time series properties like shape, duration, and variable dependencies.

These modular and configurable components collectively enrich the MLLMs' reasoning context and allow flexible adaptation to various anomaly types by providing semantic grounding, contrastive cues, and heuristic decision rules.

Expert-Guided Chain-of-Thought Inference. Next, DCoTP decomposes the anomaly verification task into five sequential reasoning steps:

1) Anomaly Type Matching: The model first compares the candidate description against textual definitions of predefined anomaly types. This step enables coarse-grained semantic category identification and domain priors.
2) False Positive Elimination: It then evaluates whether the candidate matches any predefined descriptions of common noise. This step reduces false positives by introducing contrastive knowledge in the reasoning process.
3) Judgment Rule Check: The model checks whether the observed signal satisfies key decision criteria, such as trend, amplitude, and duration, guided by predefined judgment rules.
4) Case-Based Reasoning: The model learns the most relevant true anomaly of cases across visual and textual modalities, supporting finer-grained inference. This step leverages the results from Sect. 4.2, where the model retrieves top-K templates from a curated template pool to support case-based inference.
5) Explanation Generation: Finally, the model synthesizes a human-readable justification by integrating information from the previous steps.

By integrating curated astronomical knowledge into a structured reasoning framework, DCoTP equips MLLMs with a context-aware understanding of anomalies and supports the generation of transparent, expert-aligned justifications.

4.4 Anomaly Interpretation

The proposed method MAVI generates informative anomaly reports that improve interpretability and support expert decision-making in astronomical time series anomaly detection. The report encompasses the following four key aspects:

1) Anomaly Score After Refinement. By querying the MLLM for each candidate C_i with its corresponding prompt, we obtain the anomaly score S_i^{mllm}. However, S_i^{mllm} is purely based on the image and text description, without taking into account the original numerical time series. Thus, we further refine the score by summing the original anomaly score S_i^C from the base TSAD model and S_i^{mllm} in a training-free manner [28]. The refined score $\mathcal{S}$ provides a quantitative estimate of anomaly severity and serves as a complementary form of explanation. Formally:

$$\mathcal{S}_i = (S_i^C + S_i^{mllm})/2 \tag{6}$$

2) Anomaly Flag. For each anomaly candidate, MAVI explicitly assigns an anomaly flag $\mathcal{Y}$ to indicate whether it is a true or false anomaly after validation.

3) Anomaly Type Classification. For each validated anomaly, MAVI assigns a corresponding anomaly type $\mathcal{A}$, serving as a higher-level explanation to support astronomers in subsequent verification efforts.

4) Anomaly Explanation. MAVI provides a step-by-step explanation $\mathcal{E}$ of its decision process, following the steps defined in Sect. 4.3. By mimicking expert reasoning and generating natural language reports, MAVI enhances transparency in the decision process. This, in turn, strengthens astronomers' ability to assess the results and increases their trust in the system's outcome.

5 Experiments

This section evaluates the performance of MAVI from the following aspects: **A1.** Comparison with baselines for anomaly detection in astronomical observations. **A2.** Ablation study to assess the contribution of each component in MAVI. **A3.** Interpretability Analysis. **A4.** Cost Analysis for the MAVI process.

5.1 Experimental Setup

Implementation Details. In our implementation of MAVI, we adopt GPT-4o [29] as the multimodal large language model and use AERO [4] as the base TSAD method. For the image encoder E_v, we employ ViT-B/32 [30], which has demonstrated strong performance across a range of visual tasks. For the language encoder E_{text}, we use the BERT [31], chosen for its balance between efficiency and contextual understanding in textual encoding. We implement baseline models by adhering to their original code repository and configuration settings.

Baselines. We compare the performance of our proposed MAVI with eight applicable baselines. This comprehensive comparison spans from cutting-edge deep learning models to large language models, ensuring a thorough analysis of anomaly detection in time series data. Specifically, 1) **OmniAnomaly** [11] uses a VAE framework to stochastically model inter-variable dependencies in multivariate time series. 2) **GDN** [3] and 3) **ESG** [13] employ static and dynamic graph neural networks, respectively, to capture inter-variable correlations, with anomaly scores computed from prediction errors. 4) **AnomalyTransformer** [14] introduces anomaly-attention and association discrepancy into a Transformer-based reconstruction model for anomaly detection. 5) **TranAD** [2] enhances Transformer with adversarial training and score-based self-conditioning for robust anomaly detection. 6) **AERO** [4] is a two-stage time series anomaly detection method tailored to the unique characteristics in astronomical observations. 7) **TimesNet** [21] is a foundation model for time series analysis, including anomaly detection, which applies convolutions in a transformed 2D space. 8) **LLMTime** [15] encodes time series as a string of numerical digits to forecast time series, and we use the prediction error as anomaly signals.

Datasets. In experiments, we utilize the astronomical time series anomaly detection benchmark introduced in AERO [4], which includes three real-world and three synthetic datasets. These datasets are constructed to assess model robustness under varying anomaly-to-noise (A/N) ratios. The synthetic sets simulate stellar light curves with injected anomalies and typical noise, while the real datasets, derived from telescope observations, exhibit complex astrophysical variability. Each dataset provides accurate anomaly intervals, enabling fine-grained and rigorous evaluation of anomaly detection methods in terms of generalization and interpretability under realistic scientific conditions. Dataset statistics are summarized in Table 1.

Table 1. Dataset Statistics.

Dataset	#train	#test	#variates	**A**nomaly(%)	**N**oise(%)	**A/N**	#Anomaly Segments
SyntheticMiddle	4000	4000	24	0.180	1.719	0.105	5
SyntheticHigh	4000	4000	24	0.359	1.719	0.209	10
SyntheticLow	4000	4000	24	0.180	3.438	0.052	5
AstrosetsMiddle	5540	5387	54	0.153	4.173	0.037	2
AstrosetsHigh	8000	6117	38	0.117	2.405	0.049	2
AstrosetsLow	6255	2950	40	0.190	8.419	0.023	6

*Anomaly(%) represents the proportion of anomalous data points.
*Noise(%) is the proportion of data points affected by noise.
*A/N denotes the anomaly-to-noise ratio measuring the ratio of true anomalies in candidates.

Evaluation Metrics. We use precision (Prec), recall, and F1-Score (F1) over the test datasets to evaluate the performance of all the compared methods:

$$Prec = \frac{TP}{TP + FP}, Recall = \frac{TP}{TP + FN}, F1 = \frac{2 \times Prec \times Recall}{Prec + Recall} \tag{7}$$

where TP, TN, FP, and FN are the numbers of true positives, true negatives, false positives, and false negatives, respectively. Traditional point-wise F1 evaluation assesses each timestamp independently, which may overlook the temporal continuity of anomalies and reduce practical relevance. Following prior work [2,11,32], we adopt a point-adjust strategy that aligns better with segment-level anomaly detection and real-world application needs.

5.2 Comparison with Baselines (A1)

Table 2 reports the results of precision, recall, and F1 score for all baselines and MAVI on six datasets. Based on the results, we make several observations.

First, MAVI consistently improves precision across all datasets while maintaining comparable recall, resulting in enhanced overall detection performance. Operating on a pool of anomaly candidates with mixed TP and FP, MAVI effectively filters out FP by correcting noisy inputs from FP to TN, thereby boosting

Table 2. Comparison with baselines: results on precision, recall, and F1-score(%).

Method	SyntheticMiddle			SyntheticHigh			SyntheticLow		
	Prec	Recall	F1	Prec	Recall	F1	Prec	Recall	F1
OmniAnomaly	20.37	34.78	25.70	26.86	28.26	27.54	44.54	38.27	41.17
AnomalyTransformer	29.76	14.49	19.49	90.55	50.00	64.43	14.79	24.69	18.50
TranAD	31.03	100.0	47.36	54.16	100.0	70.26	35.68	100.0	52.60
GDN	89.58	79.71	84.36	86.03	50.00	63.24	87.93	62.96	73.38
ESG	79.55	71.01	75.04	85.80	63.04	72.68	69.02	50.62	58.40
AERO	90.79	100.0	95.17	90.67	100.0	95.10	92.68	100.0	96.20
TimesNet	83.33	71.01	76.68	88.58	100.0	93.94	86.54	100.0	92.78
LLMTIME	68.52	34.78	46.14	81.24	50.00	61.90	60.45	100.0	75.35
MAVI	**91.81**	100.0	**95.73**	**91.63**	100.0	**95.63**	**93.83**	100.0	**96.82**
Method	AstrosetMiddle			AstrosetHigh			AstrosetLow		
OmniAnomaly	41.93	22.23	29.05	64.10	55.56	59.52	86.37	75.00	80.28
AnomalyTransformer	68.97	77.78	73.11	55.89	44.44	49.51	55.76	25.00	34.52
TranAD	06.47	22.23	10.03	11.61	44.44	18.42	41.61	92.86	57.47
GDN	79.72	100.0	88.71	64.94	55.56	59.88	69.20	33.33	44.99
ESG	40.24	22.23	28.63	57.47	55.56	56.50	68.18	42.86	52.63
AERO	80.72	100.0	89.33	75.36	100.0	85.95	89.00	91.67	90.31
TimesNet	41.15	22.23	28.86	68.09	55.56	61.19	85.54	91.67	88.50
LLMTIME	41.23	77.78	53.89	61.48	44.44	51.59	72.36	42.88	53.85
MAVI	**80.81**	100.0	**89.39**	**75.73**	100.0	**86.19**	**89.92**	91.67	**90.79**

precision without compromising recall. **Second**, MAVI achieves its strongest results on datasets with low anomaly-to-noise (A/N) ratios, where its noise-filtering capabilities are most beneficial. Conversely, performance declines on high A/N datasets in both synthetic and real-world data. **Third**, we observe performance degradation as the number of variates (N) increases. A larger N imposes a greater cognitive load on the model and may reduce the representational quality of individual variates, limiting overall effectiveness.

In summary, MAVI achieves the highest overall performance. AERO is the strongest baseline, benefiting from domain-specific design. GNN-based methods (e.g., GDN, ESG) also perform well by capturing inter-variate dependencies. In contrast, models such as OmniAnomaly, AnomalyTransformer, and TranAD underperform due to weaker noise filtering and relational modeling. LLMTIME, which uses LLM's prediction error as the anomaly score, exhibits low performance. These results highlight MAVI's advantage in combining multimodal in-context learning and domain-guided reasoning for scientific anomaly detection.

5.3 Ablation Study (A2)

To assess the contribution of key components in MAVI, we conduct ablation experiments focusing on two aspects: 1) multimodal in-context learning (MICL), and 2) domain-guided chain-of-thought prompting (DCoTP). Experiments are performed on one synthetic and two real-world datasets.

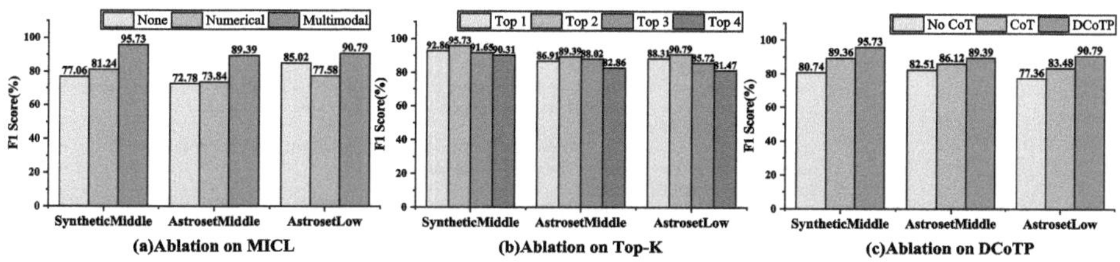

Fig. 4. Results for Ablation Study.

MICL. We examine the impact of MICL under two settings: 1) modality of retrieved samples (multimodal, numerical-only, or none), and 2) the number of retrieved templates K.

As shown in Fig. 4(a), numerical-only ICL does not consistently improve performance, and in some cases underperforms compared to no ICL. This highlights the limitations of unimodal guidance and the robustness of multimodal retrieval. Incorporating multimodal demonstrations enables MAVI to better recognize true astronomical anomalies, leading to enhanced detection accuracy.

Regarding the number of retrieved templates, performance improves up to $K = 2$, but declines when more demonstrations are used, as shown in Fig. 4(b). While increasing K provides richer contextual cues, excessive demonstrations may introduce noise and hinder reasoning. Using the setting $K = 2$ offers a favorable balance between context diversity and inference stability.

DCoTP. We evaluate the effectiveness of DCoTP by removing it and replacing it with standard CoT (only asking LLM to think step-by-step), as shown in Fig. 4(c). As expected, CoT can significantly improve the performance of TSAD, while our DCoTP, which incorporates domain knowledge and follows the logical thinking of experts, can further boost the performance to a greater level. On average, CoT improves the F1 by 7.6%, and using DCoTP can further improve CoT by 6.5%, which is a substantial boost. This improvement can be attributed to the fact that as MLLMs infer like humans, using step-wise thinking combined with injecting domain expertise leads to improved performance in TSAD.

5.4 Interpretability Analysis (A3)

To evaluate the interpretability of MAVI, we adopt a qualitative analysis approach. Specifically, we present two representative cases that demonstrate how the explanation results are consistent with human reasoning and domain knowledge.

As shown in Fig. 5, each case includes an anomaly flag, a refined anomaly score, and an anomaly type classification, followed by a stepwise explanation comprising four stages: anomaly type matching, false positive elimination, judgment rule check, and case-based reasoning.

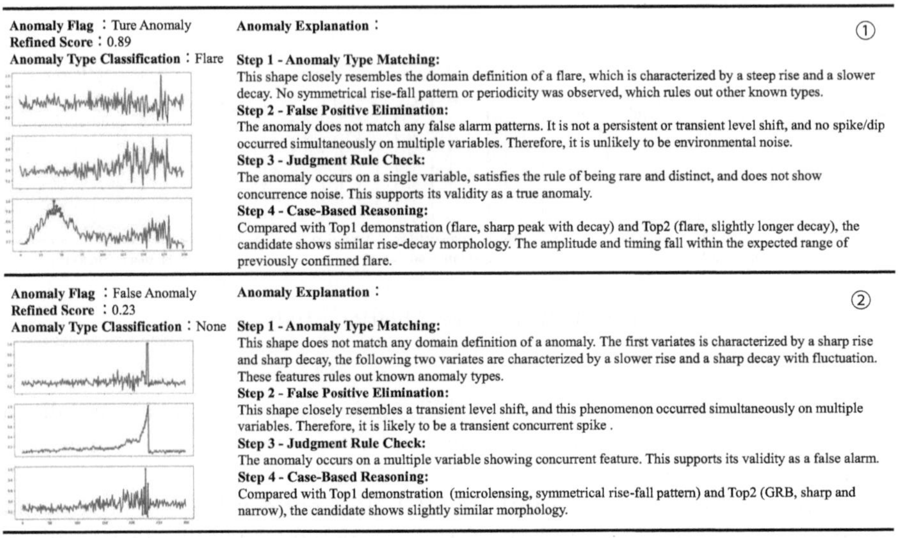

Fig. 5. Examples for MAVI explanation.

In the first case ①, the candidate is labeled as an anomaly, with a refined score of 0.89, and classified as a flare. The explanation highlights a steep rise followed by a gradual decay, which aligns well with the definition of a flare. It further rules out false alarm patterns, confirms the anomaly with rule-based criteria, and supports classification through comparison with historical flare cases.

In contrast, the second case ② is marked as noise (i.e., false anomaly), with a low refined score of 0.23 and no anomaly type assigned. The explanation notes that the observed pattern does not align with known anomaly types. Instead, concurrent transient behaviors across multiple variables suggest a transient level shift, supporting its classification as a false alarm. The case-based reasoning further confirms this by showing low similarity to known true anomalies.

These examples illustrate that MAVI can produce transparent and structured explanations, supported by both semantic reasoning and domain-aligned features, thereby enhancing its interpretability in practice.

5.5 Cost Analysis (A4)

We estimate the cost by calculating the API usage fees. For each anomaly candidate, the complete MVAI process incurs a cost of approximately \$0.017. Across all six datasets, the base model generates a total of 141 candidates, resulting in a total cost of approximately \$2.4. Furthermore, we evaluate the cost of using MLLM directly for anomaly detection, without relying on the base model. With the same setting as AERO (i.e., the base TSAD model [4]), the MLLM incurs a significantly higher cost of \$429.32, as it requires processing 25,254 samples across all datasets. This comparison highlights the substantial cost-efficiency gained by the MAVI framework. It is worth noting that the effectiveness of MVAI is inherently constrained by the base model's recall. If the base model fails to identify true anomalies as candidates, MVAI cannot subsequently detect them. We therefore recommend using a base model with high recall.

6 Conclusion and Future Work

This paper presents MAVI, a multimodal validation and interpretation framework to address the unique challenges of astronomical time series anomaly detection in a cost-efficient and post-processing manner. By designing the multimodal in-context learning mechanism and domain-guided chain-of-thought prompting, MAVI significantly reduces false positives, enhances interpretability, and aligns with domain-specific analytical workflows. Experimental results demonstrate that MAVI offers an effective and domain-adapted solution for anomaly detection in scientific discovery. This work underscores the potential of MLLM as a validator and interpreter for scientific anomaly detection. For future work, we plan to incorporate expert feedback to enable human-in-the-loop optimization.

Acknowledgments. This work is supported by the National Natural Science Foundation of China (Grant No: 62172423).

References

1. Siffer, A., Fouque, P.A., Termier, A. et al.: Anomaly detection in streams with extreme value theory. In: SIGKDD 2017, pp. 1067–1075 (2017)
2. Tuli, S., Casale, G., Jennings, N.R.: Tranad: deep transformer networks for anomaly detection in multivariate time series data. Proc. VLDB Endow. **15**(6), 1201–1214 (2022)
3. Deng, A., Hooi, B.: Graph neural network-based anomaly detection in multivariate time series. In: AAAI, vol. 35, pp. 4027–4035 (2021)
4. Hao, X., Chen, Y., Yang, C., et al.: From chaos to clarity: time series anomaly detection in astronomical observations. In: IEEE ICDE, pp. 570–583 (2024)
5. Chen, X., Qiu, Q., Li, C. et al.: Graphad: a graph neural network for entity-wise multivariate time-series anomaly detection. In: ACM SIGIR, pp. 2297–2302 (2022)
6. Malhotra, P., Vig, L., Shroff, G. et al.: Long short term memory networks for anomaly detection in time series. In: ESANN, vol. 2015, p. 89 (2015)
7. Li, D., Chen, D., Jin, B., Shi, L., Goh, J., Ng, S.-K.: MAD-GAN: multivariate anomaly detection for time series data with generative adversarial networks. In: Tetko, I.V., Kůrková, V., Karpov, P., Theis, F. (eds.) ICANN 2019. LNCS, vol. 11730, pp. 703–716. Springer, Cham (2019). https://doi.org/10.1007/978-3-030-30490-4_56

8. Angeloudi, E., Audenaert, J., Bowles, M., et al.: The multimodal universe: enabling large-scale machine learning with 100 TB of astronomical scientific data. Adv. Neural. Inf. Process. Syst. **37**, 57841–57913 (2024)
9. Rizhko, M., Bloom, J.S.: Self-supervised multimodal model for astronomy. In: 2rd Foundation Models for Science: Progress, Opportunities, and Challenges@Neurips
10. An, J., Cho, S.: Variational autoencoder based anomaly detection using reconstruction probability. Special lecture on IE **2**(1), 1–18 (2015)
11. Su, Y., Zhao, Y., Niu, C., et al.: Robust anomaly detection for multivariate time series through stochastic recurrent neural network. In: ACM SIGKDD (2019)
12. Scarselli, F., Gori, M., Tsoi, A.C., et al.: The graph neural network model. IEEE Trans. Neural Networks **20**(1), 61–80 (2009)
13. Ye, J., Liu, Z., Du, B. et al.: Learning the evolutionary and multi-scale graph structure for multivariate time series forecasting. In: ACM SIGKDD (2022)
14. Xu, J., Wu, H., Wang, J. et al.: Anomaly transformer: time series anomaly detection with association discrepancy. arXiv preprint arXiv:2110.02642 (2021)
15. Gruver, N., Finzi, M., Qiu, S., et al.: Large language models are zero-shot time series forecasters. NeurIPS **36**, 19622–19635 (2023)
16. Liu, Y., Qin, G., Huang, X., et al.: Autotimes: autoregressive time series forecasters via large language models. In: NeurIPS, vol. 37, pp. 122154–122184 (2024)
17. Jin, M., Wang, S., Ma, L. et al.: Time-LLM: time series forecasting by reprogramming large language models. arXiv preprint arXiv:2310.01728 (2023)
18. Liu, J., Zhang, C., Qian, J. et al.: Large language models can deliver accurate and interpretable time series anomaly detection. arXiv preprint arXiv:2405.15370
19. Li, Z., Li, S., Yan, X.: Time series as images: vision transformer for irregularly sampled time series. In: NeurIPS, vol. 36, pp. 49187–49204 (2023)
20. Wang, Z., Oates, T., et al.: Encoding time series as images for visual inspection and classification using tiled convolutional neural networks. In: AAAI, pp. 1–7 (2015)
21. Wu, H., Hu, T., Liu, Y. et al.: TimesNet: temporal 2D-variation modeling for general time series analysis. arXiv preprint arXiv:2210.02186 (2022)
22. Xu, X., Wang, H., Liang, Y. et al.: Can multimodal LLMs perform time series anomaly detection? arXiv preprint arXiv:2502.17812 (2025)
23. Faraway, J., Mahabal, A., Sun, J., et al.: Modeling lightcurves for improved classification of astronomical objects. Statist. Anal. Data Min. ASA Data Sci. J. **9**(1), 1–11 (2016)
24. Xu, B., Yao, J., Yi, X., et al.: Towards better value principles for large language model alignment: a systematic evaluation and enhancement. In: Proceedings of the 63rd Annual Meeting of the Association for Computational Linguistics (Volume 1: Long Papers), pp. 28991–29010 (2025)
25. Fei, Y., Hou, Y., Chen, Z. et al.: Mitigating label biases for in-context learning. In: Proceedings of the 61st Annual Meeting of the Association for Computational Linguistics (Volume 1: Long Papers), pp. 14014–14031 (2023)
26. Salvador, S., Chan, P.: Toward accurate dynamic time warping in linear time and space **11**(5), 561–580 (2007)
27. Wei, J., Wang, X., Schuurmans, D., et al.: Chain-of-thought prompting elicits reasoning in large language models. NeurIPS **35**, 24824–24837 (2022)
28. Zanella, L., Menapace, W., Mancini, M. et al.: Harnessing large language models for training-free video anomaly detection. In: CVPR, pp. 18527–18536 (2024)
29. Hurst, A., Lerer, A., Goucher, A.P., et al.: GPT-4o system card. arXiv preprint arXiv:2410.21276 (2024)
30. Dosovitskiy, A., Beyer, L., Kolesnikov, A., et al.: An image is worth 16x16 words: transformers for image recognition at scale. In: ICLR (2021)

31. Kenton, J.D.M.W.C., Toutanova, L.K.: Bert: pre-training of deep bidirectional transformers for language understanding. In: NAACL-HLT, vol. 1, p. 2 (2019)
32. Xu, H., Chen, W., Zhao, N., et al.: Unsupervised anomaly detection via variational auto-encoder for seasonal KPIS in web applications. In: WWW 2018, pp. 187–196 (2018)

Joint Multi-level Attention and Consistency Aligned for Multimodal Emotion Recognition Based on Physiological Signals

Yuanbo Zeng[1(✉)], Yao Yao[2], Shuaiqi Fu[1], Ganbo Cao[1], Jing Li[1], Liu Yi[1], Xiangdong Peng[1], and Shuqiang Guo[1]

[1] Jiangxi University of Finance and Economics, Nanchang, China
2202320786@stu.jxufe.edu.cn
[2] Huazhong Agricultural University, Wuhan, China

Abstract. Emotion recognition is crucial for human-computer interaction, enhancing user experience and safety. Although Electroencephalogram (EEG)-based methods have made progress, they often capture redundant information and fail to extract key emotional features from frequency and spatial domains. They also overlook the use of richer multimodal data that could enhance model performance. Physiological signals such as EEG, Electrocardiogram (ECG), and Galvanic Skin Response (GSR) exhibit inherent differences in signal properties and feature distributions. These heterogeneities lead to cross-modal feature inconsistency, making it difficult to achieve precise alignment and effective fusion. To this end, we propose a joint multi-level attention and consistency aligned method (MLACA) for multimodal emotion recognition. A multi-level channel-spatial attention to mine important emotional information across multiple layers in both the spatial and frequency domains of the EEG modality, enabling the extraction of rich emotional features. To further achieve fine-grained cross-modal alignment of physiological signals, we design a multimodal cross-attention alignment method that effectively captures local dynamic associations between modalities, facilitating consistent alignment and interaction of inter-modal sequence features. Extensive experimental results on three datasets demonstrate that our method outperforms the state-of-the-art methods.

Keywords: Multimodal emotion recognition · Multi-level attention · Consistency aligned · Physiological signal · Deep learning

1 Introduction

Emotions can be understood as a unique cognitive experience that occurs when individuals process external information and is influenced by personal preferences, knowledge, and intentions. In many fields, such as medical robots, intelligent education systems, and self-driving cars, quality of human-computer interaction has a significant impact on the emotional state of the user, which can either enhance or weaken the overall happiness of the user. Therefore, to improve user satisfaction and minimize stress, it is particularly important to design human-centered systems that can effectively identify and interpret emotional cues [1].

M. Yoshikawa et al. (Eds.): ADMA 2025, LNAI 16197, pp. 68–82, 2026.
https://doi.org/10.1007/978-981-95-3453-1_5

Changes in emotions often trigger significant physical and physiological reactions [2]. Participants can conceal their inner emotions by controlling physical reactions such as facial expressions, language, or gestures, which to some extent affects the accuracy of emotion recognition models based on physical signals. Internal data patterns reflect physiological signals of emotional responses in the central or peripheral nervous system, such as EEG [3], eye movements [4] electromyography [5], body temperature [6], skin conductance [7], respiration [8], and ECG [9]. Since these internal data are directly connected to the nervous system and can reveal the cognitive experience of the individual, their bias is relatively small. Although unimodal recognizers have achieved some results through advanced deep learning methods, they still have certain limitations in representing complex emotional states [10]. In addition, a single physiological feature may only cover a narrow range of emotional responses in time and space, which limits the ability of the model to distinguish subtle emotion categories [11]. Physiological signals like EEG, GSR, and ECG each capture only one aspect of emotional states—EEG reflects brain activity, while GSR measures skin conductance. Relying on a single signal may lead to incomplete information. Thus, integrating multimodal physiological data is crucial, as it leverages their complementarity to enhance signal quality and better represent complex emotional dynamics.

The challenges of emotion recognition through physiological signal fusion can be summarized as follows: (i) Effectively handling the heterogeneity of multimodal signals, especially EEG remains a major challenge in emotion recognition, as it involves extracting key emotional features from noisy and redundant data while avoiding additional redundancy or noise introduced by overlapping information across modalities. (ii) To date, the inconsistency in the representation of emotional states between different physiological signals has not been fully studied or resolved.

To address modality inconsistency and information redundancy in multimodal emotion recognition, we propose the MLACA method based on physiological signals. First, we introduce a multi-level channel-spatial attention (MCSA) network that uses channel attention to capture the importance of EEG channels across frequency bands (alpha, beta, gamma, theta), and spatial attention to focus on emotion-related brain regions (e.g., prefrontal cortex), enhancing relevant features and suppressing noise. Then, to align modalities, we design a cross-modal transformer (CMT) with cross-attention and integrate a low-rank fusion (LRF) module to reduce redundancy during fusion. Overall, MLACA effectively aligns and fuses multimodal signals for robust emotional feature extraction. The main contributions are: (1) We propose a novel MLACA framework for multimodal emotion recognition based on physiological signals, which effectively addresses issues of inconsistency and information redundancy, enabling the learning of efficient and discriminative multimodal emotional features. (2) A MCSA network is proposed to tackle the challenge of extracting emotional features from EEG signals. By integrating spatial and channel attention, the model effectively captures critical spatial and frequency-domain information while reducing redundancy and heterogeneity, thereby enhancing multimodal emotion recognition performance. (3) We propose a MCAF network to address inconsistencies in emotional state representation across peripheral physiological signals. A CMT is introduced to optimize feature interactions between modalities, while a LRF strategy reduces redundancy during feature fusion.

Together, these components enhance the consistency and effectiveness of multimodal emotion representation. (4) We conducted extensive experiments on three public multimodal datasets using two validation methods to evaluate MLACA. Its performance was compared against various fusion strategies and single-modality baselines.

2 Method

The architecture of MLACA is shown in Fig. 1. First, we preprocess the EEG and peripheral signals separately to extract features. Then, we propose a multi-level channel-spatial attention network to extract feature representations of EEG signals in both spatial and frequency domains. After that, the cross-domain affective network and the multimodal consistency alignment fusion network are applied to the statistic-complexity EEG features and the multimodal peripheral features, respectively. Finally, the network is trained using scene adaptation and contrastive alignment techniques.

2.1 Multi-level Channel-Spatial Attention Network for the EEG Modality

To reduce observation noise, MLACA applies exponential smoothing across all physiological modalities, where each current feature vector $\boldsymbol{x}_n$ and its previous smoothed version $\widetilde{\boldsymbol{x}}_{n-1}$ are combined using a fixed coefficient $\lambda_1 \in [0, 1]$, Then we get the smoothed features $\widetilde{\boldsymbol{x}}_n$, effectively suppressing high-frequency noise while preserving signal trends. For EEG signals, features are organized across spatial, frequency, complexity, and statistical domains. Frequency features $\tilde{X}_f^{(eeg)} \in R^{L \times P_f}$ are mapped into a spatial-frequency tensor $\boldsymbol{X}_f^{(eeg)} \in R^{s \times s \times P_f}$ via interpolation based on EEG channel positions, where $s \times s$ represents the spatial map size. Complexity and statistical features are represented as matrices $X_{co}^{(eeg)} \in R^{L \times P_{co}}$ and $X_{sta}^{(eeg)} \in R^{L \times P_{sta}}$, respectively.

To extract emotional features from the spatial-frequency domain of EEG signals $\boldsymbol{X}_f^{(eeg)}$, , we embed an MCSA network into MLACA. MCSA combines full-dimensional dynamic convolution (ODConv) with self-norm to enhance feature learning. It consists of three layers, each using 3 × 3 ODConv, self-norm, ReLU, and an Attention-Enhanced Feature Fusion (AEFF) module. The motivation behind implementing AEFF is to effectively capture and represent both local and global patterns in $\boldsymbol{X}_f^{(eeg)}$ in the spatial-frequency domain by aggregating multi-scale context from the EEG spatial-frequency features. First, the output tensor of the ResNet block $\widetilde{\boldsymbol{X}}_{fo}^{(eeg)}$ is computed using the following formula:

$$\widetilde{\boldsymbol{X}}_{fo}^{(eeg)} = \tilde{X}_{fi}^{(eeg)} + F_{resb}\left(\tilde{X}_{fi}^{(eeg)}\right) \tag{1}$$

where $\widetilde{\boldsymbol{X}}_{fi}^{(eeg)}$ represents the block input, and F_{resb} is the feedforward operator designed to learn the high-level representation of $\widetilde{\boldsymbol{X}}_{fi}^{(eeg)}$.

Then, AEFF uses a local context learner through stacked ODConv operations to discover the local spatial patterns $\widetilde{\boldsymbol{X}}_{fol}^{(eeg)} \in R^{s \times s \times P_f}$. Simultaneously, the spatial mean is

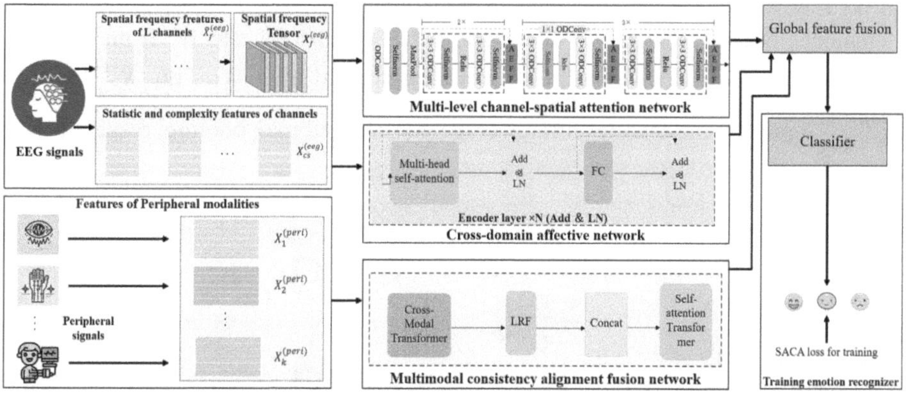

Fig. 1. The architecture of MLACA consists of several components: FC (fully-connected layer), LN (layer normalization), SACA (scenario adaptation with contrastive alignment), ODConv (Omni-Dimensional Dynamic Convolution), LRF (Low Rank Fusion), and AEFF (Attention-Enhanced Feature Fusion Module).

computed:

$$\tilde{x}_{fog}^{(eeg)} = \frac{1}{\tilde{s}^2} \cdot \sum_{i=1}^{\tilde{s}} \sum_{j=1}^{\tilde{s}} \widetilde{\boldsymbol{X}}_{fo}^{eeg}(i, j, :) \quad (2)$$

To enhance EEG-based emotion recognition, we introduce an MCSA module that adaptively weights both frequency and spatial features. The fused spatial-frequency pattern is first processed by a global context learner to extract $\tilde{x}_{fog}^{\prime(eeg)}$, then passed through the MCSA. The channel attention mechanism uses average and max pooling with a two-layer convolution to assign weights to each frequency band, emphasizing emotion-related features while suppressing redundancy. The spatial attention mechanism generates a spatial attention map to highlight emotion-relevant brain regions. By combining both, MCSA effectively captures key emotional features and improves model accuracy and generalization (Fig. 2).

The input to MCSA is $\widetilde{\boldsymbol{X}}_{fo}^{(eeg)} \in R^{s \times s \times P_f}$. The channel attention mechanism can be expressed as:

$$M_c\left(\widetilde{\boldsymbol{X}}_{fo}^{(eeg)}\right) = sigmoid((MLP(MLP(Maxpool(\widetilde{\boldsymbol{X}}_{fo}^{(eeg)})) + AvgPool(\widetilde{\boldsymbol{X}}_{fo}^{(eeg)}))))$$

$$= sigmoid\left(\left(W_1\left(W_2\left({\widetilde{\boldsymbol{X}}_{fo}^{(eeg)}}^{c}_{max}\right)\right) + W_2\left(W_1\left({\widetilde{\boldsymbol{X}}_{fo}^{(eeg)}}^{c}_{avg}\right)\right)\right)\right) \quad (3)$$

where sigmoid is the function that compresses the dynamic range of the input activation vector to [0, 1]. $W_1 \in R^{L/(L \times r)}$ assigns high weights to channels with critical information, and $W_2 \in R^{\frac{L\times L}{r}}$ assigns lower weights to other channels. Here, L represents the number of channels. Notably, these two inputs share the weights W_1 and W_2 from the multi-layer perceptron (MLP), with ReLU activation applied after W_1. $W_1.\widetilde{X}_{fo\ \ max}^{(eeg)^c}$ and $\widetilde{X}_{fo\ \ avg}^{(eeg)^c}$ represent the features after average pooling and max pooling, respectively.

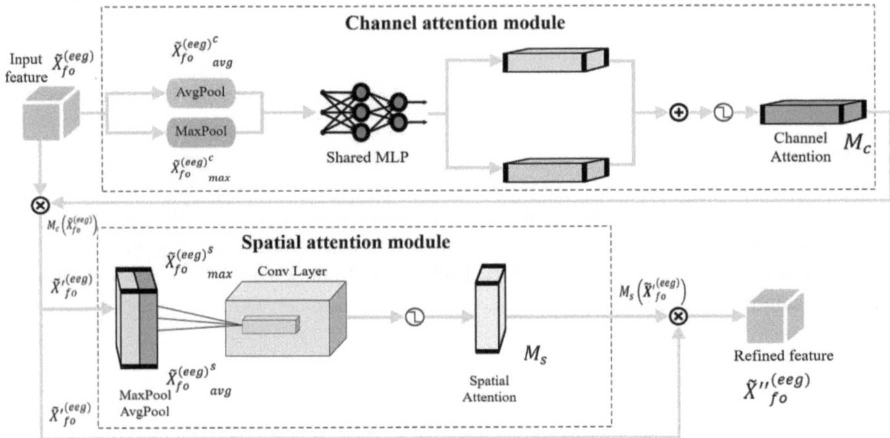

Fig. 2. MCSA based on channel-spatial attention mechanism.

The channel attention module compresses global temporal information using max-pooling and average-pooling to generate channel descriptors. These are processed through a gating mechanism to model inter-channel dependencies, producing feature tensors $\widetilde{X}_{fo\ \ max}^{(eeg)^c}$ and $\widetilde{X}_{fo\ \ avg}^{(eeg)^c}$. Both tensors are passed through a shared MLP, summed, and activated with a sigmoid function to generate the final channel attention weights $M_c\left(\widetilde{X}_{fo}^{(eeg)}\right)$, which are then element-wise multiplied with the original input features. The sigmoid function compresses the dynamic range of the input activation vector to [0,1]. $W_1 \in R^{L/(L\times r)}$ assigns high weights to channels containing key information, while $W_2 \in R^{\frac{L\times L}{r}}$ assigns lower weights to other channels. Here, L represents the number of channels. It is important to note that these two inputs share the weights W_1 and W_2 from a multi-layer perceptron (MLP), and the ReLU activation function is applied after W_1. $W_1.\widetilde{X}_{fo\ \ max}^{(eeg)^c}$ and $\widetilde{X}_{fo\ \ avg}^{(eeg)^c}$ represent the average pooling and max pooling features, respectively.

The spatial attention mechanism captures spatial dependencies by computing attention weights that reflect the importance of different positions in the feature map. It begins by applying average and max pooling along the channel dimension to obtain $\widetilde{X}_{fo\ \ avg}^{(eeg)^s} \in R^{1\times s\times s}$ and $\widetilde{X}_{fo\ \ max}^{(eeg)^s} \in R^{1\times s\times s}$.. These are concatenated along the channel axis and passed through a convolution layer, preserving the spatial size. A sigmoid activation is then applied to generate the final spatial attention weights, which are multiplied with the input feature map to enhance key spatial features. The spatial attention mechanism can be formulated as:

$$M_s\left(\widetilde{X}_{fo}^{(eeg)}\right) = sigmoid((f^{7\times7}\left(\left[AvgPool\left(\widetilde{X}_{fo}^{(eeg)}\right)\right); MaxPool\left(\widetilde{X}_{fo}^{(eeg)}\right)\right] \\ = sigmoid(\left(f^{7\times7}\left(\left[\widetilde{X}_{fo}^{(eeg)^s}{}_{avg}; \widetilde{X}_{fo}^{(eeg)^s}{}_{max}\right]\right)\right), \tag{4}$$

where $f^{7\times7}$ denotes a convolution operation with a 7×7 filter.

Then, based on the 1D channel attention map $M_c \in R^{L\times1\times1}$, and the 2D spatial attention map $M_s \in R^{1\times s\times s}$, the enhanced features $\widetilde{X}_{fo}^{\prime\prime(eeg)}$ are expressed as:

$$\widetilde{X}_{fo}^{\prime(eeg)} = M_c\left(\widetilde{X}_{fo}^{(eeg)}\right) \otimes \widetilde{X}_{fo}^{(eeg)}, \tag{5}$$

$$\widetilde{X}_{fo}^{\prime\prime(eeg)} = M_s\left(\widetilde{X}_{fo}^{\prime(eeg)}\right) \otimes \widetilde{X}_{fo}^{\prime(eeg)}, \tag{6}$$

where $\otimes$ denotes element-wise multiplication.

the output of the AEFF module $Z_{fo}^{(eeg)}$ is calculated as:

$$Z_{fo}^{(eeg)} = \widetilde{X}_{fo}^{\prime\prime(eeg)} \otimes \widetilde{X}_{fi}^{(eeg)} + \left(1 - \widetilde{X}^{\prime}(eeg)_{fo}\right) \otimes F_{resb}\left(\widetilde{X}_{fi}^{(eeg)}\right). \tag{7}$$

Finally, the spatial-frequency affective information Z'^{eeg}_{sf} can be obtained in the output of the final block, i.e.,

$$Z_{sf}^{\prime}(eeg) = MCSA\left(X_f^{(eeg)} | \theta_{MCSAN}\right), Z_{sf}^{\prime}(eeg) \in R^{\tilde{s}^{\prime}\times\tilde{s}^{\prime}\times\tilde{P}_f^{\prime}} \tag{8}$$

In the equation, $\tilde{s}^{\prime} \times \tilde{s}^{\prime}$ represents the spatial dimension, $\tilde{P}_f^{\prime}$ represents the dimension of the learned features in each spatial region, and θ_{MCSA} represents the set of all learnable parameters of the MCSA network.

To overcome the limitations of MCSA network in capturing multi-domain EEG features caused using fixed convolution kernel sizes, we introduce a parallel self-attention mechanism that fuses complexity and statistical features. We construct feature matrices $X_{co}^{(eeg)} \in R^{L\times P_{co}}$ and $X_{sta}^{(eeg)} \in R^{L\times P_{sta}}$, The terms L, P_{co} and P_{sta} represent the number of channels and feature dimensions in the complexity and statistical domains, respectively. These are merged into $X_{cs}^{(eeg)} \in R^{L\times(P_{co}+P_{sta})}$ and linearly projected as $\tilde{X}_{cs}^{(eeg)} = X_{cs}^{(eeg)} W_{ae}^{(C)}$ with $W_{ae}^{(C)} \in R^{(P_{co}+P_{sta})\times d_1}$. Self-attention is then applied: Attention$(\tilde{X}_{cs}^{(eeg)} | \theta^{(Q,K,V)}, L)$. To enhance representation diversity, we use Multi-Head Self-Attention (MHSA), where each head is $H_i = (\tilde{X}_{cs}^{(eeg)} | \theta_i{}^{(Q,K,V)})$ and the outputs are combined as:

$$MHSA\left(\tilde{X}_{cs}^{(eeg)}\right) = Concat(H_1, H_2, \cdots H_h)W^{(H)}. \tag{9}$$

The final result passes through an encoder layer $EL(\tilde{X}_{cs}^{(eeg)} | \phi_{EL})$, and after stacking $N^{(AL)}$ layers, we obtain the integrated representation $\widehat{X}_{cs}^{(eeg)} \in R^{L\times d_1}$.

Multimodal Consistency Alignment Fusion Network for the Peripheral Features. Existing multimodal fusion methods often face issues like modality inconsistency and information redundancy due to poor alignment and naive fusion strategies, which can degrade performance. To address this, we propose the MCAF network, designed to enhance consistency and reduce redundancy during fusion. MCAF first employs a Cross-Modal Transformer (CMT) for bidirectional alignment between modalities, followed by a Low-Redundancy Fusion (LRF) module to integrate and filter the transferred features. Finally, a Self-Attention Transformation (SAT) refines the combined features, capturing inter-modality correlations and producing enhanced peripheral representations.

We use the cross-modal transformer (CMT) to refine peripheral features and resolve inter-modal inconsistencies, followed by low-rank fusion (LRF) to eliminate redundancy introduced during fusion. A self-attention–based modality encoder is then applied to capture emotional correlations across modalities. For K peripheral physiological modalities, each represented as a vector $X_k^{(peri)} \in R^{P_k}$ at each time step (with P_k denoting the feature dimension), the combined multimodal peripheral feature is $X^{(peri)} \in R^{d_P}$, where $d_P = \sum_{k=1}^{K} P_k$ is the concatenated dimension.

CMT enables information transfer between peripheral modalities by treating each modality as a query and others as key-value pairs. This helps reduce inter-modal inconsistencies. Information from one modality is transferred to another using CMT. For instance, transferring information from modality 1 to modality 2 is defined as:

$$X_{12}^{(peri)} = CMT\left(X_1^{(peri)} \rightarrow X_2^{(peri)}\right) \quad (10)$$

where $X_{12}^{(peri)}$ denotes the features transferred from the first modality generate the features of the second modality.

CMT consists of N encoder layers. Each encoder layer contains a Multi-Head Cross-Modal Attention (MHCMA) module, which has J Cross-Modal Attention (CMA) blocks, and the latent features learned by the jth head ($j = 1, \ldots, J$) are denoted as $\ddot{X}_{12}^{(peri)(j)} \in R^{P_v}$, which can be calculated as:

$$\ddot{X}_{12}^{(peri)(j)} = CMA\left(\ddot{X}_1^{(peri)} \rightarrow \ddot{X}_2^{(peri)}\right) = softmax\left(\frac{\left(Q_2^{(j)} \times (K_1^{(j)})^T\right)}{\sqrt{d_k}} \times V_1^{(j)}\right) \quad (11)$$

We project $\ddot{X}_2^{(peri)(j)}$ into a query using $W_{Q_2}^{(j)}$, and $\ddot{X}_1^{(peri)(j)}$ into key and value representations using $W_{K_1}^{(j)}$ and $W_{V_1}^{(j)}$, respectively. The attention weights are calculated through a scaled dot-product between query and key, followed by SoftMax normalization, which are then applied to the value to obtain the attended representation. This enables adaptive focus on cross-modal cues for better feature alignment and emotional representation. The output of the MHCMA block is:

$$\ddot{X}_{12}^{(peri)} = Concat\left(\ddot{X}_{12}^{(peri)(1)}, \ldots, \ddot{X}_{12}^{(peri)(j)}\right) \quad (12)$$

Then, the output of the nth encoder layer in the CMT is given by:

$$\tilde{X}_{12}^{[n](peri)} = MHCMA\left(LN\left(X_{12}^{[n-1](peri)}\right), LN\left(X_{1}^{(peri)}\right)\right) + LN\left(X_{12}^{[n-1](peri)}\right)$$

$$X_{12}^{[n](peri)} = f_\theta\left(LN\left(\tilde{X}_{12}^{[n](peri)}\right)\right) + LN\left(\tilde{X}_{12}^{[n](peri)}\right) \tag{13}$$

In CMT, each encoder layer consists of Layer Normalization (LN) and a position-wise feed-forward network f_θ. After N encoder layers, in CMT, feature transfers between modalities are represented as $X_{21}^{(peri)} X_{23}^{(peri)}$, ... To reduce redundancy between modalities, we adopt the Low-Rank Fusion (LRF) strategy [12], which integrates multi-modal signals by modeling their interactions via a low-rank tensor. This reduces both computational cost and redundancy. The optimized embedding of the k-th modality $\overline{X}_k^{(peri)} \in R^{J \times P_v}$ is expressed as:

$$\overline{X}_k^{(peri)} = LRF\left(X_k^{(peri)}, X_{1k}^{(peri)}, \ldots, X_{(k-1)k}^{(peri)}, X_{(k+1)k}^{(peri)} \ldots, X_{Kk}^{(peri)}\right) \tag{14}$$

By leveraging CMT for information transfer and LRF for fusion, inter-modal inconsistencies and redundancy are effectively reduced. To further capture shared and complementary information, a Self-Attention Transformer (SAT) is applied to the concatenated optimized features, producing the enhanced Peripheral-Modality Affective Pattern $\widehat{X}^{(peri)} \in R^{J \times d_P}$:

$$\widehat{X}^{(peri)} = SAT\left(Concat\left(\overline{X}_1^{(peri)}, \overline{X}_2^{(peri)}, \ldots, \overline{X}_k^{(peri)}\right)\right) \tag{15}$$

Multi-modal Global Feature Fusion and Training of MLACA. To integrate emotional features from the spatial-frequency EEG pattern $Z_{sf}^{\prime(eeg)}$, cross-peripheral modality pattern $\widehat{X}^{(peri)}$, and cross-domain EEG pattern $\widehat{X}_{cs}^{(eeg)}$, we propose a Global Feature Fusion Module in MLACA. While CMT focuses on cross-modal interactions [14]. The operation can be defined as:

$$CRA\left(Y_\alpha, Y_\beta | \theta_{ca}^{(Q,K,V)}, L_\alpha \times L_\beta\right) = f_{softmax}\left(\frac{Y_\alpha W_{ca}^{(Q)} {W_{ca}^{(K)}}^T Y_\beta^T}{\sqrt{d_{ca}}}\right) Y_\beta W_{ca}^{(V)}. \tag{16}$$

where $Y_\alpha \epsilon R^{L_\alpha \times d_{ca}}$ represents the query, and $Y_\beta \epsilon R^{L_\beta \times d_{ca}}$ represents the key and value.

After that, We using the spatial-frequency EEG pattern $Z_{sf}^{\prime(eeg)} \in R^{\tilde{s}' \times \tilde{s}' \times \tilde{P}_f'}$ as the query and the cross-domain pattern $\widehat{X}_{cs}^{(eeg)}$ as key and value. Both are linearly projected into a shared space:

$$E_{sf}^{(eeg)} = Z_{sf}^{\prime(eeg)} W_{sf}^{(ML)}, E_{cs}^{(eeg)} = \widehat{X}_{cs}^{(eeg)} W_{cs}^{(ML)} \tag{17}$$

Then, the representations between $E_{sf}^{(eeg)}$ and $E_{cs}^{(eeg)}$ are computed using the multi-head cross-attention mechanism with the parameter set θ_{sfcs}:

$$X_{sfcs}^{(eeg)} = MHCA\left(E_{sf}^{(eeg)}, E_{cs}^{(eeg)} | \theta_{sfcs}\right). \tag{18}$$

at the same time, we also use $E_{sf}^{(eeg)}$ as the query and learn the symmetric pattern $X_{sfcs}^{\prime(eeg)} \in R^{L \times d_2}$ with the learnable parameter set $\theta_{sfcs}^{\prime}$,

$$X_{sfcs}^{\prime(eeg)} = MHCA\left(E_{sf}^{(eeg)}, E_{cs}^{(eeg)} | \theta_{sfcs}^{\prime}\right). \tag{19}$$

For modality-level fusion, cross-attention is performed between $\widehat{X}^{(peri)}$ and $X_{sfcs}^{\prime(eeg)}$:

$$X^{(phy)} = MHCA\left(\widehat{X}^{(peri)}, X_{sfcs}^{\prime(eeg)} | \theta_{phy}\right) + \widehat{X}^{(peri)}, \tag{20}$$

$$X^{\prime(phy)} = MHCA\left(X_{sfcs}^{\prime(eeg)}, \widehat{X}^{(peri)} | \theta_{phy}^{\prime}\right) + X_{sfcs}^{\prime(eeg)}. \tag{21}$$

where $X^{(phy)} \in R^{d_P \times d_2}$ and $X^{\prime(phy)}$ represent the symmetric modality fusion pattern from external features with a dimension of d_P and the L channels in the EEG feature representation. The learnable parameter sets are defined by θ_{phy} and $\theta_{phy}^{\prime}$, respectively.

Finally, global fusion is achieved by mapping $X_{sfcs}^{(eeg)}$, $X^{(phy)}$, and $X^{\prime(phy)}$ into a unified space via FC layers and concatenating the outputs to obtain:

$$x^{(fused)} = GF\left(X_{sfcs}^{\prime(eeg)}, X^{(phy)}, X^{\prime(phy)} | \theta_{GF}\right). \tag{22}$$

To mitigate distributional variations across emotional scenes, we apply a scene adaptation pre-training strategy [15] using center loss to promote scene-discriminative feature clustering.

3 Experiment

3.1 Data Description and Preprocessing

The summary of the three databases used for validating the MLACA is listed in Table 1 For all three datasets, EEG signals were re-referenced to the common mean and bandpass filtered (4–45 Hz). Artifacts from eye movement, muscle activity, and motion were removed using ICA [16]. For SEED-IV and SEED-V, PCA was applied to eye movement signals to suppress brightness and environmental noise [17]. Signals were segmented into non-overlapping 3-s intervals (4 s for SEED-IV and SEED-V for better frequency resolution). From EEG, features including differential entropy, frequency bands were extracted [18]. Peripheral features included 55 for DEAP, and 22 and 24 from eye movement signals in SEED-IV and SEED-V, respectively [19, 20].

Performance Comparison and Performance of MLACA Using Different Modalities. We compare the proposed MLACA framework with several representative multimodal integration methods, including JE [21], SAN [22], ATTEN [23], DFAF [24], R2GAN [25], DCCA [26], CMHSA [27], and RHPRNet [15]. Experimental results on DEAP, SEED-IV, and SEED-V datasets under participant-generic settings (Tables 2) show that MLACA consistently achieves superior performance. It records the highest accuracy and macro-F1 scores on SEED-IV and SEED-V, and performs competitively on DEAP, particularly in valence recognition, demonstrating strong generalization and robustness across different scenarios.

Table 1. Summary of the DEAP, SEED-IV, and SEED-V databases.

Databases	DEAP	SEED-IV	SEED-V
Participants	32	15	16
Emotional stimuli	Musical videos	Movie clips	Movie clips
Modalities	7	2	2
EEG channels	32	62	62
Sessions for a participant	1	3	3
Trials for a session	40	24	15
Sampling rate of signals	128 Hz	200 Hz	200 Hz
Target emotional classes	Binary	Four	Five

Table 2. We evaluate MLACA against other multimodal methods using accuracy and macro-F1 (%) under the participant-generic setup, reporting both the average and standard deviation across participants.

Databases	DEAP-valence	DEAP-arousal
JE	58.12 $\pm$ 3.98/59.87 $\pm$ 4.12	58.12 $\pm$ 3.98/59.87 $\pm$ 4.12
SAN	54.63 $\pm$ 2.76/51.44 $\pm$ 6.03	54.63 $\pm$ 2.76/51.44 $\pm$ 6.03
ATTEN	53.78 $\pm$ 3.64/52.09 $\pm$ 5.82	53.78 $\pm$ 3.64/52.09 $\pm$ 5.82
DFAF	60.42 $\pm$ 6.01/61.53 $\pm$ 5.27	56.42 $\pm$ 6.01/61.53 $\pm$ 5.27
R^2 GAN	55.21 $\pm$ 3.43/49.67 $\pm$ 7.13	55.21 $\pm$ 3.43/49.67 $\pm$ 7.13
DCCA	56.89 $\pm$ 3.55/56.32 $\pm$ 3.18	56.89 $\pm$ 3.55/56.32 $\pm$ 3.18
CMHA	60.37 $\pm$ 5.60/61.76 $\pm$ 5.83	56.38 $\pm$ 4.31/56.45 $\pm$ 4.19
RHPRNet	61.37 $\pm$ 4.20/60.74 $\pm$ 4.15	57.85 $\pm$ 3.79/**59.41** $\pm$ 4.68
Ours	**63.05** $\pm$ 4.41/**62.94** $\pm$ 4.64	**58.47** $\pm$ 2.94/58.44 $\pm$ 3.78
Databases	SEED-IV	SEED-V
JE	65.50 $\pm$ 6.75/64.90 $\pm$ 6.38	66.70 $\pm$10.10/66.00 $\pm$ 9.90
SAN	67.50 $\pm$ 7.10/67.20 $\pm$ 7.05	63.50 $\pm$ 9.30/62.40 $\pm$ 9.15
ATTEN	66.00 $\pm$ 4.25/65.70 $\pm$ 4.15	62.20 $\pm$ 9.80/61.90 $\pm$ 9.80
DFAF	65.80 $\pm$ 7.50/64.60 $\pm$ 7.40	63.90 $\pm$ 11.80/62.50 $\pm$ 12.50
R^2 GAN	65.20 $\pm$ 5.55/64.80 $\pm$ 5.50	65.60 $\pm$ 10.10/64.50$\pm$ 10.10
DCCA	67.20 $\pm$ 7.00/67.00 $\pm$ 6.95	64.80 $\pm$ 11.50/62.90 $\pm$ 11.20
CMHA	66.40 $\pm$ 6.50/65.80 $\pm$ 6.60	66.40 $\pm$ 9.90/65.70 $\pm$ 9.80
RHPRNet	68.30 $\pm$ 7.20/67.80 $\pm$ 7.11	68.30 $\pm$ 12.30/67.60 $\pm$ 12.20
Ours	**69.23** $\pm$ 5.33/**68.70** $\pm$ 5.72	**68.76** $\pm$ 10.96/68.34 $\pm$ 11.56

MLACA outperforms existing methods thanks to its hierarchical design and two core modules: MCSA and MCAF. MCSA captures emotion-relevant EEG features by attending to both spatial and spectral dimensions, while MCAF uses a Cross-Modal Transformer and Low-Rank Fusion to align peripheral signals and reduce inter-modality inconsistencies. Existing methods like JE, DCCA, DFAF, and RHPRNet lack targeted mechanisms for handling modality-specific inconsistencies and heterogeneous physiological signals. In contrast, MLACA introduces adaptive attention and alignment tailored to such data, leading to improved performance and interpretability.

As shown in Table 3, combining EEG with peripheral modalities significantly enhances MLACA's performance across all datasets, outperforming single-modality models. In DEAP, both MCSA and MCAF are vital—MCSA captures fine-grained EEG features, while MCAF aligns modalities and reduces inconsistencies via cross-modal transformers and low-rank fusion. For SEED-IV and SEED-V, the gains mainly come from MCSA, highlighting EEG's dominant role. Its attention mechanism effectively focuses on key brain regions and frequencies, improving recognition accuracy.

Table 3. The emotion recognition performance of MLACA is evaluated on the DEAP, SEED-IV, SEED-V databases, using accuracy and macro-F1 score (%) across various combinations of physiological data modalities. The average value across participants and the associated standard deviation are presented.

Modality	DEAP-valence	DEAP-arousal	SEED-IV	SEED-V
EEG	62.36 ± 3.00/62.20 ± 3.08	57.91 ± 3.04/57.88 ± 3.41	37.67 ± 3.51/37.40 ± 3.74	36.58 ± 4.93/35.77 ± 5.12
Peri	61.38 ± 6.13/61.36 ±6.37	55.97 ± 3.28/55.81 ± 3.54	63.94 ± 7.02/63.52 ± 6.97	67.26 ± 10.03/67.74 ± 8.73
EEG-Peri	**63.05** ± 4.41/**62.94** ± 4.64	**58.47** ± 2.94/**58.44** ± 3.78	**69.23** ± 5.33/**68.70** ± 5.72	**68.76** ±10.96/**68.34** ± 11.56

Note: The maximum value in each column is highlighted in bold

Ablation Study. We conducted an ablation study under the participant-generic setting (Table 4) to evaluate the contributions of MCSA, CMT, and LRF. Removing MCSA reduced performance significantly (e.g., SEED-IV accuracy dropped from 69.23% to 65.02%), showing its importance in capturing spatial-spectral EEG features. Replacing its hierarchical fusion with a parallel strategy offered slight gains but still underperformed the full model. Excluding CMT caused the largest drop, especially on DEAP, confirming its role in semantic alignment. Removing LRF also led to a performance decline, though less severe, indicating its effect in reducing redundancy. Overall, the full MLACA model performed best, validating the effectiveness of all components.

Table 4. The effectiveness of each key module of the proposed framework under participant-generic configurations.

Databases	DEAP-valence	DEAP-arousal	SEED-IV	SEED-V
MLACA-MCSA	60.42 ± 4.38/59.80 ± 4.12	54.91 ± 3.26/54.33 ± 3.45	65.02 ± 5.15/65.61 ± 5.87	65.19 ± 10.20/65.88 ± 10.42
MLACA-MCSA(Parallel)	62.11 ± 4.51/61.73 ± 4.76	56.86 ± 3.01/57.19 ± 3.37	68.35 ± 4.90/66.94 ± 5.43	67.85 ± 9.87/66.79 ± 10.01
MLACA -CMT	59.87 ± 4.65/59.31 ± 4.33	53.60 ± 3.44/52.98 ± 3.62	68.93 ± 5.28/6788 ± 5.97	66.77 ± 9.66/67.01 ± 9.92
MLACA -LRF	60.73 ± 4.12/59.95 ± 4.48	55.01 ± 2.83/54.44 ± 3.12	68.22 ± 5.09/67.61 ± 5.76	67.42 ± 10.05/66.88 ± 10.28
MLACA	**63.05** ± 4.41/**62.94** ± 4.64	**58.47** ± 2.94/**58.44** ± 3.78	**69.23** ± 5.33/**68.70** ± 5.72	**68.76** ±10.96/**68.34** ± 11.56

4 Conclusion

We propose MLACA, a novel joint attention and alignment framework addressing feature inconsistency and redundancy in multimodal physiological signals for emotion recognition. MLACA employs a multi-level channel-spatial attention mechanism to extract emotionally relevant features from EEG signals and integrates a cross-domain emotion encoder with a multimodal consistency alignment module based on a cross-modal transformer and low-rank fusion. This design enhances semantic coherence across modalities while reducing redundancy. Experiments on DEAP, SEED-IV, and SEED-V datasets show MLACA outperforms state-of-the-art methods in both participant-dependent and -independent settings, with significant gains in accuracy and macro F1-score.

Disclosure of Interests. The authors declare that there are no conflicts of interest that could be construed as affecting the findings.

References

1. Liu, R., Jia, Z., Bao, F., Li, H.: Retrieval-augmented dialogue knowledge aggregation for expressive conversational speech synthesis. Inf. Fusion **118**, 102948 (2025). https://doi.org/10.1016/j.inffus.2025.102948
2. Jia, Z., Liu, Y., Wang, H., et al.: Cross-modal knowledge distillation for enhanced unimodal emotion recognition. IEEE Trans. Affect. Comput. (2025)
3. Wang, J., Feng, Z., Ning, X., et al.: Two-stream dynamic heterogeneous graph recurrent neural network for multi-label multi-modal emotion recognition. IEEE Trans. Affect. Comput. (2025)
4. Bao, J., Tao, X., Zhou, Y.: An emotion recognition method based on eye movement and audiovisual features in MOOC learning environment. IEEE Trans. Comput. Soc. Syst. 1–13 (2022). https://doi.org/10.1109/TCSS.2022.3221128
5. Kim, H., Zhang, D., Kim, L., Im, C.-H.: Classification of Individual's discrete emotions reflected in facial microexpressions using electroencephalogram and facial electromyogram. Expert Syst. Appl. **188**, 116101 (2022). https://doi.org/10.1016/j.eswa.2021.116101
6. Rahman, Md.M., et al.: Recognition of human emotions using EEG signals: a review. Comput. Biol. Med. **136**, 104696 (2021). https://doi.org/10.1016/j.compbiomed.2021.104696
7. Shukla, J., Barreda-Angeles, M., Oliver, J., Nandi, G.C., Puig, D.: Feature extraction and selection for emotion recognition from electrodermal activity. IEEE Trans. Affect. Comput. **12**, 857–869 (2021). https://doi.org/10.1109/TAFFC.2019.2901673
8. Zhang, Q., Chen, X., Zhan, Q., Yang, T., Xia, S.: Respiration-based emotion recognition with deep learning. Comput. Ind. **92–93**, 84–90 (2017). https://doi.org/10.1016/j.compind.2017.04.005
9. Zontone, P., et al.: Car driver's sympathetic reaction detection through electrodermal activity and electrocardiogram measurements. IEEE Trans. Biomed. Eng. **67**, 3413–3424 (2020). https://doi.org/10.1109/TBME.2020.2987168
10. Cheng, C., Liu, W., Wang, X., et al.: DISD-Net: a dynamic interactive network with self-distillation for cross-subject multi-modal emotion recognition. IEEE Trans. Multimedia (2025)
11. Jia, Z., Zhao, F., Guo, Y., et al.: Multi-level disentangling network for cross-subject emotion recognition based on multimodal physiological signals. In: Proceedings of the Thirty-Third International Joint Conference on Artificial Intelligence, pp. 3069–3077 (2024)

12. Liu, C., Zhou, X., Zhu, Z., et al.: VBH-GNN: variational Bayesian heterogeneous graph neural networks for cross-subject emotion recognition. In: The Twelfth International Conference on Learning Representations (2024)
13. Liu, Z., Shen, Y., Lakshminarasimhan, V.B., Liang, P.P., Zadeh, A., Morency, L.-P.: Efficient low-rank multimodal fusion with modality-specific factors. arXiv preprint arXiv:1806.00064 (2018)
14. Li, J., et al.: Hybrid fusion with intra- and cross-modality attention for image-recipe retrieval. In: Proceedings of the 44th International ACM SIGIR Conference on Research Development Information Retr., ACM, New York, NY, USA, 2021, pp. 244–254 (2021). https://doi.org/10.1145/3404835.3462965
15. Tang, J., Ma, Z., Gan, K., Zhang, J., Yin, Z.: Hierarchical multimodal-fusion of physiological signals for emotion recognition with scenario adaption and contrastive alignment. Inf. Fusion **103**, 102129 (2024)
16. Yin, Z., Zhao, M., Wang, Y., Yang, J., Zhang, J.: Recognition of emotions using multimodal physiological signals and an ensemble deep learning model. Comput. Methods Programs Biomed. **140**, 93–110 (2017). https://doi.org/10.1016/j.cmpb.2016.12.005
17. Soleymani, M., Pantic, M., Pun, T.: Multimodal emotion recognition in response to videos. IEEE Trans. Affect. Comput. **3**(2), 211–223 (2012). https://doi.org/10.1109/TAFFC.2011.37
18. Duan, R.-N., Zhu, J.-Y., Lu, B.-L.: Differential entropy feature for EEG-based emotion classification. In: Proceedings of the 6th International IEEE/EMBS Conference on Neural Eng. (NER), San Diego, CA, USA, Nov. 2013, pp. 81–84 (2013). https://doi.org/10.1109/NER.2013.6695876
19. Atkinson, J., Campos, D.: Improving BCI-based emotion recognition by combining EEG feature selection and kernel classifiers. Expert Syst. Appl. 47, 35–41 (2016). https://doi.org/10.1016/j.eswa.2015.10.049
20. Lu, Y., Zheng, W.-L., Li, B., Lu, B.-L.: Combining eye movements and EEG to enhance emotion recognition. In: Proceedings of the 24th International Conference on Artificial Intelligent (IJCAI), Buenos Aires, Argentina, Jul. 2015, pp. 1170–1176 (2015)
21. Salvador, A., et al.: Learning cross-modal embeddings for cooking recipes and food images. In: Proceedings of the IEEE Conference on Computing Vision Pattern Recognition (CVPR), Honolulu, HI, USA, Jul. 2017, pp. 3068–3076 (2017). https://doi.org/10.1109/CVPR.2017.327
22. Chen, J., Pang, L., Ngo, C.-W.: Cross-modal recipe retrieval: How to cook this dish?" In: MultiMedia Modeling: Proceedings 23rd International Conference (MMM), Reykjavik, Iceland, Jan. 2017, pp. 588–600 (2017). https://doi.org/10.1007/978-3-319-51811-4_48
23. Chen, J.-J., Ngo, C.-W., Feng, F.-L., Chua, T.-S.: Deep understanding of cooking procedure for cross-modal recipe retrieval. In: Proceedings of the 26th ACM International Conference on Multimedia, Association for Computing Machinery, New York, NY, USA, 2018, pp. 1020–1028 (2018). https://doi.org/10.1145/3240508.3240627
24. Gao, P., et al.: Dynamic fusion with intra- and inter-modality attention flow for visual question answering. In: Proceedings of the IEEE/CVF Conference on Computing Vision Pattern Recognition (CVPR), Long Beach, CA, USA (2019)
25. Zhu, B., Ngo, C.-W., Chen, J., Hao, Y.: R2GAN: cross-modal recipe retrieval with generative adversarial network. In: Proc. IEEE/CVF Conf. Comput. Vis. Pattern Recognit. (CVPR), Long Beach, CA, USA, Jun. 2019, pp. 11469–11478 (2019). https://doi.org/10.1109/CVPR.2019.01174

26. Liu, W., Qiu, J.-L., Zheng, W.-L., Lu, B.-L.: Comparing recognition performance and robustness of multimodal deep learning models for multimodal emotion recognition". IEEE Trans. Cogn. Dev. Syst. **14**(3), 715–729 (2022). https://doi.org/10.1109/TCDS.2021.3071170
27. Tzirakis, P., Chen, J., Zafeiriou, S., Schuller, B.: End-to-end multimodal affect recognition in real-world environments. Inf. Fusion **68**, 46–53 (2021). https://doi.org/10.1016/j.inffus.2020.10.011

A Novel Historical-Meteorology-Informed Approach for One-Week Air Quality Forecasting

Xiang Li[1], Huihui Zheng[2], and Zhewei Wei[3](✉)

[1] Data Science Institute, Shandong University, Jinan, China
li_xiang@mail.sdu.edu.cn

[2] China National Environmental Monitoring Centre, Beijing, China
zhenghuihui@cnemc.cn

[3] Gaoling School of Artificial Intelligence, Renmin University of China, Beijing, China
zhewei@ruc.edu.cn

Abstract. In recent years, sustained efforts to mitigate air pollution have significantly improved overall air quality. However, infrequent pollution events pose new challenges for accurate forecasting. Concurrently, the growing availability of monitored datasets offers opportunities for data-driven deep learning approaches in this field. Most deep learning models rely solely on time series methods, neglecting the influence of meteorological factors. Although some studies incorporate these factors with complicated structures, they often fail to capture the intrinsic relationships between meteorological conditions and pollution outcomes, especially for week-ahead predictions. This study presents an adaptive method for modeling the impact of evolving meteorological conditions on pollution outcomes. Inspired by the annual periodicity of meteorological patterns and their consistent impact on pollution, we align historical pollution events with the current window, adaptively shifting them in response to the corresponding meteorological patterns. Specifically, we represent meteorological variations as changes in the coefficients of orthogonal polynomials and employ a gating mechanism to dynamically transfer the influence of historical meteorological data to the current window, thereby indicating potential pollution process under current conditions. Experiment Results demonstrate that our method can enhance a linear model to outperform existing pure time series approaches on week-ahead prediction, emphasizing the importance of considering meteorological conditions and pollution results in air quality forecasting. Moreover, our findings align with expert domain knowledge and achieve further performance improvements when incorporating forecasted meteorological data.

Keywords: Air Quality Forecasting · Meteorologically-Informed · Time Series Forecasting

M. Yoshikawa et al. (Eds.): ADMA 2025, LNAI 16197, pp. 83–97, 2026.
https://doi.org/10.1007/978-981-95-3453-1_6

1 Introduction

Air quality is a critical determinant of human health and environmental sustainability. Over the past decade, China has made substantial progress in air pollution control, supported by the nationwide deployment of state-controlled air quality monitoring networks and a series of environmental regulations. As a result, the frequency and duration of severe pollution events have declined significantly. However, this progress also brings new challenges to air quality forecasting. With fewer pollution episodes and more transient pollution processes, predicting future pollutant concentrations has become increasingly complex.

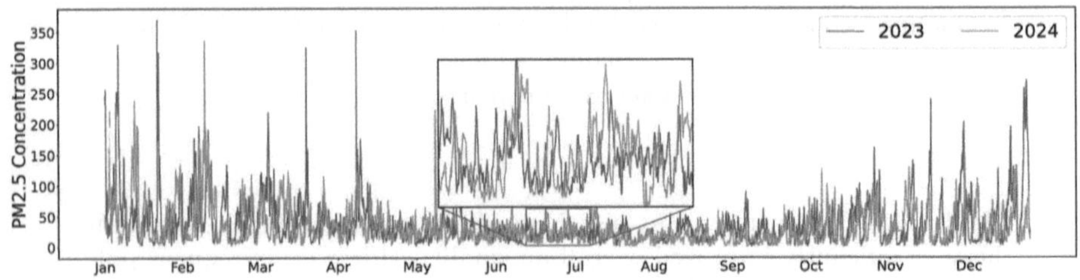

Fig. 1. Pollutant concentrations at a given station exhibit strong annual periodicity. However, temporal misalignments caused by shifts in meteorological patterns lead to variations in pollution behavior, resulting in either a lead or a lag.

Traditionally, numerical models have dominated the field of air pollution forecasting [30]. These models are grounded in physical and chemical principles, simulating pollutant dynamics by solving partial differential equations. Despite their interpretability and strong theoretical foundation, such models are computationally expensive and slow to adapt to scenarios beyond predefined reaction mechanisms. These limitations hinder their applicability in fast-response or data-scarce environments, motivating a shift toward data-driven approaches, particularly deep-learning models. Data-driven models offer several advantages. The increasing density of monitoring stations has enabled the collection of high-resolution, large-scale data on pollutants and meteorological conditions. Deep-learning models can leverage this data to learn complex temporal patterns without explicitly modeling physical mechanisms [31]. However, most deep-learning models treat pollutant concentrations as univariate or multivariate time series, relying on learning trends and periodicities from historical data. These purely sequential models often underperform in real-world air quality forecasting, where meteorological factors dominate in shaping the accumulation and dispersion of pollution.

To address this, several recent works attempt to incorporate meteorological variables [3,4]. One class of methods focuses on wind-based pollutant transport modeling, grounded in physical intuition [5,19]. However, such methods are limited to small-scale regions and become increasingly ineffective as heavy pollution events become less frequent. Another class uses cross-attention mechanisms to model the influence of meteorological inputs on pollutant sequences [23]. While flexible, these methods often lack physical interpretability. They rely on token similarity between meteorological and pollutant sequences within the same input window, making the model highly sensitive to sequence length and unable to generalize when using forecasted meteorological inputs. Moreover, there is a growing demand for forecasts exceeding seven days, yet most existing studies continue to focus on short-term predictions of three days or less.

In this work, we revisit the physical influence of meteorological variation on pollutant evolution by leveraging the temporal similarity between pollution patterns in different years. As illustrated in Fig. 1, pollutant concentrations at a given station exhibit strong annual periodicity, with similar monthly values of peaks and troughs. However, temporal misalignments caused by shifts in meteorological patterns lead to variations in pollution behavior, resulting in either a lead or a lag. Assuming that the underlying mechanisms by which meteorological variables influence pollution remain stable over time, we propose learning how current meteorological conditions differ from historical ones and using this difference to adapt historical pollution trajectories to the current forecast window.

Our Contributions. The main contributions of this paper can be summarized as follows:

- **We propose a novel method that captures how meteorological variation influences pollution processes.** By leveraging the annual regularity in pollutant evolution and modeling the difference in orthogonal polynomial coefficients, our method adaptively transfers the historical meteorology-pollution relationship to the current prediction window.
- **We provide a benchmark for air quality forecasting, highlighting the importance of aligning historical meteorological conditions in one-week predictions.** Experimental results highlight the essential role of adaptively aligning historical meteorological factors with current conditions in deep learning-based air quality forecasting, especially for a one-week-ahead forecasting.
- **Our methods are physically interpretable.** We reveal interpretable meteorological influences on pollutant prediction: precipitation emerges as the most impactful factor in multi-day forecasts, while wind speed becomes dominant when using simulated forecasted inputs.

2 Preliminary

2.1 Problem Formulation

Given a sequence of historical pollutant concentrations $\mathbf{X}_t \in \mathbb{R}^{P \times N}$ and historical meteorological data $\mathbf{M_t} \in \mathbb{R}^{P \times N}$, where P is the input window length and N denotes the number of stations, the goal is to forecast pollutant concentrations $\mathbf{Y_t} \in \mathbb{R}^{F \times N}$ for the next F time steps. Thus, the air quality prediction problem is generally defined as:

$$\mathcal{F}(\mathbf{X}_t, \mathbf{M}_t) \longrightarrow \mathcal{F}(\mathbf{Y}_t)$$

2.2 Deep Learning-Based Time Series Forecasting

Deep learning-based time series forecasting models are generally categorized into two main types. The first focuses on Long-term Time Series Forecasting (LTSF), aiming to extract periodic and trend patterns from historical data for accurate future predictions [2]. Transformer-based models have achieved notable success in this domain. For example, PatchTST [14] converts point-wise inputs into patch-wise representations to better capture local context and employs channel-wise independence to reduce variable interference. iTransformer [12] maintains the core Transformer architecture but redefines its usage by treating the input window as a series-wise patch, enabling effective modeling of inter-variable dependencies via attention. CATS [8] replaces self-attention with cross-attention, using future time steps as queries and incorporating parameter sharing to enhance long-term forecasting accuracy while reducing model complexity. Fredformer [15] addresses the frequency bias often observed in Transformers—where low-frequency components are overemphasized—by promoting balanced learning across frequency bands, thereby preserving crucial high-frequency, low-amplitude signals. Beyond Transformers, linear models such as DLinear and CycleNet [10,29] offer efficient and competitive performance, making them standard baselines in the literature.

The second category comprises Spatio-Temporal Graph Neural Networks (STGNNs), which combine graph-based and sequential modeling to capture dependencies among variables explicitly. These models are widely applied to tasks such as traffic flow forecasting [18,25].

Despite their strengths, both LTSF models and STGNNs primarily rely on endogenous patterns and often overlook the influence of exogenous factors. Addressing this gap, TimeXer [23] introduces cross-attention between endogenous and exogenous variables to incorporate external information. However, it only considers exogenous effects within the input window and models them based on sequence similarity, lacking further physical interpretability.

2.3 Air Quality Forecasting

Air quality forecasting is a representative spatiotemporal prediction task, with the core challenge lying in accurately modeling the propagation of pollutants over time and space while adequately accounting for the critical influence of meteorological factors [9,26]. Early studies commonly adopted standard spatiotemporal graph neural network (STGNN) architectures, where graph convolutional networks (GCNs) were used to capture spatial dependencies among monitoring stations [11], and recurrent neural networks (e.g., LSTM) were employed to model temporal dynamics [7,20]. For instance, GAGNN [1] integrates wind speed and direction to construct pollutant transmission pathways between stations or cities, thereby enhancing the modeling of spatial interactions. AirPhyNet [5] and Air-DualODE [19] incorporate physical mechanisms into the learning process, leveraging two classical principles governing the movement of airborne particles—diffusion and advection—to build physics-constrained spatiotemporal models. PM2.5GNN [22] and MDSTNet [27] go a step further by incorporating meteorological forecast data, thereby simulating prediction scenarios that more closely resemble real-world applications and demonstrating strong practical utility (Fig. 2) .

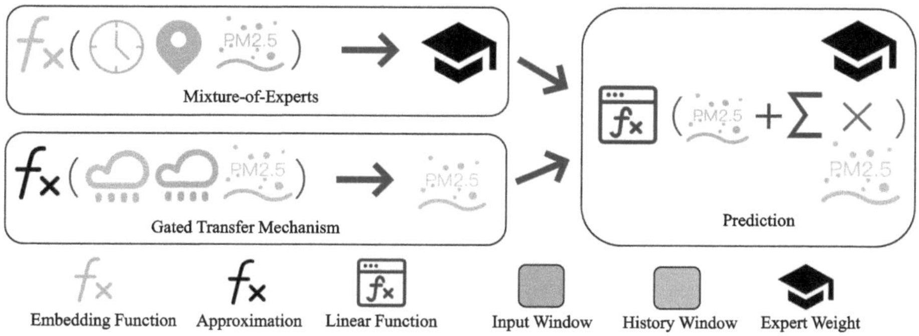

Fig. 2. Illustration of the modeling process using precipitation as an example. Polynomial approximation captures its impact on pollutants; spatial-temporal context determines expert weights; a linear model generates the final prediction.

However, limitations remain in how these models integrate meteorological factors. In most cases, such information is either used as auxiliary input for graph construction or embedded directly as input features. The former approach may become less effective as pollution events become more sporadic due to improved environmental regulation. At the same time, the latter, being a "black-box" treatment, may improve predictive accuracy but lacks interpretability. Specifically, it fails to explicitly model or explain the mechanisms through which meteorological variables influence pollution dynamics.

3 Methodology

In this section, we introduce our meteorology-aware air quality forecasting framework, which aims to leverage historical relationships between meteorology and pollution to enhance prediction accuracy within the current forecast window. Specifically, we use orthogonal polynomials to approximate the meteorological sequences and pollutant processes. Differences in polynomial coefficients capture different components variation, and a gated mechanism adaptively transfers the influence of historical meteorology on pollution to the current time window. We introduce a mixture-of-experts structure that learns diverse mappings from meteorological changes to pollutant behavior, thereby better modeling the heterogeneous effects of meteorological changes on pollutant behavior.

Based on the annual periodicity of pollution processes, we use the previous year's data as historical information. We formalized the mapping process as:

$$k \times \varDelta \mathbf{M}_{t \leftarrow hist} = \varDelta \mathbf{X}_{t \leftarrow hist},$$

where the transformation from historical meteorological changes to pollutant variations is assumed to be parameterized by k. There are two correspondences between the historical window and the current window: one between the input window $\mathbf{X}_t$ and its historical counterpart $\mathbf{X}_{hist}$, and the other between the target window $\mathbf{Y}_t$ and $\mathbf{Y}_{hist}$. From a practical perspective, the former corresponds to forecasting based on observed meteorological data, while the latter corresponds to forecasting using meteorological predictions.

Our framework comprises three key components: (1) an orthogonal polynomial approximation module for input sequences, (2) a gated transfer mechanism to adapt historical meteorologyâĂŞpollution dynamics, and (3) a mixture-of-experts module for capturing diverse meteorological impacts.

3.1 Orthogonal Polynomial Approximation

To capture temporal structure, ensure consistency across time windows, and improve numerical stability, we approximate sequences using Chebyshev polynomials of the first kind evaluated at Chebyshev nodes, where the rescaled time points for a sequence of length L are defined as:

$$x_t = \cos\left(\frac{(2t+1)\pi}{2L}\right), \quad t = 0, 1, \ldots, L-1. \tag{1}$$

Let $T_k(x)$ denote the k-th Chebyshev polynomial:

$$T_0(x) = 1, \quad T_1(x) = x, \quad T_k(x) = 2xT_{k-1}(x) - T_{k-2}(x), \quad \text{for } k \geq 2. \tag{2}$$

We construct the design matrix $\mathbf{\Phi} \in \mathbb{R}^{L \times (K+1)}$, where each entry is:

$$\varPhi_{t,k} = T_k(x_t), \quad \text{for } t = 1, \ldots, L, \ k = 0, \ldots, K. \tag{3}$$

The sequence $\mathbf{z}$ is then approximated by a linear combination of basis functions:

$$\hat{\mathbf{z}} = \mathbf{\Phi}\boldsymbol{\theta}, \quad \text{where } \boldsymbol{\theta} \in \mathbb{R}^{K+1}. \tag{4}$$

The coefficient vector $\boldsymbol{\theta}$ is obtained via the closed-form least squares solution:

$$\boldsymbol{\theta} = (\mathbf{\Phi}^\top \mathbf{\Phi})^{-1} \mathbf{\Phi}^\top \mathbf{z}. \tag{5}$$

We apply this approximation independently to historical pollutant sequence $\mathbf{X}_{\text{hist}}$, current meteorological sequence $\mathbf{M}_t$ and historical meteorological sequence $\mathbf{M}_{\text{hist}}$. The three kinds of sequences at each station share the same polynomial basis but have different coefficients. The resulting coefficient vectors $\boldsymbol{\phi}_{\text{hist}}, \boldsymbol{\theta}_t$ and $\boldsymbol{\theta}_{\text{hist}} \in \mathbb{R}^{K+1}$ are used downstream to quantify meteorological variation and its effect on pollution dynamics.

3.2 Gated Transfer Mechanism

To adaptively transfer the influence of historical meteorological conditions to the current forecast window, we employ a Gated Linear Unit (GLU) that modulates the historical pollutant representation ϕ_{hist} based on the difference between current and historical meteorological conditions.

Let $\Delta\boldsymbol{\theta} = \boldsymbol{\theta}_t - \boldsymbol{\theta}_{\text{hist}} \in \mathbb{R}^{K+1}$ denote the difference in Chebyshev polynomial coefficients between current and historical meteorology. We compute the transferred representation as:

$$\hat{\boldsymbol{X}}_{i,t} = \boldsymbol{\Phi} \times (\boldsymbol{\phi}_{\text{hist}} + \text{GLU}(\Delta\boldsymbol{\theta})), \tag{6}$$

where GLU is defined as:

$$\text{GLU}(\Delta\boldsymbol{\theta}) = \Delta\boldsymbol{\theta} \odot \sigma(\mathbf{W}\Delta\boldsymbol{\theta}), \tag{7}$$

with $\odot$ denoting element-wise multiplication, σ the sigmoid activation, and $\mathbf{W} \in \mathbb{R}^{(K+1)\times(K+1)}$ as learnable parameter matrices.

This formulation enables the model to learn which components of the historical pollutant trajectory should be enhanced or suppressed under current meteorological variation. The resulting vector $\hat{\mathbf{X}}_{i,t} \in \mathbb{R}^L$ represents a physically-adjusted latent representation of pollution dynamics and is used as input to the downstream forecasting module (Fig. 3).

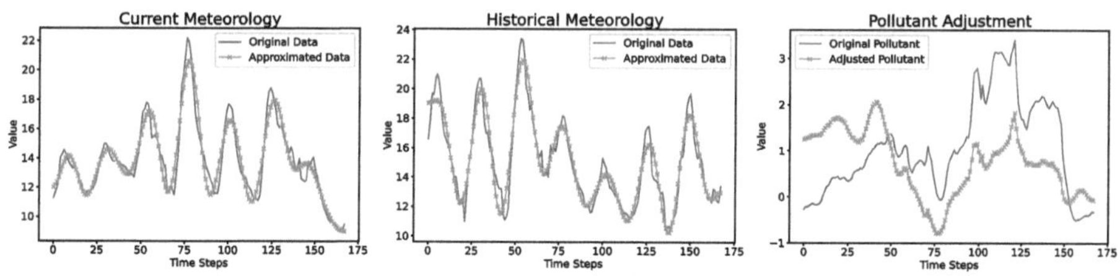

Fig. 3. Example of polynomial approximation and the adjusted historical pollutant sequence. The adjusted pollutant sequence reflects variations attributable to changes in meteorological factors.

3.3 Mixture-of-Experts for Meteorological Influence

To enhance the model's capacity for handling diverse and nonlinear meteorology-pollution interactions, we adopt a mixture-of-experts (MoE) architecture that dynamically selects among multiple expert models based on both spatial and temporal context. Empirically, pollution patterns vary across regions and seasons. For instance, springtime pollution in Beijing is often driven by dust storms, whereas winter pollution is more closely associated with emissions related to heating.

We incorporate station-specific spatial embeddings $\mathbf{H}_s \in \mathbb{R}^{N \times d}$ and temporal embeddings $\mathbf{H}_t \in \mathbb{R}^{C \times d}$, where N is the number of stations, and C is the number of calendar positions (e.g., day-of-year, hour-of-day). For each input sample (i, t), we extract the corresponding embeddings $\mathbf{h}_s = \mathbf{H}_s[i]$ and $\mathbf{h}_t = \mathbf{H}_t[t]$, and combine them with a projection of the input feature $\mathbf{X}_{i,t} \in \mathbb{R}^P$ to compute the gating weights:

$$\mathbf{G} = \mathrm{softmax}\left(\mathbf{W}_p\left(\mathbf{W}_e\mathbf{X}_{i,t} + \mathbf{h}_s + \mathbf{h}_t\right)\right), \quad \mathbf{G} \in \mathbb{R}^e, \tag{8}$$

where $\mathbf{W}_e \in \mathbb{R}^{d \times L}$, $\mathbf{W}_p \in \mathbb{R}^{e \times d}$ are learnable weight matrices.

Each expert $f^{(j)}(\cdot)$, $j = 1, \ldots, e$, receives the GLU-modulated historical pollutant representation $\hat{\boldsymbol{X}}_{i,t} \in \mathbb{R}^L$ and outputs an expert-specific forecast vector $\hat{\boldsymbol{X}}_{i,t}^{(j)} \in \mathbb{R}^L$. The aggregated pollutant representation is computed as:

$$\hat{\boldsymbol{X}}_{i,t}^{\mathrm{MoE}} = \sum_{j=1}^{e} G_j \cdot \hat{\boldsymbol{X}}_{i,t}^{(j)}. \tag{9}$$

Finally, the expert-modulated historical estimate is combined with the current pollutant observations and passed through a linear transformation $\mathcal{F}(\cdot)$ to obtain the prediction:

$$\hat{\mathbf{Y}}_{i,t} = \mathcal{F}\left(\mathbf{X}_{i,t} + \hat{\boldsymbol{X}}_{i,t}^{\mathrm{MoE}}\right). \tag{10}$$

This MoE structure allows the model to adaptively focus on different types of meteorological-pollutant relationships depending on the region and time, improving both flexibility and predictive accuracy.

4 Experiments

4.1 Experimental Setup

The dataset was provided by the China National Environmental Monitoring Centre (CNEMC) and is organized at the city level. It includes hourly PM2.5 concentration monitoring data for 41 major cities in China, spanning from January 1, 2021, to December 31, 2024, along with the corresponding meteorological data for these cities, including wind speed, wind direction, temperature, humidity, precipitation, and pressure. Since our proposed method focuses on modeling the impact of meteorological factor variations on pollution processes rather than pollutant transport mechanisms, wind direction is excluded from our experiments. However, we retain wind speed, as its influence on local pollutant dynamics at individual stations is non-negligible. Our experiments were conducted on a 40G NVIDIA A100 GPU with the open-source platform BasicTS [16] to provide a unified comparison.

4.2 Baselines and Metrics

A number of existing models for air pollution forecasting rely on non-public or specially processed datasets, and some only provide partially released code, posing challenges for reproducibility and fair performance comparison. Given that the primary objective of this work is to develop a novel method for capturing the influence of meteorological factors on pollution processes, we mainly compare our approach against pure time-series models to emphasize the importance of incorporating meteorological information in air quality forecasting. As a result, we conduct a comprehensive evaluation of our method alongside several state-of-the-art deep learning architectures for forecasting, categorizing these baselines into transformer-based and non-transformer-based models.

Transformer-Based: Autoformer [24], PatchTST [14], CATS [8], iTransformer [12], Fredformer [15] and MGSFformer. Notably, MGSFformer [28] is explicitly developed for the task of air pollution forecasting, incorporating domain-specific considerations into its design.

Non transformer-Based: DLinear [29] and CycleNet [10], TimeMixer [21], CrossGNN [6], Koopa [13], FiLM [32] and STID [17].

We also compared a newly proposed **Physics-Guided** Neural Network designed for air quality prediction [5] within our unified framework.

Our evaluation is conducted on re-normalized data, employing metrics such as Mean Absolute Error (MAE), Root Mean Squared Error (RMSE), and Weighted Mean Absolute Percentage Error (WAPE) to achieve an extensive evaluation of forecasting performance. Assuming y_j denotes the j-th actual sample $\mathbf{Y}_{i,t}$, and $\hat{y}_j$ is the corresponding prediction, these metrics are formulated as follows:

$$MAE(\hat{y_j}, y_j) = \frac{1}{F}\sum_{j=1}^{F} |y_j - \hat{y_j}|,$$

$$RMSE(\hat{y_j}, y_j) = \sqrt{\frac{1}{F}\sum_{j=1}^{F}(y_j - \hat{y_j})^2},$$

$$WAPE(\hat{y_j}, y_j) = \frac{\sum_{j=1}^{n} |y_j - \hat{y_j}|}{\sum_{j=1}^{n} |y_j|},$$

where F is the forecasting horizon, MAE and RMSE metrics reflect prediction accuracy, while WAPE normalizes the error to account for differences in data scale.

5 Discussion

In this section, we analyze the experimental results. It is important to note that in real-world air quality forecasting applications, meteorological forecast data is typically used to predict whether pollution events will occur. Some prior works have also considered this practical constraint. To reflect this aspect, we denote two settings in our experiments: **F**(uture)**Linear** and **P**(ast)**Linear**, which indicate whether future meteorological information is used when modeling the relationship between meteorological variables and pollutant dynamics. **Linear** refers to a single-layer linear transformation without considering meteorological factors.

To isolate the impact of inaccurate meteorological forecasts on pollution predictions, we use future meteorological observations as a proxy for forecasted data, enabling a more precise evaluation of our method's effectiveness. Specifically, assuming accurate meteorological forecasts, we evaluate whether our model effectively translates meteorological changes into more accurate pollutant predictions. We then analyze the sensitivity of hyperparameters to the most influential meteorological factors based on the experimental results, and explore the impact of other meteorological variables on the pollution process when optimal results are achieved.

Table 1. Experimental results. The best-performing results for each metric are bolded, and the second-best results are underlined.

Methods	Horizon=72			Horizon=168			Overall		
	MAE	RMSE	WAPE	MAE	RMSE	WAPE	MAE	RMSE	WAPE
Autoformer	23.49	37.24	70.12%	23.56	36.18	70.00%	23.27	36.79	58.61%
PatchTST	22.00	35.75	60.78%	21.55	34.05	62.19%	20.70	33.62	50.98%
iTransformer	20.50	36.65	62.19%	22.10	34.99	63.30%	22.11	35.71	53.81%
CATS	21.99	35.75	60.78%	21.71	34.56	61.64%	20.80	33.96	50.99%
Fredformer	23.13	38.18	62.11%	22.44	36.29	63.25%	21.38	35.35	52.05%
MGSFformer	22.22	35.82	62.24%	22.16	34.82	64.27%	21.06	34.22	51.76%
DLinear	21.39	35.07	60.87%	20.82	33.04	62.50%	20.21	33.26	50.55%
CycleNet	22.22	36.01	60.71%	22.02	34.87	63.28%	20.98	34.13	51.34%
FiLM	22.07	35.76	61.53%	21.81	34.64	62.62%	20.92	34.10	51.24%
Koopa	22.46	36.12	63.89%	22.42	35.32	65.69%	21.55	34.63	53.25%
STID	20.93	34.18	60.71%	<u>20.23</u>	32.01	60.31%	<u>19.71</u>	32.18	50.00%
CrossGNN	<u>20.92</u>	<u>33.88</u>	60.40%	20.30	<u>31.81</u>	61.49%	19.73	<u>32.00</u>	<u>49.63%</u>
TimeMixer	21.08	35.09	<u>58.89%</u>	20.36	32.71	**58.63%**	20.04	33.00	49.82%
AirPhyNet	21.14	34.06	62.56%	20.74	32.42	62.86%	20.65	33.26	52.55%
Linear	21.45	35.24	61.11%	20.93	33.18	63.15%	20.27	33.23	51.48%
PLinear	**20.27**	**33.38**	**57.79%**	**19.52**	**30.97**	<u>59.50%</u>	**19.28**	**31.74**	**48.45%**
FLinear	**19.47**	**32.03**	**54.91%**	**19.04**	**30.29**	**56.33%**	**18.64**	**30.43**	**47.18%**

5.1 Performance Study

The experimental results are summarized in Table 1. For practical relevance, we selected two time horizons for comparison: horizon=72 (representing a 3-day forecast) and horizon=168 (representing a 7-day forecast). These time points reflect the model's performance in a rolling forecast context. Additionally, we compared the model's performance for a one-time 7-day forecast, as shown in the "Overall" column.

The results reveal two key insights. First, incorporating meteorological factors is crucial for 7-day predictions. Our proposed method significantly enhances the performance of the linear model, surpassing existing purely time-series models. We also observed that while most pure time-series models performed reasonably well at the 24-hour horizon, their performance deteriorated substantially as the forecast horizon lengthened. This suggests that although pure time-series models can capture historical patterns, the relationship between pollutant changes and prior time steps is weaker in air pollution forecasting tasks.

In contrast, CrossGNN, which explicitly constructs a graph structure, significantly outperforms models that implicitly capture relationships between variables, such as iTransformer, which uses attention mechanisms. This warrants further investigation in future research. Additionally, CrossGNN outperforms AirPhyNet, which incorporates physical diffusion processes. We believe this indicates that correlations between monitoring stations can extend beyond geographic constraints. Relying on physical diffusion processes in model design may limit the learning of these correlations, as their significance can be obscured by physical distance.

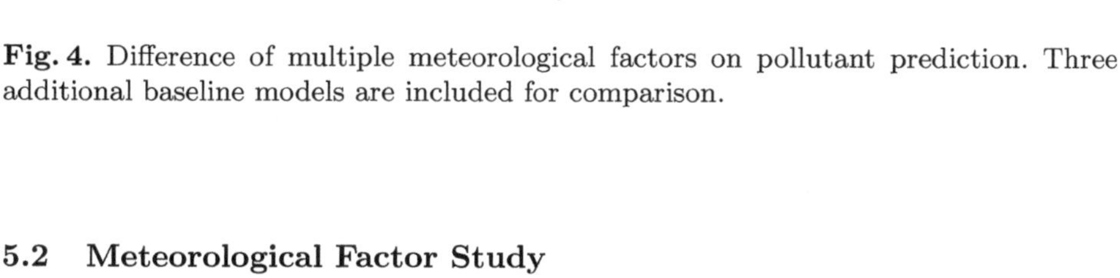

Fig. 4. Difference of multiple meteorological factors on pollutant prediction. Three additional baseline models are included for comparison.

5.2 Meteorological Factor Study

In Fig. 4 we assessed the impact of various meteorological factors on the PM2.5 pollution process using the Plinear and FLinear settings, yielding inconsistent results. When only current meteorological factors were considered, precipitation had the greatest impact on prediction accuracy, followed by wind speed. However, when meteorological forecasts were included, wind speed significantly improved the pollution process prediction, with humidity as the second most influential factor. This finding aligns with current human forecasting practices and intuitive understanding. For instance, air quality typically remains good during and after rainfall, while strong winds are likely to create favorable conditions for pollutant dispersion. This supports the effectiveness of our proposed method, which is also grounded in clear physical reasoning.

5.3 Hyperparameter Study

We analyze the sensitivity of our method to hyperparameters under the meteorological conditions where it achieves the best performance in both the PLinear (precipitation) and FLinear (wind speed) settings. Our method involves two main hyperparameters: the number of experts in the mixture-of-experts module and the degree of the polynomial used in the approximation. To facilitate comparison, we also present the results of three baseline models. AirPhyNet is a physics-driven model specifically designed for air pollution forecasting. MGSF-former is another domain-specific model developed for pollutant prediction, though it does not account for meteorological factors. CrossGNN, in contrast, serves as a strong non-domain-specific baseline. Despite not being explicitly tailored for air quality tasks, it demonstrates robust generalization ability and achieves competitive performance on our dataset.

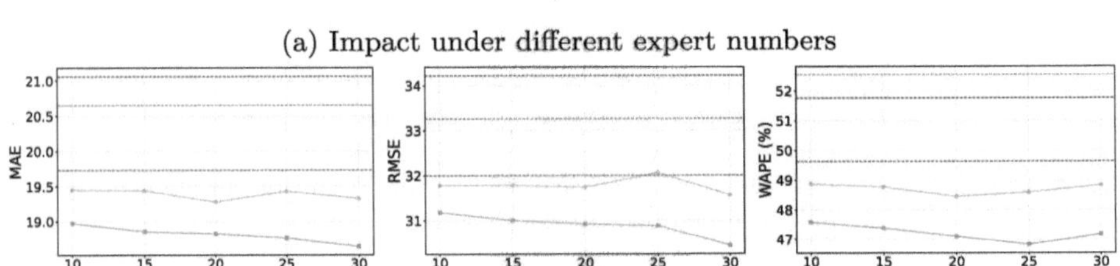

(a) Impact under different expert numbers

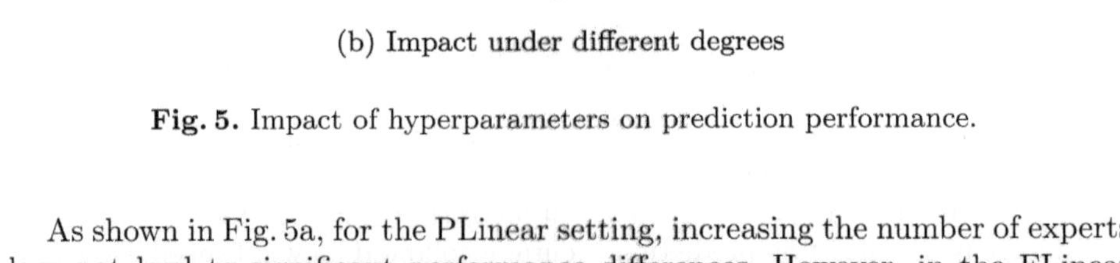

(b) Impact under different degrees

Fig. 5. Impact of hyperparameters on prediction performance.

As shown in Fig. 5a, for the PLinear setting, increasing the number of experts does not lead to significant performance differences. However, in the FLinear setting, more experts result in a lower RMSE, suggesting that accurate meteorological forecasts can help improve predictions of more extreme pollution events.

In addition, we modified the formulation in Equation (8). Interestingly, we found that excluding the current pollutant sequence $\mathbf{X_{i,t}}$ when computing expert weights slightly improves performance. While this phenomenon may be attributed to the seasonal patterns inherent in temporal features, we believe it is worth further investigation in future work, especially in scenarios involving the joint influence of multiple pollutants.

The polynomial degree controls the approximation capacity, effectively reflecting the model's ability to capture sharp fluctuations such as peaks and troughs in the sequence. A higher degree enables more precise fitting. As shown in Fig. 5b, both the PLinear and FLinear settings exhibit a consistent decline in prediction error as the degree increases. This trend aligns with intuition, indicating that meteorological variation exerts a meaningful influence on pollutant dynamics—and the more accurately this variation is approximated, the more effectively the model can learn its impact.

6 Conclusion

In this work, we introduce a novel meteorologically informed method for air quality forecasting that effectively leverages the interplay between meteorological variations and pollutant dynamics. By modeling meteorological changes through orthogonal polynomial approximations and employing a gated transfer mechanism alongside a mixture-of-experts architecture, our approach adaptively transfers historical meteorology-pollution relationships to current forecasting windows. This results in a physically interpretable framework that significantly enhances the performance of linear models, surpassing state-of-the-art pure time-series models, particularly for longer-term forecasts such as 7-day horizons. Our experimental results on the China Air Quality dataset (2021âĂŞ2024) demonstrate the critical role of meteorological factors, with precipitation and wind speed emerging as dominant influences in multi-day and forecast-driven predictions, respectively. These findings align with physical intuition and underscore the importance of accurate meteorological forecasts in enhancing the accuracy of deep-learning-based air quality predictions. Furthermore, our method's interpretability and scalability make it a practical solution for real-world applications, offering a robust foundation for future research into integrating exogenous factors in spatiotemporal forecasting tasks.

Acknowledgments. This study was funded by the National Key R&D Program of China (2022YFC3701201) and we sincerely thank the China National Environmental Monitoring Centre for providing the data. Code is available at https://github.com/superarthurlx/APM

Disclosure of Interests. The authors have no competing interests to declare that are relevant to the content of this article.

References

1. Chen, L., Xu, J., Wu, B., Qian, Y., Du, Z., Li, Y., Zhang, Y.: Group-aware graph neural network for nationwide city air quality forecasting (arXiv:2108.12238), August 2021. [cs]
2. Guang-Yu, W., Hui-Chuan, H., Zhi-Qing, Z., Wen-Long, S., Yong-Hao, W., Ai-Min, F.: Dualmamba: a patch-based model with dual mamba for long-term time series forecasting. Front. Comp. Sci. **20**(2), 2002315 (2026)
3. Han, J., Liu, H., Zhu, H., Xiong, H.: Kill two birds with one stone: a multi-view multi-adversarial learning approach for joint air quality and weather prediction. IEEE Trans. Knowl. Data Eng. **35**(11), 11515–11528 (2023). https://doi.org/10.1109/TKDE.2023.3236423
4. Han, J., Liu, H., Zhu, H., Xiong, H., Dou, D.: Joint air quality and weather predictions based on multi-adversarial spatiotemporal networks
5. Hettige, K.H., Ji, J., Xiang, S., Long, C., Cong, G., Wang, J.: Airphynet: Harnessing physics-guided neural networks for aír quality prediction (arXiv:2402.03784), February 2024. [cs]
6. Huang, Q., Shen, L., Zhang, R., Ding, S., Wang, B., Zhou, Z., Wang, Y.: Crossgnn: Confronting noisy multivariate time series via cross interaction refinement
7. Jin, X.B., Wang, Z.Y., Kong, J.L., Bai, Y.T., Su, T.L., Ma, H.J., Chakrabarti, P.: Deep spatio-temporal graph network with self-optimization for air quality prediction. Entropy **25**(2), 247 (2023). https://doi.org/10.3390/e25020247
8. Kim, D., Park, J., Lee, J., Kim, H.: Are self-attentions effective for time series forecasting? (arXiv:2405.16877) (Oct 2024)
9. Liang, Y., Xia, Y., Ke, S., Wang, Y., Wen, Q., Zhang, J., Zheng, Y., Zimmermann, R.: Airformer: Predicting nationwide air quality in China with transformers (arXiv:2211.15979), November 2022. [eess]
10. Lin, S., Lin, W., Hu, X., Wu, W., Mo, R., Zhong, H.: Cyclenet: Enhancing time series forecasting through modeling periodic patterns (arXiv:2409.18479), October 2024
11. Liu, B., Qi, Z., Gao, L.: Enhanced air quality prediction through spatio-temporal feature sxtraction and fusion: a self-tuning hybrid approach with gcn and gru. Water, Air, Soil Pollution **235**(8), 532 (2024)
12. Liu, Y., et al.: itransformer: inverted transformers are effective for time series forecasting (arXiv:2310.06625), October 2023
13. Liu, Y., Li, C., Wang, J., Long, M.: Koopa: Learning non-stationary time series dynamics with koopman predictors (arXiv:2305.18803), October 2023 [cs]
14. Nie, Y., Nguyen, N.H., Sinthong, P., Kalagnanam, J.: A time series is worth 64 words: Long-term forecasting with transformers (2023)
15. Piao, X., Chen, Z., Murayama, T., Matsubara, Y., Sakurai, Y.: Fredformer: Frequency debiased transformer for time series forecasting. In: Proceedings of the 30th ACM SIGKDD Conference on Knowledge Discovery and Data Mining, pp. 2400–2410, August 2024. https://doi.org/10.1145/3637528.3671928, http://arxiv.org/abs/2406.09009
16. Shao, Z., et al.: Exploring progress in multivariate time series forecasting: comprehensive benchmarking and heterogeneity analysis. IEEE Trans. Knowl. Data Eng. **37**(1), 291–305 (2024)
17. Shao, Z., Zhang, Z., Wang, F., Wei, W., Xu, Y.: Spatial-temporal identity: a simple yet effective baseline for multivariate time series forecasting. arXiv, August 2022. http://arxiv.org/abs/2208.05233

18. Shao, Z., Zhang, Z., Wei, W., Wang, F., Xu, Y., Cao, X., Jensen, C.S.: Decoupled dynamic spatial-temporal graph neural network for traffic forecasting. arXiv preprint arXiv:2206.09112 (2022)
19. Tian, J., et al.: Air quality prediction with physics-guided dual neural odes in open systems (2025)
20. Wang, J., Li, J., Wang, X., Wang, J., Huang, M.: Air quality prediction using CT-LSTM. Neural Comput. Appl. **33**(10), 4779–4792 (2020). https://doi.org/10.1007/s00521-020-05535-w
21. Wang, S., Wu, H., Shi, X., Hu, T., Luo, H., Ma, L., Zhang, J.Y., Zhou, J.: Timemixer: Decomposable multiscale mixing for time series forecasting (2024)
22. Wang, S., Li, Y., Zhang, J., Meng, Q., Meng, L., Gao, F.: Pm2.5-gnn: a domain knowledge enhanced graph neural network for pm2.5 forecasting. In: Proceedings of the 28th International Conference on Advances in Geographic Information Systems, pp. 163–166, November 2020. https://doi.org/10.1145/3397536.3422208 [eess]
23. Wang, Y., et al.: TimeXer: Empowering Transformers for Time Series Forecasting with Exogenous Variables, February 2024. http://arxiv.org/abs/2402.19072 [cs]
24. Wu, H., Xu, J., Wang, J., Long, M.: Autoformer: decomposition transformers with auto-correlation for long-term series forecasting. In: NIPS. arXiv (2021). http://arxiv.org/abs/2106.13008
25. Wu, Z., Pan, S., Long, G., Jiang, J., Chang, X., Zhang, C.: Connecting the dots: Multivariate time series forecasting with graph neural networks. In: KDD. arXiv (2020). http://arxiv.org/abs/2005.11650, [cs, stat]
26. Yi, X., Zhang, J., Wang, Z., Li, T., Zheng, Y.: Deep distributed fusion network for air quality prediction. In: Proceedings of the 24th ACM SIGKDD International Conference on Knowledge Discovery & Data Mining, pp. 965–973 (2018)
27. Yin, H., Zhang, Y.M., Xu, J., Chang, J.L., Li, Y., Liu, C.L.: Air quality prediction with a meteorology-guided modality-decoupled spatio-temporal network (arXiv:2504.10014), April 2025 [cs]
28. Yu, C., Wang, F., Wang, Y., Shao, Z., Sun, T., Yao, D., Xu, Y.: Mgsfformer: a multi-granularity spatiotemporal fusion transformer for air quality prediction. Inf. Fusion **113**, 102607 (2025). https://doi.org/10.1016/j.inffus.2024.102607
29. Zeng, A., Chen, M., Zhang, L., Xu, Q.: Are transformers effective for time series forecasting? In: Proceedings of the AAAI Conference on Artificial Intelligence, vol. 37, pp. 11121–11128 (2023)
30. Zhang, B., Zhang, H., Zhao, G., Lian, J.: Constructing a pm2.5 concentration prediction model by combining auto-encoder with bi-lstm neural networks. Environ. Modelling Softw. **124**, 104600 (2020). https://doi.org/10.1016/j.envsoft.2019.104600
31. Zhang, Z., Zhang, S., Chen, C., Yuan, J.: A systematic survey of air quality prediction based on deep learning. Alex. Eng. J. **93**, 128–141 (2024). https://doi.org/10.1016/j.aej.2024.03.031
32. Zhou, T., et al.: Film: frequency improved legendre memory model for long-term time series forecasting. arXiv, September 2022. http://arxiv.org/abs/2205.08897 [cs, stat]

Enchancing Time Series Forecasting: A Time-Frequency Analysis Perspective

Shang Zeng[1,3(✉)], Yiyang Fan[1,3], Shaobing Zhang[1,2,3], and Zhe Cui[1,3]

[1] Chengdu Institute of Computer Application, Chinese Academy of Sciences, Chengdu 610041, China
zengshang19@mails.ucas.ac.cn
[2] Shenzhen CBPM-KEXIN Banking Technology Company Limited, Shenzhen 518206, Guangdong, China
[3] University of Chinese Academy of Sciences, Beijing 10049, China

Abstract. Time series forecasting models often either focus on time-domain or frequency-domain, or alternatively, create network architectures to combine both. While these models achieve performance improvements, modeling time-domain and frequency-domain either solely or separately tends to underestimate the complementary relationship and intrinsic connections between them. In this paper, we explore a novel time-frequency perspective, propose a unified time-frequency analysis framework that models temporal dynamics and spectral components simultaneously via a spectrogram-based time-frequency representation (TFR). We introduce Spectrogram Band-wise Normalization to balance the distribution of frequency bands, and then introduce a Time-Frequency Enhanced (TFE) Block with two lightweight, channel-independent models, SpecLinear and SpecMLP, that learn directly from the normalized STFT spectrogram. Our models follow mainstream architectural patterns but with significantly simplified structures. Extensive experiments on multiple real-world datasets demonstrate that our TFE models outperform traditional methods, highlighting the effectiveness of a unified time-frequency perspective for designing simple yet powerful forecasting approaches.

Keywords: Time Series Forecasting · Spectrogram Band-wise Normalization · Time-Frequency Analysis

1 Introduction

Time series forecasting is a fundamental task and widely applied in areas such as weather forecasting [15], electricity demand forecasting, medical time series analysis [5] and decision making [8].

Both time-domain and frequency-domain are crucial for understanding time series data [1]. Time domain analysis focuses on the sequential order of series and models the dependency on local features, effectively capturing characteristics that change over time. Conversely, frequency domain analysis views series as a

M. Yoshikawa et al. (Eds.): ADMA 2025, LNAI 16197, pp. 98–112, 2026.
https://doi.org/10.1007/978-981-95-3453-1_7

superposition of signals at different frequencies, revealing the multi-periodicity and frequency characteristics.

However, real-world time series often exhibit intricate temporal patterns (see Fig. 1). The cycles and magnitudes vary, and there are some transient anomalies. Both time domain and frequency domain methods have their limitations when dealing with such variations. Time domain methods rely on local features, which makes them more sensitive to anomalies and causes them to struggle with time series that exhibit multiple periodicities [9]. On the other hand, frequency domain methods, such as Discrete Fourier Transform (DFT), presuppose that the series is time-invariant [1], therefore lose the local temporal context. Even brief transient anomalies can influence all frequency components (see Fig. 1(a)(b)), thereby undermining the accuracy of frequency analysis. These limitations restrict the effectiveness of using either time domain or frequency domain methods solely for time series forecasting.

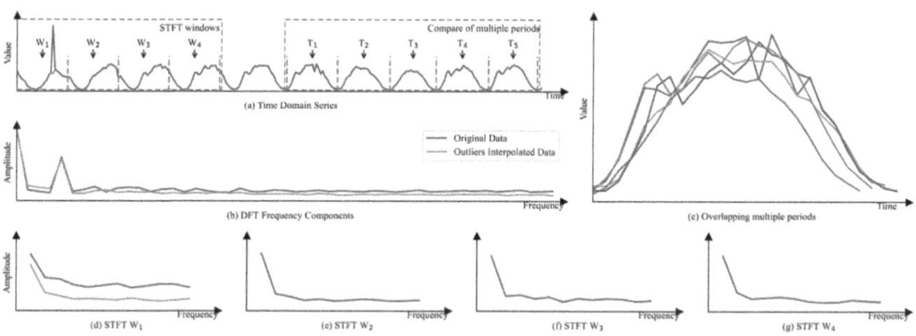

Fig. 1. Example of real-world time-varying series in the Traffic dataset.

Consequently, recent research attempts to combine time and frequency modeling. However, direct integration of time domain and frequency domain features faces challenges, as the time domain uses real-valued representations, while the frequency domain employs complex-valued representations. This hinders their effective integration. Existing methods typically model the time domain and frequency domain separately, use different networks to extract features from both domains, and then design a network structure to integrate them. Although carefully designed models that integrate time domain and frequency domain features can achieve good performance, we question whether such approaches are sufficiently effective and efficient. Meanwhile, some recent lightweight models show that simpler architectures can outperform complex ones. Particularly, DLinear [13] demonstrated that a single linear layer can beat Transformers in long-horizon forecasting, highlighting the importance of inductive bias and data representation over model complexity.

In this paper, we propose rethinking the integration of time-domain and frequency-domain from a Time-Frequency Analysis perspective to enhance the performance of time series forecasting. Time-Frequency Analysis is capable of simultaneously modeling temporal and spectral features, enabling a more comprehensive analysis of the series dynamics, while keeping the model structure simple. Our approach is built on the Short-Time Fourier Transform (STFT), which provides a time-frequency representation (TFR) of the time series in the form of a spectrogram (signal energy distributed over time and frequency). By learning directly from this 2D time-frequency representation, our models inherently integrate both domains: they can learn how frequency components evolves over time.

A key challenge in time-frequency domain modeling is that frequency bands across different windows can exhibit vastly different magnitudes, as illustrated in Fig. 1. This issue is compounded by the non-stationary nature of real-world time series, where distribution shifts between input and output series often occur (e.g., abrupt magnitude changes in specific frequency bands). Consequently, the frequency domain model may tend to focus more on the typically larger magnitude low-frequency components [12], which may limit effective learning on frequency components. To address this, we propose a novel Spectrogram Band-wise Normalization (SBN). SBN normalizes each frequency band of the spectrogram across time windows by dividing it by its mean magnitude, and later reverses this normalization on the model's output. Intuitively, this reduces the discrepancies in scale among different frequency components and between input and output distributions, thereby mitigating the effects of non-stationarity (distribution shift). By ensuring the model operates on a spectrally normalized input, we make the forecasting task more about predicting changes and patterns rather than dealing with raw scale differences.

We integrate SBN into a Time-Frequency Enhanced (TFE) Block, which is the core of our proposed framework. In addition to SBN, the TFE block includes an adaptive complex-valued filter that learns to emphasize or attenuate specific time-frequency components of the spectrogram. We also incorporate an optional Gaussian smoothing filter along the frequency axis to suppress high-frequency components, which are often attributed to anomalies and tend to be less predictive. Using the TFE block, we build two extremely simple models: SpecLinear, a single-layer linear model operating on the complex spectrogram, and SpecMLP, a three-layer feed-forward network (MLP).

In summary, our contributions are:

- We propose a novel unified time-frequency analysis framework for time series forecasting. By leveraging STFT to obtain a time-frequency representation (spectrogram) of the time series, our approach simultaneously captures temporal dynamics and spectral characteristics in a single model.
- We design a Time-Frequency Enhanced (TFE) Block that introduces a specialized Spectrogram Band-wise Normalization (SBN) module to alleviate spectral magnitude distribution shift, as well as an adaptive spectrogram filter and optional Gaussian smoothing to intelligently modulate frequency components. We also include ablation studies and analysis to validate the contribution of each component.

- We present two lightweight models, SpecLinear and SpecMLP, built using the TFE block. Extensive experiments on real-world datasets demonstrate show that our approach matches or outperforms state-of-the-art (SOTA) models while keeping a simple model architecture. This highlights that appropriate representations (time-frequency) and normalizations can unlock strong performance without a carefully designed architectures.

2 Related Work

Considerable progress has been achieved in the time-domain methods, including models like Multi-Layer Perceptrons (MLP) [2] and Transformer-based models [10,14,16]. For instance, DLinear [13] employs seasonal-trend decomposition, and applies linear autoregression to the seasonal and trend components separately. The iTransformer [4] leverages attention mechanisms and feed-forward networks in inverted dimensions. In the frequency domain, FreTS [12] transform time series data into frequency domain spectra using the Discrete Fourier Transform (DFT) and utilize an MLP specifically designed for complex numbers to project these components onto the output. Similarly, FITS [11] implements interpolation techniques within the complex-valued frequency domain. However, the distribution of frequency components in the series are highly unbalanced, as illustrated in Fig. 1. Consequently, the frequency domain model may tend to focus more on the typically larger magnitude low-frequency components, as shown in the results of the FreTS [12], which may limit effective learning on frequency components.

Some efforts have been made to integrate time-domain and frequency-domain modeling. For instance, FEDformer [17] first decomposes the input sequence using seasonal-trend decomposition, applying autoregression to the trend component and converting the seasonal component to a frequency-domain representation to replace the time-domain attention block. TimesNet [9] selects k key frequency components based on the magnitude of frequency-domain components, folds the sequence into a 2D space by periods, and applies convolution algorithms. Both approaches extract time-domain and frequency-domain features through complex model designs.

3 Methodology

3.1 Problem Definition and Time-Frequency Representation

Time series forecasting refers to the task of predicting future values of variates based on historical observations. Consider a time series $X = \{X_1, X_2, \ldots, X_L\} \in \mathbb{R}^{L \times C}$ with L historical timesteps (a.k.a. look-back window) and C variates. The goal is to learn a model f_θ to generate predictions $\hat{Y} = \{X_{L+1}, X_{L+2}, \ldots, X_{L+H}\} \in \mathbb{R}^{H \times C}$ for a future timesteps (a.k.a. forecast horizon) of length H.

Rather than learning f_θ directly in the time domain, we propose to transform the input into a time-frequency domain using the Short-Time Fourier Transform (STFT). The STFT of a time series window produces a complex-valued spectrogram that captures how the signal's frequency content evolves over time. Specifically, for each variable c, we compute:

$$S_c(f, w) = \sum_{n=0}^{N-1} x[n + wR]\omega[n]e^{-j\frac{2\pi fn}{N}} \tag{1}$$

where $\omega(n)$ is a window function of length N. The result $S_c(f, w)$ is the complex spectrogram, with $f \in [0, F-1]$ indexing frequency bins and $w \in [0, W-1]$ indexing the time (window) position. We stack the spectrograms of all variables to obtain $S(f, w) \in \mathbb{C}^{C\times F\times W}$ as the time-frequency representation of the input series. In practice, N and R are hyperparameters, a common setting is $R = \frac{N}{4}$.

One key advantage of using the STFT is that it is an invertible transform. We can apply the inverse STFT (ISTFT) on a spectrogram to reconstruct the time-domain series. All operations on the spectrogram can be made differentiable, enabling end-to-end training of our models with standard practice.

3.2 Time-Frequency Enhanced Block

Spectrogram Band-Wise Normalization. Real-world time series often exhibit varying power across different frequency bands and over different time periods, which can lead to distribution shift between input series and output series. Directly training a model on the raw spectrogram could force it to learn these scale differences, which can hurt generalization.

To alleviate this, we introduce a Spectrogram Band-wise Normalization, SBN, applied to each frequency in each spectrogram instance independently. Formally, let $S_c(f, w)$ be the complex spectrogram for variable c. Define $A_c(f, w) = |S_c(f, w)|$ as the magnitude, and $\Phi_c(f, w) = \angle S_c(f, w)$ as the phase. Our normalization focuses on the magnitudes, as phase carries temporal alignment information that we do not want to distort by normalization. We compute the mean of the magnitude for each frequency band f across all time windows of the spectrogram $S_c(f, w)$:

$$\mu_{(c,f)} = \frac{1}{W} \sum_{w=1}^{W-1} A_c(f, w) \tag{2}$$

Then we normalize the magnitude:

$$\tilde{S}_c(f, w) = \frac{S_c(f, w)}{\mu_{(c,f)} + \epsilon} \tag{3}$$

where ϵ is a small constant to avoid division by zero. Because each frequency band is divided by its own mean $\mu_{(c,f)}$ (a real number), its magnitudes are effectively scaled to lie around 1, which promotes numerical stability during training while leaving the phase $\Phi_c(f, w)$ unchanged.

Adaptive Spectrogram Filtering. Besides normalization, we introduce an adaptive spectrogram filter. For each frequency bin f and window w, we learn a complex-valued weight (which can be represented by two real parameters for real and imaginary parts, or amplitude and phase). Let $H(f, w)$ be the adaptive spectrogram filter. The filtered spectrogram $\hat{S}(f, w)$ is obtained as:

$$\hat{S}(f, w) = \tilde{S}(f, w) \odot H(f, w) \tag{4}$$

Here, $\tilde{S}(f, w)$ is the normalized spectrogram from SBN, $\odot$ denotes element-wise multiplication. This operation amplifies or attenuates each frequency component and can also adjust phase via the complex multiplication. If $|H(f, w)| > 1$, the filter amplifies frequency f at window w; if $|H(f, w)| < 1$, it attenuates it. The angle of $H(f, w)$ will shift the phase of that frequency component. Because $H(f, w)$ is learned, the model can discover which frequencies are important for forecasting future values and which are less predictive.

Gaussian Filtering (Optional). As previous work FEDformer [17] and FITS [11] have shown that low-pass filters can cause performance drops, we employ a Gaussian filter in our models. Unlike low-pass filters, Gaussian filters preserve the essential features of the series, maintaining the integrity of important frequency components while eliminating unwanted noise. Formally, let $\hat{S}(f, w)$ be the filtered spectrogram from adaptive filters, and $g(0, \sigma)$ be the Gaussian filter, the filtered spectrogram $S'(f, w)$ is obtained by the following equation:

$$S'(f, w) = \hat{S}(f, w) \odot g(0, \sigma) \tag{5}$$

3.3 Model Architectures: SpecLinear and SpecMLP

Using the TFE block described above, we instantiate two lightweight, channel-independent model SpecLinear and SpecMLP, the overall architecture is shown in Fig. 2.

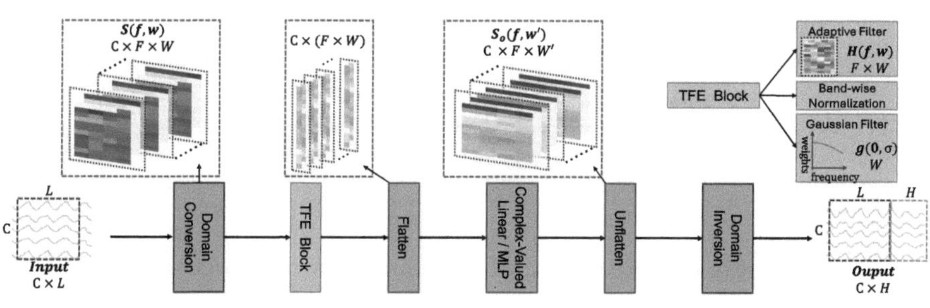

Fig. 2. The overall architecture of time-frequency enhanced model SpecLinear and SpecMLP.

SpecLinear. This model is a single-layer linear forecaster in the time-frequency domain. Unlike FITS, which implements interpolation in the frequency domain, SpecLinear generates predictions of arbitrary lengths by forecasting future STFT windows in the time-frequency domain. This means that the number of frequency components of each window in the output spectrogram $S_o(f, w')$ match those in the input spectrogram $S(f, w)$. In other words, by forecasting the evolution of each frequency component (spectrogram) and outputting a sequence of continuous STFT windows (spectrogram), SpecLinear is able to predict time-varying series.

Let $S(f, w) \in \mathbb{C}^{C \times F \times W}$ be the input spectrogram, the prediction of SpecLinear $\hat{Y}$ is obtained by:

$$\hat{Y} = \mathrm{ISTFT}(\mathrm{ComplexLinear}(\mathrm{TFEBlock}(S(f, w)))) \tag{6}$$

SpecMLP. SpecMLP share the same architecture with SpecLinear but adds nonlinear modeling. As illustrated in Fig. 3, after applying TFE Block to the input spectrogram, SpecMLP processes each spectrogram through a multi-layer perceptron (three-layer MLP with ComplexReLU and ComplexDropout). SpecMLP can be considered a simplified version of mainstream model architectures. While conventional models focus on the carefully designed network, SpecMLP simplifies this approach by employing just a single linear layer. This design choice not only aligns with the fundamental principles of mainstream architectures but also emphasizes minimalism and efficiency, thereby enabling a more straightforward comparison with the performance of mainstream models.

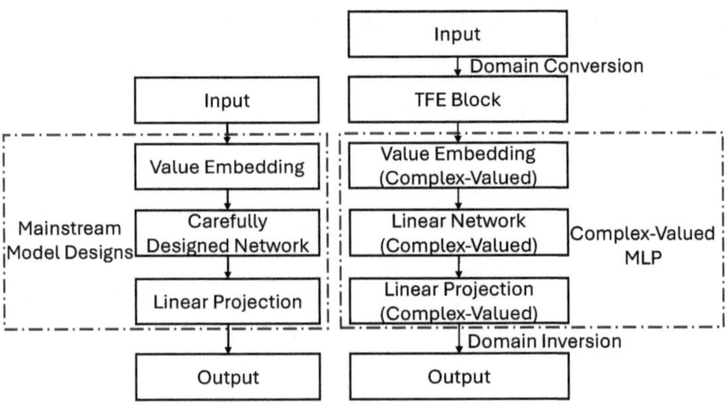

Fig. 3. Comparison between SpecMLP and mainstream model architectures.

4 Experiments

4.1 Dataset and Implementation Details

We conduct extensive experiments on nine real-world datasets—ETT (h1, h2, m1, m2) [16], Solar, Electricity, Weather [10], Exchange [3] and Traffic [7]—and on two synthetic univariate datasets we constructed: Fre-vary, which concatenates semi-sine waves with fixed amplitude but varying periods, and Amp-vary, which concatenates semi-sine waves with fixed period but varying amplitudes.

All the experiments are implemented with PyTorch, and conducted on a single NVIDIA RTX 3090 24GB GPU. We take the Adam as optimizer and MSE as the loss function. For all experiments, we follow the setting of previous works [9,12], fix the look-back window $L = 96$, report the MSE and MAE for multivariate tasks, and the RMSE and MAE for univariate tasks. All experiments are conducted at least 5 times with different random seeds across 3 different learning rates.

4.2 Compared Methods

For univariate forecasting, we compare SpecLinear with FITS [11]. FITS includes a low-pass filter (LPF) to remove high-frequency components, but experiments have shown that this reduces performance. Therefore we implement a version of FITS without frequency truncation. For multivariate forecasting, we include iTransformer [4], TimesNet [9], PatchTST [6], DLinear [13], FEDformer [17].

4.3 Main Results

In Table 1, FITS represent frequency-domain approaches, whereas our implementation of PatchLinear employs a folding strategy analogous to STFT's sliding windows and uses a single linear layer for output projection. PatchLinear can thus be seen as a time-domain ablation of SpecLinear.

Although all these models are one layer linear network, SpecLinear consistently achieves superior results in most cases. In particular, it delivers substantial gains on datasets such as Traffic, Solar, and Weather, which feature complex temporal patterns and unbalanced frequency components—thereby validating the effectiveness of Spectrogram Band-wise Normalization. Furthermore, SpecLinear demonstrates pronounced advantages on our two synthetic datasets, indicating that the TFE Block effectively captures variations in both cycle length and amplitude.

Multivariate Time Series Forecasting. As shown in Table 2, despite its architectural simplicity, SpecMLP wins or ties for the best MSE/MAE on almost every dataset and horizon. It leads unambiguously on ETTm1/ETTm2 (all four horizons), and maintains first place on Exchange and Weather across the board. On ETTh1/ETTh2 it either ranks first or a close second, and on Solar it claims three of four top spots while narrowly trailing on the fourth. Electricity is

Table 1. "Full results of the univariate forecasting task. We fix the look-back window $L = 96$. The best results are in **bold** and the second best are underlined.

Models		SpecLinear		PatchLinear		FITS		DLinear	
Metrics		RMSE	MAE	RMSE	MAE	RMSE	MAE	RMSE	MAE
ETTm1	96	0.0292	0.1268	0.0292	**0.1263**	0.0292	0.1273	**0.0289**	0.1263
	192	0.0435	0.1582	0.0435	**0.1577**	0.0437	0.1586	**0.0432**	0.1578
	336	0.0566	0.1831	0.0565	**0.1826**	0.0569	0.1836	**0.0563**	0.1826
	720	0.0791	0.2164	0.0793	0.2161	0.0795	0.2170	**0.0790**	**0.2161**
ETTm2	96	**0.0650**	**0.1824**	0.0691	0.1894	0.0694	0.1883	0.0693	0.1888
	192	**0.0968**	**0.2293**	0.1001	0.2344	0.1004	0.2340	0.0999	0.2336
	336	**0.1256**	**0.2678**	0.1280	0.2713	0.1296	0.2736	0.1276	0.2707
	720	**0.1768**	**0.3249**	0.1782	0.3272	0.1788	0.3277	0.1774	0.3258
ETTh1	96	**0.0545**	**0.1764**	0.0561	0.1790	0.0563	0.1794	0.0555	0.1777
	192	**0.0737**	**0.2069**	0.0753	0.2096	0.0757	0.2101	0.0749	0.2088
	336	**0.0885**	**0.2329**	0.0903	0.2359	0.0912	0.2373	0.0901	0.2355
	720	**0.0936**	**0.2409**	0.0964	0.2447	0.0957	0.2441	0.0964	0.2444
ETTh2	96	**0.1252**	**0.2680**	0.1271	0.2712	0.1266	0.2706	0.1257	0.2685
	192	**0.1764**	**0.3250**	0.1779	0.3268	0.1776	0.3267	0.1788	0.3252
	336	**0.2205**	**0.3727**	0.2221	0.3744	0.2216	0.3740	0.2219	0.3735
	720	0.2451	0.3976	0.2527	0.4049	**0.2426**	**0.3963**	0.2551	0.4062
Electricity	96	**0.3341**	**0.4147**	0.3842	0.4441	0.4107	0.4539	0.3912	0.4472
	192	**0.3343**	**0.4089**	0.3658	0.4294	0.3814	0.4335	0.3709	0.4312
	336	**0.3774**	**0.4373**	0.4015	0.4540	0.4180	0.4567	0.4063	0.4548
	720	**0.4287**	**0.4819**	0.4543	0.4962	0.4715	0.5024	0.4594	0.4977
Exchange	96	**0.0904**	0.2274	0.0936	0.2278	0.0943	0.2274	0.0905	**0.2215**
	192	**0.1945**	0.3354	0.2026	0.3394	0.2198	0.3511	0.2005	**0.3340**
	336	**0.3817**	**0.4765**	0.4099	0.4806	0.4735	0.5125	0.4186	0.4809
	720	**0.9929**	**0.7525**	1.0848	0.7974	1.1059	0.8070	1.1113	0.8022
Traffic	96	**0.2027**	**0.2778**	0.2366	0.3114	0.2522	0.3118	0.2405	0.3145
	192	**0.1769**	**0.2484**	0.2054	0.2765	0.2097	0.2697	0.2074	0.2786
	336	**0.1696**	**0.2452**	0.1963	0.2706	0.2026	0.2644	0.1972	0.2730
	720	**0.1891**	**0.2657**	0.2172	0.2910	0.2240	0.2866	0.2186	0.2938
Weather	96	**0.0011**	**0.0238**	0.0013	0.0259	0.0013	0.0264	0.0013	0.0253
	192	**0.0014**	**0.0267**	0.0015	0.0285	0.0016	0.0295	0.0015	0.0286
	336	**0.0015**	**0.0283**	0.0016	0.0301	0.0017	0.0307	0.0017	0.0306
	720	**0.0021**	**0.0345**	0.0021	0.0348	0.0022	0.0354	0.0022	0.0351
Solar	96	**0.2299**	**0.2981**	0.3294	0.3323	0.3725	0.3854	0.3028	0.3266
	192	0.2486	**0.3076**	0.3738	0.3485	0.3747	0.3673	**0.2419**	0.3460
	336	**0.2674**	**0.3211**	0.4025	0.3610	0.4144	0.3764	0.3835	0.3617
	720	**0.2691**	**0.3187**	0.3996	0.3548	0.4143	0.3622	0.3871	0.3555
Fre-Vary	96	**0.2400**	**0.3806**	0.4800	0.5320	0.5088	0.5522	0.4800	0.5316
	192	**0.1914**	**0.3391**	0.3476	0.4398	0.3646	0.4508	0.3475	0.4397
	336	**0.1883**	**0.3357**	0.3275	0.4242	0.3397	0.4323	0.3275	0.4242
	720	**0.1936**	**0.3409**	0.3487	0.4391	0.3622	0.4481	0.3486	0.4390
Amp-Vary	96	**0.1147**	**0.2670**	0.1926	0.3264	0.1982	0.3328	0.1924	0.3262
	192	**0.0997**	**0.2492**	0.1500	0.2873	0.1533	0.2909	0.1499	0.2871
	336	**0.0956**	**0.2440**	0.1425	0.2797	0.1449	0.2823	0.1424	0.2796
	720	**0.0968**	**0.2456**	0.1502	0.2865	0.1532	0.2894	0.1500	0.2863

Table 2. Full results of the multivariate forecasting task. We fix the length of the look-back window $L = 96$. The rest of the numbers are taken from the results from iTransformer. The best results are in **bold** and the second best are underlined.

Models		SpecMLP (Ours)		iTransformer (ICLR2024)		TimesNet (ICLR2023)		PatchTST (ICLR2023)		DLinear (AAAI2023)		FEDformer (ICML2022)	
Metrics		MSE	MAE	MSE	MAE	MSE	MAE	MSE	MAE	MSE	MAE	MSE	MAE
ETTm1	96	**0.311**	**0.341**	0.334	0.368	0.338	0.375	0.329	0.367	0.345	0.372	0.346	0.388
	192	**0.362**	**0.368**	0.377	0.391	0.374	0.387	0.367	0.385	0.380	0.389	0.426	0.441
	336	**0.391**	**0.390**	0.426	0.420	0.410	0.411	0.399	0.410	0.413	0.413	0.445	0.459
	720	**0.450**	**0.427**	0.491	0.459	0.478	0.450	0.454	0.439	0.474	0.453	0.543	0.490
ETTm2	96	**0.170**	**0.250**	0.180	0.264	0.187	0.267	0.175	0.259	0.193	0.292	0.203	0.287
	192	**0.235**	**0.293**	0.250	0.309	0.249	0.309	0.241	0.302	0.284	0.362	0.269	0.328
	336	**0.295**	**0.331**	0.311	0.348	0.321	0.351	0.305	0.343	0.369	0.427	0.325	0.366
	720	**0.392**	**0.390**	0.412	0.407	0.408	0.403	0.402	0.400	0.554	0.522	0.421	0.415
ETTh1	96	**0.375**	**0.390**	0.386	0.405	0.384	0.402	0.414	0.419	0.386	0.400	0.376	0.396
	192	0.431	**0.417**	0.441	0.436	0.436	0.429	0.460	0.445	0.437	0.432	**0.420**	0.424
	336	0.476	**0.438**	0.487	0.458	0.491	0.469	0.501	0.466	0.481	0.459	**0.458**	0.466
	720	**0.478**	**0.459**	0.503	0.491	0.521	0.500	0.500	0.488	0.519	0.516	0.501	0.501
ETTh2	96	**0.283**	**0.330**	0.297	0.349	0.340	0.374	0.302	0.348	0.333	0.387	0.358	0.397
	192	**0.358**	**0.379**	0.380	0.400	0.402	0.414	0.388	0.400	0.477	0.476	0.429	0.439
	336	**0.402**	**0.413**	0.428	0.432	0.452	0.452	0.426	0.433	0.594	0.541	0.496	0.487
	720	**0.410**	**0.430**	0.427	0.445	0.462	0.468	0.431	0.446	0.831	0.657	0.463	0.474
Electricity	96	0.160	0.247	**0.148**	**0.240**	0.168	0.272	0.181	0.270	0.197	0.282	0.193	0.308
	192	0.174	0.259	**0.162**	**0.253**	0.184	0.289	0.188	0.274	0.196	0.285	0.201	0.315
	336	0.190	0.277	**0.178**	**0.269**	0.198	0.300	0.204	0.293	0.209	0.301	0.214	0.329
	720	0.229	0.311	**0.217**	**0.305**	0.220	0.320	0.246	0.324	0.245	0.333	0.246	0.355
Exchange	96	**0.082**	**0.199**	0.086	0.206	0.107	0.234	0.088	0.205	0.088	0.218	0.148	0.278
	192	**0.162**	**0.297**	0.177	0.299	0.226	0.344	0.176	0.299	0.176	0.315	0.271	0.315
	336	**0.293**	0.406	0.331	0.417	0.367	0.448	0.301	**0.397**	0.313	0.427	0.460	0.427
	720	0.862	0.699	0.847	**0.691**	0.964	0.746	0.901	0.714	**0.839**	0.695	1.195	0.695
Weather	96	**0.160**	**0.203**	0.174	0.214	0.172	0.220	0.177	0.218	0.196	0.255	0.178	0.223
	192	**0.201**	**0.245**	0.221	0.254	0.219	0.261	0.225	0.259	0.237	0.296	0.226	0.262
	336	**0.253**	**0.288**	0.278	0.296	0.280	0.306	0.278	0.297	0.283	0.335	0.289	0.309
	720	**0.325**	**0.340**	0.358	0.347	0.365	0.359	0.354	0.348	0.345	0.381	0.362	0.353
Solar	96	**0.199**	**0.223**	0.203	0.237	0.250	0.292	0.234	0.286	0.290	0.378	0.242	0.342
	192	0.234	**0.245**	**0.233**	0.261	0.296	0.318	0.267	0.310	0.320	0.398	0.285	0.380
	336	0.258	**0.259**	**0.248**	0.273	0.319	0.330	0.290	0.315	0.353	0.415	0.282	0.376
	720	0.262	**0.262**	**0.249**	0.275	0.338	0.337	0.289	0.317	0.356	0.413	0.357	0.427

the only case where iTransformer is marginally better at short horizons, yet SpecMLP stays within 3–5% error. These results confirm that our lightweight time-frequency framework with TFE Block yield robust, state-of-the-art multivariate forecasts with minimal model complexity.

4.4 Ablation Study

Table 3 compares the SpecMLP TFE Block against two ablated variants across four forecast horizons on ETTm1, Electricity and Exchange datasets. Omitting the Gaussian Filter ("w/o Gaussian Filter") yields a small but consistent performance loss, demonstrating that frequency-axis smoothing modestly suppresses high-frequency noise. Removing the Adaptive Spectrogram Filter ("w/o Adaptive Filter") results in a moderate performance drop, indicating that the learnable Adaptive Spectrogram Filter effectively captures specific filtering patterns to enhance time-frequency representation quality. Removing Spectrogram Band-wise Normalization ("w/o SBN") incurs a much larger degradation, particularly at longer horizons (for example, Exchange-720 MSE jumps from 0.862 to 1.114, a 29% rise), confirming that SBN is critical for stabilizing per-band statistics and maintaining accuracy over extended forecasts.

4.5 Spectrogram Band-Wise Normalization Visualization

As shown in Fig. 4, the magnitudes of frequency components exhibit substantial scale differences across bands; after Spectrogram Band-wise Normalization, these magnitudes are brought onto a more unified scale, making them considerably more consistent. This normalization enables the model to focus on learning the true variations of frequency components rather than being distracted by scale differences caused by distribution shifts.

4.6 Hyperparameter Sensitivity

We conduct hyperparameter sensitivity experiments with SpecLinear on multiple datasets. As shown in Fig. 5, varying the STFT window size across a practical

Table 3. Ablation study of SpecMLP TFE Block components on multivariate time series forecasting.

Dataset		ETTm1				Electricity				Exchange			
Prediction Length		96	192	336	720	96	192	336	720	96	192	336	720
SpecMLP	MSE	**0.311**	**0.362**	**0.391**	**0.450**	**0.160**	**0.174**	**0.190**	**0.229**	**0.082**	**0.162**	**0.293**	**0.862**
	MAE	**0.341**	**0.368**	**0.390**	**0.427**	**0.247**	**0.259**	**0.277**	0.311	**0.199**	**0.297**	**0.406**	**0.699**
w/o Gaussian Filter	MSE	0.311	0.362	0.392	0.453	0.162	0.176	0.191	0.231	0.083	0.177	0.341	0.865
	MAE	0.342	0.370	0.391	0.430	0.250	0.261	0.277	**0.310**	0.201	0.298	0.421	0.700
w/o Adaptive Filter	MSE	0.311	0.365	0.393	0.454	0.163	0.179	0.197	0.237	0.087	0.164	0.319	0.915
	MAE	0.342	0.370	0.391	0.428	0.249	0.261	0.278	0.311	0.204	0.300	0.426	0.721
w/o SBN	MSE	0.318	0.368	0.401	0.482	0.173	0.180	0.203	0.237	0.088	0.179	0.346	1.114
	MAE	0.347	0.375	0.400	0.451	0.255	0.263	0.283	0.311	0.207	0.299	0.424	0.769

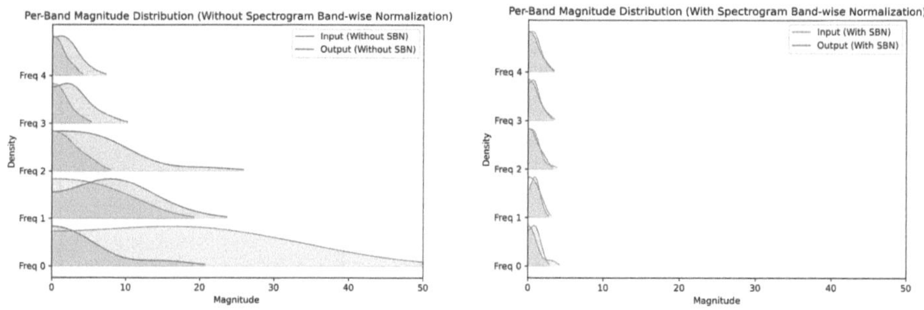

Fig. 4. Per-Band Magnitude Distribution Comparison (With vs. Without SBN).

range causes almost no change in MSE, suggesting that this hyperparameter does not require meticulous selection. For learning-rate sensitivity, SpecLinear shows similarly negligible variation in MSE across different learning rates, confirming that our time-frequency enhanced framework is robust and does not rely on precise tuning.

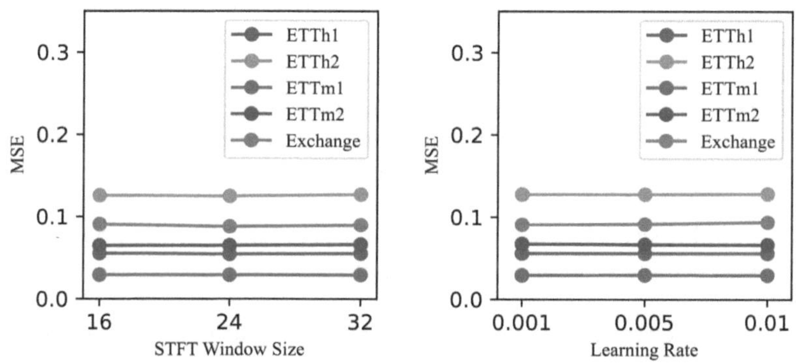

Fig. 5. Hyperparameter sensitivity of SpecLinear on univariate forecasting. The prediction length $H = 96$.

4.7 Efficiency Analysis

Figure 6 report per-batch training time and GPU memory usage. SpecMLP's efficiency is close the Linear model in both plots, showing that the added TFE Block incurs only a marginal overhead—training latency per batch increases, and peak memory consumption remains within a few percent of Linear models. Altogether, these efficiency measurements confirm that SpecMLP delivers the forecasting accuracy of more complex models while preserving the high throughput and low memory footprint characteristic of linear predictors.

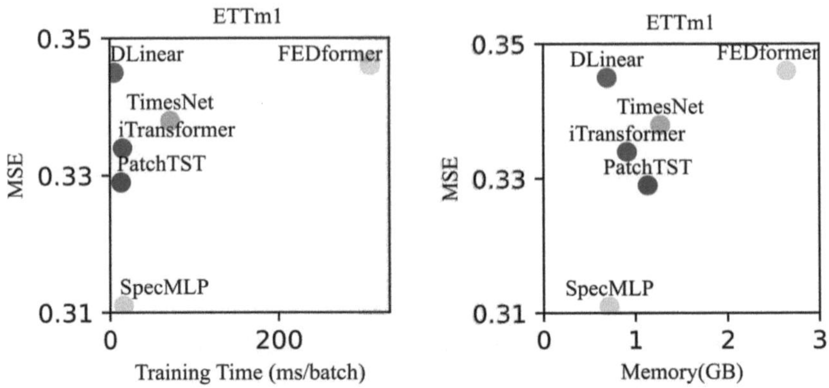

Fig. 6. Model efficiency comparison of different methods on ETTm1 with prediction length $H = 96$.

5 Conclusion

We present a novel time-frequency approach to time series forecasting that leverages the time-frequency analysis (based on STFT) to provide a unified representation, time-frequency representation of time series. Central to our approach is the Time-Frequency Enhanced (TFE) Block, which combines Spectrogram Band-wise Normalization (SBN), adaptive spectrogram filtering, and optional Gaussian smoothing to address unbalanced distribution and noise in time-frequency components. Integrating this block into two lightweight architectures—SpecLinear and SpecMLP—we demonstrated that even extremely simple models can capture complex temporal and spectral dynamics.

Extensive experiments on eleven datasets show that our TFE models consistently match or surpass state-of-the-art methods, often with fewer parameters and faster training times. In particular, other models achieves up to 15% MSE reduction on datasets with highly non-stationary frequency content, validating the effectiveness of SBN for stable and generalizable learning.

By unifying time-domain and frequency-domain modeling within a streamlined architecture, our approach challenges the notion that improved forecasting accuracy requires ever-deeper or more complex networks. We believe this work opens new avenues for leveraging classic signal-processing tools in modern deep learning pipelines.

References

1. Box, G.E.P., Jenkins, G.M., Reinsel, G.C., Ljung, G.M.: Time Series Analysis: Forecasting and Control. Wiley Series in Probability and Statistics, John Wiley & Sons, Inc, Hoboken, New Jersey, fifth edition edn. (2016)
2. Chen, S.A., Li, C.L., Yoder, N., Arik, S.O., Pfister, T.: TSMixer: An All-MLP Architecture for Time Series Forecasting, September 2023. https://doi.org/10.48550/arXiv.2303.06053
3. Lai, G., Chang, W.C., Yang, Y., Liu, H.: Modeling Long- and Short-Term Temporal Patterns with Deep Neural Networks. In: The 41st International ACM SIGIR Conference on Research & Development in Information Retrieval. pp. 95–104. SIGIR '18, Association for Computing Machinery, New York, NY, USA, June 2018. https://doi.org/10.1145/3209978.3210006
4. Liu, Y., et al.: iTransformer: inverted transformers are effective for time series forecasting, October 2023. https://doi.org/10.48550/arXiv.2310.06625
5. Matsubara, Y., Sakurai, Y., van Panhuis, W.G., Faloutsos, C.: FUNNEL: Automatic mining of spatially coevolving epidemics. In: Proceedings of the 20th ACM SIGKDD International Conference on Knowledge Discovery and Data Mining, pp. 105–114. KDD '14, Association for Computing Machinery, New York, August 2014. https://doi.org/10.1145/2623330.2623624
6. Nie, Y., Nguyen, N.H., Sinthong, P., Kalagnanam, J.: A Time Series is Worth 64 Words: Long-term Forecasting with Transformers, March 2023. https://doi.org/10.48550/arXiv.2211.14730
7. Sen, R., Yu, H.F., Dhillon, I.S.: Think globally, act locally: a deep neural network approach to high-dimensional time series forecasting. In: Advances in Neural Information Processing Systems, vol. 32. Curran Associates, Inc. (2019)
8. Seyedan, M., Mafakheri, F.: Predictive big data analytics for supply chain demand forecasting: methods, applications, and research opportunities. J. Big Data **7**(1), 1–22 (2020). https://doi.org/10.1186/s40537-020-00329-2
9. Wu, H., Hu, T., Liu, Y., Zhou, H., Wang, J., Long, M.: TimesNet: Temporal 2D-Variation Modeling for General Time Series Analysis, April 2023. https://doi.org/10.48550/arXiv.2210.02186
10. Wu, H., Xu, J., Wang, J., Long, M.: Autoformer: decomposition transformers with auto-correlation for long-term series forecasting. In: Advances in Neural Information Processing Systems, vol. 34, pp. 22419–22430. Curran Associates, Inc. (2021)
11. Xu, Z., Zeng, A., Xu, Q.: FITS: Modeling Time Series with $10k$ Parameters. In: The Twelfth International Conference on Learning Representations, October 2023
12. Yi, K., et al.: Frequency-domain MLPs are More Effective Learners in Time Series Forecasting, November 2023. https://doi.org/10.48550/arXiv.2311.06184
13. Zeng, A., Chen, M., Zhang, L., Xu, Q.: Are Transformers Effective for Time Series Forecasting? Proceedings of the AAAI Conference on Artificial Intelligence 37(9), pp. 11121–11128 (2023). https://doi.org/10.1609/aaai.v37i9.26317
14. Zhang, Y., Yan, J.: Crossformer: transformer utilizing cross-dimension dependency for multivariate time series forecasting. In: The Eleventh International Conference on Learning Representations, September 2022
15. Zheng, Y., Yi, X., Li, M., Li, R., Shan, Z., Chang, E., Li, T.: Forecasting fine-grained air quality based on big data. In: Proceedings of the 21th ACM SIGKDD International Conference on Knowledge Discovery and Data Mining, pp. 2267–2276. KDD '15, Association for Computing Machinery, New York, August 2015. https://doi.org/10.1145/2783258.2788573

16. Zhou, H., Zhang, S., Peng, J., Zhang, S., Li, J., Xiong, H., Zhang, W.: Informer: Beyond Efficient Transformer for Long Sequence Time-Series Forecasting. Proceedings of the AAAI Conference on Artificial Intelligence 35(12), pp. 11106–11115 (2021). https://doi.org/10.1609/aaai.v35i12.17325
17. Zhou, T., Ma, Z., Wen, Q., Wang, X., Sun, L., Jin, R.: FEDformer: frequency enhanced decomposed transformer for long-term series forecasting. In: Proceedings of the 39th International Conference on Machine Learning, pp. 27268–27286. PMLR, June 2022

RPGCN: Relational Probabilistic Graphs for EEG-Based Emotion Mining

Xinliang Zhou[1], Jianheng Zhou[1], Jiaping Xiao[1], Yingwei Zhang[2], Xiaoshuai Hao[3], Jing Wang[4], Badong Chen[5(✉)], and Qingsong Wen[6(✉)]

[1] Nanyang Technological University, Singapore, Singapore
[2] Institute of Computing Technology, Chinese Academy of Sciences, Beijing, China
[3] Xiaomi EV, Beijing, China
[4] Beijing Jiaotong University, Beijing, China
[5] Xi'an Jiaotong University, Xi'an, China
[6] Squirrel Ai Learning, Bellevue, WA, USA
qingsongedu@gmail.com

Abstract. Effectively mining emotional states from multi-channel electroencephalography (EEG) signals presents significant data mining challenges. Prevailing methods often limit recognition accuracy by treating signals as independent time-series. This approach neglects crucial inter-channel correlations and results in non-interpretable 'black-box' models. To overcome these deficiencies, we propose the Relational Probabilistic Graph Convolutional Network (RPGCN), a novel two-stage framework. First, RPGCN transforms raw EEG signals into probabilistic graphs that explicitly model dynamic, inter-channel dependencies to capture latent emotional variations. Subsequently, a Graph Convolutional Network (GCN) learns discriminative patterns directly from these structured representations for classification. Extensive experiments demonstrate that RPGCN significantly outperforms state-of-the-art methods. Crucially, our framework provides inherent model interpretability, with learned relational patterns aligning with established cognitive neuroscience findings. Our work establishes a new, interpretable graph-based paradigm for EEG analysis, paving the way for more robust and intelligent human-computer interaction.

Keywords: Emotion Recognition · EEG · Probabilistic Graphs · Relational Thinking · GCN

1 Introduction

Traditional emotion recognition, relying on behavioral cues like facial expressions, is often unreliable. These signals can be concealed, misleading, and context-dependent, limiting the robustness of models for human-computer interaction (HCI) [18]. This unreliability poses challenges for increasingly vital applications, such as personalized healthcare for mental well-being and adaptive cognitive tools [27].

M. Yoshikawa et al. (Eds.): ADMA 2025, LNAI 16197, pp. 113–128, 2026.
https://doi.org/10.1007/978-981-95-3453-1_8

Electroencephalography (EEG), which records the brain's electrical activity, provides a more direct and objective physiological data source for emotion recognition. However, many existing data mining methods fail to fully leverage the potential of EEG. They either reduce complex, multi-channel signals to independent time-series for generic CNN or RNN architectures, or, when using Graph Neural Networks (GNNs) [21], they fail to capture the true relational structure of neural activity through effective graph construction. A critical drawback of these prevailing methods is their tendency to produce 'black-box' models, whose decision-making processes lack transparency. In domains like healthcare, where insight into the relationship between neural patterns and emotional states is required, this lack of interpretability constitutes a significant impediment.

Cognitive neuroscience indicates that emotions arise from dynamic interactions between distributed brain regions, not from isolated neural activity [5]. This highlights the need for models that explicitly capture these inter-channel relationships, a feature often neglected in current pipelines that treat channels uniformly or fail to separate signal from noise.

To address these limitations, we propose the Relational Probabilistic Graph Convolutional Network (RPGCN), a novel framework designed to model these crucial inter-channel EEG relationships and enhance interpretability. Inspired by relational thinking theory [11,35], RPGCN employs a two-stage strategy. First, it transforms raw multi-channel EEG signals into a structured, probabilistic "emotion relation graph," where nodes are channels and weighted edges represent their functional interactions. This stage produces a summary graph for overarching emotional patterns alongside variant graphs for individual differences. Second, a Graph Convolutional Network (GCN) mines these structured representations to learn discriminative, emotion-relevant features. Extensive experiments demonstrate that RPGCN significantly improves recognition accuracy while providing inherent model interpretability, eliminating the need for post-hoc explanation modules.

2 Related Work

2.1 Behavior-Based Emotion Recognition

Behavior-based emotion recognition relies on the visual cues (*e.g.*, facial expressions, posture and textual communication) or auditory cues (*e.g.*, vocal patterns and speech rate) of individuals. For example, Xie *et al.* [40] analyze acoustic features for emotion detection, while Pranav *et al.* [32] classify facial emotions using CNNs. Piana *et al.* [31] use 3D motion data to infer emotions, and Xu *et al.* [39] explore speech-text alignment for multi-modal emotion recognition. A common drawback of the above behavior-based methods is that they do not directly take into account the brain signals, the direct reflection of emotions, resulting in a less accurate and objective emotional assessment.

2.2 EEG-Based Emotion Mining

Compared with behavior-based methods, EEG-based emotion mining has received increasing attention. Manual feature extraction from EEG data is challenging and risks losing crucial signal details. Thus, feeding raw EEG signals into CNNs or RNNs for automatic feature learning is a better choice. However, EEG signals are highly variable across individuals. To address domain shifts between training and testing, domain discriminators are employed to learn domain-invariant features [20]. For robust EEG-based emotion recognition, variational Bayesian graph neural networks are also used to model complex dependencies and signal uncertainty [22,23].

Neuroscience studies highlight strong links between emotions and specific cortical areas [26]. Given EEG's multichannel structure, modeling inter-signal relations is vital. Using an adjacency matrix to encode electrode topology and applying dynamic GCNs helps capture these dependencies [36]. Alternatively, grouping electrodes by spatial location and fusing features enhances inter-regional correlation modeling [6]. Recent work also incorporates spatio-temporal patterns via specialized spatial and temporal modules [2,12].

Despite their effectiveness, these models function as black boxes, hindering their adoption in practical settings where interpretability is indispensable, especially in healthcare or commercial applications. Hence, developing interpretable EEG-based emotion mining models is essential.

3 Relational Probabilistic Graph Convolutional Network

3.1 Preliminary

Emotion Modeling and Brain Interactions. Emotional states are commonly described along two axes: valence (positive-negative) and arousal (high-low intensity) [9]. These dimensions enable continuous emotion modeling and are widely used in affective neuroscience. In parallel, neuroscience models view the brain as a hierarchical network of interconnected regions [30], where cognitive tasks like emotion generation arise from distributed, whole-brain patterns rather than localized activity [29].

3.2 Motivation and Problem Formulation

Relational thinking refers to the cognitive process of understanding and analyzing relationships between different elements or entities [1]. It involves identifying connections, patterns and dependencies among various pieces of information or concepts. This motivates us to use the relational thinking to reveal the hidden mechanisms underlying the emotional activities in various brain regions.

To apply relational thinking to EEG signals, we represent the EEG signals as a graph structure. The reasons are as follows. First, relational thinking focuses on relationships, specifically how entities connect. The graph structure explicitly captures these relations as connections between nodes, aligning closely with the

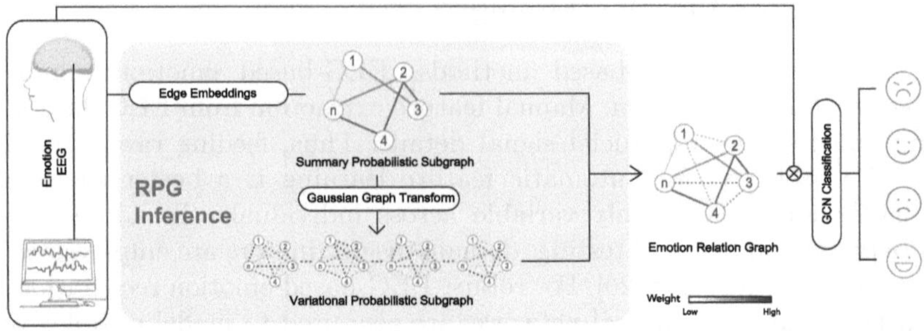

Fig. 1. Architecture of RPGCN. First, RPG inference generates a summary probabilistic subgraph by extracting edge embeddings between signal nodes. Second, this summary graph is transformed via Gaussian graph transforms into variant subgraphs representing individual emotion states. These two sets of subgraphs are then merged to create a unified 'emotion relation graph'. Finally, a Graph Convolutional Network (GCN) uses this emotion relation graph and the original EEG signals to extract features, which a classifier then uses to predict the emotion labels.

core focus of relational thinking. Second, EEG measures electrical activity from neuronal interactions in the brain. These interactions can be effectively modeled as a graph, with nodes denoting EEG channels (*i.e.*, brain regions) and edges denoting functional connections, as demonstrated in [41].

Figure 1 shows the overall architecture of the Relational Probabilistic Graph Convolutional Network (RPGCN). The first stage is the RPG inference, which obtains the emotion relation graphs that represent the relationships of EEG channels. Let $X = \{X_t \in \mathbb{R}^{C\times N}|t \in T\}$ denotes the EEG signals in T time segments and $G^{Emo} = \{G_t^{Emo}|t \in T\}$ denotes a set of emotion relation graphs obtained from X, in which C is the number of EEG channels and N is the number of sampling points within a time segment. It can be formalized as:

$$G^{Emo} = \text{RPG}(X), \tag{1}$$

where RPG denotes the RPG inference function. See Sect. 3.3 for details of the RPG inference process.

The second stage is the classification, which uses a GCN to map the input EEG signals to the corresponding emotion categories with the guidance of emotion relation graphs. It can be formalized as:

$$\hat{y} = \text{GCN}(X, G^{Emo}), \tag{2}$$

where GCN denotes a GCN model and $\hat{y}$ denotes the label of the emotion category. See the GCN details in Sect. 3.4.

Finally, the two stages are progressively learned with carefully designed loss functions, which are detailed in Sect. 3.5.

3.3 RPG Inference

As shown in Fig. 1, the RPG inference aims to model the emotion relation as the posterior graphs' edge distribution and obtain a emotion relation graph.

First, we transform EEG signals into two initial graph structures, the summary probabilistic graph and the variant probabilistic graph, based on relational thinking. The former is generated to capture general emotional relations. From this, variant probabilistic graph is created to represent possible variations of these relations. These two graph types are then merged to form an emotion relation graph that contains both overall and variant emotional activities, thereby simulating a realistic human emotional state.

Node Embedding and Edge Embedding. The input EEG segment $X_t \in \mathbb{R}^{C\times N}$ is transformed into a graph by encoding each EEG channel as a node and computing edge embeddings between node pairs. A linear layer $\boldsymbol{f}_{node}(\cdot)$ maps each channel's data to a node embedding, and a convolutional network $\boldsymbol{f}_{edge}(\cdot)$ generates edge embeddings from node pairs:

$$v_{i,t} = \boldsymbol{f}_{node}(x_{i,t}), \tag{3}$$

$$\boldsymbol{e}_{i,j,t} = \boldsymbol{f}_{edge}(v_{i,t}, v_{j,t}), \tag{4}$$

where $i, j \in [1, C]$ and $i \neq j$, $x_{i,t}$ is the EEG data of channel i at time t, $v_{i,t} \in \mathbb{R}^{1\times D_n}$ is the node embedding, and $\boldsymbol{e}_{i,j,t} \in \mathbb{R}^{1\times D_e}$ is the edge embedding. D_n and D_e are the dimensions of node and edge embeddings, respectively.

Summary Probabilistic Graph. To effectively model the global dependencies among brain regions for emotion recognition, we construct a summary probabilistic graph, G^{Sum}, to capture general emotion relations. This graph is based on multi-regional brain interactions, where a node's connectivity is influenced by both direct and numerous indirect nodes, making each edge dependent on the broader network structure.

Let $\lambda_{i,j} = \Pr(e_{i,j})$ be the probability of an edge's existence and $\alpha_{i,j}$ denote the edge following a specific distribution. For simplicity, we distinguish $\alpha_{i,j}$ from the initial time-stamped edge $e_{i,j,t}$ and omit the subscript t. Since neural interactions involve binary spike transmissions, we assume:

$$\alpha_{i,j}^{n} \sim \mathbf{Bern}(\lambda_{i,j}), \tag{5}$$

where $n \in [0, +\infty]$, and $\lambda_{i,j} \to 0$ indicates minimal neural activity [8]. The edge $\alpha_{i,j}^{Sum}$ in G^{Sum} aggregates all $\alpha_{i,j}^{n}$:

$$\alpha_{i,j}^{Sum} = \sum_{n=1}^{\infty} \alpha_{i,j}^{n}. \tag{6}$$

Direct computation of Eq. (6) is infeasible. Instead, we approximate it with a binomial distribution:

$$\alpha_{i,j}^{Sum} \sim \lim_{n\to\infty, \lambda_{i,j}\to 0} \mathbf{Bin}(n, \lambda_{i,j}), \tag{7}$$

where $n \to \infty$ reflects the influence of countless nodes.

Following VRNN [4], posterior parameters can be estimated via an RNN encoder. However, due to the infinite n and near-zero $\lambda_{i,j}$, direct inference is intractable. According to the De Moivre-Laplace theorem [34], the binomial distribution can be approximated by a Gaussian distribution:

$$\alpha_{i,j}^{Sum} \sim \mathcal{N}(\mu_{i,j}^{Sum}, \mu_{i,j}^{Sum}(1 - \mu_{i,j}^{Sum})), \tag{8}$$

where $\mu_{i,j}^{Sum}$ is:

$$\mu_{i,j}^{Snm} = \frac{(1 - 2\mu_{i,j}) + 2\sigma_{i,j}^2}{1 - 2\mu_{i,j}} - \frac{\left|1 - 2\mu_{i,j} + \sqrt{(1 - 2\mu_{i,j})^2 + 4\sigma_{i,j}^2}\right|}{1 - 2\mu_{i,j}}. \tag{9}$$

To stabilize the term $1/(1 - 2\mu_{i,j})$, we use an approximation:

$$q_{i,j} = \frac{1}{1 - 2\mu_{i,j}} \sim \text{Softplus}(\mu_{i,j}) + \epsilon_{i,j}, \tag{10}$$

then, $\mu_{i,j}^{Sum}$ simplifies to:

$$\mu_{i,j}^{Sum} \sim \frac{1 + 2q_{i,j}\sigma_{i,j}^2 - \sqrt{1 + 4q_{i,j}^2\sigma_{i,j}^4}}{2}. \tag{11}$$

Before computing $\alpha_{i,j}^{Sum}$, we obtain $\mu_{i,j}$ and $\sigma_{i,j}$ using two convolutional layers f_{mean} and f_{std}. This re-parameterization enables sampling from the posterior distribution $Q(\alpha_{i,j}^{Sum})$ using the Gaussian approximation $\mathcal{N}(\mu_{i,j}^{Sum}, \mu_{i,j}^{Sum}(1 - \mu_{i,j}^{Sum}))$ [15].

Variant Probabilistic Graph. The variant probabilistic subgraph depends on the summary probabilistic subgraph. Based on the sample $z_{i,j}^{Sum}$ drawn from $Q(\alpha_{i,j}^{Sum})$, we define a conditional Gaussian distribution with parameters generated via a linear layer, forming the variant subgraph:

$$z_{i,j}^{Sum} = \sqrt{\mu_{i,j}^{Sum}(1 - \mu_{i,j}^{Sum})}\, \varepsilon_{i,j}^{Sum} + \mu_{i,j}^{Sum}, \tag{12}$$

$$\alpha_{i,j,t}^{Var} \sim \mathcal{N}(z_{i,j}^{Sum}\, \tilde{\mu}_{i,j,t},\ z_{i,j}^{Sum}\, \tilde{\sigma}_{i,j,t}^2). \tag{13}$$

Here, $\mathcal{N}(\tilde{\mu}_{i,j,t}, \tilde{\sigma}_{i,j,t}^2)$ is produced by a linear layer following [25]. The final edge sampling is:

$$z_{i,j,t}^{Var} = \sqrt{z_{i,j}^{Sum}}\, \sigma_{i,j,t}^{Var}\, \varepsilon_{i,j,t}^{Var} + z_{i,j}^{Sum}\, \mu_{i,j,t}^{Var}. \tag{14}$$

Emotion Relation Graph. In the final step of RPG inference, to extract emotional feature representation, based on two subgraphs generated in the summary probabilistic graph and the variant probabilistic graph, we define the edge of the emotion relation graph as follows:

$$\alpha_{i,j,t}^{Emo} = z_{i,j,t}^{Var} \, z_{i,j}^{Sum}, \tag{15}$$

where $\alpha_{i,j,t}^{Emo}$ describes the edges of the final emotion relation graph generated from the summary probabilistic subgraph and variant probabilistic subgraph.

3.4 Graph Convolutional Network Emotion Classification

Given a graph $G = (V, E)$ with $V = \{v_{i,t}\}$ and $E = \{e_{i,j,t}\}$, GCNs update node features by aggregating neighbor information. The propagation rule is:

$$H^{(l+1)} = \sigma\left(\tilde{D}^{-\frac{1}{2}}\tilde{A}\tilde{D}^{-\frac{1}{2}}H^{(l)}W^{(l)}\right), \tag{16}$$

where $H^{(l)}$ is the node feature matrix at layer l, $\tilde{A}$ is the adjacency matrix with self-loops, $\tilde{D}$ is its degree matrix, $W^{(l)}$ is the trainable weight, and σ is the ReLU activation. After RPG inference, we apply GCN for emotion classification. The final representation is passed through a Sigmoid function to obtain class probabilities Z:

$$Z = \text{Sigmoid}\left(H^{(l+1)}W^{(l+1)}\right), \tag{17}$$

$$Z = \text{Sigmoid}\left(\sigma\left(\hat{D}^{-\frac{1}{2}}\hat{A}\hat{D}^{-\frac{1}{2}}H^{(l)}W^{(l)}\right)W^{(l+1)}\right). \tag{18}$$

3.5 Learning of Relational Probabilistic Graphs

We apply variant inference [10] to optimize probabilistic modeling in RPG inference. The optimization target is the evidence lower bound (ELBO):

$$ELBO = \min\left\{\text{KL}\left[Q(G^{Emo} \mid X)||P(G^{Emo} \mid X)\right] + CE(Y, Y'|G^{Emo}, X)\right\}, \tag{19}$$

$$\begin{aligned} &\text{KL}\left[Q(G^{Emo} \mid X)||P(G^{Emo} \mid X)\right] = \\ &\sum_{(i,j)\in E}\left\{KL[Q(\alpha_{i,j}^{Sum})||P(\alpha_{i,j}^{Sum})] + KL[Q(\alpha_{i,j}^{Var})||P(\alpha_{i,j}^{Var})]\right\}. \end{aligned} \tag{20}$$

Here, the ELBO consists of the KL term and cross-entropy loss. The KL divergence captures the difference between the posterior and prior over the emotion graph G^{Emo}, including both the summary ($\alpha_{i,j}^{Sum}$) and variant ($\alpha_{i,j}^{Var}$) edge variables. The cross-entropy term models the emotional inference given EEG input.

Term 1: Summary KL Term. We define the summary probabilistic subgraph as the foundation for RPG inference across time, capturing dynamic changes in the variant probabilistic subgraph. The summary subgraph is computed as:

$$E_{i,j}^{Sum} = f_{edge}\left(\sum_{t=1}^{T} v_{i,t}, \sum_{t=1}^{T} v_{j,t}\right), \quad (i,j \in [1,C],\ i \neq j), \tag{21}$$

$$\lambda_{i,j} = \text{Softplus}(f_{\lambda}(E_{i,j}^{Sum})), \tag{22}$$

$$\mu_{i,j} = f_{\mu}(E_{i,j}^{Sum}), \tag{23}$$

$$\sigma_{i,j} = \text{Softplus}(f_{\sigma}(E_{i,j}^{Sum})). \tag{24}$$

Here, f_{edge}, f_{λ}, f_{μ}, and f_{σ} are neural networks used to compute the edge embedding $E_{i,j}^{Sum}$ and the distribution parameters. $\mu_{i,j}$ and $\sigma_{i,j}$ define the Gaussian approximation in Eq. (9), while $\lambda_{i,j}$ corresponds to the prior of the binomial distribution in Eq. (7).

Since the posterior $Q(\alpha_{i,j}^{Sum}) = \mathbf{BIN}(n, \lambda_{i,j})$ involves an infinite n, the KL term is approximated in closed-form as:

$$KL[Q(\alpha_{i,j}^{Sum})||P(\alpha_{i,j}^{Sum})] \sim$$
$$\sum_{(i,j)\in E} \left\{ \mu_{i,j}^{Sum} \log \frac{\mu_{i,j}^{Sum} + \epsilon}{\lambda_{i,j} + \epsilon} + (1 - \mu_{i,j}^{Sum}) \log \frac{1 - \mu_{i,j}^{Sum} + (\mu_{i,j}^{Sum})^2/2 + \epsilon}{1 - \lambda_{i,j} + (\lambda_{i,j})^2/2 + \epsilon} \right\}. \tag{25}$$

Here, ϵ is a small constant to avoid numerical instability.

Term 2: Variant KL Term. In Step 3, we define the posterior and prior distributions of the variant probabilistic subgraph edge $\alpha_{i,j,t}^{Var}$ as:

$$Q(\alpha_{i,j,t}^{Var}) \sim \mathcal{N}(z_{i,j}^{Sum} \cdot f_{\tilde{\mu}}(E_{i,j,t}),\ z_{i,j}^{Sum} \cdot \zeta(f_{\tilde{\sigma}}(E_{i,j,t}))^2), \tag{26}$$

$$P(\alpha_{i,j,t}^{Var}) \sim \mathcal{N}(z_{i,j}^{Sum} \cdot f_{\bar{\mu}}(E_{i,j,:t}),\ z_{i,j}^{Sum} \cdot \zeta(f_{\bar{\sigma}}(E_{i,j,:t}))^2). \tag{27}$$

Here, $f_{\tilde{\mu}}$, $f_{\tilde{\sigma}}$, $f_{\bar{\mu}}$, and $f_{\bar{\sigma}}$ are networks used to generate the Gaussian parameters. $E_{i,j,t}$ and $E_{i,j,:t}$ denote the edge features at time t and before t, respectively.

We abbreviate the two distributions as $Q \sim \mathcal{N}(\mu_{i,j,t}^{q}, \sigma_{i,j,t}^{q\,2})$ and $P \sim \mathcal{N}(\mu_{i,j,t}^{p}, \sigma_{i,j,t}^{p\,2})$. The KL divergence is computed as:

$$KL\left(Q\left(\alpha_{i,j}^{Var}\right) || P\left(\alpha_{i,j}^{Var}\right)\right) =$$
$$\sum_{t\in T,(i,j)\in E^t} \left\{ 2\log\left(\frac{\sigma_{i,j,t}^{p} + \epsilon}{\sigma_{i,j,t}^{q} + \epsilon}\right) + \frac{\sigma_{i,j,t}^{q\,2} + (\mu_{i,j,t}^{q} - \mu_{i,j,t}^{p})^2}{(\sigma_{i,j,t}^{p} + \epsilon)^2} - 1 \right\}, \tag{28}$$

where ϵ is a small constant for numerical stability.

Term 3: Emotion Relation Term. According to Eq. (15), we define an adjacency matrix $A_t^{Emo} = [\alpha_{i,j,t}^{Emo}]$, and we define the Emotion Relation Inference process as follows:

$$\bar{V}_t = GCN(V_t, A_t^{Emo}), \tag{29}$$

$$\bar{Y} = f_{n2y}(\sum_{t \in T}(\bar{V}_t)), \tag{30}$$

where $GCN(\cdot)$ denotes a GCN, $V_t = \{v_{c,t}|c \in C\}$ denotes the node in the tth time segment, and $\bar{V}_t$ denotes the node based on the emotion relation graph after updating the weights by the GCN. Finally, we combine the signals of the nodes in different time segments and apply a classification network f_{n2y} to obtain the input signal's emotion classification result $\bar{Y}$. This result will be used with the ground truth Y to compute the CE loss in Eq. (19) as the emotion relation term in the RPG optimization objective.

3.6 Model Interpretability of RPGCN

Interpretability of RPGCN is inherent to its design. The first stage, RPG inference, models inter-channel relationships by generating and fusing probabilistic graphs into a final 'emotion relation graph'. This graph explicitly captures the importance and emotional connectivity of each EEG channel. The second stage then uses this interpretable graph to guide a GCN for robust classification. Crucially, after training, the learned edge weights of the emotion relation graph offer direct insight into the model's logic, eliminating the need for external interpretability tools. This simplifies the process of understanding how the model utilizes connections between EEG channels to detect emotional states.

4 Experiment and Discussion

4.1 Datasets

The Database for Emotion Analysis using Physiological Signals (Deap) dataset [16] is a multimodal dataset for emotion analysis, containing EEG and peripheral physiological signals (ECG, GSR, etc.) from 32 participants watching 40 one-minute music videos. After each clip, participants rated their emotional states in terms of valence and arousal. The dataset provides 32-channel EEG and several peripheral signals, including ECG, GSR, respiration, and skin temperature. Similarly, the Dreamer dataset [14] serves as a benchmark for emotion recognition using EEG and ECG signals. It includes recordings from 23 participants watching emotion-eliciting video clips, with self-reported valence and arousal ratings. The dataset provides 14-channel EEG and 14-channel ECG data for analyzing brain and cardiac responses to emotional stimuli.

4.2 Implementation Details

Experimental Setup. We performed subject-specific 10-fold cross-validation, with an 80%/20% train-validation split applied to non-test folds. To prevent data leakage, signals were cropped into non-overlapping segments. Final results are the average accuracy and F1-score over 10 runs, using an ensemble of the 5 best-performing models from the validation set. All hyperparameters and implementation details are listed in Table 1.

Table 1. Implementation Details and Hyperparameter Settings.

Parameter	Value	Parameter	Value	Parameter	Value
Optimizer	Adam	Learning Rate	3×10^{-5}	Weight Decay	1×10^{-3}
Dropout	0.5	Epochs	500	Segment Length	4s
Hidden Dim.	128	Sequence Length	512	Time Steps	32
Attention Heads	4	Conv. Channels	8	Stability ($\epsilon, \epsilon_{i,j}$)	$10^{-6}, 10^{-5}$

Table 2. Comparison (in %) of RPGCN and Baselines.

Method	Deap				Dreamer			
	Arou Acc	Arou F1	Vale Acc	Vale F1	Arou Acc	Arou F1	Vale Acc	Vale F1
DGCNN [36]	75.23	75.63	70.38	73.45	74.28	73.88	75.30	75.82
RGNN [41]	70.56	69.18	69.94	68.11	68.82	67.45	66.51	67.77
EEGNet [17]	75.65	74.88	70.61	66.08	70.65	71.17	69.43	67.11
TSception [7]	65.34	64.91	67.39	52.18	66.32	65.43	67.27	65.48
SpikingNN [37]	63.15	60.28	66.23	66.21	62.72	64.39	64.29	63.85
3DCANN [24]	72.15	69.81	65.23	64.18	70.23	71.81	61.67	62.89
MSDTT [3]	60.23	59.94	61.08	62.84	61.12	59.03	59.88	60.37
SS-STANN [33]	75.56	76.18	72.94	76.11	72.82	71.45	71.51	70.77
MTGNN [38]	74.76	73.54	72.84	74.13	70.28	70.55	72.71	73.12
ASTG-LSTM [19]	75.72	74.15	76.34	76.22	74.85	75.47	72.82	72.33
ST-GCLSTM [13]	76.47	76.53	75.83	75.17	74.27	74.78	74.01	73.26
RPGCN (Ours)	**80.46**	**79.13**	**79.28**	**77.21**	**80.68**	**79.88**	**77.28**	**77.73**

4.3 Experiment Results and Comparison with Prior Art

The experimental results include per-subject scores on the DEAP and DREAMER datasets (see Fig. 2), overall performance metrics, and a comparison against ten existing baselines (see Table 2).

Per-Subject Result: On the DEAP dataset, standard deviations (SD) for accuracy were 4.77 (arousal) and 5.48 (valence); for F1-scores, they were 3.46 and 3.74, respectively. On the DREAMER dataset, the accuracy SDs were 4.12 (arousal) and 3.78 (valence), with corresponding F1-score SDs of 2.65 and 2.97. Notably, the prediction difficulty differs across datasets: valence is harder to predict on DEAP, while arousal is more challenging on DREAMER.

Comparison Results: We compare RPGCN with various baselines:

EEGNet [17] captures time-frequency features, but its local receptive fields limit the modeling of global spatio-temporal relationships.

Spiking NN (SNN) [37] infers temporal relations but lacks spatial awareness. Models with separate components like TSception [7] and MSDTT [3] also have limited temporal inference, capturing only inter-segment correlations while neglecting complex intra-segment spatial dependencies.

3DCANN [24] and SS-STANN [33] use spatio-temporal encoders but show limited generalization as they ignore brain topology and lack physiological interpretability.

Graph-based models such as DGCNN [36], MTGNN [38], ASTG-LSTM [19], and ST-GCLSTM [13] learn relations from model parameters without incorporating brain structure. While RGNN [41] introduces a spatial prior based on physical distance, it still fails to capture task-specific relations. In contrast, RPGCN uses self-inferred, physiologically-aligned relational graphs, providing simultaneous gains in both performance and interpretability.

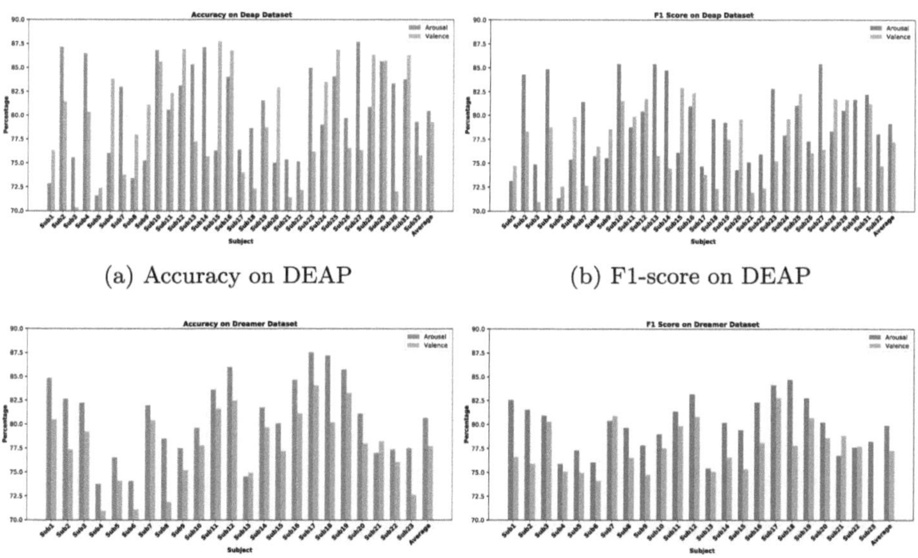

(a) Accuracy on DEAP

(b) F1-score on DEAP

(c) Accuracy on Dreamer

(d) F1-score on Dreamer

Fig. 2. Accuracy and F1-score of RPGCN on DEAP and Dreamer datasets.

4.4 Relation Feature Visualization Comparison

To validate the interpretability of RPG-inferred relations, we compare RPGCN against a baseline GCN with a randomly initialized adjacency matrix. For both models, we visualize their input adjacency weights (as emotion relations) for selected DEAP subjects under high-arousal/high-valence (positive) and low-arousal/low-valence (negative) conditions.

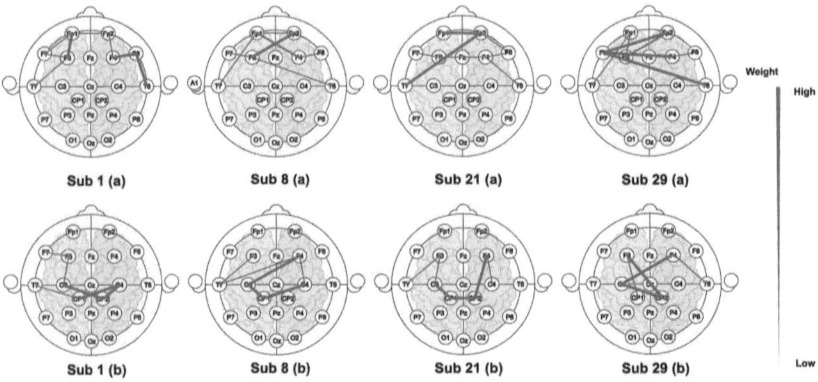

Interpretable Emotion Relation Feature Visualization on DEAP Dataset. Subfigures (a) indicate that the emotion relationship of positive emotions varies in different subjects. In contrast, the emotion relationship is located in the prefrontal cortex and anterior insula regions. Subfigures (b) indicate that the emotion relationship of negative emotions varies in different subjects, while the emotion relationship is located in the postcentral gyrus region.

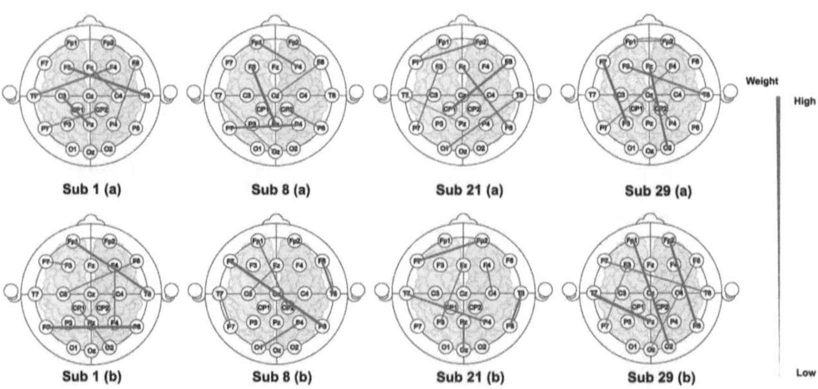

Comparison Emotion Feature Visualization on DEAP Dataset. The visualization illustrates that traditional GCN with a random-initiated adjacency matrix cannot capture the relation between specific brain regions (EEG channels) and emotion.

Fig. 3. Interpretable and Comparison Feature Visualization on DEAP.

As shown in Fig. 3, the RPGCN-inferred relations for positive emotions (row 1) are concentrated in the prefrontal cortex and anterior insula (Fp1, Fp2, F3, F4, F7, F8, T7, T8), aligning with findings in [28]. In contrast, negative emotion relations (row 2) are focused on mid-posterior regions like the postcentral gyrus (C3, C4, CP1, CP2). The visualizations also distinguish between universal emotion relations across all subjects (row 3) and subject-specific variability (row 4), demonstrating that RPG inference captures both common and individual emotion patterns.

Conversely, the graph from the baseline GCN (second graph in Fig. 3), which lacks RPG guidance, reveals no meaningful correlations between channels and emotions. Its uninformative adjacency weights underscore the critical role of RPG inference in providing the interpretability that our framework offers.

4.5 Ablation Study

To verify the advantages of RPG inference, we replace it with other traditional feature extractors and do the ablation studies on the Deap and Dreamer datasets, respectively. The results of the ablation study show that our RPG inference has an advantage over the typical feature extractors (Table 3). We applied the same experiment setting as the main experiment. The experiment result in Table 3 demonstrates that our RPG inference outperforms the traditional feature extractors.

Table 3. Comparison of RPG Inference and other Feature Extractor.

Method	Deap				Dreamer			
	Arou Acc	Arou F1	Vale Acc	Vale F1	Arou Acc	Arou F1	Vale Acc	Vale F1
CNN-Spa	68.61	67.82	67.61	66.28	60.12	60.08	61.28	61.96
CNN-Temp	68.37	66.91	67.89	64.88	60.74	60.91	60.19	62.81
TF-Spa	61.23	60.11	62.18	56.21	61.61	60.22	60.23	60.89
TF-Temp	62.33	59.17	60.12	60.01	61.33	61.01	61.88	62.21
CNN-Joint	75.66	73.82	74.71	72.13	71.01	67.92	70.08	69.92
TF-Joint	76.92	77.18	75.04	73.41	70.56	70.11	69.94	68.11
RPG Inference	**80.46**	**79.13**	**79.28**	**77.21**	**80.68**	**79.88**	**77.28**	**77.73**

5 Conclusion

In this paper, we propose RPGCN, a novel method for EEG emotion recognition based on relational thinking. By inferring emotion relations using summary and variant probabilistic subgraphs, our model achieves state-of-the-art performance on the DEAP and DREAMER datasets. Ablation studies confirm the importance of our RPG inference module, which produces interpretable emotion relation graphs consistent with known physiological phenomena. This combination of SOTA performance and clear interpretability makes RPGCN a promising tool for developing personalized applications in healthcare, education, product design, and interactive entertainment (*e.g.* VR/AR).

References

1. Alexander, P.A.: Relational thinking and relational reasoning: harnessing the power of patterning. NPJ Sci. Learn. **1**(1), 1–7 (2016)
2. Cheng, C., Liu, W., Feng, L., Jia, Z.: Emotion recognition using hierarchical spatial–temporal learning transformer from regional to global brain. Neural Netw. **179**, 106624 (2024). https://doi.org/10.1016/j.neunet.2024.106624, https://www.sciencedirect.com/science/article/pii/S0893608024005483
3. Cheng, C., Zhang, Y., Liu, L., Liu, W., Feng, L.: Multi-domain encoding of spatiotemporal dynamics in EEG for emotion recognition. IEEE JBHI **27**(3), 1342–1353 (2022)
4. Chung, J., Kastner, K., Dinh, L., Goel, K., Courville, A.C., Bengio, Y.: A recurrent latent variable model for sequential data. NeurIPS **28** (2015)
5. Dalgleish, T.: The emotional brain. Nat. Rev. Neurosci. **5**(7), 583–589 (2004)
6. Ding, Y., Robinson, N., Tong, C., Zeng, Q., Guan, C.: Lggnet: learning from local-global-graph representations for brain–computer interface. IEEE TNNLS (2023)
7. Ding, Y., et al.: Tsception: a deep learning framework for emotion detection using EEG. In: IJCNN, pp. 1–7. IEEE (2020)
8. Fox, P.T., Raichle, M.E., Mintun, M.A., Dence, C.: Nonoxidative glucose consumption during focal physiologic neural activity. Science **241**(4864), 462–464 (1988)
9. Gupta, K., et al.: Total vrecall: using biosignals to recognize emotional autobiographical memory in virtual reality. IMWUT **6**(2), 1–21 (2022)
10. Huang, H., Xue, F., Wang, H., Wang, Y.: Deep graph random process for relational-thinking-based speech recognition. In: ICML, pp. 4531–4541. PMLR (2020)
11. Hudson, D., Manning, C.D.: Learning by abstraction: The neural state machine. NeurIPS **32** (2019)
12. Jia, Z., Lin, Y., Cai, X., Chen, H., Gou, H., Wang, J.: Sst-emotionnet: spatial-spectral-temporal based attention 3d dense network for eeg emotion recognition. In: Proceedings of the 28th ACM International Conference on Multimedia, MM 2020, pp. 2909–2917. Association for Computing Machinery, New York (2020). https://doi.org/10.1145/3394171.3413724, https://doi.org/10.1145/3394171.3413724
13. Jiang, W.B., Yan, X., Zheng, W.L., Lu, B.L.: Elastic graph transformer networks for EEG-based emotion recognition. In: ICASSP, pp. 1–5. IEEE (2023)
14. Katsigiannis, S., Ramzan, N.: DREAMER: a database for emotion recognition through EEG and ECG signals from wireless low-cost off-the-shelf devices. IEEE JBHI **22**(1), 98–107 (2017)
15. Kingma, D.P., Welling, M.: Auto-encoding variational bayes. ArXiv Preprint arXiv:1312.6114 (2013)
16. Koelstra, S., et al.: Deap: a database for emotion analysis; using physiological signals. IEEE TAFFC **3**(1), 18–31 (2011)
17. Lawhern, V.J., Solon, A.J., Waytowich, N.R., Gordon, S.M., Hung, C.P., Lance, B.J.: EEGNet: a compact convolutional neural network for EEG-based brain-computer interfaces. JNE **15**(5), 056013 (2018)

18. Li, W., et al.: Cogemonet: a cognitive-feature-augmented driver emotion recognition model for smart cockpit. IEEE TCSS **9**(3), 667–678 (2021)
19. Li, X., Zheng, W., Zong, Y., Chang, H., Lu, C.: Attention-based spatio-temporal graphic lstm for EEG emotion recognition. In: IJCNN, pp. 1–8. IEEE (2021)
20. Li, Y., Zheng, W., Cui, Z., Zhang, T., Zong, Y.: A novel neural network model based on cerebral hemispheric asymmetry for EEG emotion recognition. In: IJCAI, pp. 1561–1567 (2018)
21. Liu, C., et al.: Graph neural networks in eeg-based emotion recognition: a survey (2025). https://arxiv.org/abs/2402.01138
22. Liu, C., et al.: Vsgt: variational spatial and gaussian temporal graph models for eeg-based emotion recognition. In: Proceedings of the Thirty-Third International Joint Conference on Artificial Intelligence, pp. 3078–3086 (2024)
23. Liu, C., Zhou, X., Zhu, Z., Zhai, L., Jia, Z., Liu, Y.: Vbh-gnn: variational bayesian heterogeneous graph neural networks for cross-subject emotion recognition. In: The Twelfth International Conference on Learning Representations (2024)
24. Liu, S., Wang, X., Zhao, L., Li, B., Hu, W., Yu, J., Zhang, Y.D.: 3DCANN: a spatio-temporal convolution attention neural network for EEG emotion recognition. IEEE JBHI **26**(11), 5321–5331 (2021)
25. Liu, Y., Jia, Z.: BSTT: A bayesian spatial-temporal transformer for sleep staging. In: ICLR (2022)
26. Lotfi, E., Akbarzadeh-T, M.R.: Practical emotional neural networks. Neural Netw. **59**, 61–72 (2014)
27. Mao, S., et al.: Time series analysis for education: Methods, applications, and future directions. arXiv preprint arXiv:2408.13960 (2024)
28. Min, J., et al.: Emotion downregulation targets interoceptive brain regions while emotion upregulation targets other affective brain regions. J. Neurosci. **42**(14), 2973–2985 (2022)
29. Pang, J.C., et al.: Geometric constraints on human brain function. Nature pp. 1–9 (2023)
30. Perich, M.G., Rajan, K.: Rethinking brain-wide interactions through multi-region 'network of networks' models. Curr. Opin. Neurobiol. **65**, 146–151 (2020)
31. Piana, S., Staglianò, A., Odone, F., Camurri, A.: Adaptive body gesture representation for automatic emotion recognition. ACM TiiS **6**(1), 1–31 (2016)
32. Pranav, E., Kamal, S., Chandran, C.S., Supriya, M.: Facial emotion recognition using deep convolutional neural network. In: ICACCS, pp. 317–320. IEEE (2020)
33. Sartipi, S., Torkamani-Azar, M., Cetin, M.: A hybrid end-to-end spatio-temporal attention neural network with graph-smooth signals for EEG emotion recognition. IEEE TCDS (2023)
34. Sheynin, O.B.: Laplace's theory of errors. Arch. Hist. Exact Sci. **17**(1), 1–61 (1977)
35. Smolensky, P.: Connectionist AI, symbolic AI, and the brain. Artif. Intell. Rev. **1**(2), 95–109 (1987)
36. Song, T., Zheng, W., Song, P., Cui, Z.: EEG emotion recognition using dynamical graph convolutional neural networks. IEEE TAFFC **11**(3), 532–541 (2018)
37. Tan, C., Šarlija, M., Kasabov, N.: Neurosense: short-term emotion recognition and understanding based on spiking neural network modelling of spatio-temporal EEG patterns. Neurocomputing **434**, 137–148 (2021)
38. Wu, Z., Pan, S., Long, G., Jiang, J., Chang, X., Zhang, C.: Connecting the dots: Multivariate time series forecasting with graph neural networks. In: KDD, pp. 753–763 (2020)
39. Xu, H., Zhang, H., Han, K., Wang, Y., Peng, Y., Li, X.: Learning alignment for multimodal emotion recognition from speech. arXiv preprint arXiv:1909.05645 (2019)

40. Ying, X., Yizhe, Z.: Design of speech emotion recognition algorithm based on deep learning. In: IEEE AUTEEE, pp. 734–737. IEEE (2021)
41. Zhong, P., Wang, D., Miao, C.: EEG-based emotion recognition using regularized graph neural networks. IEEE TAFFC **13**(3), 1290–1301 (2020)

Federated Spatio-Temporal Attention for Time Series Anomaly Detection

Weicheng Wang[1], Yue He[1], Xiaoliang Chen[1,2,5](✉), Duoqian Miao[2], Hongyun Zhang[2], Xiaolin Qin[3], Shangyi Du[4], and Peng Lu[5]

[1] School of Computer and Software Engineering, Xihua University, Chengdu 610039, People's Republic of China
chenxl@mail.xhu.edu.cn, chexiaol@iro.umontreal.ca

[2] College of Electronic and Information Engineering, Tongji University, Shanghai 201804, People's Republic of China

[3] Chengdu Institute of Computer Applications, Chinese Academy of Sciences, Chengdu 610041, People's Republic of China

[4] Department of Computer Science, McGill University, Montreal, QC H3A0G4, Canada

[5] Department of Computer Science and Operations Research, University of Montreal, Montreal, QC H3C3J7, Canada

Abstract. The accelerated expansion of Industrial Internet of Things (IIoT) systems has concurrently precipitated the generation of substantial multivariate time series data, where the implementation of precise anomaly detection mechanisms is indispensable for ensuring operational safety, and mitigating risks in complex industrial ecosystems.

Current methodologies predominantly leverage local spatial and temporal representations derived from adjacent nodes and recent time points. This approach, which focuses on local processing, frequently overlooks global topological relationships and temporal patterns—both crucial for precise anomaly detection. This limitation arises from insufficient modeling of global sensor-time dependencies and inadequate integration of spatial-temporal interdependencies, resulting in high false-positive rates.

To mitigate these issues, we introduce **FL-STAM**, a federated learning-enhanced framework with a spatio-temporal attention mechanism for unsupervised MTS anomaly detection.

First, a parallel graph attention architecture independently extracts global sensor dependencies and temporal patterns through serial-oriented and time-oriented modules. Second, a dual-branch transformer with Wasserstein distance quantification explicitly models spatial-temporal association discrepancies, amplifying discriminative features between normal and anomalous patterns. Third, this paper adopts a privacy-preserving federated learning paradigm, which can realize collaborative model training among distributed devices while effectively preventing the exposure of local data.

Extensive experiments on five IIoT datasets (SMAP, MSL, SMD, PSM, SWaT) demonstrate FL-STAM's superiority, achieving state-of-the-art (SOTA) F1 scores of 98.52% on PSM and 97.86% on SWaT. Abla-

M. Yoshikawa et al. (Eds.): ADMA 2025, LNAI 16197, pp. 129–141, 2026.
https://doi.org/10.1007/978-981-95-3453-1_9

tion studies verify component effectiveness, notably the parallel graph mechanism enhancing accuracy by 6.86% over baselines.

Keywords: Time series · Anomaly detection · Federated learning · Graph attention mechanism · Transformer

1 Introduction

A time series refers to a chronological arrangement of data points, commonly recorded at consistent intervals and organized in a sequential manner. Time series analysis involves applying various analytical methods to examine these data, with the goal of identifying significant statistical patterns and extracting meaningful insights.

The IIoT infrastructure widely collects multivariate time series (MTS) data, which possess complex topological relationships and significant temporal dynamic characteristics [13]. However, industrial networks and devices are vulnerable to potential threats and attacks, potentially leading to severe repercussions. Consequently, the development of efficient and precise anomaly detection systems for MTS is of paramount importance.

While temporal models like Recurrent Neural Networks (RNNs) [6], Autoencoders (AEs) [17], and Generative Adversarial Networks (GANs) [8] effectively capture temporal dependencies, they often neglect inter-series spatial correlations.

To mitigate these challenges, we introduce a distributed training framework, termed FL-STAM, grounded in the federated mean algorithm. This framework augments the STAM approach to enable the distributed training of anomaly detection models, improving their accuracy while safeguarding data privacy.

The contributions of this study are delineated as follows:

1. This research constructs a sequence-time parallel graph attention framework, designed to simultaneously handle global sequence dependencies and temporal dependencies in MTS data. The framework integrates two graph attention mechanisms, which precisely allocate weights to the influence of global nodes across both sequence and time dimensions, thereby significantly enhancing the optimization of time series feature representation.
2. Transformers are employed to extract spatial and temporal correlations. Furthermore, based on the extracted spatial and temporal correlations, the discrepancies among these correlations are computed through a spatio-temporal multi-head attention mechanism.
3. To ascertain the efficacy of FL-STAM, this study undertook a thorough experimental evaluation. Across five publicly available datasets, FL-STAM was benchmarked against eight prominent anomaly detection methods. The results illustrate that FL-STAM substantially surpasses other SOTA methods in performance.

2 Related Work

ADMTS has garnered substantial interest owing to its pivotal applications across diverse domains, including industrial monitoring, finance, and healthcare. The traditional approaches used in ADMTS are neatly sorted into three different types: clustering-based approaches [7], distance-based techniques [2], and isolation-based methods [9]. These methods have laid the foundational framework for identifying anomalies by leveraging statistical and machine learning techniques. However, with the advent of deep learning, more sophisticated and powerful models have emerged, significantly enhancing the feature representation and generalization capabilities for anomaly detection [15]. Deep learning methodologies predominantly encompass two intrinsic paradigms: reconstruction-centric and prediction-oriented models [3].

Reconstruction-based frameworks endeavor to encapsulate the intrinsic distribution of MTS data through latent variables, subsequently facilitating the reconstitution of original data samples from the acquired distribution [5].

AnomalyTrans [14] augments anomaly detection capabilities by leveraging a transformer architecture fortified with an anomaly attention mechanism. GLAD [16] extends AnomalyTrans by integrating a Gumbel-Softmax-based graph structure learning technique. These reconstruction-based methods have shown remarkable capabilities in automatically learning high-level data features, thereby enhancing the detection of complex anomaly patterns. Nevertheless, a significant drawback of these methods is their tendency to overlook sudden fluctuations in time series data, which may represent normal variations, leading to an increased false positive rate.

Prediction-based frameworks [11] center on anticipating forthcoming temporal instances or intervals based on historical MTS data, with anomalies detected through the evaluation of prediction errors. SCNN [4] is an adaptive, interpretable, and scalable prognostic architecture designed to independently model each constituent of spatio-temporal dynamics. By operating based on a predefined MTS generative process, SCNN captures the latent structure of spatio-temporal patterns, offering improved traceability and predictability compared to traditional methods. THOC [10] introduces a classification framework that captures temporal dynamics across various scales through the use of a dilated RNN with skip connections, and integrates multiscale time features by means of hierarchical clustering. This methodology efficiently addresses the complex temporal dependencies inherent in MTS data. However, a key limitation is that they primarily capture correlations between different timestamps in MTS, often neglecting the interdependencies between data from different sensors. This limitation highlights the need for more comprehensive approaches capable of efficaciously modeling both spatial and temporal dependencies within MTS data.

3 Methodology

In this chapter, we formally delineate the conceptual domain of unsupervised ADMTS problems and provide a detailed exposition of the Federated Learning framework for Spatio-Temporal Anomaly Detection (FL-STAM). The objective is to identify temporal patterns that markedly diverge from normal behavior, independent of labeled anomaly data during the training phase.

3.1 Overview of Proposed Model

This study proposes a privacy-preserving anomaly detection architecture for MTS data within IIoT systems. The FL-STAM architecture identifies anomalies by analyzing discrepancies between spatial correlations and temporal interactions while integrating federated learning to enable distributed computation and preserve data privacy. FL-STAM comprises two principal components:

- **FL-STAM Model for ADMTS.** As depicted in Fig. 1, the process commences with data preprocessing to address discrepancies in values across different dimensions of the MTS. Subsequently, a parallel graph structure learning approach is deployed, encompassing a serial-oriented graph attention layer to capture sensor dependencies and a time-oriented graph attention layer to model temporal interdependencies within the time series. A transformer architecture incorporating an anomaly attention mechanism is then employed, enabling simultaneous modeling of spatial and temporal associations. Beyond reconstruction loss, the anomaly transformer is refined through a minimax optimization strategy, aimed at sharpening the distinction of association discrepancies and thereby improving the detection of time series anomalies.
- **FL-STAM Distributed Training Framework.** Sensor systems within distinct manufacturing areas stream temporal operational data to local edge devices. These edge devices train local FL-STAM models using resident data, enabling timely edge-based anomaly detection. This distributes computational load, preventing cloud server overload.

3.2 FL-STAM Model for ADMTS

Serial-Time Parallel Graph Attention Mechanism. In the context of MTS, interdependencies among nodes are captured through correlation analysis in both serial and time dimensions. To address this, we employ two parallel graph attention layers, each tasked with the distinct extraction of features from the serial and temporal components of the multivariate time series data. Specifically, we use GATv2 [1], a modified version of the standard Graph Attention Network, to perform computations in both serial and time domains.

(1) Serial-Oriented Graph Attention Layer
The objective is to identify correlations among MTS without prior knowledge. To achieve this, the MTS is represented as a complete graph, wherein each node corresponds to a unique time series, exemplified by sensor data. The edges of the graph capture the relationships between pairs of time series. Specifically, the time series i is represented as a vector $\boldsymbol{x}^i = \{\nu_1^i, \nu_2^i, \ldots, \nu_s^i\}$, where s denotes the number of timestamps within a sliding window. The set of adjacent nodes

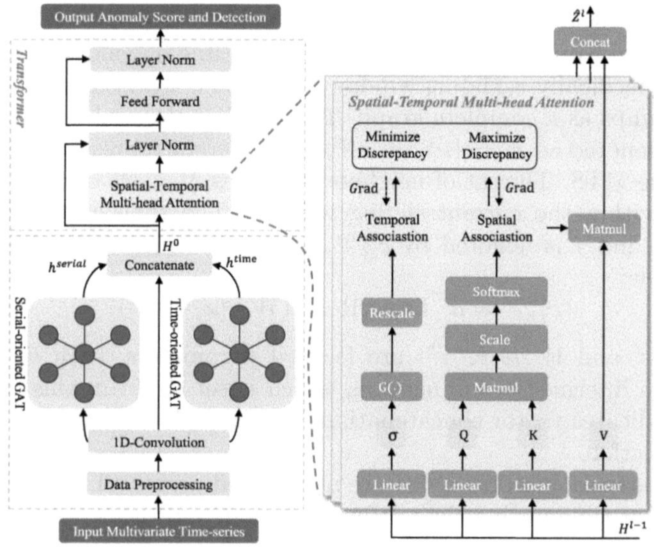

Fig. 1. The framework of FL-STAM.

$\mathcal{N}_i \in \mathbb{R}^n$ includes all other time series within the MTS. The serial correlation between time series i and j is denoted by c_{ij}^{serial}, with the corresponding formula provided below:

$$c_{ij}^{serial} = \boldsymbol{a}^\top \text{LeakyReLU}(W \cdot [x^i \parallel x^j]) \tag{1}$$

where $a \in \mathbb{R}^d$ and $W \in \mathbb{R}^{d \times 2s}$ are learned parameters, with d representing a configurable intermediate dimension, which is set to $2s$ in this context. The operator $||$ indicates vector concatenation, and LeakyReLU refers to a nonlinear activation function.

The serial attention scores α_{ij}^{serial} are normalized across all neighboring time series $j \in \mathcal{N}_i$ using the softmax function. The corresponding attention mechanism is then defined as follows:

$$\alpha_{ij}^{serial} = \frac{\exp\left(c_{ij}^{serial}\right)}{\sum_{j \in \mathcal{N}_i} \exp\left(c_{ij}^{serial}\right)} \tag{2}$$

Finally, the serial graph attention layer determines the output representation $\boldsymbol{h}^i$ for each node using the following process, ensuring that the dimensionality of $\boldsymbol{h}^i$ matches that of the input $\boldsymbol{x}^j$.

$$\boldsymbol{h}^i = \sigma\left(\sum_{j \in \mathcal{N}_i} \alpha_{ij}^{serial} \boldsymbol{x}^j\right) \tag{3}$$

where σ represents the sigmoid activation function.

(2) Time-Oriented Graph Attention Layer
Graph Attention Networks are utilized to model time dependencies within time series data. Specifically, a sliding window approach is utilized, representing each set of timestamps as a complete graph. To formalize the problem, let the vector at time i be denoted as $\boldsymbol{x}_i = \{\nu_i^1, \nu_i^2, \cdots, \nu_i^n\}$, where n represents the number of features in the MTS. The set of neighboring nodes $\mathcal{N}_i \in \mathbb{R}^s$ consists of all other timestamps within the current sliding window. The time correlation between timestamps i and j is denoted by c_{ij}^{time}, and the corresponding formulation is provided below:

$$c_{ij}^{time} = \boldsymbol{a}^\top \text{LeakyReLU}(\boldsymbol{W} \cdot [\boldsymbol{x}_i \parallel \boldsymbol{x}_j]) \tag{4}$$

where $a \in \mathbb{R}^d$ and $W \in \mathbb{R}^{d \times 2n}$ are learned parameters, with d representing a configurable intermediate dimension, which is set to $2n$ in this context. The operator $||$ indicates vector concatenation, and LeakyReLU refers to a nonlinear activation function.

The time attention scores α_{ij}^{time} are normalized across all neighboring timestamps $j \in \mathcal{N}_i$ at time i using the softmax function, and the attention mechanism is defined as follows:

$$\alpha_{ij}^{time} = \frac{\exp\left(c_{ij}^{time}\right)}{\sum_{j \in \mathcal{N}_i} \exp\left(c_{ij}^{time}\right)} \tag{5}$$

Finally, the time graph attention layer generates the output representation $\boldsymbol{h}_i$ for each node through the following process, ensuring that the dimensionality of $\boldsymbol{h}_i$ is consistent with that of the input $\boldsymbol{x}_i$.

$$\boldsymbol{h}_i = \sigma\Big(\sum_{j \in \mathcal{N}_i} \alpha_{ij}^{time} \boldsymbol{x}_i\Big) \tag{6}$$

where σ represents the sigmoid activation function.

The output of the serial graph attention layer, h^{serial}, is represented as a matrix of dimensions $n \times s$, while the output of the time graph attention layer, denoted as h^{time}, is represented as a matrix of dimensions $s \times n$. To integrate information from multiple sources, the outputs from both the serial and time graph attention layers, along with the output from the one-dimensional convolutional neural network, are concatenated. This combined process is referred to as Embedding(X).

Transformer with Spatial and Temporal Association Discrepancies. The Transformer model introduced in this study, which accounts for spatial and temporal association discrepancies, is distinguished by its interleaved arrangement of spatial-temporal multi-head attention blocks and feedforward neural network layers. Let the model consist of L layers, with the input time series denoted as $H \in \mathbb{R}^{s \times d}$, where s indicates the series length and d its dimensionality. The general formulation for the l-th layer can be formally expressed as:

$$Z^l = \text{LayerNorm}(\text{ST-Attention}(H^{l-1}) + H^{l-1}) \tag{7}$$

$$H^l = \text{LayerNorm}(\text{FeedForward}(Z^l) + Z^l) \tag{8}$$

Here, $H^l \in \mathbb{R}^{s\times d}$, for $l \in \{1, \ldots, L\}$, signifies the output of the l-th layer with d channels. The initial input, $H^0 = \text{Embedding}(X)$, denotes the embedded form of the original sequence. $Z^l \in \mathbb{R}^{s\times d}$ is the hidden representation at the l-th layer. The function ST-Attention($\cdot$) is employed to calculate the spatial and temporal association discrepancies.

Spatial-Temporal Multi-head Attention. The single-branch self-attention mechanism [12] is limited in concurrently modeling spatial and temporal associations. To mitigate this limitation, we introduce a ST-Attention mechanism with a dual-branch structure [14].

Spatial-association is designed to examine the relationships within the original sequence, adaptively identifying the most relevant associations across the entire spatiotemporal domain, irrespective of temporal distance.

$$\mathrm{Q}, \mathrm{K}, \mathrm{V}, \sigma = H^{l-1}W_{\mathrm{Q}}^{l}, H^{l-1}W_{\mathrm{K}}^{l}, H^{l-1}W_{\mathrm{V}}^{l}, H^{l-1}W_{\sigma}^{l} \tag{9}$$

$$SA^l = \mathrm{Softmax}\left(\frac{\mathrm{QK}^{\mathrm{T}}}{\sqrt{d_{\mathrm{model}}}}\right) \tag{10}$$

$$\widehat{Z}^l = SA^l\mathrm{V} \tag{11}$$

In the self-attention mechanism, $\mathrm{Q}, \mathrm{K}, \mathrm{V} \in \mathbb{R}^{s\times d}$ and $\sigma \in \mathbb{R}^{s\times 1}$ correspond to the Query, Key, Value, and the learned scale parameter, respectively. $W_{\mathrm{Q}}^l, W_K^l, W_V^l \in \mathbb{R}^{d\times d}$ and $W_\sigma^l \in \mathbb{R}^{d\times 1}$ denote the weights associated with $\mathrm{Q}, \mathrm{K}, \mathrm{V}$, and σ at the l-th layer, respectively.

The spatial-association matrix $SA^l \in \mathbb{R}^{s\times s}$ is normalized along its last dimension by the Softmax function, ensuring that each row of SA^l forms a discrete probability distribution. $\widehat{Z}^l \in \mathbb{R}^{s\times d}$ represents the hidden state obtained after the l-th layer of the ST-Attention mechanism.

Output Anomaly Score and Detection. The discrepancy between spatial and temporal associations is assessed through the symmetric Wasserstein Distance, a metric that quantifies the divergence between two probability distributions, wherein a reduced Wasserstein Distance signifies a lesser disparity between the distributions. The association discrepancy across various layers is aggregated to consolidate multi-level feature associations into a holistic measure, as delineated below:

$$\mathrm{Dis}(TA, SA; X) = \left(\frac{1}{L}\sum_{l=1}^{L}\left[\mathcal{W}(TA_{i,:}^l, SA_{i,:}^l) + \mathcal{W}(SA_{i,:}^l, TA_{i,:}^l)\right]\right)_{i=1}^{s}. \tag{12}$$

where Wasserstein($\cdot$) denote the Wasserstein Distance between two discrete probability distributions. The vector $\mathrm{Dis}(TA, SA; X) \in \mathbb{R}^{s\times 1}$ quantifies the point-wise association discrepancy of the time series X with respect to the temporal associations TA and spatial associations SA across multiple layers. The i-th element of Dis corresponds to the association discrepancy at the i-th time point of X. Empirical findings suggest that anomalous time points tend to exhibit lower $\text{Dis}(TA, SA; X)$ values than normal time points, thereby rendering Dis a distinctive and effective measure for anomaly detection.

As an unsupervised task, the model is optimized using reconstruction loss during the training phase. This loss function directs the spatial association component to discern the most relevant correlations. To effectively distinguish between normal and anomalous time points, an auxiliary loss term is introduced to accentuate the association discrepancy. Given the unimodal nature of the temporal association, the discrepancy loss encourages the spatial association to emphasize non-adjacent regions, thereby intensifying the challenge of reconstructing anomalies and thereby enhancing their detectability. The loss function for the input series $X \in \mathbb{R}^{s \times d}$ is formulated as follows:

$$\mathcal{L}_{\text{Total}} = \|X - \widehat{X}\|_{\text{F}}^2 - \lambda \times \|\text{Dis}(TA, SA; X)\|_1 \tag{13}$$

where $\widehat{X} \in \mathbb{R}^{s \times n}$ represents the reconstructed matrix of X. $\|\cdot\|_{\text{F}}, \|\cdot\|_1$ indicate the Frobenius norm and 1-norm. The parameter λ serves to balance the various terms within the loss function. For $\lambda > 0$, the optimization process aims to amplify the association discrepancy.

3.3 FL-STAM Distributed Training Framework

This section implements a FL framework for collaborative anomaly detection model training across distributed edge devices, enhancing detection accuracy while ensuring data privacy preservation. Each edge device operates as a client node running the FL-STAM anomaly detection model (Sect. 3.2), with parameter aggregation performed at the central server. The framework maintains data privacy by restricting transmission to model parameters while retaining raw data locally. The proposed framework's workflow comprises seven sequential stages:

(1) Sensor data acquired by the edge device constitutes the local dataset for model training.
(2) The edge device performs local model training following the methodology described in Sect. 3.2, utilizing the collected local dataset.
(3) Upon completion of local training, the edge device transmits the updated model parameters to the global server.
(4) The global server executes a model aggregation process, integrating parameters from all participating clients to produce an updated global model.
(5) The aggregated global model parameters are subsequently disseminated to all edge devices.
(6) Steps (2) through (5) are repeated until the global model converges.
(7) Th edge device performs anomaly detection using the converged global model.

4 Experiments

This segment delineates the experimental framework, including the specification of the data set, evaluation metrics, and baseline comparisons. Comprehensive experiments substantiate the efficacy of the proposed model across multiple performance dimensions (Table 1).

Table 1. Comprehensive statistical summary of the experimental datasets.

Dataset	Applications	Dimensions	Train	Test	Anomaly rate(%)
MSL	Space	55	58,317	73,729	10.72
SMAP	Space	25	135,183	427,617	13.13
PSM	Server	25	132,481	87,841	27.75
SMD	Server	38	708,405	708,420	4.16
SWaT	Water	51	496,800	449,919	11.98

4.1 Experimental Setup

Computationally, all experiments were conducted on Ubuntu 20.04 with an RTX 4090 GPU (24 GB VRAM), utilizing PyTorch 2.0 and CUDA 11.8. Model configuration included a non-overlapping sliding window of size 100, dataset-specific anomaly thresholds (SWaT: 0.1%, SMD: 0.5%, others: 1%), and a 3-layer transformer (8 heads, 512 hidden units). Optimization used the Adam optimizer ($\beta_1 = 0.9, \beta_2 = 0.999$) with a learning rate of 10^{-4} and a loss balance coefficient $\lambda = 3$ (Eq. 13).

4.2 Experimental Results

This segment presents the experimental outcomes that validate the effectiveness of the FL-STAM model in anomaly detection across five diverse datasets. The assessment is grounded in the comparison of FL-STAM with multiple SOTA anomaly detection techniques, employing precision, recall, and F1 score as key performance metrics. Each dataset was evaluated through ten test iterations to ensure dependability and robustness of the reported performance metrics.

Baseline Comparisons. Table 2 provides a detailed comparison of the FL-STAM model with eight SOTA anomaly detection methods across five distinct datasets. The best results are highlighted in bold, with rankings provided in parentheses. Both average F1 scores and their corresponding rankings are reported in the table. The technique recognized as the best performer is the one that captures the highest average F1 scores and achieves top rankings. If a method secures high average F1 scores but falls short in rankings, it might perform especially well on certain datasets while showing inconsistent results. On the other hand, a method with excellent rankings but lower average F1 scores could suggest stable performance across all datasets, even if it doesn't excel particularly on any single dataset.

Table 2. Experimental results for all methods on five public datasets, measured by F1 score.

Method	MSL			SMAP			PSM		
	Pre	Rec	F1	Pre	Rec	F1	Pre	Rec	F1
IF (2008)	53.94	86.54	66.45(9)	52.39	59.07	55.53(9)	75.91	57.36	65.34(9)
OmniAnomaly (2019)	89.02	86.37	87.67(6)	92.49	81.99	86.92(7)	88.39	74.46	80.83(8)
MTAD-GAT (2020)	87.54	94.40	90.84(4)	89.06	91.23	90.13(4)	95.28	75.65	84.34(5)
InterFusion (2021)	81.28	92.70	86.62(7)	89.77	88.52	89.14(5)	83.61	83.45	83.52(6)
GDN (2021)	91.35	86.12	88.66(5)	89.32	88.72	89.02(6)	80.65	83.57	82.09(7)
AnomalyTrans (2022)	92.09	95.15	93.59(3)	94.13	99.40	96.69(3)	96.91	98.90	97.89(3)
Dcdetector (2023)	93.69	99.69	**96.60(1)**	95.63	98.92	97.02(2)	97.14	98.74	97.94(2)
RWKV-TS (2024)	78.11	77.74	77.92(8)	89.04	55.94	68.71(8)	98.32	95.92	97.10(4)
FL-STAM	92.79	98.13	95.39(2)	96.16	99.45	**97.78(1)**	98.67	98.37	**98.52(1)**

Method	SMD			SWaT			Avg F1	Avg F1 Ranking
	Pre	Rec	F1	Pre	Rec	F1		Rank (value)
IF (2008)	42.31	73.29	53.64(9)	49.42	44.95	47.02(9)	57.60(9)	9(9)
OmniAnomaly (2019)	83.68	86.82	85.22(6)	81.42	84.30	82.83(7)	84.69(7)	6.8(7)
MTAD-GAT (2020)	88.28	84.92	86.57(4)	92.46	75.12	82.89(6)	86.95(4)	4.6(4)
InterFusion (2021)	84.02	88.41	86.16(5)	80.59	85.58	83.01(5)	85.69(5)	5.6(5)
GDN (2021)	71.70	99.74	83.42(8)	99.35	68.12	80.82(8)	84.80(6)	6.8(7)
AnomalyTrans (2022)	89.40	95.45	92.33(2)	91.55	96.73	94.07(3)	94.91(3)	2.8(3)
Dcdetector (2023)	83.59	91.10	87.18(3)	93.11	99.77	96.33(2)	95.01(2)	2(2)
RWKV-TS (2024)	87.45	81.43	84.33(7)	88.20	94.85	91.40(4)	83.89(8)	6.2(6)
FL-STAM	93.79	95.29	**94.53(1)**	97.33	98.41	**97.86(1)**	**96.82(1)**	**1.2(1)**

The FL-STAM model achieves an average F1 score of 96.82 across all datasets, surpassing all competing methods and establishing itself as the top performer. With an average F1 ranking of 1.2, FL-STAM occupies the leading position. In contrast, the InterFusion, AnomalyTrans, and Dcdetector models represent the SOTA methods for 2021, 2022, and 2023, respectively, serving as the primary benchmarks in this comparison. On the SWaT dataset, FL-STAM exhibits a 17.89% improvement over InterFusion, a 4.03% advantage over AnomalyTrans, and a 1.59% edge over Dcdetector. In the PSM dataset, FL-STAM achieves an F1 score 17.96% higher than InterFusion, 0.64% superior to AnomalyTrans, and 0.59% better than Dcdetector. Across all five datasets, FL-STAM demonstrates a 12.99% increase in F1 score over InterFusion, a 2.01% improvement over AnomalyTrans, and a 1.91% gain over Dcdetector. In all other datasets assessed, FL-STAM consistently outperforms the baseline models.

The analysis of traditional models such as Isolation Forest (IF) reveals their limitations; they consistently underperform relative to deep learning models across all datasets, primarily due to their inability to effectively capture complex patterns and spatiotemporal correlations in MTS data. For example, while OmniAnomaly demonstrates improvements in temporal modeling, it fails to capture the spatial dependencies critical for accurate anomaly detection. The GDN methodology, encompassing node embeddings and graph attention mechanisms, underscores the inter-nodal correlations while disregarding temporal dependencies. In contrast, both InterFusion and MTAD-GAT effectively address both time and spatial dependencies, leading to improved detection precision. Nonetheless, these models still rely on basic judgment criteria. AnomalyTrans pioneers anomaly detection criterion that integrates association discrepancy, facilitating the differentiation between anomalous and normal data. However, it overlooks the spatial dependencies within MTS, which limits the enhancement of anomaly detection performance. In contrast, DCdetector accounts for both spatial and time dependencies, leading to moderate improvements.

The proposed FL-STAM distinguishes itself by minimizing reconstruction errors while simultaneously considering both spatial and temporal correlations in the data. This methodological enhancement is pivotal in enabling the model to achieve more precise anomaly detection. For example, FL-STAM not only integrates association discrepancy to differentiate anomalous from normal data but also strengthens spatial and temporal dependency modeling, resulting in its superior performance.

5 Conclusion

This study proposes FL-STAM, a novel transformer-based framework for MTS anomaly detection, enhanced by a parallel graph structure learning mechanism. Our key contributions are: (1) a parallel graph attention mechanism capturing global serial and temporal dependencies independently, overcoming the common limitation of isolated spatial or temporal modeling; (2) a spatial-temporal multi-head attention mechanism within the transformer, simultaneously modeling spatiotemporal associations and utilizing their Wasserstein Distance discrepancy for nuanced anomaly understanding, significantly boosting accuracy; (3) the integration of federated learning, enabling privacy-preserving distributed training across devices, enhancing scalability and addressing IIoT data security. Evaluations on five benchmark datasets confirm FL-STAM's consistent superiority over SOTA methods in F1-score and computational efficiency, with ablation studies underscoring the critical role of the parallel graph structure. While demonstrating strong performance, ongoing challenges in dynamic IIoT environments warrant future research into adaptive continuous learning mechanisms, domain knowledge integration, and framework extension to diverse time-series data.

Acknowledgment. This work is supported by the National Key R&D Plan "Key Special Project of Cyberspace Security Governance" (No. 2022YFB3104700), the National Natural Science Foundation (Grant nos. 62402395, 62376198, 62076182), the Science and Technology Program of Sichuan Province (Grant no. 2023YFS0424), the Science and Technology Service Network Initiative (No. KFJ-STS-QYZD-2021-21-001), and the Talents by Sichuan provincial Party Committee Organization Department, and Chengdu - Chinese Academy of Sciences Science and Technology Cooperation Fund Project (Major Scientific and Technological Innovation Projects).

References

1. Brody, S., Alon, U., Yahav, E.: How attentive are graph attention networks? arXiv preprint arXiv:2105.14491 (2021)
2. Chaovalitwongse, W.A., Fan, Y.J., Sachdeo, R.C.: On the time series k-nearest neighbor classification of abnormal brain activity. IEEE Trans. Syst. Man Cybernetics-Part A: Syst. Humans **37**(6), 1005–1016 (2007)
3. Darban, Z.Z., Webb, G.I., Pan, S., Aggarwal, C., Salehi, M.: Deep learning for time series anomaly detection: A survey. ACM Comput. Surv. **57**(1), 15:1–15:42 (2025). https://doi.org/10.1145/3691338
4. Deng, J., Chen, X., Jiang, R., Yin, D., Yang, Y., Song, X., Tsang, I.W.: Disentangling structured components: Towards adaptive, interpretable and scalable time series forecasting. IEEE Trans. Knowl. Data Eng. **36**(8), 3783–3800 (2024). https://doi.org/10.1109/TKDE.2024.3371931
5. Goodge, A., Hooi, B., Ng, S.K., Ng, W.S.: Robustness of autoencoders for anomaly detection under adversarial impact. In: Bessiere, C. (ed.) Proceedings of the twenty-ninth international conference on international joint conferences on artificial intelligence, pp. 1244–1250. International Joint Conferences on Artificial Intelligence Organization (2021). https://doi.org/10.24963/IJCAI.2020/173
6. Hundman, K., Constantinou, V., Laporte, C., Colwell, I., Soderstrom, T.: Detecting spacecraft anomalies using lstms and nonparametric dynamic thresholding. In: Guo, Y., Farooq, F. (eds.) Proceedings of the 24th ACM SIGKDD International Conference on Knowledge Discovery & Data Mining, pp. 387–395. ACM, New York (2018). https://doi.org/10.1145/3219819.3219845
7. Kiss, I., Genge, B., Haller, P., Sebestyén, G.: Data clustering-based anomaly detection in industrial control systems. In: 2014 IEEE 10th International Conference on Intelligent Computer Communication and Processing (ICCP), pp. 275–281. IEEE (2014)
8. Li, D., Chen, D., Jin, B., Shi, L., Goh, J., Ng, S.K.: Mad-gan: Multivariate anomaly detection for time series data with generative adversarial networks. In: International Conference on Artificial Neural Networks, pp. 703–716. Springer (2019)
9. Liu, F.T., Ting, K.M., Zhou, Z.: Isolation forest. In: Proceedings of the 8th IEEE International Conference on Data Mining (ICDM 2008), December 15-19, 2008, Pisa, Italy, pp. 413–422. IEEE Computer Society (2008).https://doi.org/10.1109/ICDM.2008.17
10. Shen, L., Li, Z., Kwok, J.: Timeseries anomaly detection using temporal hierarchical one-class network. Adv. Neural. Inf. Process. Syst. **33**, 13016–13026 (2020)
11. Tealab, A.: Time series forecasting using artificial neural networks methodologies: a systematic review. Future Comput. Inform. J. **3**(2), 334–340 (2018). https://doi.org/10.1016/j.fcij.2018.10.003

12. Vaswani, A.: Attention is all you need. Advances in Neural Information Processing Systems (2017)
13. Wang, H.z., Li, G.q., Wang, G.b., Peng, J.c., Jiang, H., Liu, Y.t.: Deep learning based ensemble approach for probabilistic wind power forecasting. Appl. Energy **188**, 56–70 (2017)
14. Xu, J., Wu, H., Wang, J., Long, M.: Anomaly transformer: Time series anomaly detection with association discrepancy. CoRR abs/2110.02642 (2021). https://arxiv.org/abs/2110.02642
15. Zhang, C., et al.: A deep neural network for unsupervised anomaly detection and diagnosis in multivariate time series data. In: Proceedings of the AAAI Conference on Artificial Intelligence, vol. 33, pp. 1409–1416 (2019)
16. Zhou, X., Dai, C., Wang, W., Qiu, T.: Global-local association discrepancy for multivariate time series anomaly detection in iiot. IEEE Internet Things J. **11**(7), 11287–11297 (2024). https://doi.org/10.1109/JIOT.2023.3330696
17. Zong, B., et al.: Deep autoencoding gaussian mixture model for unsupervised anomaly detection. In: International Conference on Learning Representations (2018)

Machine Learning for Data Mining

Domain Graph-Structured Multi-source Domain Adaptation with Dual Integration

Jiayi Wang[1], Xin Zheng[2], Yi Li[1], and Yanqing Guo[1(✉)]

[1] Dalian University of Technology, Dalian, China
jiayiwang@mail.dlut.edu.cn, {liyi, guoyq}@dlut.edu.cn
[2] Griffith University, Brisbane, Australia
xin.zheng@griffith.edu.au

Abstract. Multi-source unsupervised domain adaptation (MUDA) is a challenging research problem, which aims to leverage the knowledge from multiple labeled source domains to adapt to an unlabeled target domain for better inference performance. Despite many methods having made efforts to extract common domain-invariant representations across all domains to improve adaptation ability, diverse cross-domain distribution discrepancies and imprecise decision boundaries have not been well addressed. To deal with such challenges, in this work, we propose a **D**omain **G**raph-**S**tructured Multi-Source **D**omain **A**daptation with Dual Integration named **DGSDA**. Specifically, the proposed method contains three core modules: (1) Input integration, where the combined source domain is integrated by the multi-source domains to reduce the distribution discrepancies; (2) Domain graph structure module, constructed through feature extraction and domain graph embedding, focusing on aligning domain-specific distributions to learn multiple domain-invariant representations for optimal category decision boundaries; (3) Decision integration, aiming to assign different weights to multiple classification results to efficiently explore useful knowledge. Extensive experiments on real-world image classification tasks demonstrate that the proposed DGSDA achieves expressive domain adaptation performance.

Keywords: Unsupervised domain adaptation · Multi-source domains · Domain graph construction · Dual integration

1 Introduction

In recent years, deep neural networks have shown great ability in a variety of visual learning tasks, especially in image classification [28], object detection [4], and object recognition [26]. These achievements mainly come from the availability of large-scale labeled data for supervised learning. Nevertheless, obtaining abundant labeled data is costly and time-consuming in the real world [8]. Domain adaptation (DA) mainly solves this problem by establishing knowledge transfer from the labeled source domain to the unlabeled target domain. By

The original version of the chapter has been revised. A correction to this chapter can be found at https://doi.org/10.1007/978-981-95-3453-1_37

M. Yoshikawa et al. (Eds.): ADMA 2025, LNAI 16197, pp. 145–159, 2026.
https://doi.org/10.1007/978-981-95-3453-1_10

boosting domain-invariance of feature representations from different domains, DA has been an expert at pattern recognition [13], computer vision [6], and image processing [23] in recent years.

Unsupervised domain adaptation (UDA) methods [12,20] seek to bridge the performance gap due to domain shift via adaptation of the model on small amounts of unsupervised data from the target domain. Single-source unsupervised domain adaptation (SUDA) methods assume that only one source domain and one target domain are available for training and testing. SUDA methods have been proposed [1,19,24] in the past decade. In practice, the labeled data are not drawn from a single data distribution, and the samples are always collected from different deployment environments. To consider different knowledge from different domains, some Multi-source unsupervised domain adaptation (MUDA) methods have been proposed [14,27,30], which are dedicated to transferring the knowledge learned from multiple source domains to an unlabeled target domain. Due to potential complementary information, MUDA is both feasible in practice and more valuable in performance improvement and has received considerable attention in real-world application fields.

Graph convolutional network (GCN) for UDA [32] demonstrates that data structure, domain labels, and category labels play a crucial role in establishing connections between the labeled source domain and the unlabeled target domain. The data structure typically reflects the inherent characteristics of a dataset. Category labels are essential in facilitating greater semantic consistency across different domains. In general, integrating them has been verified to be effective in reducing domain shift. It can enhance the model's generalization capability across diverse domains, rendering it more robust to variations in data distribution, feature representation, and other factors contributing to domain shift.

The majority of current MUDA approaches optimize to minimize the empirical risk on the source data and to make the target and source features indistinguishable from each other. However, there are some significant challenges in solving the MUDA task. **Challenge 1:** A clear discrepancy exists among multiple source domains, hampering the effectiveness of mainstream single-source domain adaptation methods. **Challenge 2:** Without precise learning of decision boundaries, the model cannot make full use of the effective features in the data to optimize the classification decision.

Motivated by these limitations, we propose a novel multi-source unsupervised domain adaptation algorithm called **D**omain **G**raph-**S**tructured Multi-Source **D**omain **A**daptation with Dual Integration named **DGSDA**. According to abstract, with (1)(2)(3) modules, being consistent. Dual integration consists of input integration and decision integration. The combined source domain is firstly integrated by the multi-source domains. Through this input integration, we can generally reduce the distribution discrepancies between each source and target domain, as well as the discrepancies between different source domains. Then, we respectively map each source domain and the target domain into the domain graph structure module composed by the feature extraction based on

Convolutional neural network (CNN) and domain graph embedding with GCN. By aligning the distribution of the target domain with each source domain, the model can further optimize the utilization of each source domain's knowledge. After that, we concatenate the domain-specific CNN features and GCN features to feed into the multiple domain-specific classifiers. We utilize decision integration to combine all predictions by assigning different weights to multiple classification results, and obtain the final results. Our main contributions are as follows:

- **Input integration.** The combined domain is integrated by the multi-source domains to align the distribution with the target domain, which can minimize the distribution discrepancies.
- **Domain graph structure module.** We extract domain-specific features by mapping into the domain graph structure module, which can align domain-specific distributions to learn multiple domain-invariant representations and optimize category decision boundaries.
- **Decision integration.** We assign different weights to obtain the final results, which can take the reliabilities of the different results produced by the multiple domain-specific classifiers into consideration.
- Extensive experiments on two available datasets demonstrate the effectiveness of our proposed method for multi-source domain adaptation.

2 Related Work

In this section, we will introduce the related work for Single-source unsupervised domain adaptation (SUDA) and Multi-source unsupervised domain adaptation (MUDA).

2.1 Single-Source Unsupervised Domain Adaptation (SUDA)

SUDA methods focus on transferring knowledge from a single labeled source domain to an unlabeled target domain. Recent research in SUDA has focused on three primary strategies: adversarial learning, self-supervised learning, and discrepancy minimization.

Adversarial learning, inspired by generative adversarial networks, aligns feature distributions by reducing the gap between the source and target domains through domain discriminators [3]. Methods such as domain-adversarial neural networks and conditional domain adversarial networks have shown promising results in reducing domain shifts [12]. Self-supervised learning has emerged as another effective approach, integrating tasks such as contrastive learning and pseudo-labeling to improve target domain generalization. Techniques like consistency learning [29] enhance feature robustness in the target domain. Additionally, discrepancy minimization approaches such as Maximum mean discrepancy (MMD) and Correlation alignment (CORAL) explicitly measure and minimize the statistical distance between domains. Recent method proposes hybrid models integrating MMD with adversarial learning to achieve better adaptation [20].

2.2 Multi-source Unsupervised Domain Adaptation (MUDA)

MUDA extends the adaptation process to multiple source domains, introducing complexities in handling diverse data distributions. Domain-invariant representation learning methods, such as deep subspace alignment and multi-source domain adversarial networks, project multiple source domains into a shared feature space, aligning them with the target domain [31]. Dynamic source weighting techniques, such as importance weighted adversarial networks, assign different importance levels to source domains based on their relevance to the target domain, improving adaptation performance [15]. Recent research has proposed a prototype aggregation method that utilizes a group of prototypes as representative feature embeddings [7]. Another approach involves a bi-level adaptation framework for graphs, incorporating a graph-modeling-based domain selector and a sub-graph node selector to capture variations in transferability within and across domains [33]. Moreover, meta-learning techniques are increasingly employed to enhance MUDA, allowing models to generalize across diverse source distributions.

3 Method

In this section, we introduce the details of the proposed model (see Fig. 1). In multi-source unsupervised domain adaptation, there are N different source domains and the target domain T, and their underlying distributions are respectively denoted as $\{p_{S_i}(X,Y)\}_{i=1}^{N}$ and $p_T(X,Y)$. The labeled source domain data $\{(X_{S_i}, Y_{S_i})\}_{i=1}^{N}$ and the unlabeled target domain $X_T = \{x_j^T\}_{j=1}^{|X_T|}$ are sampled from the distributions respectively, where $X_{S_i} = \{x_j^{S_i}\}_{j=1}^{|X_{S_i}|}$ represents samples from j-th source domain distribution and $Y_{S_i} = \{y_j^{S_i}\}_{j=1}^{|Y_{S_i}|}$ represents its corresponding labels.

In our framework, the combined source domain is integrated by the multi-source domains, then we have $N+1$ source domains. Only one source domain and the target domain are fed into the network at each iteration during the training process. The combined source domain and the original N source domains are used iteratively. The mini-batch we used to train are denoted as $\{(\hat{X}_{S_i}, \hat{Y}_{S_i})\}_{i=1}^{N+1}$and $\hat{X}_T$. After that, we combine feature extraction based on CNN and domain graph embedding with GCN to form the domain graph structure module, where each source domain and the target domain data are respectively mapped. Then, we concatenate the domain-specific CNN features and GCN features to feed into the multiple domain-specific classifiers. To leverage complementary knowledge of the multiple classification results as much as possible, we introduce decision integration strategy to combine all predictions from multiple domain-specific classifiers to obtain the final results. By assigning different weights to different classifiers, the classifiers with high reliability play a key role in decision making and reducing the risk of misclassification.

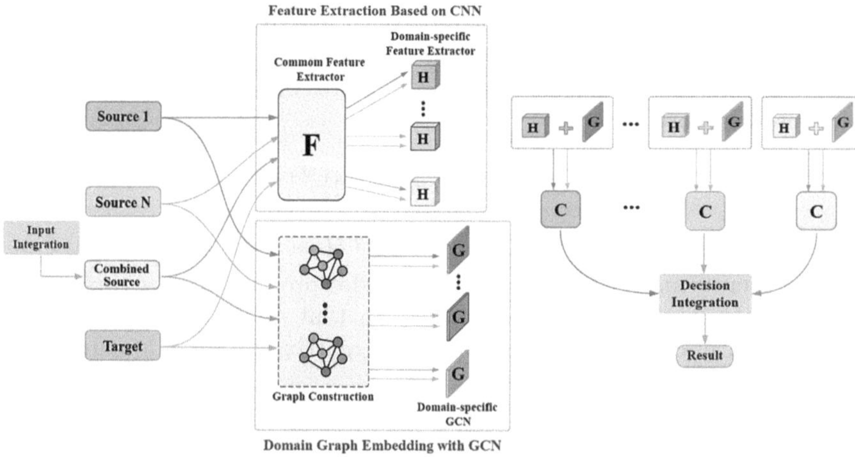

Fig. 1. The architecture of the proposed framework. In the input integration, the combined source domain is integrated by the multi-source domains. Domain graph structure module is formed by the feature extraction based on CNN and domain graph embedding with GCN, where each source domain and the target domain data are respectively mapped. Then, we concatenate the domain-specific CNN features and GCN features to feed into the multiple domain-specific classifiers. In the decision integration, we assign different weights to multiple classification results to obtain the final results.

3.1 Domain Graph Structure Module

Feature Extraction Based on CNN. To learn the common representations for all domains, we use a common feature extractor $F(\cdot)$ to map the input data from the original feature space into a common feature space. In order to take domain-specific decision boundaries in consideration and learn domain-invariant representations, we utilize domain-specific feature extractors $\{H_i\}_{i=1}^{N+1}$ following the common feature extractor. These are used to map data into their corresponding domain-specific feature spaces. The parameters of these CNN are not shared. The CNN features of source domains and target domain can be represented by

$$\{CF_{S_i}(\hat{X}_{S_i}) = H_i(F(\hat{X}_{S_i}))\}_{i=1}^{N+1} \tag{1}$$

$$\{CF_T^{S_i}(\hat{X}_T) = H_i(F(\hat{X}_T))\}_{i=1}^{N+1} \tag{2}$$

Domain Graph Embedding with GCN. Appling to GCN is a graph embedding technique to acquire the feature representation. We first construct a graph for each domain, and then embed each domain graph into the corresponding domain-specific GCN to obtain GCN features.

Each node of the domain graph represents each image in the domain. By treating the image as a graph node, we can take advantage of the structure-processing capabilities of GCNs to classify images more effectively. The weights of the edges of the graph are determined by the structural similarities between the nodes and are represented by the adjacency matrix. We deliver each source domain and the target domain data into domain-specific data structure analyzers $\{A_i\}_{i=1}^{N+1}$, and we can obtain the structural features of source domain and target domain represented by

$$\{SF_{S_i}(\hat{X}_{S_i}) = A_i(\hat{X}_{S_i})\}_{i=1}^{N+1} \tag{3}$$

$$\{SF_T^{S_i}(\hat{X}_T) = A_i(\hat{X}_T)\}_{i=1}^{N+1} \tag{4}$$

The adjacency matrix is generated by the dot product of structural features and reflects the similarity between nodes, so that GCN can learn the inter-domain structural consistency. The adjacency matrix of source domain and target domain are represented by

$$\{A_{S_i} = SF_{S_i}(\hat{X}_{S_i})SF_{S_i}(\hat{X}_{S_i})^T\}_{i=1}^{N+1} \tag{5}$$

$$\{A_T^{S_i} = SF_T^{S_i}(\hat{X}_T)SF_T^{S_i}(\hat{X}_T)^T\}_{i=1}^{N+1} \tag{6}$$

The features extracted by domain-specific CNNs and the structural features extracted by domain-specific data structure analyzers jointly constitute the features of nodes, which enhances the category discrimination ability and ensures that the graph construction captures both global distribution patterns and local structural dependencies.

In this case, every source domain graph and target domain graph for GCN can be constructed. In an undirected graph $\overline{G}$, $\tilde{A} = A + I$, where A stands for the adjacency matrix and I represents the identity matrix. In the lth layer, the parameter matrix and activation matrix are respectively denoted as $W^{(l)} \in \mathbb{R}^{C\times F}$ and $H^{(l)} \in \mathbb{R}^{N\times C}$. The degree matrix can be computed as $\widetilde{D}_{ii} = \sum_j \tilde{A}_{ij}$. When $l^{th} = 0$, H is the input node matrix. Consequently, the graph convolution network can be represented by

$$H^{(l+1)} = \sigma\left(\widetilde{D}^{-\frac{1}{2}}\widetilde{A}\widetilde{D}^{-\frac{1}{2}}H^{(l)}W^{(l)}\right) \tag{7}$$

where σ indicates the activation layer function. We map constructed domain graphs into domain-specific GCNs $\{G_i\}_{i=1}^{N+1}$.

$$\{GF_{S_i}(\hat{X}_{S_i}) = G_i(CF_{S_i}(\hat{X}_{S_i}), A_{S_i})\}_{i=1}^{N+1} \tag{8}$$

$$\{GF_T^{S_i}(\hat{X}_T) = G_i(CF_T^{S_i}(\hat{X}_T), A_T^{S_i})\}_{i=1}^{N+1} \tag{9}$$

where $GF_{S_i}(\widehat{X}_{S_i})$ denotes the GCN features of source domain i and $GF_T^{S_i}(\widehat{X}_T)$ denotes GCN features of target domain with respect to source domain i. Then, we concatenate the domain-specific CNN features and GCN features so that domain labels and data structure information can mutually restrict each other.

3.2 Decision Integration

The domain-specific features are received by the domain-specific classifier $\{C_i\}_{i=1}^{N+1}$. To leverage complementary knowledge of the multiple classification results as much as possible, we introduce a more flexible strategy.

Suppose there exist N labeled source domains and a combined source domain, we can obtain $N+1$ classification results. Each output of the domain-specific classifiers is represented by

$$\hat{y}_j(\cdot) = \{r_{1j}, ..., r_{Cj}\}_{j=1}^{N+1} \tag{10}$$

where C denotes the number of the class.

Then, we allocate a weight vector $\{w_{ij}\}_{i=1,j=1}^{i=N+1,j=C}$ to each classifier. When dealing with two source domains, we set the weight of the combined domain as 0.6. We define a learnable parameter as the weight of the first source domain, and the weight of the second source domain is calculated as 0.4 minus the value of this learnable parameter. In the case of three source domains, we set the weight of the combined domain as 0.4. We utilize a learnable parameter to determine the weight of the first source domain. The weight of the second source domain is calculated as 0.4 minus the value of this learnable parameter, and the third source domain has a fixed weight of 0.2. These learnable parameters are initialized with an initial value 0.2 and are updated automatically through backpropagation based on the loss function. The final results are formulated as

$$\hat{y}_f(\cdot) = \left\{ \sum_{i=1}^{N+1} r_{1i} w_{1i}, ..., \sum_{i=1}^{N+1} r_{Ci} w_{Ci} \right\} \tag{11}$$

where the weights are adjusted according to the weight assignment for each source domain. This approach enables the model to dynamically adapt to diverse source domains during training via learnable parameters, optimizing weight combinations to maximize classification performance.

3.3 Objective Function

Maximum mean discrepancy (MMD) is regarded as the most typical measurement method in the deep domain adaptation networks, which measures the discrepancy of two distributions with their mean embeddings in the Reproducing Kernel Hilbert space $\mathcal{H}$. It is formulated as

$$MMD(X_S, X_T) = \left\| \frac{1}{|X_S|} \sum \phi(x_S^i) - \frac{1}{|X_T|} \sum \phi(x_T^i) \right\|_{\mathcal{H}}^2 \tag{12}$$

where $\phi(\cdot)$ denotes a mapping function from the original samples space to $\mathcal{H}$ space. In order to focus on the samples that are beneficial to transfer knowledge from the source domains to the target domain, we use a weighted strategy represented by

$$MMD(\hat{X}_S, \hat{X}_T) = \left\| \sum_{x_S^i \in \mathcal{S}} \frac{\left\|\hat{y}(x_S^i) - \max(\hat{y}(x_S^i)) \cdot y(x_S^i)\right\|_2}{\sum_{x_S^j \in \mathcal{S}} \left\|\hat{y}(x_S^j) - \max(\hat{y}(x_S^j)) \cdot y(x_S^j)\right\|_2} \phi(x_S^i) \right.$$
$$\left. - \sum_{x_T^i \in \mathcal{T}} \frac{\left\|\hat{y}(x_T^i) - \max(\hat{y}(x_T^i)) \cdot \bar{y}(x_T^i)\right\|_2}{\sum_{x_T^j \in \mathcal{T}} \left\|\hat{y}(x_T^j) - \max(\hat{y}(x_T^j)) \cdot y(x_T^j)\right\|_2} \phi(x_T^i) \right\|_{\mathcal{H}}^2 \tag{13}$$

where $y(\cdot)$ denote the one-hot labels of the source domain samples, $\widehat{y}(\cdot)$ denote the predictions of the classifiers, and $\bar{y}(\cdot)$ denote the pseudo labels predicted by the network. Motivated by the entropy minimization principle, we reduce the conditional entropy of the category distribution, which is represented by

$$\mathcal{L}_{con} = -\frac{1}{n} \sum_{i=1}^{n} \hat{y}(x_T^i) \log\left(\hat{y}(x_T^i)\right) \tag{14}$$

To make the discrepancy measure precise, we employ the intraclass and interclass MMD represented by

$$MMD(\hat{X}_S, \hat{X}_T)_{intra} = \frac{\sum_{m=1}^{K} \left\| \frac{1}{n_S^m} \sum_{i=1}^{n_S^m} \phi(x_S^{mi}) - \frac{1}{n_T^m} \sum_{j=1}^{n_T^m} \phi(x_T^{mj}) \right\|_{\mathcal{H}}^2}{K}$$
$$MMD(\hat{X}_S, \hat{X}_T)_{inter} = \frac{\sum_{z=1}^{K} \sum_{m=1, m \neq z}^{K} \left\| \frac{1}{n_S^z} \sum_{i=1}^{n_S^z} \phi(x_S^{zi}) - \frac{1}{n_T^m} \sum_{j=1}^{n_T^m} \phi(x_T^{mj}) \right\|_{\mathcal{H}}^2}{K(K-1)} \tag{15}$$

where K denotes the same categories, x_S^{mi} denotes the i-th source domain sample belonging to the m-th category, n_S^m denotes the number of the source domain samples belonging to the m-th category, x_T^{mj} denotes the j-th target domain sample belonging to the m-th category, n_T^m denotes the number of the target domain samples belonging to the m-th category. Intraclass MMD can minimize the cross-domain distribution difference of similar samples and enhance intra-class compactness. Interclass MMD can maximize the cross-domain distribution difference of different types of samples and enhance the separability between classes. In this case, the loss function of the MMD is represented by

$$\begin{aligned} \mathcal{L}_{MMD} &= MMD(\hat{X}_S, \hat{X}_T) + \mathcal{L}_{con} \times \lambda \\ &+ (MMD(\hat{X}_S, \hat{X}_T)_{intra} - MMD(\hat{X}_S, \hat{X}_T)_{inter}) \times \delta \end{aligned} \tag{16}$$

where λ and δ are hyper-parameters to control the corresponding weights.

Then, we utilize $\mathcal{L}_{cross}$ to denote the cross-entropy loss used to train the network with labeled source data, and we utilize $\mathcal{L}_{cls}$ to reduce the discrepancies among different classifiers. The overall objective is formulated as follows,

$$\mathcal{L}_{cross} = -\frac{1}{n} \sum_{i=1}^{n} y(x_S^i) \log\left(\hat{y}(x_S^i)\right) \tag{17}$$

$$\mathcal{L}_{cls} = \frac{2}{N(N+1)} \sum_{a=1}^{N} \sum_{b=a+1}^{N+1} \left(\frac{1}{n} \sum_{i=1}^{n} |\hat{y}_a(x_T^i) - \hat{y}_b(x_T^i)| \right) \tag{18}$$

$$\mathcal{L}_{overall} = \mathcal{L}_{cross} + (\mathcal{L}_{MMD} + \mathcal{L}_{cls}) \times \gamma \tag{19}$$

where γ denotes the weight of the discrepancy penalty term, and n denotes the number of training batch samples.

4 Experiment

4.1 Datasets

Office-31 Dataset [17]. It contains 4110 images in 31 categories collected from the Amazon website, Web camera, and Digital SLR camera, respectively. It consists of three domains: Amazon (A) with 2817 images, Webcam (W) with 795 images, and DSLR (D) with 498 images. We utilize all domain combinations and establish three MDA tasks: A, W → D; A, D → W; D, W → A.

Office-Home Dataset [22]. It contains 15,500 images in 65 categories collected from the offices and homes. It consists of four domains: Art (A) with 2427 images, Clipart (C) with 4365 images, Product (P) with 4439 images, and Real world (R) with 4357 images. We utilize all domain combinations and establish four MDA tasks: C, P, R → A; A, P, R → C; A, C, R → P; A, C, P → R.

4.2 Baselines

We compare our method with MFSAN [34], DCA [9], CASR [25], PMSDAN [10], which are regarded as MUDA baselines. In addition, some SUDA methods are also compared, including ResNet [5], DDC [21], DAN [11], D-CORAL [18], RevGrad [2]. Specifically, three standards for MUDA are employed: 'Single Best' means the top performance in SUDA to assess whether the best outcome can be further enhanced by incorporating other source domains; 'Source Combine' means the performance achieved by SUDA after all source domains are combined into one domain; 'Multi-Source' stands for MUDA performance.

4.3 Implementation Details

We execute all the deep methods described above using Pytorch. We apply mini-batch stochastic gradient descent with a momentum of 0.9 and a batch-size of 32 to train the network. The learning rate is computed by the formula: lr $= (\eta_0/(1+\alpha p)^\beta)$, where p increases linearly from 0 to 1, varying with the training completion rate. η_0 is an initial constant which is set to 0.001 for the common feature extractor and the decision integration layer. For other layers, the default values are $\eta_0 = 0.01$, $\alpha = 10$, and $\beta = 0.75$ to speed up convergence. Since the network's predictions are imprecise at the start of training, the hyper-parameter γ is gradually adjusted from 0 to 1 following a progressive

Table 1. Performance comparison of classification accuracy (%) on Office-31 dataset.

Standards	Method	A, W → D	A, D → W	D, W → A	Avg.
Single Best	ResNet	99.3	96.7	62.5	86.2
	DDC	98.2	95.0	67.4	86.9
	DAN	99.5	96.8	66.7	87.7
	D-CORAL	99.7	98.0	65.3	87.7
	RevGrad	99.1	96.9	68.2	88.1
Source Combine	DAN	99.6	97.8	67.6	88.3
	D-CORAL	99.3	98.0	67.1	88.1
	RevGrad	99.7	98.1	67.6	88.5
Multi-Source	MFSAN	99.5	98.5	72.7	90.2
	DCA	99.6	98.9	75.1	91.2
	CASR	**99.8**	**99.8**	76.2	91.9
	PMSDAN	**99.8**	98.7	76.8	91.8
	DGSDA (ours)	**99.8**	99.0	**77.6**	**92.1**

schedule: $\gamma = (2/(1+\exp(-\theta p))) - 1$, with $\theta = 10$. We use the ResNet [5] model pretrained on the ImageNet dataset [16] as the common feature extractor. The AlexNet model pretrained on ImageNet is used as the domain-specific structure analyzer to extract structure features. We adopt the (Conv1× 1, Conv3×3, Conv1×1) structure as domain-specific extractors, and single-layer GCN networks act as domain-specific GCNs. Domain-specific feature extractors and classifiers are trained through backpropagation, and their learning rate is set to be ten times that of the other layers.

4.4 Results

Table 1 and Table 2 present the experimental results of our proposed model compared with other algorithms. Our model significantly improves upon existing domain adaptation approaches. Through comprehensive experiments on the baselines, we draw the following meaningful conclusions from the experimental results.

- Most of the methods in the source combined standard perform better than those in the single best standard. The primary reason is that source combined methods have more training data than single best methods.
- The results of most multi-source methods are superior to those of single best and source combined methods. This demonstrates that multi-source methods take into account the domain shift among multiple source domains, and exploring information from multiple sources can effectively enhance the DA performance.

Table 2. Performance comparison of classification accuracy (%) on Office-Home dataset.

Standards	Method	C, P, R → A	A, P, R → C	A, C, R → P	A, C, P → R	Avg
Single Best	ResNet	65.3	49.6	79.7	75.4	67.5
	DDC	64.1	50.8	78.2	75.0	67.0
	DAN	68.2	56.5	80.3	75.9	70.2
	D-CORAL	67.0	53.6	80.3	76.3	69.3
	RevGrad	67.9	55.9	80.4	75.8	70.0
Source Combine	DAN	68.5	59.4	79.0	82.5	72.4
	D-CORAL	68.1	58.6	79.5	82.7	72.2
	RevGrad	68.4	59.1	79.5	82.7	72.4
Multi-Source	MFSAN	72.1	62.0	80.3	81.8	74.1
	DCA	72.1	63.6	80.5	81.4	74.4
	CASR	72.2	61.1	82.8	82.8	74.7
	PMSDAN	75.0	65.8	83.7	83.8	77.1
	DGSDA (ours)	**75.5**	**66.1**	**83.9**	**84.4**	**77.5**

- In the multi-source standard, our proposed model achieves higher accuracy on most adaptation tasks. This indicates that combining all source domains into one domain may benefit from the sufficient training samples. Compared with baseline PMSDAN, the results demonstrate the effectiveness of GCN. Using the domain graph embedding with GCN to extract structural features of data in different domains may enable the model to utilize data structure information and improve performance. In the decision integration, the use of the learnable parameter enables the model to capture the unique features of each source domain more accurately, leading to better generalization and a more robust classification model in multi-source domain adaptation tasks.

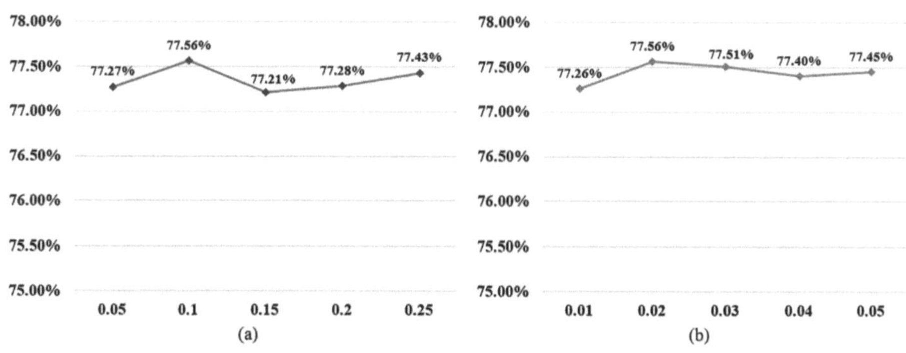

Fig. 2. The sensitivity of λ and δ on task D, W → A. (a) Accuracy with respect to λ. (b) Accuracy with respect to δ.

4.5 Parameter Sensitivity

Our model involves two hyper-parameters λ and δ, which regulate the loss function served as auxiliary elements, so the values of the hyper-parameters should be small. To prove that the proposed method is stable regarding hyper-parameters, we select values from 0.05, 0.1, 0.15, 0.2, 0.25 and 0.01, 0.02, 0.03, 0.04, 0.05 to respectively investigate the sensitivity of λ and δ on task D, W $\rightarrow$ A. All the results are displayed in Fig. 2. It can be noticed that the accuracy rate varies within a narrow range. This indicates that our model is robust to the hyper-parameters λ and δ. In fact, setting $\lambda = 0.1$ and $\delta = 0.02$ often leads to good performance, thus it is advisable to set $\lambda = 0.1$ and $\delta = 0.02$ as the default values.

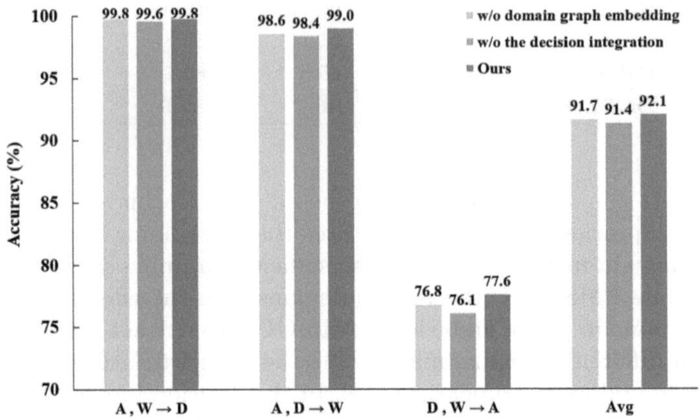

Fig. 3. Ablation results on Office-31 dataset.

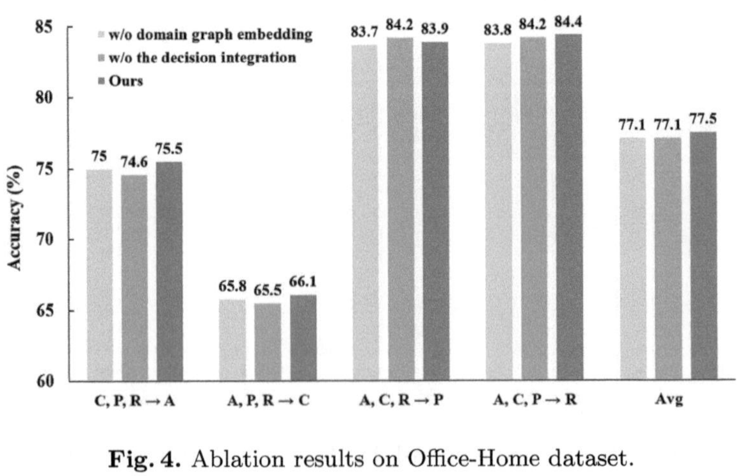

Fig. 4. Ablation results on Office-Home dataset.

4.6 Ablation Analysis

We reserved the domain graph embedding and decision integration provided by existing studies and compared the adaptation performance of our model without (w/o) domain graph embedding and decision integration. The results are shown in Fig. 3 and Fig. 4. This proves that the introduction of domain graph embedding with GCN plays an active role in cross-domain classification. Since the structure information reflects the inherent properties of the data, GCN can extract the structural features in different domains to provide richer information, so that the model can better adapt to the structural changes between different domains and improve the adaptive performance of the domain. Using the strategy of decision integration can take the reliabilities of the different results and efficiently explore useful knowledge. Thus, adopting domain graph embedding and decision integration in a unified MDA network can effectively enhance the performance.

5 Conclusion

In this paper, we propose a novel multi-source unsupervised domain adaptation algorithm called **DGSDA**, which contains three core modules: (1) Input integration; (2) Domain graph structure module; and (3) Decision integration. Specifically, the design of input integration could effectively merge all source domains into a single domain to minimize the distribution discrepancies. And the domain graph structure module, which is constructed through feature extraction and domain graph embedding, is capable of aligning domain-specific distributions and optimizing category decision boundaries. By leveraging domain graph embedding to extract the structural features of data across different domains, our proposed DGSDA can effectively utilize data structure information and enhance its adaptation performance. The decision integration strategy for obtaining the final results renders the classification more accurate. Experimental results on two available datasets demonstrate that the proposed DGSDA achieves expressive performance for multi-source domain adaptation.

Acknowledgments. This work is supported by the Major Program of the National Social Science Foundation of China under Grant No. 19ZDA127.

References

1. Chai, Z., Zhao, C.: A fine-grained adversarial network method for cross-domain industrial fault diagnosis. IEEE Trans. Autom. Sci. Eng. **17**(3), 1432–1442 (2020)
2. Ganin, Y., Lempitsky, V.: Unsupervised domain adaptation by backpropagation. In: International Conference on Machine Learning, pp. 1180–1189. PMLR (2015)
3. Ganin, Y., Ustinova, E., Ajakan, H., Germain, P., Larochelle, H., Laviolette, F., March, M., Lempitsky, V.: Domain-adversarial training of neural networks. J. Mach. Learn. Res. **17**(59), 1–35 (2016)

4. He, H., Wang, S., Yang, D., Wang, S.: Sar target recognition and unsupervised detection based on convolutional neural network. In: 2017 Chinese Automation Congress (CAC), pp. 435–438. IEEE (2017)
5. He, K., Zhang, X., Ren, S., Sun, J.: Deep residual learning for image recognition. In: Proceedings of the IEEE Conference on Computer Vision and Pattern Recognition, pp. 770–778 (2016)
6. Huang, L.Q., Liu, Z.G., Dezert, J.: Cross-domain pattern classification with distribution adaptation based on evidence theory. IEEE Trans. Cybern. **53**(2), 718–731 (2021)
7. Huang, M., Xie, Z., Sun, B., Wang, N.: Multi-source unsupervised domain adaptation with prototype aggregation. Mathematics **13**(4), 579 (2025)
8. Jagabathula, S., Mitrofanov, D., Vulcano, G.: Personalized retail promotions through a directed acyclic graph-based representation of customer preferences. Oper. Res. **70**(2), 641–665 (2022)
9. Li, K., Lu, J., Zuo, H., Zhang, G.: Dynamic classifier alignment for unsupervised multi-source domain adaptation. IEEE Trans. Knowl. Data Eng. **35**(5), 4727–4740 (2022)
10. Liu, Z.G., Ning, L.B., Zhang, Z.W.: A new progressive multisource domain adaptation network with weighted decision fusion. IEEE Trans. Neural Networks and Learning Systems **35**(1), 1062–1072 (2022)
11. Long, M., Cao, Y., Wang, J., Jordan, M.: Learning transferable features with deep adaptation networks. In: International Conference on Machine Learning, pp. 97–105. PMLR (2015)
12. Long, M., Cao, Z., Wang, J., Jordan, M.I.: Conditional adversarial domain adaptation. Advances in neural information processing systems **31** (2018)
13. Lu, Y., Zhu, Q., Zhang, B., Lai, Z., Li, X.: Weighted correlation embedding learning for domain adaptation. IEEE Trans. Image Process. **31**, 5303–5316 (2022)
14. Ren, C.X., Liu, Y.H., Zhang, X.W., Huang, K.K.: Multi-source unsupervised domain adaptation via pseudo target domain. IEEE Trans. Image Process. **31**, 2122–2135 (2022)
15. Ren, M., Zeng, W., Yang, B., Urtasun, R.: Learning to reweight examples for robust deep learning. In: International Conference on Machine Learning, pp. 4334–4343. PMLR (2018)
16. Russakovsky, O., et al.: Imagenet large scale visual recognition challenge. Int. J. Comput. Vision **115**, 211–252 (2015)
17. Saenko, K., Kulis, B., Fritz, M., Darrell, T.: Adapting visual category models to new domains. In: Daniilidis, K., Maragos, P., Paragios, N. (eds.) ECCV 2010. LNCS, vol. 6314, pp. 213–226. Springer, Heidelberg (2010). https://doi.org/10.1007/978-3-642-15561-1_16
18. Sun, B., Saenko, K.: Deep CORAL: correlation alignment for deep domain adaptation. In: Hua, G., Jégou, H. (eds.) ECCV 2016. LNCS, vol. 9915, pp. 443–450. Springer, Cham (2016). https://doi.org/10.1007/978-3-319-49409-8_35
19. Teng, W., Wang, N., Shi, H., Liu, Y., Wang, J.: Classifier-constrained deep adversarial domain adaptation for cross-domain semisupervised classification in remote sensing images. IEEE Geosci. Remote Sens. Lett. **17**(5), 789–793 (2019)
20. Tzeng, E., Hoffman, J., Saenko, K., Darrell, T.: Adversarial discriminative domain adaptation. In: Proceedings of the IEEE Conference on Computer Vision and Pattern Recognition, pp. 7167–7176 (2017)
21. Tzeng, E., Hoffman, J., Zhang, N., Saenko, K., Darrell, T.: Deep domain confusion: Maximizing for domain invariance. arXiv preprint arXiv:1412.3474 (2014)

22. Venkateswara, H., Eusebio, J., Chakraborty, S., Panchanathan, S.: Deep hashing network for unsupervised domain adaptation. In: Proceedings of the IEEE Conference on Computer Vision and Pattern Recognition, pp. 5018–5027 (2017)
23. Wang, F., Han, Z., Gong, Y., Yin, Y.: Exploring domain-invariant parameters for source free domain adaptation. In: Proceedings of the IEEE/CVF Conference on Computer Vision and Pattern Recognition, pp. 7151–7160 (2022)
24. Wang, M., Deng, W.: Deep visual domain adaptation: a survey. Neurocomputing **312**, 135–153 (2018)
25. Wang, S., Wang, B., Zhang, Z., Heidari, A.A., Chen, H.: Class-aware sample reweighting optimal transport for multi-source domain adaptation. Neurocomputing **523**, 213–223 (2023)
26. Wen, Z., Liu, Z., Zhang, S., Pan, Q.: Rotation awareness based self-supervised learning for sar target recognition with limited training samples. IEEE Trans. Image Process. **30**, 7266–7279 (2021)
27. Wong, W.K., Lu, Y., Lai, Z., Li, X.: Graph correlated discriminant embedding for multi-source domain adaptation. Pattern Recogn. **153**, 110538 (2024)
28. Xie, F., Fan, H., Li, Y., Jiang, Z., Meng, R., Bovik, A.: Melanoma classification on dermoscopy images using a neural network ensemble model. IEEE Trans. Med. Imaging **36**(3), 849–858 (2016)
29. Xie, Q., Luong, M.T., Hovy, E., Le, Q.V.: Self-training with noisy student improves imagenet classification. In: Proceedings of the IEEE/CVF Conference on Computer Vision and Pattern Recognition, pp. 10687–10698 (2020)
30. Xu, M., Wang, H., Ni, B.: Graphical modeling for multi-source domain adaptation. IEEE Trans. Pattern Anal. Mach. Intell. **46**(3), 1727–1741 (2022)
31. Zhao, H., Zhang, S., Wu, G., Costeira, J.P., Moura, J.M., Gordon, G.J.: Multiple source domain adaptation with adversarial training of neural networks. arXiv preprint arXiv:1705.09684 (2017)
32. Zhao, S., Li, B., Xu, P., Yue, X., Ding, G., Keutzer, K.: Madan: multi-source adversarial domain aggregation network for domain adaptation. Int. J. Comput. Vision **129**(8), 2399–2424 (2021)
33. Zhao, T., Luo, D., Zhang, X., Wang, S.: Multi-source unsupervised domain adaptation on graphs with transferability modeling. In: Proceedings of the 30th ACM SIGKDD Conference on Knowledge Discovery and Data Mining, pp. 4479–4489 (2024)
34. Zhu, Y., Zhuang, F., Wang, D.: Aligning domain-specific distribution and classifier for cross-domain classification from multiple sources. In: Proceedings of the AAAI conference on artificial intelligence. vol. 33, pp. 5989–5996 (2019)

CrossFM: Cross-City Fine-Grained Urban Flow Inference with Incomplete Data

Wenchao Wu[1,2] and Yuanbo Xu[1,2](✉)

[1] MIC Lab, College of Computer Science and Technology, Jilin University, Changchun, China
wuwc23@mails.jlu.edu.cn

[2] Key Laboratory of Symbolic Computation and Knowledge Engineering of Ministry of Education, Changchun, China
yuanbox@jlu.edu.cn

Abstract. Fine-grained urban flow inference provides important insights for smart city applications such as urban planning and traffic management, but its accuracy is often hindered by incomplete observations due to sparse sensor deployment. While existing methods can handle minor data gaps, their performance degrades significantly under high missing rates, particularly in newly developed urban areas. Along these lines, we propose a novel cross-city super-resolution data map inference framework (CrossFM), designed to transform incomplete coarse-grained urban flows into accurate fine-grained data maps by harnessing cross-city spatio-temporal dynamics. Specifically, we first perform temporal alignment between the source city and the target city data using timestamps. Then, guided by Point-of-Interest (POI) similarity to identify similar regions, we impute missing values in the target city's coarse-grained flow maps. This completion adaptively leverages information from both the source and target city data, resulting in an enhanced coarse-grained representation. Finally, a super-resolution module processes the spatial patterns within the completed coarse data to generate the high-resolution urban flow maps. The framework components are trained jointly end-to-end within a multi-task setup. We conduct extensive experiments on two real-world datasets and demonstrate that CrossFM significantly outperforms the state-of-the-art methods, especially under severe data scarcity.

Keywords: Transfer Learning · Super-Resolution · Fine-Grained Inference · Spatio-Temporal Data

1 Introduction

Fine-grained urban flows, like taxi and bike flows, depict human mobility patterns crucial for smart city applications such as urban planning, traffic management [7,22]. Acquiring such data requires dense sensor networks, incurring substantial operational and maintenance costs. However, sensor deployment is often sparse and uneven due to budget and logistical constraints, resulting in

M. Yoshikawa et al. (Eds.): ADMA 2025, LNAI 16197, pp. 160–174, 2026.
https://doi.org/10.1007/978-981-95-3453-1_11

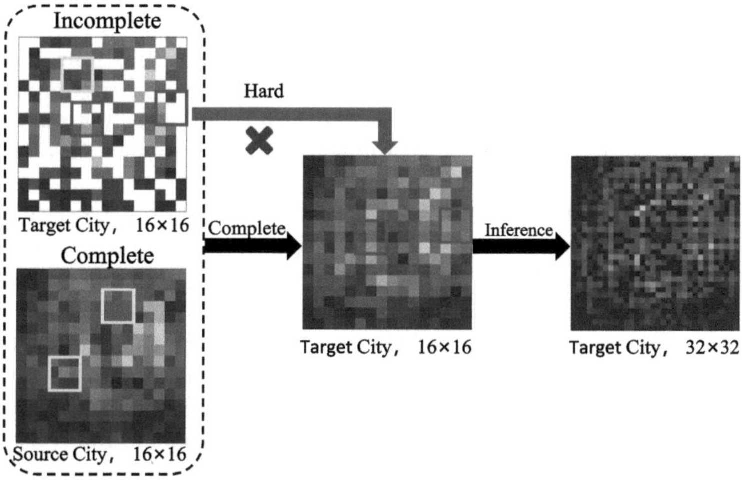

Fig. 1. Fine-grained urban flow inference with incomplete data. The white cell regions denote that the urban flows are unavailable. With high data scarcity, it is challenging to complete the flow map using only the target city's data.

coarse-grained and incomplete observations [19]. This necessitates methods for inferring high-resolution urban flows from limited sensor data, a research problem that has garnered significant attention [20].

Traditional urban flow inference relied on statistical methods like interpolation and tensor factorization [9]. However, these approaches often struggle with urban data's scale and complexity and typically ignore external factors (e.g., weather, holidays), limiting accuracy. More recently, inspired by computer vision super-resolution (SR) [21], researchers framed urban flow inference as a spatio-temporal SR task [20], treating flow snapshots as images. Building on early image SR methods (e.g., SRCNN [3], VDSR [10]), UrbanFM first adapted SR to urban flows, incorporating external factors [12]. UrbanPy introduced a pyramidal approach for higher upscaling rates [16]. Addressing the common issue of incomplete coarse data, MT-CSR proposed a multi-task framework for joint data completion and super-resolution [11].

However, a critical limitation remains: existing methods addressing data incompleteness heavily rely on intra-city context. Consequently, their performance degrades significantly with high missing rates, common in newly developed areas or regions with exceptionally sparse sensor coverage. Accurately inferring fine-grained flows under such severe data scarcity remains a significant challenge [4,5].

The challenge arises primarily from: (1) Failure of Intra-City Completion under High Scarcity: When the available data within the target city is extremely limited, completion methods relying solely on local neighborhood correlations or intra-city semantic similarities (like POI) become unreliable and insufficient for accurate completion. (2) Effectively Leveraging External Data Sources: While using external data holds promise, naively incorporating it is problematic. Utilizing data from a different city, for instance, requires robust mechanisms to ensure relevance. This involves precise temporal alignment to compare corresponding time periods and effective spatial correspondence methods to identify functionally similar, though geographically distinct, regions across different urban environments.

To tackle these challenges, we propose CrossFM, a novel Cross-City Fine-Grained Urban Flow Inference framework. CrossFM introduces a strategy to leverage data from an auxiliary source city to aid inference in the target. Specifically, CrossFM operates sequentially by first performing temporal alignment between the auxiliary source city and the target city using timestamps to ensure that comparisons and data borrowing respect temporal dynamics. Guided by Point-of-Interest (POI) similarity, which helps identify functionally similar urban regions across the two cities, our Cross-city Completion module (CrossCMP) completes missing values in the target city's coarse-grained flow map. CrossCMP adaptively leverages information from both the temporally-aligned source and the available target city data, creating an enhanced, more complete coarse-grained representation. Finally, this enhanced representation is fed into a super-resolution module that processes the spatial patterns to generate the final high-resolution, fine-grained urban flow maps for the target city. The entire CrossFM framework is designed to be trained end-to-end(Fig 1).

The contributions of this paper are summarized as follows:

- To the best of our knowledge, we are the first to address the problem of fine-grained urban flow inference through cross-city transfer learning under conditions of high data scarcity.
- We propose CrossFM, a novel cross-city transfer learning framework that leverages a data-rich source city to enhance fine-grained urban flow inference for a data-scarce target city.
- We design the CrossCMP module to perform completion by considering cross-city spatio-temporal dependencies and global POI similarity between the source and target city.
- We conduct extensive experiments on two real-world datasets, demonstrating that CrossFM significantly outperforms state-of-the-art methods.

2 Related Work

2.1 Fine-Grained Urban Flow Inference

Inferring fine-grained urban flows often employs super-resolution (SR) techniques adapted from computer vision, with recent advancements leveraging diffusion models and other deep learning approaches [23]. Urban-specific SR models, such as UrbanFM and UrbanPy [12,16], were developed to incorporate domain knowledge like external factors or handle high upscaling rates. Recognizing data incompleteness, MT-CSR proposed joint intra-city data completion and SR [11]. However, these methods primarily rely on information within the target city and struggle significantly when coarse data suffers from high missing rates. CrossFM is specifically designed to address such high scarcity scenarios where intra-city information is insufficient.

2.2 Spatio-Temporal Data Completion

Spatio-temporal data completion aims to complete missing values in sparse datasets like urban flows. Early approaches included statistical algorithms, exemplar-based inpainting seeking similar patches [2], efficient patch-matching techniques [1], and offset fusion methods [6], though these often struggled with complex spatio-temporal correlations. More recently, deep learning (DL) has gained prominence, frequently treating gridded spatio-temporal data as images to leverage image completion advances [8]. Notable DL examples include Context Encoders using encoder-decoder structures [17], Partial Convolutions designed to handle missing data explicitly [13], and edge-focused models like EdgeConnect aimed at reducing blurriness [15]. Despite these methodological advances, a key limitation persists across many approaches, from traditional to deep learning. They fundamentally rely on sufficient surrounding context within the same domain [14]. This dependence causes performance to degrade significantly under high global scarcity or when extensive regions are missing âĂŞ precisely the challenging conditions CrossFM targets. Consequently, CrossFM introduces a novel completion methodology specifically designed to overcome this reliance on intra-city data by leveraging external information from an auxiliary cross-city source, enabling more robust completion even under severe data scarcity.

3 Nations and Problem Definition

We will first give some definitions to help state the studied problem, and then present a formal problem definition.

Definition 1. (Region). We divide a city into a grid map based on latitude and longitude, consisting of $I \times J$ cell regions. The set of all regions is denoted as $\mathcal{R} = \{r_{i,j} | 1 \leq i \leq I, 1 \leq j \leq J\}$, where $r_{i,j}$ represents the cell region at the i-th row and j-th column.

Definition 2. (Urban Flow Map). For a given time interval t, the urban flow map captures the movement intensity. For each region $r_{i,j}$, we define inflow $X^t_{in,i,j}$ and outflow $X^t_{out,i,j}$ based on trajectories $\mathcal{T}$:

$$X^t_{in,i,j} = \sum_{f \in \mathcal{T}} \{(f_{t-1} \notin r_{i,j} \wedge f_t \in r_{i,j})\} \tag{1}$$

$$X^t_{out,i,j} = \sum_{f \in \mathcal{T}} \{f_t \in r_{i,j} \wedge f_{t+1} \notin r_{i,j}\} \tag{2}$$

where f_t is the location of urban flow trajectory f at time t. We represent the inflow and outflow for all regions at time t as an urban flow map tensor $X_t \in \mathbb{R}^{2 \times I \times J}$.

Definition 3. (Coarse- and Fine-Grained Flow Maps). A coarse-grained urban flow map represents the observed urban flows derived from the flow sensors. It is generated by integrating neighboring grids within an $N \times N$ range from a fine-grained urban flow map, where N is the upscaling factor. We denote the coarse-grained and fine-grained urban flow maps at time t as $X_{cg}^t \in \mathbb{R}^{2\times I\times J}$ and $X_{fg}^t \in \mathbb{R}^{2\times NI\times NJ}$, respectively. In practice, the observed coarse-grained map $\hat{X}_{cg}^t \in \mathbb{R}^{2\times I\times J}$ is often incomplete.

Definition 4. (Point-of-Interest Features). To capture functional characteristics essential for our cross-city approach, we use Point-of-Interest (POI) features. These are represented as a tensor $P \in \mathbb{R}^{K\times I\times J}$ for a given city, where K is the number of POI categories (e.g., commercial, residential) aggregated within each coarse-grained region $r_{i,j}$. Both the target city D and source city D' have associated POI features, denoted as P^D and $P^{D'}$, respectively.

Problem Statement. Given the upscaling factor N, the sequence of observed incomplete coarse-grained flow maps $\{\hat{X}_{cg,D}^t\}_{t\in T_D}$ from the data-scarce target city D, the sequence of relatively complete coarse-grained flow maps $\{X_{cg,D'}^{t'}\}_{t'\in T_{D'}}$ from the data-rich source city D', and the POI features P^D and $P^{D'}$ for both cities, our goal is to infer the complete fine-grained urban flow map $X_{fg,D}^t \in \mathbb{R}^{2\times NI\times NJ}$ for the target city D at time t.

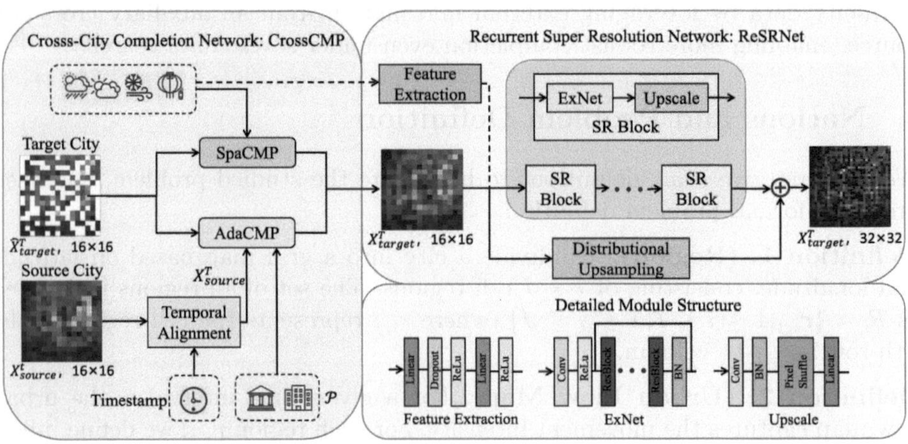

Fig. 2. Framework of the proposed CrossMF model.

4 Methodology

4.1 Cross-City Completion Network

The CrossCMP network completes missing values in coarse-grained urban flow maps of a target city by leveraging data from potentially richer source city and local spatio-temporal dynamics from the limited data in the target city, as shown in Fig. 2. Let the input be a time series $\hat{X}_{target} \in \mathbb{R}^{T \times C \times I \times J}$. To conduct data completion over the regions where the data are unavailable, we first define the mask operation as follows:

$$\mathbf{M}_{cg}(r_{i,j}) = \begin{cases} 1 & \text{if } r_{i,j} > 0 \\ 0 & \text{otherwise} \end{cases}, \tag{3}$$

where $M_{cg}()$ is a mask function,where marks the regions without observations as 0 and the regions with data as 1. CrossCMP produces a completed coarse-grained map $X_{cmp} \in \mathbb{R}^{C \times I \times J}$ through two complementary branches: Local Spatio-Temporal Completion (SpaCMP) and Adaptive Cross-City Completion (AdaCMP).

Local Spation-Temporal Completion. This module aims to capture local spatial correlations and their temporal evolution inherent in urban flow data. By modeling how flow patterns in nearby regions influence each other over time, SpaCMP can effectively complete missing values based on observed spatio-temporal context within the target city. This is achieved using a recurrent architecture employing convolutional layers to process the spatial map at each time step while propagating temporal information via a hidden state.

Let $\hat{X}_{cg}^{t}$ be the input map at time t and H^{t-1} be the hidden state from the previous step. The hidden state H^t is computed as follows:

$$H^t = \sigma(\mathcal{F}(\hat{X}_{cg}^{t}, W_x, b_x) + \mathcal{F}(H^{t-1}, W_h, b_h)). \tag{4}$$

Here, σ represents a non-linear activation function, and $\mathcal{F}$ denotes the standard 2D convolution operation with learnable weight kernels (W_x, W_h) and bias terms (b_x, b_h). The application of $\mathcal{F}$ captures local spatial dependencies. By recurrently applying Eq 4 for $t = 1, ..., T$, the network integrates information across both space and time. The final hidden state after processing all time steps, $X_{spa} = H^T$, encapsulates the learned local spatio-temporal representation and serves as the output of this branch.

Adaptive Cross-City Completion. However, SpaCMP's reliance on local spatio-temporal patterns fails under extensive target city data scarcity, potentially leaving large gaps. To ensure robust completion even with severe scarcity, we introduce the Adaptive Cross-City Completion (AdaCMP) module. AdaCMP complements SpaCMP by using Point-of-Interest (POI) guided spatial similarity and adaptively leveraging a data-rich source city when target information is insufficient. Meaningful regional similarity is needed to leverage spatial relationships with missing data; we use POI distributions for this. Since regional POI distributions reflect urban function (e.g., commercial, residential), which shapes mobility and flow patterns (volume, direction, timing), comparing POI profiles identifies functionally similar regions expected to exhibit similar flows. This provides a reliable basis for knowledge transfer, even across different cities. We first define a function $\mathrm{Sim}(r_a, r_b)$ to compute the similarity between the POI feature vectors P_{r_a} and P_{r_b} of any two regions r_a, r_b. Cosine similarity is employed:

$$\mathrm{Sim}(r_a, r_b) = \frac{P_{r_a} \cdot P_{r_b}}{||P_{r_a}||_2 ||P_{r_b}||_2}. \tag{5}$$

Adaptive Completion Logic: For each target region r_k where the data is missing ($M(r_k) = 0$), AdaCMP adaptively selects the best available information source based on POI similarity using a prioritized strategy:

Intra-city Similarity: It identifies the region r^*_{target} within the target city's observed regions $\mathcal{R}_{target} = \{r_j | M(r_j) = 1\}$ that is most similar to r_k based on POI data:

$$r^*_{target} = \arg\max_{r_j \in \mathcal{R}_{target}} \mathrm{Sim}(r_k, r_j). \tag{6}$$

If the similarity $\mathrm{Sim}(r_k, r^*_{target})$ meets or exceeds a predefined threshold θ, the flow value from this most similar observed region within the target city is used for completion:

$$X^T_{fill}(r_k) = X^T_{target}(r^*_{target}). \tag{7}$$

Cross-City Similarity: If the intra-city similarity is below the threshold θ, indicating insufficient guidance from within the target city, the module attempts to leverage information from the source city. It identifies the region r^*_{source} in the source city $\mathcal{R}_{source}$ most similar to r_k:

$$r^*_{source} = \arg\max_{r_l \in \mathcal{R}_{source}} \mathrm{Sim}(r_k, r_l). \tag{8}$$

If the source data $X_{source}(r^*_{source})$ at the corresponding location is considered valid, its value is used:

$$X^T_{fill}(r_k) = X^T_{source}(r^*_{source}). \tag{9}$$

Let $X_{fill} \in \mathbb{R}^{C \times H \times W}$ be the map containing the completed values $X_{fill}(r_k)$ for all originally missing locations. The final output of the AdaCMP branch, X_{ada}, is constructed by combining the original observed values with the newly completed values:

$$X_{ada} = M \odot \hat{X}^t_{cg} + (1 - M) \odot X_{fill}, \tag{10}$$

where $\odot$ denotes element-wise multiplication. This adaptive cross-city strategy significantly enhances completion capabilities, especially in data-scarce target regions, by intelligently borrowing information from a data-richer source city based on functional similarity.

Module Combination. The CrossCMP network integrates the complementary information captured by two branches. The output from the local spatio-temporal branch X_{spa} and the adaptive cross-city completion branch X_{ada} are fused using a learnable weight ω:

$$X_{cmp} = \omega \cdot X_{spa} + (1 - \omega) \cdot X_{ada}. \tag{11}$$

This learnable weight ω allows the model to dynamically balance the contributions from local spatio-temporal patterns and POI-based spatial similarities during end-to-end training, yielding the final conpleted coarse-grained map X_{cmp}.

4.2 Recurrent Super-Resolution Network

Feature Extraction. The ReSRNet first processes the input completed coarse map X_{cmp} to extract an initial set of features suitable for the subsequent enhancement and upscaling tasks. If optional external features E_{ext} (e.g., weather data, holidays) are provided, they are processed by a separate feature extraction sub-network, and the resulting embeddings are typically fused with the features derived from X_{cmp}. Let the output feature map after the extraction and fusion be denoted as F:

$$F^0 = \mathcal{F}_{extract}(X_{cmp}, E_{ext}), \tag{12}$$

where $\mathcal{F}_{extract}$ represents the combined operations of convolution and external feature integration.

Recurrent Super-Resolution Blocks. The core of the ReSRNet comprises a sequence of L stacked Recurrent Super-Resolution blocks. These blocks are designed to progressively refine the feature representation while simultaneously increasing the spatial resolution. Each block takes the feature map F^{t-1} from the preceding block and produces a higher-resolution feature map F^t. The total upscaling N is achieved across these stages.

Within each block, a recurrent mechanism iteratively refines the features over S steps ($s = 1, ..., S$). Let H_0^t be the initial state derived from the block input F^{t-1}. The refinement process uses a shared-parameter module $\mathcal{F}_{refine}$ based on residual connections:

$$H_s^t = \mathcal{F}_{refine}(H_{s-1}^t). \tag{13}$$

The refined features H_s^t from each internal step are then upscaled and potentially post-processed by *Upscale*. The final output F^t for stage t aggregates the information from all refinement steps via a weighted summation, allowing contributions from features at different refinement levels:

$$F^t = \sum_{s=1}^{S} \alpha_s \cdot \mathrm{Upscale}(H_s^t). \tag{14}$$

After the final stage (t=L), the resulting feature map F^L captures the high-frequency details learned through the staged, recurrent process. The final high-resolution ouput, X_{fg}, is then obtained via a global residual connection as follows:

$$X_{fg} = \mathcal{F}_{idisup}(X_{cmp}, N) + F^L, \tag{15}$$

where $\mathcal{F}_{disup(X_{cmp},N)}$ represents the completed coarse-grained map X_{cmp} after being upsampled by a factor of N using a distributional upsampling function. The residual designed facilitates the learning of these fine details, learning to the final high-resolution output X_{fg}.

4.3 Training Strategy

The proposed CrossFM framework, comprising the Cross-City Completion Network (CrossCMP) and the Recurrent Super-Resolution Network (ReSRNet), is trained in an end-to-end manner. The objective combines two pixel-wise loss terms. The completion loss $\mathcal{L}_{cmp}$ quantifies the difference between the completed coarse map output X_{cmp} and its corresponding ground truth X'_{cg}. The super-resolution loss $\mathcal{L}_{sr}$ measures the difference between the final fine-grained map X_{fg} and its ground truth X'_{fg}:

$$\mathcal{L}_{cmp} = ||X_{cmp} - X_{cg}||_F^2, \tag{16}$$

$$\mathcal{L}_{sr} = ||X_{fg} - X'_{fg}||_F^2. \tag{17}$$

The total loss $\mathcal{L}$ minimized during end-to-end training is a weighted combination of these two components:

$$\mathcal{L} = \lambda\mathcal{L}_{cmp} + \mu\mathcal{L}_{sr}, \tag{18}$$

where λ and μ are hyper-parameters balancing the contribution of each task.

5 Experiment

5.1 Datasets

We evaluate CrossFM using two real-world datasets: BJTaxi, the target city for completion and super-resolution, and NYCTaxi, the source city for cross-city information. Table 1 summarizes their specifications after preprocessing.

Table 1. Dataset Description

Dataset	BJTaxi	NYCTaxi
Longitude	(115.42, 117.51)	(-74.25, -73.70)
Latitude	(39.44, 41.06)	(40.50, 41.08)
Time Span	3/1/2015-6/30/2015	1/1/2016-6/30/2015
Time Interval	30 min	30 min
Coarse-grained Shape	$32 \times 32/64 \times 64$	$32 \times 32/64 \times 64$
Fine-grained Shape	128×128	128×128
Upscaling Factor	2/4	2/4
#POI	79063	177824

- **BJTaxi**: This dataset comprises taxi trip records from Beijing, China (March 1 - June 30, 2015), processed into coarse-grained urban flow maps. The data is split into training (70%), validation (20%), and test (10%) sets.
- **NYCTaxi**: This dataset contains taxi trip records from New York City, USA (January 1 - June 30, 2016). It serves as an external data source for cross-city insights and is not split for training/validation/testing.

5.2 Experimental Settings

Evaluation Metrics. We use Mean Absolute Error (MAE) and Root Mean Square Error (RMSE) to evaluate the inference performance.

Baselines. We compare CrossFM against several baseline methods. These include straightforward statistical approaches like Mean partition (**Mean**), which evenly distributes coarse-grained flow, and Historical Average (**HA**), which applies this distribution based on historical averages. We also consider advanced deep learning models for Fine-grained Urban Flow Inference (FUFI). Specifically, **UrbanFM** [12] utilizes distributional upsampling and fuses external factors. **UrbanPy** [16] employs a pyramid architecture with multiple components for upsampling and refinement. **UrbanSTC** [18] applies contrastive self-supervised learning for efficient pre-training, especially in low-resource scenarios. The most recent state-of-the-art model, **MT-CSR** [11], tackles FUFI with potentially incomplete coarse data through multi-task learning for simultaneous completion and super-resolution.

Table 2. Model Performance on BJTaxi Dataset (Best results bold, second best underlined)

Model				Mean	HA	UrbanFM	UrbanPy	UrbanSTC	MT-CSR	CrossFM	Improve
BJTaxi	2	40%	MAE	9.50	9.30	7.50	7.23	**5.99**	6.10	6.05	-1.0%
			RMSE	26.60	26.04	21.02	20.24	**17.10**	17.40	17.25	-0.9%
		60%	MAE	10.51	10.30	8.22	8.00	6.90	6.85	**6.45**	5.8%
			RMSE	29.43	28.84	22.96	22.42	19.60	19.45	**19.01**	2.3%
		80%	MAE	11.80	10.67	9.30	9.12	8.05	8.01	**7.12**	11.1%
			RMSE	33.04	28.81	26.04	25.48	22.70	22.59	**20.32**	10.1%
	4	40%	MAE	10.88	10.67	8.50	8.27	7.02	7.10	**7.01**	0.1%
			RMSE	29.38	28.81	22.86	20.99	20.56	19.21	**19.19**	0.1%
		60%	MAE	12.31	12.07	9.50	9.85	8.05	7.99	**7.56**	5.4%
			RMSE	33.24	32.59	25.45	23.73	21.95	21.83	**19.95**	8.6%
		80%	MAE	13.92	13.66	10.55	11.41	9.35	9.29	**7.85**	15.5%
			RMSE	37.58	36.88	28.14	27.19	25.00	24.81	**20.33**	18.1%

5.3 Performance Comparison

As Table 2 shows, we can observe that most methods degrade as missing data increases. Simple baselines (Mean, HA) perform poorly, unable to model complex spatio-temporal dynamics. Sophisticated methods like UrbanFM and UrbanPy, while better, still falter, potentially lacking inherent mechanisms for large missing blocks without specific completion modules or retraining. Models considering spatio-temporal correlations or advanced architectures (e.g., UrbanSTC, MT-CSR) handle missing data better than simpler methods. MT-CSR, designed for the same joint task, is a strong competitor but consistently outperformed by CrossFM, especially as scarcity increases. This suggests CrossFM's adaptive cross-city completion strategy offers benefits over MT-CSR's auxiliary completion under severe scarcity. UrbanSTC is also competitive but surpassed by CrossFM, indicating CrossFM's targeted cross-city knowledge transfer provides an edge beyond capturing local spatio-temporal patterns alone.

Amidst these trends, CrossFM consistently achieves the best performance across all evaluated scenarios. More importantly, while baseline accuracy declines sharply with increasing data scarcity, CrossFM demonstrates significantly greater robustness via a much more gradual performance decrease. For instance, comparing results at 80% versus 40% missing data, the increase for CrossFM is considerably less pronounced than for most baseline methods.

CrossFM's robustness stems from its unique design. The key is the AdaCMP module, which mitigates data scarcity by adaptively borrowing information from a source city based on POI similarity. Furthermore, the joint end-to-end training of the completion (CrossCMP) and super-resolution (ReSRNet) modules creates a powerful synergy, enhancing overall accuracy and outperforming less integrated approaches.

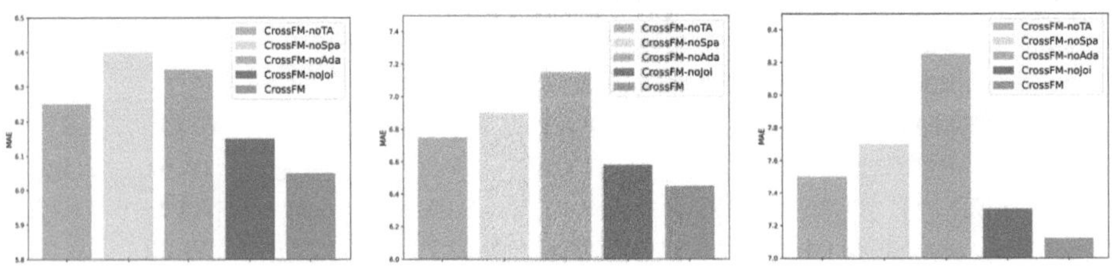

(a) mr=40%, N=2 (b) mr=60%, N=2 (c) mr=80%, N=2

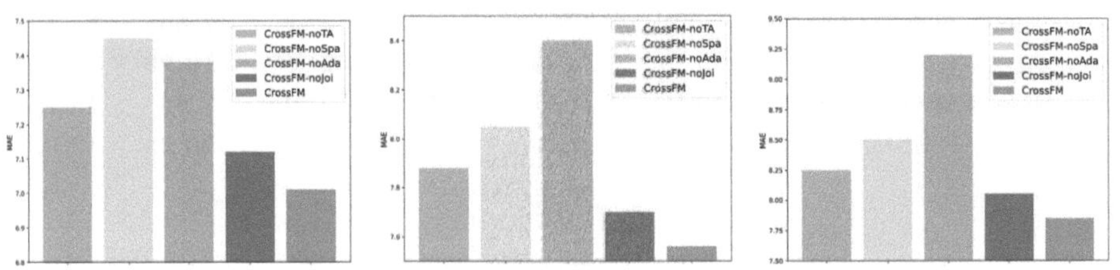

(d) mr=40%, N=4 (e) mr=60%, N=4 (f) mr=80%, N=4

Fig. 3. Ablation study of CrossFM on BJTaxi. 'mr' means 'missing rate'

5.4 Ablation Study

We compare the full CrossFM model against variants where specific components are ablated: **CrossFM-noSpa**. This variant removes the local Spatio-temporal completion module. **CrossFM-noAda**. Removing the Adaptive Cross-City completion. **CrossFM-noTA** This variant removes Time Alignment. **CrossFM-noJoi**. This variant uses separate training instead of joint optimization. The performance under different missing rates (mr) and scale factor (N) is illustrated in Fig. 3. The results consistently demonstrate that CrossFM achieves the best performance across all variants, indicating that all ablated components contribute positively to the final performance. Among the variants, CrossFM-noAda results in the most significant performance degradation, yielding the highest MAE in all depicted cases. Conversely, CrossFM-noJoi shows the smallest performance drop compared to the full model, suggesting that while joint end-to-end optimization provides benefits, the core components function effectively even when trained sequentially. The CrossFM-noTA and CrossFM-noSpa also led to performance decreases, further confirming their positive contributions to the framework. These findings collectively underscore the effectiveness of each component.

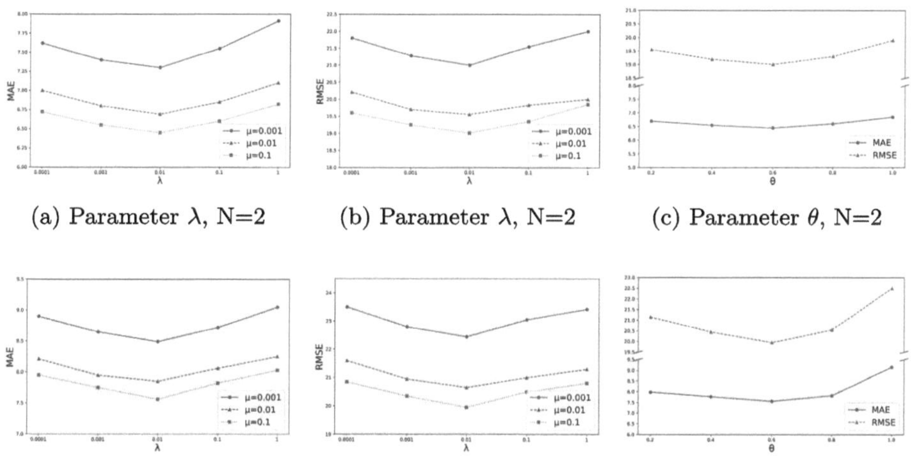

(a) Parameter λ, N=2 (b) Parameter λ, N=2 (c) Parameter θ, N=2

(d) Parameter λ, N=4 (e) Parameter λ, N=4 (f) Parameter θ, N=4

Fig. 4. Parameter study of CrossFM on BJTaxi. Missing rate = 60%

5.5 Parameter Analysis

We analyze the impact of the completion loss weight λ in the final objective function (Eq. 18) and the POI similarity threshold θ used within the AdaCMP module (described in Sect. 4). Experiments were conducted under a 60% missing rate setting, and the results are summarized in Fig. 4. For the loss weights, we varied λ within the range 0.0001, 0.001, 0.01, 0.1 and μ in the range 0.001,0.01,0.1. As observed, performance initially improves as λ increases from very low values, but degrades slightly when λ is set equal to μ (0.1), suggesting a slight emphasis on the super-resolution task yields better overall results. For the POI similarity threshold θ, we tested values in 0.2, 0.4, 0.6, 0.8, 1.0. The results indicate that performance suffers when the threshold is too low (potentially accepting less relevant intra-city matches) or too high (under-utilizing intra-city information or forcing reliance on cross-city data, with $\theta = 1.0$ performing worst). Based on these empirical results, we determined the optimal settings for our experiments to be $\lambda = 0.01$ alongside $\mu = 0.1$ and $\theta = 0.6$, which achieved the best performance.

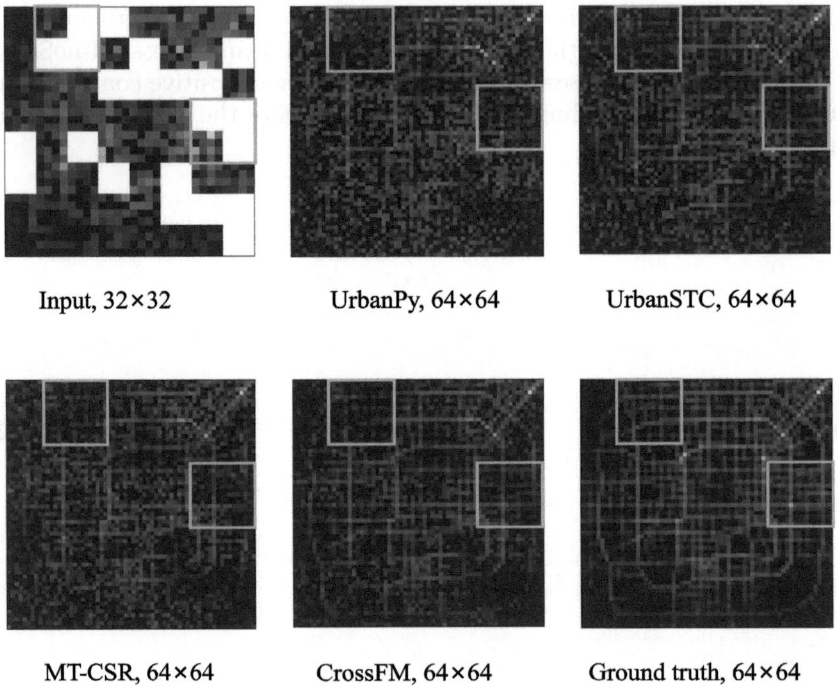

Fig. 5. Visualization of the urban flows inference with different methods on the BJTaxi. The cell regions in white color denote that the urban flow data are unavailable

5.6 Visualization

To further intuitively demonstrate the model performance, we visualize the fine-grained urban flow inference results generated by different methods alongside the ground truth. Figure 5 compares visualized heat maps (BJTaxi dataset,

32x32 input to 64x64 output, 40% missing rate) from CrossFM against baselines UrbanPy, UrbanSTC, and MT-CSR, alongside the ground truth. Visual inspection reveals that while baselines capture general patterns, UrbanPy lacks sharpness, UrbanSTC misses fine-grained hotspots, and MT-CSR shows minor inaccuracies, particularly in completed regions. In contrast, CrossFM generates maps visually closest to the ground truth, accurately reconstructing both sharp details and high-flow hotspots. This visual superiority confirms CrossFM's effectiveness in generating high-fidelity fine-grained urban flow maps.

6 Conclusion

We proposed CrossFM, a framework leveraging cross-city dynamics to infer fine-grained urban flow from sparse observations. CrossFM integrates a POI-guided cross-city completion network (CrossCMP) and a recurrent super-resolution network (ReSRNet) via end-to-end training. Experiments show CrossFM outperforms state-of-the-art methods. While our current temporal alignment is simple, future work will focus on developing more sophisticated techniques to further improve inference accuracy in complex urban environments.

Acknowledgments. This work is supported by the Natural Science Foundation of China No. 62472196, Jilin Science and Technology Research Project 202301 01067JC.

References

1. Barnes, C., Shechtman, E., Finkelstein, A., Goldman, D.B.: Patchmatch: a randomized correspondence algorithm for structural image editing. ACM Trans. Graph. **28**(3), 24 (2009)
2. Criminisi, A., Pérez, P., Toyama, K.: Region filling and object removal by exemplar-based image inpainting. IEEE Trans. Image Process. **13**(9), 1200–1212 (2004)
3. Dong, C., Loy, C.C., He, K., Tang, X.: Image super-resolution using deep convolutional networks. IEEE Trans. Pattern Anal. Mach. Intell. **38**(2), 295–307 (2015)
4. Gao, H., Su, Y., Yang, F., Yang, Y.: Fine-grained data inference via incomplete multi-granularity data. In: Proceedings of the ACM on Web Conference 2025, pp. 3377–3388 (2025)
5. Guo, W., Zhuang, F., Zhang, X., Tong, Y., Dong, J.: A comprehensive survey of federated transfer learning: challenges, methods and applications. Front. Comp. Sci. **18**(6), 186356 (2024)
6. He, K., Sun, J.: Image completion approaches using the statistics of similar patches. IEEE Trans. Pattern Anal. Mach. Intell. **36**(12), 2423–2435 (2014)
7. Hong, Y., et al.: Stkopt: automated spatio-temporal knowledge optimization for traffic prediction. In: Proceedings of the ACM on Web Conference 2025, pp. 2238–2249 (2025)
8. Ji, J., et al.: Spatio-temporal self-supervised learning for traffic flow prediction. In: Proceedings of the AAAI conference on artificial intelligence, vol. 37, pp. 4356–4364 (2023)

9. Jiang, Y., Yang, Y., Xu, Y., Wang, E.: Spatial-temporal interval aware individual future trajectory prediction. IEEE Trans. Knowl. Data Eng. **36**(10), 5374–5387 (2024). https://doi.org/10.1109/TKDE.2023.3332929
10. Kim, J., Lee, J.K., Lee, K.M.: Accurate image super-resolution using very deep convolutional networks. In: Proceedings of the IEEE conference on computer vision and pattern recognition, pp. 1646–1654 (2016)
11. Li, J., Wang, S., Zhang, J., Miao, H., Zhang, J., Yu, P.S.: Fine-grained urban flow inference with incomplete data. IEEE Trans. Knowl. Data Eng. **35**(6), 5851–5864 (2023)
12. Liang, Y., et al.: Urbanfm: inferring fine-grained urban flows. In: Proceedings of the 25th ACM SIGKDD international conference on knowledge discovery and data mining, pp. 3132–3142 (2019)
13. Liu, G., Reda, F.A., Shih, K.J., Wang, T.C., Tao, A., Catanzaro, B.: Image inpainting for irregular holes using partial convolutions. In: Proceedings of the European Conference on Computer Vision (ECCV), pp. 85–100 (2018)
14. Luo, H., Zhuang, F., Xie, R., Zhu, H., Wang, D., An, Z., Xu, Y.: A survey on causal inference for recommendation. Innovation **5**(2) (2024)
15. Nazeri, K., Ng, E., Joseph, T., Qureshi, F., Ebrahimi, M.: Edgeconnect: structure guided image inpainting using edge prediction. In: Proceedings of the IEEE/CVF International Conference on Computer Vision Workshops, pp. 0–0 (2019)
16. Ouyang, K., et al.: Fine-grained urban flow inference. IEEE Trans. Knowl. Data Eng. **34**(6), 2755–2770 (2020)
17. Pathak, D., Krahenbuhl, P., Donahue, J., Darrell, T., Efros, A.A.: Context encoders: feature learning by inpainting. In: Proceedings of the IEEE Conference on Computer Vision and Pattern Recognition, pp. 2536–2544 (2016)
18. Qu, H., Gong, Y., Chen, M., Zhang, J., Zheng, Y., Yin, Y.: Forecasting fine-grained urban flows via spatio-temporal contrastive self-supervision. IEEE Trans. Knowl. Data Eng. **35**(8), 8008–8023 (2022)
19. Wang, A., Ye, Y., Song, X., Zhang, S., Yu, J.J.: Traffic prediction with missing data: a multi-task learning approach. IEEE Trans. Intell. Transp. Syst. **24**(4), 4189–4202 (2023)
20. Wang, S., Cao, J., Philip, S.Y.: Deep learning for spatio-temporal data mining: a survey. IEEE Trans. Knowl. Data Eng. **34**(8), 3681–3700 (2020)
21. Wang, Z., Chen, J., Hoi, S.C.: Deep learning for image super-resolution: a survey. IEEE Trans. Pattern Anal. Mach. Intell. **43**(10), 3365–3387 (2020)
22. Xu, Y., Wang, E., Yang, Y., Chang, Y.: A unified collaborative representation learning for neural-network based recommender systems. IEEE Trans. Knowl. Data Eng. **34**(11), 5126–5139 (2022). https://doi.org/10.1109/TKDE.2021.3054782
23. Zheng, Y., et al.: Diffuflow: robust fine-grained urban flow inference with denoising diffusion model. In: Proceedings of the 32nd ACM International Conference on Information and Knowledge Management, pp. 3505–3513 (2023)

Make LLMs Perform Better in Knowledge Graph Completion Combined with RAG

Mengfei Xu[1], Bohan Li[1,2,3(✉)], Haofen Wang[4], Peixuan Huang[1], Chen Chen[1], and Ruilong Huang[1]

[1] College of Artificial Intelligence and Computer Science and Technology, Nanjing University of Aeronautics and Astronautics, Nanjing, China
{jensonxu,bhli}@nuaa.edu.cn

[2] Ministry of Industry and Information Technology, Collaborative Innovation Center of Novel Software Technology and Industrialization, Nanjing, China

[3] National Engineering Laboratory for Integrated Aero-Space-Ground Ocean Big Data Application Technology, Xi'an, China

[4] College of Design and Innovation, Tongji University, Shanghai, China

Abstract. Knowledge Graphs (KGs) face challenges of incompleteness, driving the requirements of Knowledge Graph Completion (KGC). The development of Large Language Models (LLMs) provides a new perspective for KGC research. Several methods instruct LLMs to conduct KGC by Prompt Engineering. However, they struggle with missing entity/relation descriptions, "text mismatch" between LLMs' responses and entities in KG, and insufficient utilization of structural information in KG. Notably, Retrieval-Augmented Generation (RAG) has emerged as a powerful paradigm, demonstrating promising potential in enhancing LLM performance across diverse tasks by grounding generation in retrieved knowledge. To capitalize on this potential and address the aforementioned KGC-specific limitations, this paper proposes a novel RAG paradigm named Ret-Gen specifically designed for KGC. It retrieves the corresponding subgraph for each query, and then generates additional knowledge to enhance LLMs' comprehension of query. Based on this paradigm, we propose a muti-stage method named RAG-KGC. It first constructs a subgraph related to query. Then, it generates abundant additional knowledge based on the query and its subgraph to enhance LLMs' cognition, thereby enabling better inference. Finally, it conducts similarity matching between LLMs' responses and the entities in knowledge graph to obtain candidate entities, with the aim to solve "text mismatch" problem. We conduct comprehensive experiments on the link prediction task using two benchmark datasets, FB15K-237-N and Wiki27K, which proved the effectiveness of RAG-KGC.

Keywords: Knowledge Graphs · Knowledge Graph Completion · Large Language Models · Retrieval Augmented Generation

M. Yoshikawa et al. (Eds.): ADMA 2025, LNAI 16197, pp. 175–190, 2026.
https://doi.org/10.1007/978-981-95-3453-1_12

1 Introduction

Knowledge Graph (KG) is a collection of massive facts, which are in the form of triples, represented as (*head entity, relation, tail entity*). KGs have been widely applied in various fields such as search [10], question answering [27], recommendation systems [22,29]. However, KGs are generally sparsely structured and incomplete [20], which motivate Knowledge Graph Completion (KGC), with the aim to predict incomplete triples in KG.

Existing KGC methods can be divided into two categories: methods based on embedding and methods based on pre-trained language models (PLM). In recent years, large language models (LLMs) demonstrate remarkable performance in several natural language processing tasks [3], which provide novel perspectives for KGC. Some methods [13,26,30] employ unified prompt templates to instruct LLMs to complete KGC. Despite demonstrating promising results, they still face several limitations. First, existing KGs generally lack textual descriptions of entities/relations, and methods mentioned above are constrained in dealing with these entities/relations, which leads to LLMs' hallucination [28]. Subsequently, the entities predicted by LLMs may exhibit semantic equivalence with existing entities in KG despite differing in literal expression. For example, the KG stores a canonical entity name such as "United States of America," whereas the LLM might output its common abbreviation, "U.S.A." Although these expressions denote the same real-world entity ("U.S.A." being a standard alias for "United States of America"), existing methods often lack robust entity alignment mechanisms, leading to incorrect classification of such predictions as erroneous. This misjudgment impacts model performance in KGC tasks, resulting in unnecessary accuracy degradation. Last but not least, conventional methods oversimplify the structural integration by converting KG queries into natural language questions while neglecting topological information, which results in diminishing LLMs' contextual awareness of entities and generate inaccurate responses. Therefore, it is necessary to consider KG data as additional knowledge to enhance LLMs' comprehension. However, conventional methods compute on the entire KG, which is inapplicable for LLMs because they impose limits on the length of input tokens [19].

Retrieval Augmented Generation (RAG) has been widely applied in tasks driven by LLMs, which retrieves relevant knowledge and integrates it into the prompt, enabling LLMs to consult corresponding knowledge and provide reasonable responses [12]. In this paper, we attempt to introduce RAG into the LLMS-based KGC, and propose a new KGC paradigm named **Ret-Gen**, which is a foundational and general framework applicable to LLMs' inference in KGC. Ret-Gen mainly consists of two parts: Graph-based Retriever and LLMs-based Generator. Graph-based Retriever is designed to retrieve and construct a subgraph relevant to query. LLMs-based Generator instructs LLMs to reason and generate answer based on the subgraph. Ret-Gen works based on the fact that only a small subgraph of the entire KG may be relevant for answering the query. In contrast to traditional KGC paradigm, Ret-Gen performs reasoning on subgraph, which can not only reduce computational cost [24], but also avoid length limits of input tokens that LLMs impose. Furthermore, the retriever and the generator are mutually independent, with lower degree of coupling, which means greater flexibility and expandability.

Furthermore, in this paper, we propose a novel method for LLMs-based KGC named **RAG-KGC**, which is considered as an implementation of Ret-Gen. It is a multi-stage framework for KGC that incorporates Retrieval, Generation, and Matching. Specifically, RAG-KGC extracts the k-hop subgraph of the entities related to the query, which constructs context of entity for inference. Then, we design a multi-perspective prompt strategy based on chain-of-thought (CoT) [18] in RAG-KGC to generate additional knowledge for LLMs. The generation is partitioned into two parts: subgraph and query. For subgraph, we introduce a KG-to-text generation task, which generates natural language text relevant to the subgraph. The textualization of structured knowledge can construct contexts of the query, thus reinforcing LLMs' cognition of entities. For query, we instruct LLMs to search for information related to the entity and relation in query, which can enormous knowledge base in LLMs to obtain additional knowledge. Finally, Similarity Matching is designed to address the "text mismatch" problem. It computes the similarity score between LLMs' response and the names of entities in KG to achieves the alignment. After that, RAG-KGC get a sorted list that represents candidate entities.

In conclusion, the main contributions of our work are as follows:

1. We conduct a preliminary investigation into KGC research utilizing RAG, and propose a foundational and general paradigm named Ret-Gen. The paradigm introduces RAG into LLMs-based KGC with the aim to enable LLMs to complete KGC better.
2. We propose RAG-KGC, a new LLMs-based KGC method, which integrates three stages of subgraph retrieval, knowledge generation, and similarity matching. It can fully exploit the rich knowledge base in LLMs to enhance the reasoning performance.
3. We conducted experiments on two benchmark datasets: FB15K-237N and Wiki27K. The results demonstrate that RAG-KGC achieves superior performance in several evaluation metrics. The code will be avaliable at https://github.com/JensonXu-NUAA/RAG-KGC.git.

2 Related Work

2.1 Knowledge Graph Completion

Knowledge Graph Completion (KGC) has emerged as a fundamental task in knowledge representation learning, aiming to infer missing facts in incomplete knowledge graphs. The methodological landscape of KGC has evolved through three distinct paradigms, each leveraging different computational approaches and representational strategies.

Embedding-based methods constitute the foundational approach, wherein entities and relations are embedded into continuous vector spaces to facilitate reasoning through geometric operations. TransE [2] pioneered this direction by modeling relations as translations in embedding space, while RotatE [23] and TuckER [1] refined this paradigm through complex rotations and tensor factorization techniques, respectively. However, these methods suffer from limited capacity to incorporate textual semantics.

Pre-trained Language Model (PLM) based methods represent a paradigm shift toward leveraging linguistic understanding by reformulating KGC as text-based tasks. KG-BERT [25] employs BERT to encode triple representations, while PKGC [11] advances this through carefully designed prompt templates. KG-ICL [4] designs a prompt-based knowledge graph foundation model with the aim to address existing methods lacking the ability to generalize and transfer knowledge across diverse KGs and reasoning settings.

LLMs are pre-trained on large-scale corpus and have stronger semantic comprehension and capability of generation than PLMs. Recent researches begin to explore the utilization of LLMs for KGC. KGLLaMA [26] employs instruction fine-tuning with entity descriptions as prompts, KG-LLM [13] converts structured knowledge into natural language for multi-hop prediction, KICGPT [19] combines embedding-based retrievers with LLM generators, and KC-GenRe [17] advances through sorting generation and constrained inference mechanisms.

In this paper, we focus on methods that instruct LLMs to conduct KGC [13, 26]. Despite these methods demonstrate promising results in KGC using LLMs, they still difficulties caused by the absence of textual descriptions for entities and relations, text mismatch between LLM-generated outputs and KG entries, and insufficient utilization of KG topological structure.

2.2 RAG-Based Knowledge Graph Research

Compared to traditional text-based RAG approaches, the integration of RAG with knowledge graphs can provide more precise and contextually relevant information retrieval compared to traditional text-based RAG approaches, leveraging graph-structured knowledge to enable sophisticated reasoning capabilities over interconnected entities and relations [5].

Recent methodologies incorporate structured knowledge graphs as primary retrieval sources. Think-on-Graph (ToG) [14] lets LLMs as agents to iteratively explore knowledge graph structures through beam search, enabling dynamic multi-hop reasoning while maintaining knowledge traceability. Paths-over-Graph (PoG) [15] employs multi-hop path extraction with graph-structured pruning for enhanced reasoning efficiency. MultiRAG [21] enhances graph retrieval performance and the accuracy of multi-hop question answering by aggregating multi-source knowledge retrieval, reconstructing knowledge subgraphs into a denser structure, and conducting certain confidence tests.

In this paper, we introduce RAG into LLMs-based KGC research, aiming to address the limitations of existing approaches and enhancing the comprehension and completion capabilities of LLMs.

3 Problem Formulation

Knowledge Graph is a directed graph data network, which can be expressed as a set of triples $G = (V, E) = \{(h, r, t)\}$, where $h \in V$, $r \in E$ and $t \in V$ indicates head entity, tail entity and relation respectively. V and E respectively represent sets of entities and relationships in knowledge graph. Link prediction is an important task of KGC. Specifically, given an incomplete triple $(h, r, ?)$ or $(?, r, t)$, the link prediction model infers the missing entity (denoted as ?) based on the known information in knowledge graph.

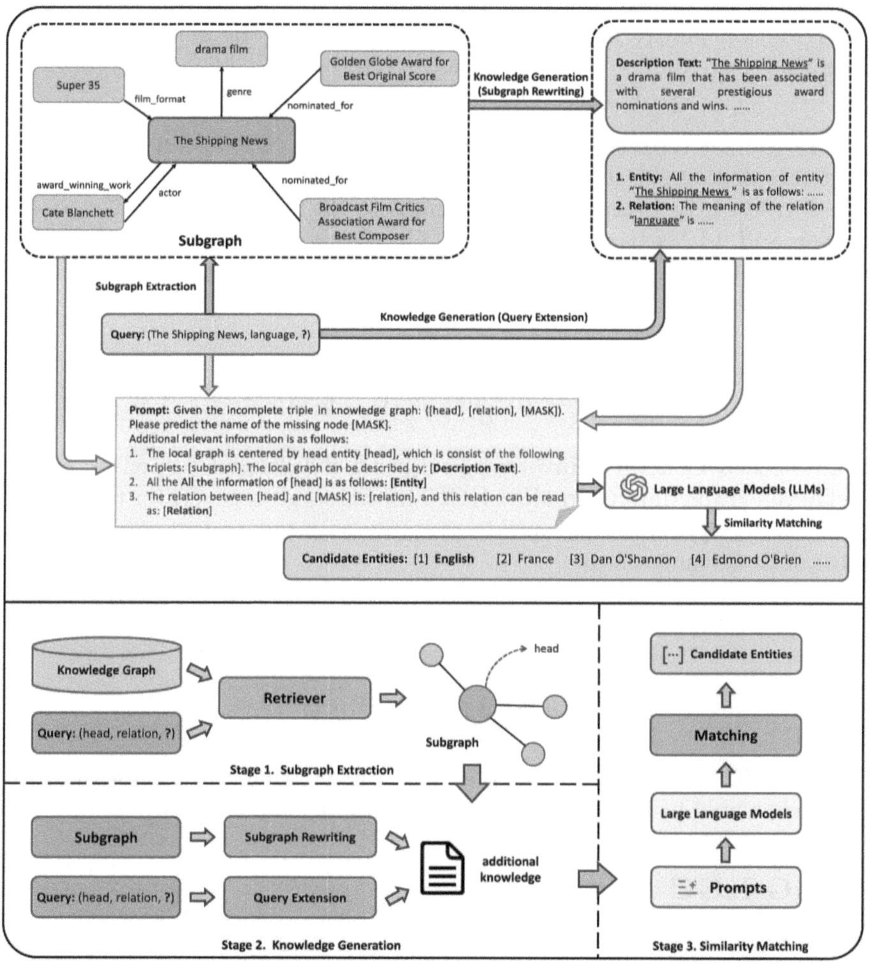

Fig. 1. The framework of RAG-KGC. It consists of three stages: subgraph retrieval, knowledge generation and similarity matching. An example (*The Shipping News*, *language*, ?) is illustrated to demonstrate its pipeline.

4 Ret-Gen Paradigm

In this section, we introduce Ret-Gen, which is applicable to LLMs' inference in KGC. Specifically, given an input query, the most relevant knowledge is retrieved from KG. Specifically, prompts are generated based on the retrieved knowledge and fed into LLMs to generate answers. The process can be defined as

$$a = arg\max_{a \in A} p\left(a \mid q, G\right), \tag{1}$$

where q represents the query input by the user, and a represents the predicted answer based on the given knowledge graph G. $p\left(a \mid q, G\right)$ represents the target distribution jointly modeled by the retriever and the generator, it can be defined as

$$p\left(a \mid q, G\right) = p_g\left(a \mid q, G^*\right) p_r\left(G^* \mid q, G\right), \tag{2}$$

where G^* represents the subgraph, and $p_g\left(a \mid q, G^*\right)$ and $p_r\left(G^* \mid q, G\right)$ respectively denote the retriever and the generator.

We introduce two components to complete the process above: Graph-based Retriever and LLMs-based Generator. Graph-based Retriever is designed to extract subgraph relevant to query from KG, providing data for the generation of answers. It can be formulated as

$$\begin{aligned} G^* &= Retriever_{KG}\left(q, G\right) \\ &= arg\max_{G^* \in G} p_r\left(G^* \mid q, G\right). \end{aligned} \tag{3}$$

LLMs-based Generator involves generating meaningful responses based on the retrieved graph data to conduct KGC. In this stage, a generator takes queries, retrieved subgraph, and appropriate prompts as input to generate a response, which can be formulated as

$$\begin{aligned} a^* &= Generator_{LLM}\left(q, G^*\right) \\ &= arg\max_{a \in A} p_g\left(a \mid q, G^*\right) \\ &= arg\max_{a \in A} p_g\left(a \mid F\left(q, G^*\right)\right), \end{aligned} \tag{4}$$

where, $F\left(q, G^*\right)$ represents a function that converts subgraph into a form that is comprehensible to LLMs, and A is the set of all possible answers.

5 RAG-KGC

5.1 Overview

In this section, we introduce RAG-KGC, which is an implementation of Ret-Gen. Figure 1 presents the overall framework of RAG-KGC, which consists mainly of three stages: subgraph retrieval, knowledge generation, and similarity matching. For subgraph retrieval, RAG-KGC extract the subgraph relevant to query. For knowledge generation, RAG-KGC generates additional knowledge to enhance

LLMs' comprehension of entities and relations. For similarity matching, RAG-KGC computes the similarity score between the responses of LLMs and the entities in KG, and ultimately outputs a list of candidate entities. In this section, we introduce the details of the implementation.

5.2 Subgraph Retrieval

Given a query triple $q : (h, r, ?)$, a retriever is designed to extract the subgraph centered on h from KG. Specifically, we use the Breadth-First Search (BFS) algorithm to traverse knowledge graph, and obtain a node set and a relation set by searching for the k-hop neighborhood centered on h, which can be formulated as

$$N_k(h) = \begin{cases} h & , k = 0 \\ N_k(h) \cup \{v \in V \mid \exists u \in N_{k-1}, r_{u,v} \in E \vee r_{v,u} \in E\} & , k > 0 \end{cases} \tag{5}$$

$$R_k(h) = \{r \in E \mid \exists u, v \in N_k(h), r_{u,v} \in E \vee r_{v,u} \in E\}, \tag{6}$$

where $N_k(h)$ and $R_k(h)$ respectively represent the node set and the relation set. Each element in $N_k(h)$ is associated with elements in $R_k(h)$ through their respective connections in the k-hop neighborhood. Therefore, the subgraph $G^*(h)$ can be represented as the Cartesian product of $N_k(h)$ and $R_k(h)$, which can be formulated as

$$G^*(h) = N_k(h) \times R_k(h). \tag{7}$$

Additionally, RAG-KGC marks each node n in $N_k(h)$ with $d(n, h)$, which represents the distance between node n and the central node h. Moreover, if $d(n, h)$ is positive, it indicates that the direction of the edge is from h to n. Otherwise, it indicates that the direction of the edge is from n to h. This operation can represent the position of each node in relation to the central node, thus better reflecting the topological structure of the subgraph.

5.3 Knowledge Generation

In our methodology, we design a multi-perspective prompt strategy based on chain-of-thought (CoT) for Knowledge Generation to guide LLMs. CoT reasoning involves providing LLMs with a series of intermediate reasoning steps or questions that help break down the problem and lead to generate knowledge for improvement on LLMs' comprehension of query. The strategy comprises two components: Subgraph Rewriting and Query Expansion.

Subgraph Rewriting. The core of Subgraph Rewriting is to transform structured triples into natural language text. Given a subgraph $G^*(h)$, the objective is to generate text sequences $X_{G^*} = (x_1, x_2, ..., x_n)$. First, we preprocess these structured triples. Specifically, combining existing research [9], we linearize the subgraph and convert it into a sequence:

$$G^*_{linear} = [<h> v_1, <h> e_{12}, <t> v_2, <r> e_{23}, <t> v_3, ..., <t> v_n] \tag{8}$$

where <h>, <r>, <t> are employed to identify entities and relations in graph, facilitating LLMs' better comprehension of the inter-entity relations and topological information in graph. Subsequently, we fine-tune LLMs and instruct them to transform the sequence G^{*}_{linear} into text via prompt template: "*Now you have a set of knowledge graph triples centered on {entity name}: {structured triples sequence* G^{*}_{linear} *}, please convert this knowledge graph into a natural language text description.*". While performing KGC, each retrieved subgraph is linearized a sequence and then transformed into textual descriptions via the designed template. The text, as query-relevant knowledge, provides entity's context for LLMs' inference in KGC to enhance the performance.

Query Extension. It is designed to enhance LLMs' understanding of entities and relations in query, by guiding LLMs to generate additional knowledge. Specifically, given a specific query $q : (h, r, ?)$, RAG-KGC employs a chain-of-thought (CoT) prompt enhancement strategy to direct LLMs to actively search and generate information relevant to entity h and relation r. In this section, we introduce the details of the implementation.

Semantic Alignment. Most existing knowledge graphs typically lack textual descriptions of entities and relations, which might result in incorrect understanding by LLMs. To cope with it, we follow PKGC [11] and design Semantic Alignment. Specifically, we provided each entity with a text segment for description. For instance, the entity *"Mutiny on the Bounty"* was defined as *"1962 film by Carol Reed, Lewis Milestone"*. For relations, we converted raw text into clearer descriptions. For example, the relation *"/people/ethnicity/people"* is replaced by *"The ethnic group of* [T] *is* [H]*."*, where [H] and [T] respectively represent placeholders for the head entity and the tail entity. In this way, we can simply establish clear and aligned textual descriptions for each entity and relation, thereby reducing the possibility of LLMs' incorrect comprehension.

Entity. RAG-KGC utilizes a unified prompt template to guide LLMs in actively searching for information relevant to entity h. We instruct LLMs to search for the information of entities from multiple perspectives in combination with CoT prompt strategy. Specifically, we initially describe the task for LLMs, then instruct LLMs to deduce the type of entity h based on its definition, and then instruct LLMs to acquire additional knowledge of h from various perspectives based on the type. For example, LLMs first determine that the entity named *"The Shipping News"* is of the type "movie", and then it will obtain additional knowledge about this film from different prospects such as language, release date, director, and awards received.

Relation. Similar to the previous practices, RAG-KGC employs CoT-based prompt strategy to enhance LLMs' comprehension of relations. Specifically, it first replaces the original text with the clear descriptions obtained in Semantic Alignment, and subsequently comprehends the relation from both global and local dimensions. In global, RAG-KGC instructs LLMs to deduce the significance of a relation, thereby inferring the domain involved in the relation. For example, the relation *"award"* can be associated with the awarded prize, and the relation *"actor"* may be related to film productions. In local, RAG-KGC guides LLMs to predict the type of the missing entity ? based on the previous understanding of the relation. For example, in the query *"(The Shipping News, actor, ?)"*, the previously inferred contents are as follows: 1) The entity *"The Shipping News"* is a film; 2) the relation *"actor"* indicates that this triple is relevant to film products. Based on the aforementioned information, LLMs deduce that the tail entity *"?"* might be relevant to a person.

All the content of prompts is not presented here due to the extent of paper, you can find it in our code.

5.4 Similarity Matching

In order to address the problem of "text mismatch", Similarity Matching is specifically designed to perform entity alignment between LLM-generated responses and the entities in KG. First, with the aim to capture semantic relevance between texts, we employ BERT model to encode contextualized word embeddings of both LLMs' responses and names of all entities' names in KG. Formally, let $\mathbf{e}_{pre} \in \mathbb{R}^d$ denotes the d-dimensional embedding vector of the LLMs' responses text, and $\mathbf{E}_{kg} = \left\{\mathbf{e}_{kg}^1, \mathbf{e}_{kg}^2, ..., \mathbf{e}_{kg}^n\right\}$ represent a set of the word embeddings of all entities in KG. Subsequently, we use cosine similarity to compute semantic similarity scores, which measures the angular distance between $\mathbf{e}_{pre}$ and $\mathbf{e}_{kg} \in \mathbf{E}_{kg}$. The calculation process can be formulated as

$$similarity\left(\mathbf{e}_{pre}, \mathbf{e}_{kg}^i\right) = \frac{\mathbf{e}_{pre} \cdot \mathbf{e}_{kg}^i}{\left\|\mathbf{e}_{pre}\right\| \cdot \left\|\mathbf{e}_{kg}^i\right\|}, \tag{9}$$

where $\mathbf{e}_{kg}^i \in \mathbb{R}^d$ represents the embedding of a single entity in KG.

After calculation, a sorted list of entities as the output of RAG-KGC is obtained ultimately by ranking the candidate entities based on the similarity scores.

5.5 Complexity Analysis

In this section, we conduct complexity analysis on each component of RAG-KGC. In Subgraph Retrieval, We adopt the BFS algorithm to extract the k-hop subgraph, so that the time complexity is $O(d^k)$, where d represents the average degree of nodes in subgraph, k represents the hop number. In Knowledge Generation, the time complexity is $O(M \cdot L)$, where M represents the number of knowledge generation steps performed by LLMs, and L denotes the length of the input token. In Similarity Matching, the time complexity of BERT's computation for encoding entities in KG is $O(e \cdot L_e)$, where e represents the number of entities in KG, L_e represents the length of text. The time complexity of encoding llms response is $O(L_r)$. Each query requires calculating the similarity between the LLMs' response and all entities in KG, where the time complexity is $O(e \cdot d)$, d represents the embedding dimension of BERT. Therefore, The overall time complexity of RAG-KGC is $O(d^k + (N + R) + M \cdot L + E \cdot d)$.

Table 1. Statistics of FB15K-237-N and Wiki27K.

Dataset	Entities	Relations	Train	Vaild	Test
FB15K-237-N	13104	93	87282	7041	8226
Wiki27K	27122	62	74793	20242	20244

6 Experiment

6.1 Settings

Datasets. We evaluate our method on two benchmark datasets, FB15k-237-N and Wiki27K. FB15k-237-N [11] is derived from FB15K-237 [16], which is more in line with the KGC tasks in real scenarios. Wiki27K [11] is a dataset for KGC based on Wikidata. The statistics of these datasets are presented in Table 1.

Baselines. We compare RAG-KGC with various existing KGC methods, which can be classified into Embedding-based, PLM-based, and LLMs-based methods. Embedding-based methods include TransE [2], TuckER [1], and RotatE [23]. PLM-based methods include KG-BERT [25], LP-RP-RR [7] and PKGC [11]. LLMs-based methods includes KG-LLaMA [26], ChatGPT [30] which is divided into zero-shot and one-shot, KG-S2S-CD [8], CSProm-KG-CD [8], and KC-GenRe [17].

Implementation Details. Our RAG-KGC constructs subgraph from train set with the hop k is set to 1. For Generator, we follow the existing work and employ default parameters provided in the original paper. In Similarity Matching, we utilize "bert-base-uncased" to encode LLMs' responses and the names of entities in KG. Moreover, we use GLM-4-Flash [6] as our LLM by invoking API, where *temperature*, *presence_penalty*, and *frequency_penalty* are set to 0.9, and *top_p* are set to 0.7. Last but not least, We adopt Mean Reciprocal Rank (MRR) and Hits@n, $n = \{1, 3, 10\}$ as metrics of the experiments under "filtered" setting [2], which is a common practice in KGC.

6.2 Main Results

As shown in Table 2, our method can achieve superior performance on most metrics in two benchmark datasets. Specifically, compared with Embedding-based and PLM-based methods, RAG-KGC achieved performance superiority

Table 2. Comparison between the proposed methods and baseline methods. The best result in terms of each metric is shown in **bold** and the second best one is underlined.

Dataset	FB15K-237N				Wiki27k			
Metric	MRR	Hits@1	Hits@3	Hit@10	MRR	Hits@1	Hits@3	Hits@10
Embedding-based Methods								
TransE [2]	0.255	0.152	0.301	0.459	0.155	0.032	0.228	0.378
TuckER [1]	0.312	0.228	0.346	0.486	0.246	0.183	0.265	0.382
RotatE [23]	0.279	0.177	0.320	0.481	0.216	0.123	0.256	0.394
PLM-based Methods								
KG-BERT [25]	0.203	0.139	0.201	0.403	0.192	0.119	0.219	0.352
LP-RP-RR [7]	0.248	0.155	0.256	0.436	0.217	0.138	0.235	0.379
PKGC [11]	0.307	0.232	0.328	0.471	0.252	0.189	0.285	0.390
LLMs-based Methods								
KG-LLaMA [26]	–	0.234	–	–	–	0.206	–	–
ChatGPT$_{zero-shot}$ [30]	–	0.197	–	–	–	0.208	–	–
ChatGPT$_{one-shot}$ [30]	–	0.225	–	–	–	0.211	–	–
KG-S2S-CD [8]	0.359	0.289	0.394	0.502	–	–	–	–
CSProm-KG-CD [8]	0.372	0.288	0.410	**0.530**	–	–	–	–
KC-GenRe [17]	0.399	0.338	0.427	0.505	0.317	0.274	0.330	0.408
RAG-KGC	**0.412**	**0.354**	**0.442**	0.495	**0.361**	**0.325**	**0.384**	**0.452**

in all metrics on two datasets. Then, we focus on comparing RAG-KGC with LLM-based methods. In contrast to LLMs-based method KC-GenRe, RAG-KGC obtains increases of 1.3% for MRR, 1.6% for Hits@1, and 1.5% for Hits@3 and a decrease of 3.5% for Hits@10 on FB15K-237-N dataset. On Wiki27K dataset, RAG-KGC gains absolute improvements of 4.4% for MRR, 5.1% for Hits@1, 5.4% for Hits@3, 4.4% for Hits@10. Our results are comparable to those of models like KC-GenRe that combine LLMs and KGE, but our approach does not require pre-training, making it more efficient. In contrast to that only employs entity and relation descriptions of a triple as prompts and utilizes the response for predictions (KG-LLaMA, ChatGPT$_{zero-shot}$, and ChatGPT$_{one-shot}$), RAG-KGC obtains significant increases of over 10% for Hits@1 on two datasets. It demonstrates that RAG-KGC employs diverse strategies for LLMs' more accurately comprehension of entities and relations within queries, thereby improving the performance of LLMs in KGC. Furthermore, we pay attention to the length of input tokens. Specifically, the average is about 6.5K on FB15k-237-N and 6.3K on Wiki27K, which is lower than the limitation of 128K set by GLM-4-Flash. It demonstrates that our method the limits on input tokens.

6.3 Ablation Study

In this section, we conduct an ablation study on Wiki27K dataset to evaluate the impact of each module. Specifically, we construct four variants by removing certain modules: "w/o Retriever", "w/o Subgraph Rewriting", "w/o Query Extension", "w/o Similarity Matching". The results are shown in Table 3.

"w/o Retriever" removes retriever and only input query triples into LLM. It can be seen that the performance declined, which demonstrates that subgraph extraction is important for RAG-KGC, because it can construct context of entity, thereby enhancing LLMs' cognition to query. "w/o Subgraph Rewriting" and "w/o Query Extension" respectively removed subgraph Rewriting and Query Extension in Generator. It can be seen that the performance is both declined, which proves the importance of Generator. Furthermore, we obverse that the performance of "w/o Query Extension" declines greater than "w/o Subgraph Rewriting", obtaining decrease of 4.6% for MRR, 5.3% for Hits@1, 6.6% for Hits@10. It demonstrates that LLM might exhibit heightened sensitivity to information about triple. "w/o Similarity Matching" removes Similarity Matching and only used LLM's responses as the final result. As shown in Table 4, there was a significant decline in performance, which demonstrates the effectiveness of Similarity Matching. In this study, we focused on observing LLM's responses and found that many answers predicted by LLM are actually aliases of entities in knowledge graph. For instance, *"United States"* is an answer predicted by LLM, and *"United States of America"* is an entity in KG. In fact, they refer to the same thing. However, after removing Similarity Matching, since there is no entity named "United States" in knowledge graph, the program will determine that the prediction predicted by LLM is incorrect. It demonstrates the necessity of Similarity Matching.

6.4 Further Study

Discussion on Subgraph. We investigate influences of the hop number k in subgraph. We conduct this study on Wiki27K dataset. As shown in Fig. 2, under different value of k, the performance of RAG-KGC remains stable, which shows that the proposed method is robust to the changes in different k-hop numbers. Furthermore, as the value of k increases, the performance of RAG-KGC slightly decreases. It might be attributed that larger value of k leads to a more extensive subgraph, which introduce more noise that interferes with the inference of LLMs.

Comparison of Different LLMs. In this section, We compare the results of RAG-KGC using various LLMs on Wiki27K dataset, as shown in Table 4. The results demonstrate that different LLMs exhibit various performance. Specifically, GLM-4-Flash achieves higher than GPT-4o mini. In terms of response time, Qwen-plus achieved the highest (6.4 s), while GLM-4-Flash was the shortest (5.3 s). Furthermore, increasing the parameter scale of LLMs may bring performance improvements. For instance, Qwen-plus (72B) gains increases of 0.4% for Hits@1, 0.6% for Hits@10, compared to GLM-4-Flash(<10B). However, the

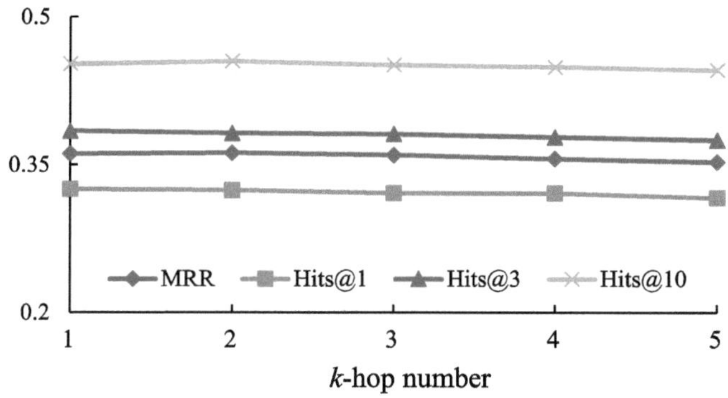

Fig. 2. Performance with different k-hop numbers

extent of the increase is slight. It may be attributed to the limited or unmemorized knowledge relevant to the query in corpus used for LLMs, which leads to no significant improvement in KGC. Therefore, a model with 7B parameters may be sufficient for KGC.

Table 3. Ablation results of RAG-KGC on Wiki27K dataset.

	MRR	Hits@1	Hits@10
RAG-KGC	**0.361**	**0.325**	**0.452**
w/o Retriever	0.345	0.301	0.423
w/o Subgraph Rewriting	0.327	0.281	0.412
w/o Query Extension	0.315	0.272	0.386
w/o Similarity Matching	–	0.226	–

Table 4. The results of the comparison of different LLMs on Wiki27K dataset.

LLM	#p	Hits@1	Hits@10	Time
GLM-4-Flash	<10B	0.325	0.452	5.3 s
Llama-3-8B	8B	0.324	0.449	5.4 s
GPT-4o mini	8B	0.320	0.450	5.6 s
Qwen-plus	72B	0.329	0.458	6.4 s

7 Conclusion

In this paper, we introduce RAG into LLMs-based KGC and propose a novel paradigm named Ret-Gen, which aims to address the limitations of LLMs in KGC. Based on this paradigm, we propose RAG-KGC. It operates through a three-stage pipeline: subgraph extraction for query-relevant knowledge retrieval, LLM-guided generation of supplementary knowledge to enhance reasoning capabilities, and similarity matching to align generated outputs with canonical knowledge graph entities, thereby mitigating text mismatch issues prevalent in prior methods. Experimental validation on link prediction tasks demonstrates the effectiveness of our approach. However, scalability remains a critical challenge for knowledge graphs exceeding one million entities, where BFS-based subgraph extraction may introduce performance bottlenecks. Future research directions include exploring more efficient retrieval mechanisms, extending the framework to additional KGC tasks such as triple classification and relation prediction, and investigating domain-specific applications in medical and financial knowledge graphs to establish broader applicability and practical utility of the Ret-Gen paradigm.

Acknowledgments. This research is supported in part by the "14th Five-Year Plan" Civil Aerospace Pre-Research Project of China under Grant No. D020101, the Natural Science Foundation of China No. 62302213, Innovation Funding of Key Laboratory of Intelligent Decision and Digital Operations No. NJ2023027, Ministry of Industrial and Information Technology Project of Hebei Key Laboratory of Software Engineering, No. 22567637H, the Natural Science Foundation of Jiangsu Province under Grant No. BK20210280.

References

1. Balažević, I., Allen, C., Hospedales, T.M.: TuckER: tensor factorization for knowledge graph completion. arXiv preprint arXiv:1901.09590 (2019)
2. Bordes, A., Usunier, N., Garcia-Duran, A., Weston, J., Yakhnenko, O.: Translating embeddings for modeling multi-relational data. Adv. Neural Inf. Process. **26** (2013)
3. Chang, Y., et al.: A survey on evaluation of large language models. ACM Trans. Intell. Syst. Technol. **15**(3), 1–45 (2024)
4. Cui, Y., Sun, Z., Hu, W.: A prompt-based knowledge graph foundation model for universal in-context reasoning. Adv. Neural. Inf. Process. Syst. **37**, 7095–7124 (2025)
5. Gao, Y., et al.: Retrieval-augmented generation for large language models: a survey. arXiv preprint arXiv:2312.10997 (2023). **2**(1)
6. GLM, T., et al.: ChatGLM: a family of large language models from GLM-130B to GLM-4 all tools. arXiv preprint arXiv:2406.12793 (2024)
7. Kim, B., Hong, T., Ko, Y., Seo, J.: Multi-task learning for knowledge graph completion with pre-trained language models. In: Proceedings of the 28th International Conference on Computational Linguistics, pp. 1737–1743 (2020)
8. Li, D., Tan, Z., Chen, T., Liu, H.: Contextualization distillation from large language model for knowledge graph completion. arXiv preprint arXiv:2402.01729 (2024)
9. Li, J., Tang, T., Zhao, W.X., Wei, Z., Yuan, N.J., Wen, J.R.: Few-shot knowledge graph-to-text generation with pretrained language models. arXiv preprint arXiv:2106.01623 (2021)
10. Li, Y., Zang, G., Song, C., Yuan, X., Ge, T.: Leveraging semantic information for enhanced community search in heterogeneous graphs. Data Sci. Eng. **9**(2), 220–237 (2024)

11. Lv, X., et al.: Do pre-trained models benefit knowledge graph completion? A reliable evaluation and a reasonable approach. In: Proceedings of the 60th Annual Meeting of the Association for Computational Linguistics: System Demonstrations, pp. 3570–3581. Association for Computational Linguistics (2022)
12. Peng, B., et al.: Graph retrieval-augmented generation: a survey. arXiv preprint arXiv:2408.08921 (2024)
13. Shu, D., Chen, T., Jin, M., Zhang, C., Du, M., Zhang, Y.: Knowledge graph large language model (KG-LLM) for link prediction. Proc. Mach. Learn. Res. **260**(1), 143 (2024)
14. Sun, J., et al.: Think-on-graph: deep and responsible reasoning of large language model on knowledge graph. arXiv preprint arXiv:2307.07697 (2023)
15. Tan, X., Wang, X., Liu, Q., Xu, X., Yuan, X., Zhang, W.: Paths-over-graph: knowledge graph empowered large language model reasoning. In: Proceedings of the ACM on Web Conference 2025, pp. 3505–3522 (2025)
16. Toutanova, K., Chen, D.: Observed versus latent features for knowledge base and text inference. In: Proceedings of the 3rd Workshop on Continuous Vector Space Models and Their Compositionality, pp. 57–66 (2015)
17. Wang, Y., Hu, M., Huang, Z., Li, D., Yang, D., Lu, X.: KC-genre: a knowledge-constrained generative re-ranking method based on large language models for knowledge graph completion. arXiv preprint arXiv:2403.17532 (2024)
18. Wei, J., et al.: Chain-of-thought prompting elicits reasoning in large language models. Adv. Neural. Inf. Process. Syst. **35**, 24824–24837 (2022)
19. Wei, Y., Huang, Q., Kwok, J.T., Zhang, Y.: KICGPT: large language model with knowledge in context for knowledge graph completion. arXiv preprint arXiv:2402.02389 (2024)
20. West, R., Gabrilovich, E., Murphy, K., Sun, S., Gupta, R., Lin, D.: Knowledge base completion via search-based question answering. In: Proceedings of the 23rd International Conference on World Wide Web, pp. 515–526 (2014)
21. Wu, W., Wang, H., Li, B., Huang, P., Zhao, X., Liang, L.: MultiRAG: a knowledge-guided framework for mitigating hallucination in multi-source retrieval augmented generation. In: 2025 IEEE 41st International Conference on Data Engineering (ICDE) (2025)
22. Xiao, S., Zhu, D., Tang, C., Huang, Z.: Combining graph contrastive embedding and multi-head cross-attention transfer for cross-domain recommendation. Data Sci. Eng. **8**(3), 247–262 (2023)
23. Yang, B., Yih, W.t., He, X., Gao, J., Deng, L.: Embedding entities and relations for learning and inference in knowledge bases. arXiv preprint arXiv:1412.6575 (2014)
24. Yang, R., Zhu, J., Man, J., Liu, H., Fang, L., Zhou, Y.: GS-KGC: a generative subgraph-based framework for knowledge graph completion with large language models. Inf. Fusion **117**, 102868 (2025)
25. Yao, L., Mao, C., Luo, Y.: KG-BERT: BERT for knowledge graph completion. arXiv preprint arXiv:1909.03193 (2019)
26. Yao, L., Peng, J., Mao, C., Luo, Y.: Exploring large language models for knowledge graph completion. arXiv preprint arXiv:2308.13916 (2023)
27. Yasunaga, M., Ren, H., Bosselut, A., Liang, P., Leskovec, J.: QA-GNN: reasoning with language models and knowledge graphs for question answering. arXiv preprint arXiv:2104.06378 (2021)
28. Zhang, Y., et al.: Siren's song in the AI ocean: a survey on hallucination in large language models. arXiv preprint arXiv:2309.01219 (2023)

29. Zhang, Y., et al.: Preference prototype-aware learning for universal cross-domain recommendation. In: Proceedings of the 33rd ACM International Conference on Information and Knowledge Management, pp. 3290–3299 (2024)
30. Zhu, Y.: LLMs for knowledge graph construction and reasoning: recent capabilities and future opportunities. World Wide Web **27**(5), 58 (2024)

cFedLoRA: Clustered Aggregation for Federated LoRA

Qi Cheng, Peng Yan, and Guodong Long(✉)

Australian Artificial Intelligence Institute, Faculty of Engineering and IT, University of Technology Sydney, Ultimo, Australia
qi.cheng-2@student.uts.edu.au, {peng.yan-1,guodong.long}@uts.edu.au

Abstract. Low-Rank Adaptation (LoRA) enables the deployment of large pre-trained models in Federated Learning (FL) by updating and communicating only lightweight adapter matrices. However, training separate LoRA adapters on non-IID client data often leads to overfitting and poor cross-client generalization. We propose cFedLoRA, a clustered aggregation framework that addresses these challenges. cFedLoRA groups clients based on the similarity of their LoRA updates and aggregates local updates within each cluster, enabling communication-efficient collaboration and cluster-wise specialization. Experiments on federated benchmarks with diverse non-IID settings show that cFedLoRA achieves higher accuracy, faster convergence, and lower computational and communication costs. These improvements underscore cFedLoRA's effectiveness and practicality for resource-constrained FL deployments.

Keywords: Federated Learning · LoRA · Pre-trained Models · Client Clustering

1 Introduction

Federated Learning (FL) is a popular machine learning paradigm that leverages decentralized training with strong privacy guarantees. However, clients in FL systems often have limited data and computational capacity, making it difficult to deploy large pre-trained foundation models [28] such as BERT [7], GPT [3], and CLIP [23], and blocking the way towards Federated Intelligence [17].

Low-Rank Adaptation (LoRA) [12] is a widely adopted technique for efficiently fine-tuning large models. By updating only lightweight adapter matrices, LoRA significantly reduces the number of trainable and transferable parameters, making it an attractive choice for federated learning, where computational and communication costs are critical constraints. However, the challenges of applying LoRA in federated learning extend beyond training efficiency. Due to the distributed and heterogeneous nature of client data [31], training separate LoRA adapters on biased and limited local datasets often results in overfitting and poor cross-client generalization.

These limitations motivate the design of cFedLoRA, a clustered aggregation framework that enhances the effectiveness of LoRA in federated learning by

M. Yoshikawa et al. (Eds.): ADMA 2025, LNAI 16197, pp. 191–205, 2026.
https://doi.org/10.1007/978-981-95-3453-1_13

leveraging collaborative training among similar clients. cFedLoRA first groups clients based on the similarity of their LoRA adapters' updates, then performs cluster-wise aggregation within each group. This targeted collaboration strategy mitigates overfitting on locally biased data by encouraging knowledge sharing and promoting the formation of specialized adapters aligned with shared patterns among similar clients. Figure 1 provides an overview of the cFedLoRA framework.

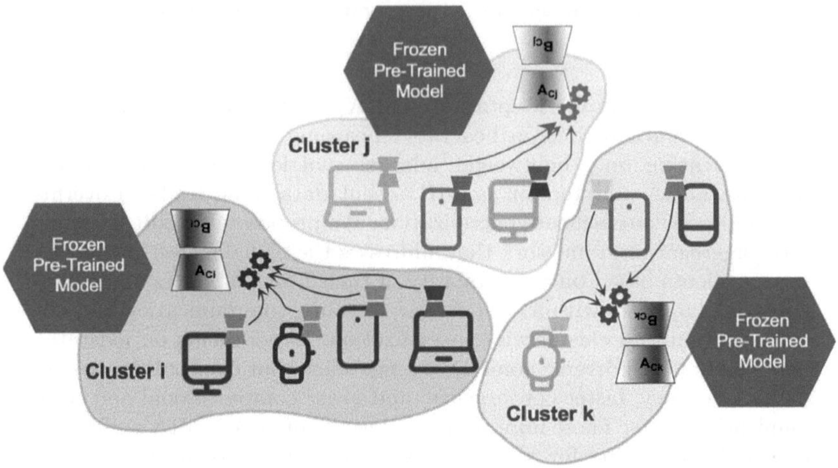

Fig. 1. Overview of the cFedLoRA framework. Each client first trains LoRA modules on local data. Clients are then clustered based on the similarity of their adapter updates. Within each cluster (indicated by color), LoRA adapters are aggregated to form a shared, cluster-specific adaptation, which is used for inference by clients in the corresponding cluster. (Color figure online)

By enabling collaboration among similar clients while decoupling adaptation across dissimilar ones, cFedLoRA extends LoRA's efficiency through a training strategy that balances collaborative learning with selective adaptation. This approach improves data efficiency and addresses the challenges posed by non-IID client distributions, making cFedLoRA particularly well-suited for deploying large pre-trained models on resource-constrained FL clients. We evaluate cFedLoRA on real-world benchmarks and compare it against a range of strong baselines, demonstrating consistent improvements in accuracy, convergence rates, and communication efficiency.

Our main contributions are summarized as follows:

- We propose **cFedLoRA**, which enables efficient deployment of large pre-trained foundation models on resource-constrained federated learning clients, bridging the gap between federated learning limitations and the scale of modern AI models.

- We design a clustered aggregation framework that enhances LoRA's effectiveness in FL by grouping clients based on adapter update similarity and allowing cluster-wise collaborative adaptation.
- We introduce a hybrid training strategy that combines cluster-wise collaboration with lightweight on-device fine-tuning, effectively balancing cluster-level specialization and client-level personalization.
- We empirically evaluate cFedLoRA on real-world federated benchmarks under varying parameter scales, showing that it significantly reduces computational and communication costs while improving accuracy and convergence over strong baselines.

2 Related Works

2.1 Federated Learning and Statistical Heterogeneity

Federated Learning [20] enables collaborative model training across decentralized clients without sharing raw data. A central challenge arises from statistical heterogeneity, where clients hold non-identically distributed data, which can degrade global model performance and training efficiency.

To address this, various approaches have been developed, including regularization based methods such as FedProx [14], which constrains local updates to reduce client drift, and optimization based strategies like FedDANE [14], which incorporates second-order information for more stable convergence. Additionally, personalization techniques have been proposed to tailor models to client-specific data, including representation-based methods [21,26,27], mixture-of-experts frameworks [19], and client-aware regularization [5,33,34].

2.2 Clustered Federated Learning

Clustered Federated Learning enhances generalization under non-IID data by grouping clients based on model update similarities and training cluster-specific models. It seeks to balance scalability and personalization while maintaining communication efficiency.

Early work such as CFL [25] clusters clients based on the cosine similarity of gradient updates, enabling cluster-wise aggregation. Briggs et al. [2] improve computational efficiency through hierarchical clustering, reducing overhead.

More recent approaches use task-driven or optimization-aware formulations. For example, ClusterFL [22] models clustering as a multi-task learning problem to jointly improve accuracy and communication efficiency. FedGroup [9] decomposes update similarity via cosine distance of optimization directions, improving FEMNIST accuracy by up to 14% points.

IFCA [10] proposes an iterative clustering algorithm that learns multiple global models and client-cluster assignments. Long et al. [18] extend this with a stochastic Expectation-Maximization framework that learns multi-center cluster models to reduce personalization error while managing overhead.

Other methods explore soft clustering (FedSPD [16]) and modular model partitioning (FedMA [29]), enabling flexibility in how clusters share and adapt models. More dynamic strategies include FedAC [38], which adjusts the number of clusters based on similarity metrics, and LCFed [39], which supports scalable re-clustering by disentangling shared and cluster-specific components.

These clustering-based methods inspire our work, though most operate on full model parameters. In contrast, cFedLoRA clusters lightweight LoRA adapters, enabling scalable adaptation for large pre-trained models under FL constraints.

2.3 Parameter Efficient Fine Tuning

Large pre-trained models such as Vision Transformers (ViT) for visual tasks, BERT for natural language understanding, and general-purpose models like LLaMA and GPT have become central to modern machine learning applications. However, fully fine-tuning these models is computationally demanding and memory-intensive, making them difficult to deploy in constrained environments such as mobile devices or FL systems.

Parameter-Efficient Fine-Tuning (PEFT) methods address these challenges by updating only a small subset of parameters, such as adapter layers or low-rank matrices, while keeping the backbone model frozen. This significantly reduces memory consumption and computational cost, enabling practical adaptation in decentralized settings. Among PEFT approaches, Low-Rank Adaptation has gained widespread use due to its straightforward implementation and strong empirical performance, making it especially suitable for FL scenarios.

2.4 PEFT in Federated Learning

Applying PEFT in FL introduces new opportunities and challenges. Recent research has explored two primary directions: computational efficiency and robustness under data heterogeneity.

Computational Efficiency. To address the high resource demands of fine-tuning large models in FL, FedPETuning [40] provides a formalized framework for tuning and aggregating lightweight PEFT modules. Their empirical study shows that several PEFT variants significantly reduce communication costs while preserving accuracy. FwdLLM [32] further reduces overhead via backpropagation-free methods, using perturbed inference for local updates. This approach accelerates convergence and minimizes memory use, allowing large models to be fine-tuned on low-resource devices. FedDPA [36] employs a dual-LoRA structure to tackle the test-time distribution shifts efficiently.

However, these methods often focus primarily on communication savings and may still struggle with performance degradation in heterogeneous FL settings.

Robustness to Data Heterogeneity. Several recent works target statistical heterogeneity in FL via improved adapter design. pFedLoRA [37] proposes a model-heterogeneous FL setting where clients train distinct models but share a homogeneous LoRA adapter to facilitate convergence. FedARA [30] employs adaptive rank allocation using SVD to match client similarity dynamically and mitigate data skew. SLoRA [1] introduces data-driven initialization for LoRA adapters, addressing performance drop under high heterogeneity. HETLoRA [6] aggregates adapter modules via client-side pruning and sparse server weighting, enhancing robustness without incurring large computational cost.

In this work, we build on LoRA to develop cFedLoRA, a clustered aggregation framework that enables scalable, efficient adaptation of large models in federated environments with heterogeneous clients.

3 Methodology

This section presents **cFedLoRA**, a framework that enhances LoRA-based adaptation in federated learning by integrating structured client clustering. While clustered FL improves personalization and communication efficiency, it seldom scaled to large pre-trained models. Conversely, per-client LoRA adapters enable efficient fine-tuning of such models but tend to overfit under highly non-IID data distributions. cFedLoRA bridges these gaps by clustering clients based on the similarity of their LoRA updates, enabling cluster-wise collaboration that improves generalization while keeping overhead low. We formalize the problem setting, introduce a hybrid training algorithm, and propose a lightweight personalization step to balance scalability and client-specific adaptation.

3.1 Problem Formulation

Let there be a set of N clients, where each client $i \in \{1, ..., N\}$ processes a private local dataset D_i. All clients share access to a frozen pre-trained foundation model Θ, and perform local adaptation by training a LoRA module denoted by Λ_i.

The goal of cFedLoRA is to improve generalization and efficiency by enabling structured collaboration among similar clients. To this end, clients are partitioned into K clusters based on the similarity of their local LoRA updates Λ_i. Within each cluster $m \in \{1, ..., K\}$, a shared cluster-level adapter $\hat{\Lambda}_m$ is computed to capture common adaptation signals. The adapted model used by client i is given by:

$$\Theta + \Lambda_i + \hat{\Lambda}_{k(i)}, \tag{1}$$

where $k(i) \in \{1, ..., K\}$ denotes the cluster assignment for client i.

The overall training objective can be formulated as a bi-level optimization problem:

$$\begin{aligned} \min_{\{\Lambda_i\}} \quad & \sum_{i=1}^{N} \sum_{m=1}^{K} \mathbb{I}[k(i) = m] \cdot L\left(\Theta + \Lambda_i + \hat{\Lambda}_m; D_i\right) \\ \text{s.t.} \quad & k(i) = \arg\min_m \left\| \Lambda_i - \hat{\Lambda}_m \right\|^2, \end{aligned} \tag{2}$$

where $L(\cdot; D_i)$ denotes the local loss function evaluated on client i's data, and $\mathbb{I}[\cdot]$ is an indicator function.

3.2 Hybrid Training

We formalize the learning procedure of cFedLoRA, which consists of local LoRA adaptation, client clustering, cluster-wise federated training, and optional local personalization.

Local Initialization and Update. Each client i initializes a LoRA module $\Lambda_i^{(0)}$. After one round of local training, the updated adapter is:

$$\Lambda_i^{(1)} = \Lambda_i^{(0)} - \eta \cdot \nabla_\Lambda L(\Theta + \Lambda_i^{(0)}; D_i), \tag{3}$$

where η is the learning rate and $L(\cdot; D_i)$ is the local loss evaluated on client i's dataset.

Client Clustering. Clients are clustered based on their first-round LoRA updates $\Lambda_i^{(1)}$ using K-means:

$$k(i) = \arg \min_{m \in 1,\ldots,K} \|\Lambda_i^{(1)} - \hat{\Lambda}_m^{(1)}\|^2, \tag{4}$$

where $k(i)$ is the cluster index for client i, and $\hat{\Lambda}_m^{(1)}$ is the centroid of cluster m in the LoRA parameter space.

Cluster-wise Federated Training. In subsequent communication rounds, clients within each cluster perform federated training using standard FedAvg updates. Specifically, the local LoRA adapter is updated as:

$$\Lambda_i^{(t+1)} = \Lambda_i^{(t)} - \eta \cdot \nabla_\Lambda L(\Theta + \Lambda_i^{(t)} + \hat{\Lambda}_m^{(t)}; D_i), \tag{5}$$

and the cluster-level adapter is updated by weighted averaging:

$$\hat{\Lambda}_m^{(t+1)} = \sum_{k(i)=m} \frac{|D_i|}{\sum_{k(j)=m} |D_j|} \cdot \Lambda_i^{(t+1)}, \tag{6}$$

for $t \geq 1$, where the summation is taken over all clients i assigned to cluster m.

Optional Local Personalization. Each client may optionally perform a final fine-tuning step using a small held-out subset $D_i^{(\mathrm{FT})} \subseteq D_i$, to personalize its adapter:

$$\Lambda_i^{(\mathrm{final})} = \Lambda_i^{(T)} - \eta \cdot \nabla_\Lambda L(\Theta + \Lambda_i^{(T)} + \hat{\Lambda}_{k(i)}^{(T)}; D_i^{(\mathrm{FT})}), \tag{7}$$

where T denotes the final communication round.

The full training procedure of cFedLoRA is summarized in Algorithm 1.

4 Experiments

We conduct experiments on real-world benchmarks to evaluate the effectiveness of cFedLoRA, focusing on accuracy, convergence rates, and parameter efficiency. Its performance is compared against strong FL baselines.

Algorithm 1. cFedLoRA Training Procedure

Require: Frozen pre-trained model Θ, local datasets $\{D_i\}_{i=1}^N$, learning rate η, number of clusters K, total communication rounds T

Initialize: Each client i initializes LoRA adapter $\Lambda_i^{(0)}$
for all clients $i = 1$ to N **in parallel do**
  Perform one round of local training by Eq. 3
end for
Perform K-means clustering over $\{\Lambda_i^{(1)}\}$ (Eq. 4) to obtain cluster assignments $k(i)$ for each client
for all clusters $m = 1$ to K **in parallel do**
  Compute initial cluster adapter by Eq. 6
end for
for $t = 1$ to T **do**
  for all clients i **in parallel do**
    Update local adapter by Eq. 5
  end for
  for all clusters $m = 1$ to K **in parallel do**
    Update cluster adapter by Eq. 6
  end for
end for
for all clients i **in parallel do**
  Fine-tune local adapter on held-out subset $D_i^{(\mathrm{FT})}$ by Eq. 7
end for

4.1 Benchmark and Data Partitioning

Data Benchmark. We evaluate cFedLoRA using the FEMNIST dataset from the LEAF benchmark suite [4], which comprises handwritten digits and characters grouped by user. This user-centric structure introduces natural data heterogeneity, making FEMNIST a widely used benchmark for testing FL methods under non-IID conditions.

Data Partitioning. We simulate a federated setting with 10 clients, each receiving a distinct, non-IID partition of the data. To induce statistical heterogeneity, we apply a Dirichlet-based partitioning scheme [11] $p_k \sim \mathrm{Dir}_J(0.5)$, where J is the number of classes. This results in imbalanced and personalized label distributions across clients, effectively modeling real-world FL challenges.

Data Transformation. As the target pre-trained backbones (MobileNetV2 [24] and ViT [8]) we select for this work take 224×224 3-channel images as input. We also transform FEMNIST data from 28×28 black-and-white images to 224×224 3-channel grayscale images with proper normalization, making the local data shape exactly matching with the pre-trained model input shape.

4.2 Base Pre-trained Models

To evaluate the effectiveness and generality of cFedLoRA, we adopt two widely used pre-trained vision models that represent distinct architectural paradigms:

convolutional neural networks and transformers. We describe their base configurations and detail how they are adapted using LoRA for efficient deployment in federated learning settings. Table 1 summarizes the number of trainable parameters under different LoRA rank settings.

Base Model Architecture.

- **MobileNetV2** [24]**:** A lightweight convolutional neural network optimized for image classification on edge devices. The original model includes 53 convolutional layers and a final classifier for 1000 output classes, totaling approximately 3.5M parameters.
- **Vision Transformer (ViT)** [8]**:** A transformer-based architecture that tokenizes images into patches and processes them via self-attention. We use the ViT-Base model with a 16×16 patch size (ViT-B/16), which comprises 12 transformer blocks and 86.7M parameters.

Base Model Adjustment.

- **LoRA Adaption:** For MobileNetV2, we freeze all 53 Conv2D layers and insert LoRA adapters into each one. For ViT-B/16, we insert LoRA adapters into the transformer encoder layers while freezing all pre-trained weights.
- **Classifier Replacement:** For MobileNetV2, We use a linear layer taking 1280 input and outputting 62 classes to replace the final classifier. For ViT-B/16 on the other hand, we construct a linear layer that takes 768 input to match the previous output shape and supplies 62 classes to replace the final classification head.
- **Trainable Parameters Reduction:**
  - The CNN model for MNIST [13] dataset has 2 Conv2D layers with 20 channels and 50 channels respectively, followed by 2×2 max pooling, then a fully connected layer, and a final softmax output layer, totaling 1.66 million parameters.
  - For MobileNetV2, the feature extraction for all 53 Conv2D layers accounts for about 2.28 million parameters. By adopting the LoRA-based strategy, the number of trainable LoRA parameters is proportional to the adapter rank R and a constant $S_{\text{conv\_factors}}$, representing the total adaptation footprint across all convolutional layers. We empirically estimate $S_{\text{conv\_factors}} \approx 8{,}480$ for MobileNetV2. Replacing the classifier(Liner(1280,62)) introduces 79,422 trainable parameters. For instance, the total trainable parameter for MobileNetV2(with LoRA rank R=4) aggregates $4 \times 8{,}480 + 79{,}422 \approx 0.11M$.
  - For ViT-B/16, the original model contains 12 transformer encoder blocks and approximately 86 million trainable parameters. To integrate LoRA, we inject adapters into three key components within each encoder block: (i) the Attention Output Projection layer, (ii) the MLP Expansion layer, and (iii) the MLP Projection layer, resulting in a total of 36 LoRA modules across the model. The classifier replacement addes $768 \times 62 + 62 \approx$

Table 1. Comparison of trainable parameters with LoRA adapters on FEMNIST.

Model	LoRA Rank	Trainable Params	Trainable %
CNN in FedAvg [20] (2 Conv2D Layers) Trainable params: 1.66M	–	1.66M	100.00%
MobileNetV2 [24] (53 Conv2D Layers) Trainable params: 2.28M	R = 4	0.11M	4.8%
	R = 8	0.14M	6.1%
	R = 16	0.21M	9.2%
ViT-B/16 [8] (12 Encoder Blocks) Trainable params: 86.73M	R = 4	0.49M	0.56%
	R = 8	0.93M	1.07%
	R = 16	1.81M	2.09%
	R = 32	3.57M	4.12%

$47K$ trainable parameters. The empirically estimation of the constant factor of the trainable parameters for LoRA-adapted ViT is approximately $110K$. Therefore, the number of trainable parameters in a LoRA-adapted($R = 4$) ViT estimates $4 \times 110K + 47K \approx 0.49M$.

These two architectures span convolutional and transformer paradigms, providing a robust testbed to assess cFedLoRA's adaptability across model families and resource constraints.

4.3 cFedLoRA Configuration

Adapter and Optimizer. Each client fine-tunes a local LoRA module with rank $R \in \{4, 8, 16, 32\}$. Training proceeds for sufficient communication rounds until convergence, with 5 local epochs per round and a batch size of 32. We use Adam optimizer with initial learning rates of 5e−3 and 5e−5 for MobileNetV2 [24] and ViT [8] backbones, respectively.

Clustering Strategy. Client-specific LoRA modules are collected after the first communication round, and clustering is performed using the K-Means. The number of clusters is fixed to $K = 3$ or $K = 4$ based on empirical observations. Once assigned, cluster memberships remain fixed for the remainder of training.

Optional Personalization Fine-Tuning. To enhance client-specific performance, we apply an additional one-epoch local fine-tuning step after cluster-wise training. This improves personalization while retaining the generalization benefits of shared cluster-level adapters.

4.4 Baselines for Comparison

To rigorously assess the effectiveness of cFedLoRA, we compare it against several representative federated learning algorithms. These baselines reflect a spectrum of strategies designed to tackle statistical heterogeneity, client drift, and personalization in distributed training.

- **FedAvg** [20]: The foundational FL algorithm that aggregates local models via weighted averaging, proportional to client dataset sizes.
- **FedCluster** [25]: An extension of FedAvg that applies hierarchical clustering to group clients based on update similarity, followed by within-cluster aggregation to better handle non-IID distributions.
- **FedDANE** [15]: A Newton-type method that augments local objectives with global gradient information to reduce client drift and improve convergence on heterogeneous data.
- **FedProx** [14]: A stabilized variant of FedAvg incorporating a proximal term in the local loss function. The coefficient is set to 0.1 in our experiments to regulate client update deviation.
- **FeSEM** [18]: A multi-center FL approach inspired by the EM algorithm. FeSEM maintains K cluster-specific models and iteratively assigns clients to clusters while refining both models and assignments.
- **FedAvg+LoRA$_{(C)}$**: FedAvg with LoRA using MobileNetV2 [24] as the backbone on FEMNIST [4]. This variant shares the setup with cFedLoRA but omits clustering. LoRA rank is fixed at 8. Subscript C denotes the convolutional backbone.
- **FedAvg+LoRA$_{(T)}$**: FedAvg with LoRA using ViT [8]. Identical to cFedLoRA except clustering is excluded. LoRA rank is set to 8. Subscript T denotes the transformer backbone.
- **cFedLoRA (ours)**: Our proposed method combining LoRA-based local adaptation with cluster-wise aggregation. We vary the LoRA rank to study parameter-efficiency trade-offs. Subscripts C and T indicate MobileNetV2 and ViT backbones, respectively.

Together, these baselines offer a comprehensive evaluation framework for measuring the performance of cFedLoRA under varying model architectures and non-IID settings.

4.5 Results and Analysis

We compare cFedLoRA against standard and state-of-the-art federated learning baselines, evaluating accuracy, convergence speed, communication efficiency, and robustness to non-IID client distributions. The results show that cFedLoRA achieves faster convergence and more effective model aggregation by reducing redundant updates and enabling cluster-wise specialization. This leads to improved generalization across diverse clients with heterogeneous data.

Overall Performance Comparison. Table 2 presents the test accuracy of all methods, covering both convolutional (MobileNetV2) and transformer (ViT) backbones. We observe that cFedLoRA consistently outperforms the baselines across both model types, demonstrating its effectiveness in handling non-IID data. The clustered adaptation strategy of cFedLoRA improves accuracy while maintaining a low computation cost and communication footprint.

Table 2. Test Accuracy Comparison of cFedLoRA and Baselines on FEMNIST. C denotes MobileNetV2 [24], T denotes ViT [8].

Method	LoRA Rank	# of Clusters	Accuracy (%)
FedAvg [20]	—	—	84.9
FedCluster [25]	—	10	84.1
FedDane [15]	—	—	40.0
FedProx [14]	—	—	72.6
FeSEM [18]	—	4	90.4
FedAvg+LoRA$_{(C)}$	8	—	80.6
cFedLoRA$_{(C)}$R = 4, K = 3	4	3	82.7
cFedLoRA$_{(C)}$R = 8, K = 3	8	3	90.5
cFedLoRA$_{(C)}$R = 16, K = 4	16	4	90.9
cFedLoRA$_{(C)}$R = 16, K = 3	16	3	**91.7**
FedAvg+LoRA$_{(T)}$	8	—	91.4
cFedLoRA$_{(T)}$R = 4, K = 3	4	3	89.2
cFedLoRA$_{(T)}$R = 8, K = 3	8	3	90.3
cFedLoRA$_{(T)}$R = 16, K = 3	16	3	90.8
cFedLoRA$_{(T)}$R = 32, K = 3	32	3	**91.6**

Convergence Behavior. Figure 2a compares the convergence of *FedAvg + LoRA* and cFedLoRA, both using LoRA adapters with rank 8 on the MobileNetV2 backbone.

The red curve shows *FedAvg + LoRA*, where clients locally fine-tune LoRA modules and aggregate them globally without clustering. Although it eventually converges, performance plateaus around 80% accuracy and convergence is slow. This suggests that naïvely averaging LoRA updates across heterogeneous clients induces parameter drift due to the highly non-IID data.

In contrast, the blue curve shows cFedLoRA, which freezes the pre-trained backbone and trains only LoRA adapters, reducing trainable parameters by 93.9%. After the first round, clients are clustered based on the similarity of their LoRA updates, and subsequent aggregation is performed within clusters. This reduces inter-client variance and provides more consistent updates, resulting in significantly faster and more stable convergence. cFedLoRA consistently achieves higher accuracy than *FedAvg + LoRA*.

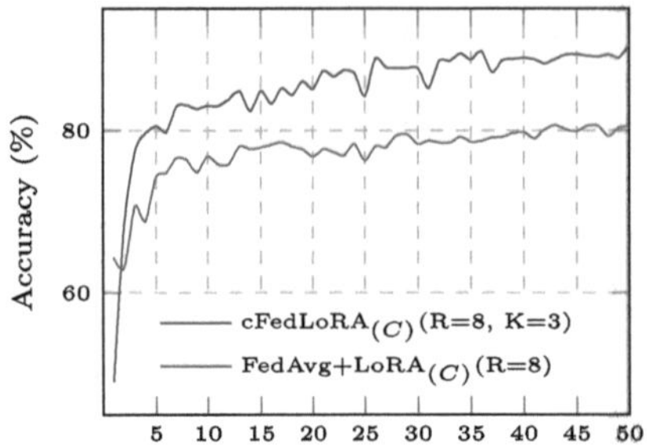

(a) Convergence speed on FEMNIST

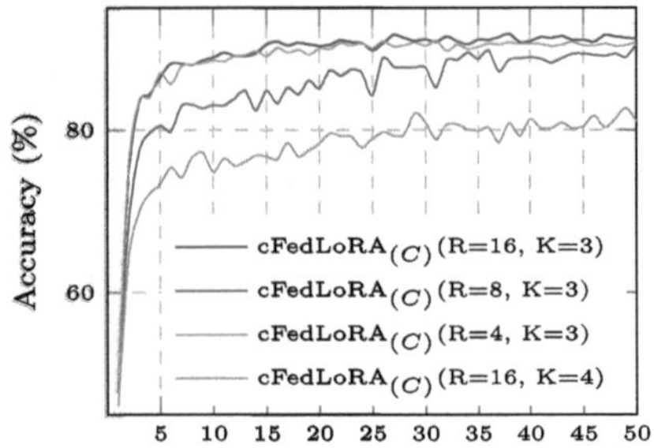

(b) cFedLoRA$_{(C)}$ Accuracy with R and K

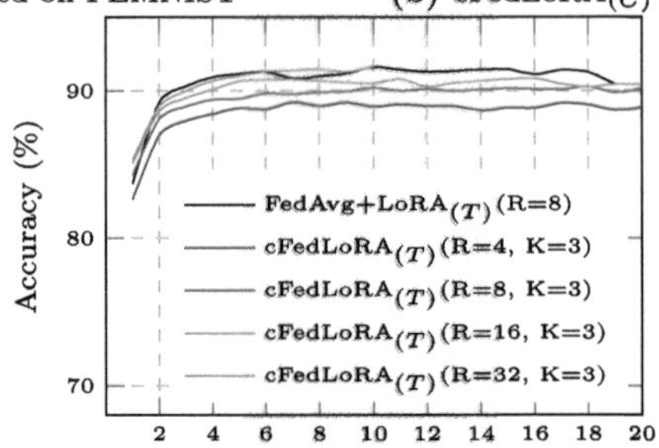

(c) cFedLoRA$_{(T)}$ Accuracy with R and K

Fig. 2. Convergence visualization: Comparison of test accuracy over communication rounds under varying LoRA ranks and numbers of clusters, highlighting the impact of cluster-wise aggregation in cFedLoRA.

Impact of LoRA Rank. To assess the expressiveness required by LoRA adapters, we vary the rank $R \in \{4, 8, 16, 32\}$ and report results in Fig. 2b and Fig. 2c. Four key observations emerge:

- **Overall Performance:** As shown in Fig. 2b, reducing R from 16 to 4 cuts the number of trainable parameters by 50% (see Table 1), but also lowers peak accuracy by around 10% points. This suggests that even low-rank adapters can capture most of the task-specific variance in heterogeneous data, though with some tradeoff in accuracy.
- **Effect of Higher Rank:** In Fig. 2c, increasing R beyond 16 yields noticeable accuracy improvements due to increased model capacity, but also incurs greater computational and communication overhead. Rank should thus be chosen based on resource availability and performance needs.
- **Clustering Impact:** A larger number of clusters generally accelerates convergence, as clients are grouped more precisely. However, extremely high K values (e.g., one client per cluster) may cause overfitting and poor generalization, while too small a K (e.g., $K = 1$) reduces the method to FedAvg + LoRA, failing to leverage personalization.
- **Rank–Efficiency Trade-off:** Higher ranks increase trainable parameters and computation cost during training and aggregation. For resource-

constrained deployments, we recommend balancing rank and accuracy by a moderate R (e.g., 8 or 16) to maintain efficiency while achieving competitive performance.

5 Conclusion and Future Work

Conclusion. We propose **cFedLoRA**, a novel clustered aggregation framework for federated learning that integrates Low-Rank Adaptation with structured client grouping. By clustering clients based on the similarity of their LoRA updates and aggregating within clusters, cFedLoRA effectively mitigates overfitting on heterogeneous data while reducing communication and computation overhead. Experimental results demonstrate that cFedLoRA consistently improves accuracy, convergence, and parameter efficiency, establishing its practicality for deploying large pre-trained models in federated environments.

Future Work. This work primarily targets image classification using LoRA-adapted backbones. Future directions include:

- extending cFedLoRA to large language models and multi-task federated settings
- incorporating the clustered LoRA with an extra global LoRA to improve generalization [35]
- exploring dynamic clustering strategies such as hierarchical or soft clustering that adapt as new clients join

We believe cFedLoRA offers a promising foundation for scalable and personalized adaptation of foundation models in real-world federated learning systems.

References

1. Babakniya, S., et al.: SLoRA: federated parameter efficient fine-tuning of language models. arXiv preprint arXiv:2308.06522 (2023)
2. Briggs, C., Fan, Z., Andras, P.: Federated learning with hierarchical clustering of local updates to improve training on non-IID data. In: 2020 International Joint Conference on Neural Networks (IJCNN), pp. 1–9. IEEE (2020)
3. Brown, T., et al.: Language models are few-shot learners. Adv. Neural. Inf. Process. Syst. **33**, 1877–1901 (2020)
4. Caldas, S., et al.: LEAF: a benchmark for federated settings. arXiv preprint arXiv:1812.01097 (2018)
5. Chen, F., Long, G., Wu, Z., Zhou, T., Jiang, J.: Personalized federated learning with graph. arXiv preprint arXiv:2203.00829 (2022)
6. Cho, Y.J., Liu, L., Xu, Z., Fahrezi, A., Joshi, G.: Heterogeneous LoRA for federated fine-tuning of on-device foundation models. arXiv preprint arXiv:2401.06432 (2024)
7. Devlin, J., Chang, M.W., Lee, K., Toutanova, K.: BERT: pre-training of deep bidirectional transformers for language understanding. In: Proceedings of the 2019 Conference of the North American Chapter of the Association for Computational Linguistics: Human Language Technologies, vol. 1 (long and short papers), pp. 4171–4186 (2019)

8. Dosovitskiy, A., et al.: An image is worth 16x16 words: transformers for image recognition at scale. arXiv preprint arXiv:2010.11929 (2020)
9. Duan, M., et al.: FedGroup: efficient clustered federated learning via decomposed data-driven measure. arXiv preprint arXiv:2010.06870 (2020)
10. Ghosh, A., Chung, J., Yin, D., Ramchandran, K.: An efficient framework for clustered federated learning. Adv. Neural. Inf. Process. Syst. **33**, 19586–19597 (2020)
11. Hsu, T.M.H., Qi, H., Brown, M.: Measuring the effects of non-identical data distribution for federated visual classification. arXiv preprint arXiv:1909.06335 (2019)
12. Hu, E.J., et al.: LoRA: low-rank adaptation of large language models. ICLR **1**(2), 3 (2022)
13. LeCun, Y., Bottou, L., Bengio, Y., Haffner, P.: Gradient-based learning applied to document recognition. Proc. IEEE **86**(11), 2278–2324 (1998)
14. Li, T., Sahu, A.K., Zaheer, M., Sanjabi, M., Talwalkar, A., Smith, V.: Federated optimization in heterogeneous networks. Proc. Mach. Learn. Syst. **2**, 429–450 (2020)
15. Li, T., Sahu, A.K., Zaheer, M., Sanjabi, M., Talwalkar, A., Smithy, V.: FedDANE: a federated Newton-type method. In: 2019 53rd Asilomar Conference on Signals, Systems, and Computers, pp. 1227–1231. IEEE (2019)
16. Lin, I.C., Yagan, O., Joe-Wong, C.: FedSPD: a soft-clustering approach for personalized decentralized federated learning (2024)
17. Long, G.: The rise of federated intelligence: from federated foundation models toward collective intelligence. In: Larson, K. (ed.) Proceedings of the Thirty-Third International Joint Conference on Artificial Intelligence, IJCAI-24, pp. 8547–8552. International Joint Conferences on Artificial Intelligence Organization (2024). https://doi.org/10.24963/ijcai.2024/980, early Career
18. Long, G., Xie, M., Shen, T., Zhou, T., Wang, X., Jiang, J.: Multi-center federated learning: clients clustering for better personalization. World Wide Web **26**(1), 481–500 (2023)
19. Mansour, Y., Mohri, M., Ro, J., Suresh, A.T.: Three approaches for personalization with applications to federated learning. arXiv preprint arXiv:2002.10619 (2020)
20. McMahan, B., Moore, E., Ramage, D., Hampson, S., y Arcas, B.A.: Communication-efficient learning of deep networks from decentralized data. In: Artificial Intelligence and Statistics, pp. 1273–1282. PMLR (2017)
21. Oh, J., Kim, S., Yun, S.Y.: FedBABU: towards enhanced representation for federated image classification. arXiv preprint arXiv:2106.06042 (2021)
22. Ouyang, X., Xie, Z., Zhou, J., Huang, J., Xing, G.: ClusterFL: a similarity-aware federated learning system for human activity recognition. In: Proceedings of the 19th Annual International Conference on Mobile Systems, Applications, and Services, pp. 54–66 (2021)
23. Radford, A., et al.: Learning transferable visual models from natural language supervision. In: International Conference on Machine Learning, pp. 8748–8763. PmLR (2021)
24. Sandler, M., Howard, A., Zhu, M., Zhmoginov, A., Chen, L.C.: MobileNetV2: inverted residuals and linear bottlenecks (2018)
25. Sattler, F., Müller, K.R., Samek, W.: Clustered federated learning: model-agnostic distributed multitask optimization under privacy constraints. IEEE Trans. Neural Netw. Learn. Syst. **32**(8), 3710–3722 (2020)
26. Tan, Y., Chen, C., Zhuang, W., Dong, X., Lyu, L., Long, G.: Is heterogeneity notorious? Taming heterogeneity to handle test-time shift in federated learning. Adv. Neural. Inf. Process. Syst. **36**, 27167–27180 (2023)

27. Tan, Y., et al.: FedProto: federated prototype learning across heterogeneous clients. In: Proceedings of the AAAI Conference on Artificial Intelligence, vol. 36, pp. 8432–8440 (2022)
28. Tan, Y., Long, G., Ma, J., Liu, L., Zhou, T., Jiang, J.: Federated learning from pre-trained models: a contrastive learning approach. Adv. Neural. Inf. Process. Syst. **35**, 19332–19344 (2022)
29. Wang, H., Yurochkin, M., Sun, Y., Papailiopoulos, D., Khazaeni, Y.: Federated learning with matched averaging. arXiv preprint arXiv:2002.06440 (2020)
30. Wu, F., Hu, J., Min, G., Wang, S.: Adaptive rank allocation for federated parameter-efficient fine-tuning of language models. arXiv preprint arXiv:2501.14406 (2025)
31. Wu, Y., et al.: A survey on federated fine-tuning of large language models. arXiv preprint arXiv:2503.12016 (2025)
32. Xu, M., Cai, D., Wu, Y., Li, X., Wang, S.: {FwdLLM}: efficient federated finetuning of large language models with perturbed inferences. In: 2024 USENIX Annual Technical Conference (USENIX ATC 2024), pp. 579–596 (2024)
33. Yan, P., Long, G.: Personalization disentanglement for federated learning. In: 2023 IEEE International Conference on Multimedia and Expo (ICME), pp. 318–323. IEEE (2023)
34. Yan, P., Long, G.: Client-supervised federated learning: towards one-model-for-all personalization. In: 2024 IEEE International Conference on Multimedia and Expo (ICME), pp. 1–6. IEEE (2024)
35. Yang, Y., Long, G., Lu, Q., Zhu, L., Jiang, J., Zhang, C.: Federated low-rank adaptation for foundation models: a survey (2025). https://arxiv.org/abs/2505.13502
36. Yang, Y., Long, G., Shen, T., Jiang, J., Blumenstein, M.: Dual-personalizing adapter for federated foundation models. In: Globerson, A., et al. (eds.) Advances in Neural Information Processing Systems, vol. 37, pp. 39409–39433. Curran Associates, Inc. (2024)
37. Yi, L., Yu, H., Wang, G., Liu, X., Li, X.: pFedLoRA: model-heterogeneous personalized federated learning with LoRA tuning. arXiv preprint arXiv:2310.13283 (2023)
38. Zhang, Y., Chen, H., Lin, Z., Chen, Z., Zhao, J.: FedAC: an adaptive clustered federated learning framework for heterogeneous data. arXiv preprint arXiv:2403.16460 (2024)
39. Zhang, Y., Chen, H., Lin, Z., Chen, Z., Zhao, J.: LCFed: an efficient clustered federated learning framework for heterogeneous data. arXiv preprint arXiv:2501.01850 (2025)
40. Zhang, Z., Yang, Y., Dai, Y., Qu, L., Xu, Z.: When federated learning meets pre-trained language models' parameter-efficient tuning methods. arXiv preprint arXiv:2212.10025 (2022)

LEAP: An LLM-Based Evidence Augmented Pipeline for Table-Based Fact Verification

Hanwen Zhang[1,2], Qingyi Si[1], Peng Fu[1(✉)], Zheng Lin[1], Zhigang Lu[3], and Weiping Wang[1]

[1] Institute of Information Engineering, Chinese Academy of Sciences, Beijing, China
{zhanghanwen,fupeng,linzheng,wangweiping}@iie.ac.cn, siqingyi@huawei.com
[2] School of Cyber Security, University of Chinese Academy of Sciences, Beijing, China
[3] Western Sydney University, Sydney, Australia
z.lu@westernsydney.edu.au

Abstract. The reasoning ability on Table-based Fact Verification (TFV) has been explored extensively on Large Language Models (LLMs). Existing TFV approaches can be broadly categorized into two main paradigms: prompting and fine-tuning. However, most prompting methods highly depend on powerful closed-sourced LLMs, which causes data breach problems, while fine-tuning methods mainly focus on generating end-to-end answers but often lack explainability. In this paper, we introduce a three-stage pipeline framework LEAP for TFV. It provides a step-by-step solution through instruction tuning by decomposing the statement into sub-questions. It consists of three sub-modules: a sentence decomposer, a table-based question answerer, and a table-based evidence-augmented verifier. To specialize the sub-modules in their respective tasks, we construct the reasoning evidence by data distillation and pseudo-labeling. Through diverse experiments on four TFV benchmarks, we demonstrate that our LEAP achieves state-of-the-art performance. In particular, it even outperforms human performance with an accuracy of 2.02% on Infotabs. As an effective method suitable for scenarios that require data privacy, our LEAP framework exhibits strong generalization and versatility on different types of backbone models and datasets. Furthermore, our experiments validate the effectiveness of each sub-module, proving its suitability as a standalone model for its respective task.

Keywords: Table-based Fact Verification · Large Language Models · Instruction Tuning · Decomposing

1 Introduction

Table-based Fact Verification (TFV), a task designed to check the veracity of statement based on table content, is a fundamental task in table-based reasoning [16,32]. It has widespread downstream applications in tasks such as rumor

M. Yoshikawa et al. (Eds.): ADMA 2025, LNAI 16197, pp. 206–221, 2026.
https://doi.org/10.1007/978-981-95-3453-1_14

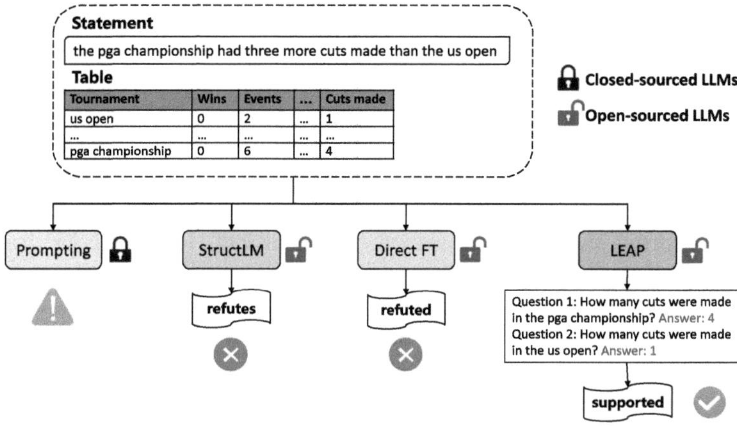

Fig. 1. A comparison between LEAP and existing LLM-based methods. Prompting closed-source LLMs is infeasible for safety reasons, while fine-tuning methods StructLM and Direct Fine-Tuning fail to provide correct answer. In contrast, LEAP delivers the correct judgment based on evidence in the form of QA pairs.

detection and fake news detection [30]. Figure 1 shows a TFV example from Tabfact [3] dataset. It requires three analytical steps to verify this statement: (1) filtering rows where the "Tournament" column is either "us open" or "pga championship", (2) retrieving the values in the "Cuts made" column for both tournaments, and (3) performing a subtraction to compare the two values. It demonstrates that the **two-dimensional structure** of tables, combined with the **complex implicit semantics** embedded in natural language statements, such as comparative expressions and logical conjunctions [3], presents significant challenges in effectively modeling the relationship between structured tables and unstructured textual statements.

To address the challenges above, previous works have made substantial efforts in advancing TFV. Traditional methods, based on small-scale pre-trained language models (SPLMs), are predominantly end-to-end and heavily rely on the quantity and quality of training data [7,9,14,28,36]. More recently, researchers have leveraged large language models (LLMs) to enhance table-based reasoning. Many studies utilize prompt engineering to decompose complex problems and generate reasoning steps [5,18,29,33], but such methods often require exposing private tables to closed-source LLMs, posing risks of data breaches and incurring high API costs. Moreover, some works try to leverage the open-source LLMs by instruction tuning. They have attempted to collect diverse table datasets and train open-source LLMs for table-related tasks [31,38]. However, these methods have limited accuracy on TFV tasks and offer only end-to-end solutions, making it difficult to understand the underlying decision-making process.

Therefore, we incorporate problem decomposition into the fine-tuning paradigm and propose **LEAP**, an **L**LM-based **E**vidence **A**ugmented **P**ipeline specifically designed for TFV. Unlike prior end-to-end approaches, LEAP explicitly separates problem decomposition from fact verification, improving reasoning transparency and factual consistency. The framework consists of three modules: 1) a Sentence Decomposer to break down complex statements, 2) a Table-based Sub-question Answerer to retrieve information and generate answers, and 3) a Table-based Evidence-augmented Fact Verifier that uses the sub-questions and answers as evidence. To address the scarcity of training data, we employ data distillation and pseudo-label generation strategy to obtain instruction data for module training. As the comparison shown in Fig. 1, our LEAP framework avoids the security risks of closed-sourced LLMs, enhances verification performance, and provides reasoning evidence in the form of question-answer pairs. Additionally, our approach exhibits strong versatility, making it applicable to not only standard relational tables but also horizontal and hierarchical tables.

Our contributions are listed as follows:

1. We propose an evidence augmented pipeline LEAP for TFV task, leveraging LLMs to decompose complex statements into sub-questions and provide answers. LEAP is especially suitable for scenarios that require data privacy, which also enhances the explainability of TFV task.
2. We introduce a novel instruction tuning dataset for sentence decomposition tailored to TFV task.
3. We fine-tune the sub-modules of LEAP, which can also be flexibly utilized as standalone models for sentence decomposition and table-based question answering (TQA).
4. Our method outperforms most LLM-based baselines that rely on end-to-end fine-tuning or prompting, achieving state-of-the-art (SOTA) performance on several in-domain and out-domain TFV datasets.

2 Related Work

An increasing number of studies are focusing on fact verification based on tabular-form evidence, creating many TFV benchmarks such as Tabfact [3], Infotabs [8], Semtabfact [23]. Early methods for TFV are developed in two main routes: program-based and pretrain-based. Program-based methods leverage semantic parsing to generate latent programs that incorporate numerical and logical semantics [3,21,28,36], which can be further leveraged to enhance the joint representation of sentences and tables. Besides, pretrain-based methods attempt to perform table-based pretraining or fine-tuning textual language models [7,9,14,34]. However, these methods highly rely on sufficient training data and still struggle to generalize on unseen datasets.

Recently, the emergence of LLMs with strong reasoning abilities has further advanced research on table-based reasoning tasks. Prompting [26] and instruction tuning [25] are two primary paradigms to utilize LLMs. While prompting uses crafted instructions and examples to prompt LLMs without additional training, instruction tuning enables LLMs to follow instructions by learning from annotated data [22]. Both paradigms have been effectively applied to table-based reasoning tasks, leading to notable progress.

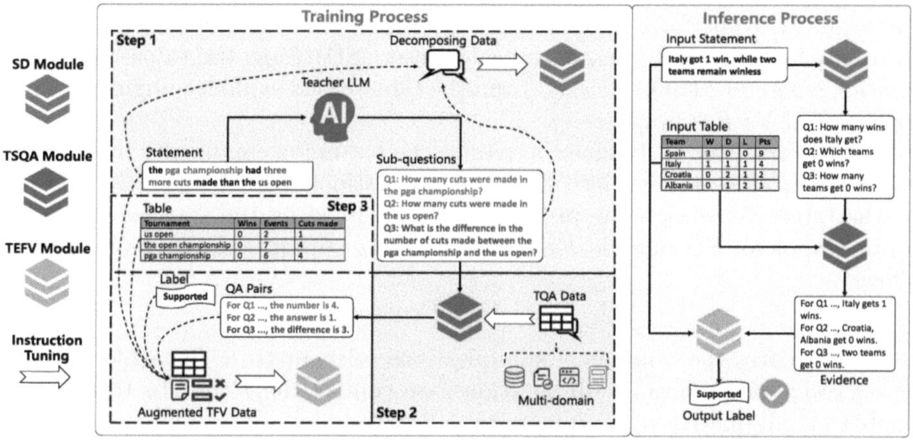

Fig. 2. The overall framework of LEAP, where the left and right parts show the training and inference processes, respectively.

For LLM prompting, most studies focus on designing few-shot prompts and reasoning steps to improve table understanding. TableCoT [2] enhances TQA and TFV tasks by adding explanations, while Binder [5] introduces symbolic languages for commonsense problems. Other methods, like Dater [29] and Tab-SQLify [18], simplify tables by prompting extraction of sub-tables. Chain-of-table [24] and ReAcTable [33] extend this approach to iterative table transformations. However, their outstanding performances often rely on large-scale closed-source LLMs, which raises privacy and cost concerns.

On the other hand, instruction tuning open-source LLMs provides another kind of solutions for table-based reasoning, with research mainly following two directions. One direction focuses on developing general-purpose table understanding models via instruction datasets and fine-tuning strategies. For example, TableLLaMA [31], fine-tuned on 14 table datasets with LongLoRA [4], performs well on both in-domain and out-domain tasks. The other one enhances task-specific reasoning by distilling knowledge from larger models. For table-to-text, [19] effectively distills GPT-3.5's generation capabilities to Flan-T5, while HeLM [1] fine-tunes a two-step framework. Nevertheless, table generalist LLMs still struggle with accuracy and explainability in TFV, and no existing methods are tailored for it. To address this, our proposed pipeline for TFV adopts the instruction tuning with data distillation and pseudo-labeling to fill the gaps.

3 Methodology

3.1 LEAP Framework

The overall framework of LEAP is shown in Fig. 2. Given a structured table T and a natural language statement S, the goal of LEAP is to determine the veracity label V of S with respect to T. Instead of end-to-end reasoning, LEAP enables LLMs to perform different roles within TFV process. It consists of three LLM-based modules: the sentence decomposer (SD) F_{SD}, the table-based sub-question answerer (TSQA) F_{TSQA}, and the table-based evidence-augmented fact verifier (TEFV) F_{TEFV}.

The **SD module** is designed to reduce the reasoning complexity by breaking down a complex statement into simpler sub-questions, without directly referencing the table. We design the prompt $Prompt_S$ filled by the statement to form the full input for SD module F_{SD}. The resulting sub-questions are obtained as follows:

$$\{SQ_i\}_{i=1}^n = F_{SD}(Prompt_S(S)), \tag{1}$$

where n denotes the number of sub-questions relevant to S. To avoid generating excessive or irrelevant sub-questions, we control the $Prompt_S$ to limit the number of sub-questions to no more than 4.

The **TSQA module** is responsible for answering the decomposed sub-questions. We prompt TSQA module F_{TSQA} to answer the sub-questions using the static strategy proposed in [27], where sub-questions are answered sequentially. The answer to each sub-question SQ_i is generated as follows:

$$SA_i = F_{TSQA}(Prompt_Q(SQ_i, T)), \tag{2}$$

where $Prompt_Q$ is the prompt employed to generate answers from the table T. These question-answer (QA) pairs $\{SQ_i, SA_i\}_{i=1}^n$ summarize the key information relevant to the statement S, which can serve as the evidence for TFV.

The **TEFV module**, guided by the prompt $Prompt_F$ shown in Fig. 3, leverages the evidence in the form of QA pairs to verify the truthfulness of the original statement. The final veracity label V is assigned through an augmented process:

$$V = F_{TEFV}(Prompt_F(S, T, \{SQ_i, SA_i\}_{i=1}^n)) \tag{3}$$

Table-based Evidence Augmented Fact Verification

Instruction
Your task is to determine the verdict about a statement towards the given table.
The verdict labels include: [label candidates]
In addition, we have broken down the statement into sub-questions and given the answers to those sub-questions. These Q&A pairs can be used as evidence in your judgment of verdict. Please note that the evidence is not always available, and the answers are sometimes wrong.
Input
Table Caption: {table caption}\nTable: {table}\nStatement: {statement}\nEvidence:\nQ1: {sub-question 1}\nAnswer to Q1: {answer 1}\nQ2: {sub-question 2}\nAnswer to Q2: {answer 2} ...
Output
{label}

Fig. 3. The prompt for table-based evidence-augmented fact verifier. The "Output" section is left empty during the inference stage.

3.2 Training Data Collection

Collecting high-quality instruction datasets is essential for developing LLM-based frameworks that rely on the fine-tuning paradigm. To train the sub-modules of LEAP, we employ data distillation and pseudo-label generation to obtain labels for unlabeled seed data, and integrate existing annotated datasets from various domains to construct instruction datasets. Specifically, we use the training sets of existing TFV datasets $W_{TFV} = \{t_j, s_j, v_j\}_{j=1}^{m_f}$ as seed data, where t_j, s_j and v_j represent the table, statement, veracity label, respectively, of the jth sample out of a total of m_f samples. For possible extension to private data, we assume that the tables in TFV datasets are also inaccessible to online closed-source LLMs.

Annotated Data for Training SD Module. The decomposition of the statement sentences can be performed without tables. Therefore, we distill the answer labels (i.e., sub-questions) of the statements in TFV datasets from an intelligent teacher LLM. This process can be symbolically represented as:

$$\{s\hat{q}_{j_i}\}_{i=1}^{n_j} = TM(Prompt_{DS}(s_j)), \tag{4}$$

where TM is the teacher LLM, $Prompt_{DS}$ is the prompt for distilling sentence decomposition labels, $\{s\hat{q}_{j_i}\}_{i=1}^{n_j}$ is a list of sub-questions broken down from s_j. The distilled data $W_{SD} = \{s_j, \{s\hat{q}_{j_i}\}_{i=1}^{n_j}\}_{j=1}^{m_f}$ is prepared for further training. After the distillation process, we apply data filtering operation to remove samples with unexpected annotations.

Annotated Data for Training TSQA Module. Given that answering the sub-questions still involves some logical or numerical operations, we collect several representative public TQA datasets covering different scenarios and domains. A detailed description of these datasets can be found in Sect. 4.1. We integrate the collected datasets for TSQA, denoted as $W_{TSQA} = \{t_{0,k}, q_{0,k}, a_{0,k}\}_{k=1}^{m_q}$, where $t_{0,k}$, $q_{0,k}$, $a_{0,k}$ are denoted as table, question and answer of the kth sample, respectively, while m_q represents the total number of TQA samples.

Annotated Data for Training TEFV Module. When distilling data for the SD module, the sub-question set $\{s\hat{q}_{j_i}\}_{i=1}^{n}$ related to the statement s_j from the seed data W_{TFV} is available, while the corresponding answer set $\{s\hat{a}_{j_i}\}_{i=1}^{n}$ remains inaccessible. To address this limitation, we adopt pseudo-labeling, a technique demonstrated to be effective for expanding training sets in scenarios with limited labeled data [13]. Specifically, we leverage pseudo-label generation to construct our training data. Once the TSQA module F_{TSQA} is fine-tuned, we perform inference on W_{TFV} to generate the answer for sub-questions $s\hat{q}_{j_i}$:

$$s\hat{a}_{j_i} = F_{TSQA}(Prompt_Q(s\hat{q}_{j_i}, t_j)) \tag{5}$$

Thus, we construct the evidence-augmented TFV training set $W_{TEFV} = \{t_j, s_j, \{s\hat{q}_{j_i}, s\hat{a}_{j_i}\}_{i=1}^{n_j}, v_j\}_{j=1}^{m_f}$, which is used to train the TEFV module in a semi-supervised manner. For further instruction tuning, we need to convert these annotated datasets into task-specific instruction-tuning formats, comprising instruction, input, and output, as illustrated in Fig. 3.

Algorithm 1: Training Process of LEAP Modules

Input: TFV dataset $W_{TFV} = \{t_j, s_j, v_j\}_{j=1}^{m_f}$, TQA dataset $W_{TSQA} = \{t_{0,k}, q_{0,k}, a_{0,k}\}_{k=1}^{m_q}$, Teacher model TM
Output: Fine-tuned module F_{SD}, F_{TSQA}, F_{TEFV}
`// Step 1: Training SD module`
for $j \leftarrow 1$ **to** m_f **do**
  $\{s\hat{q}_{j_i}\}_{i=1}^{n_j} \leftarrow TM(Prompt_{DS}(s_j))$;
end for
Define dataset $W_{SD} = \{s_j, \{s\hat{q}_{j_i}\}_{i=1}^{n_j}\}_{j=1}^{m_f}$;
Fine-tune module F_{SD} on W_{SD} with instructions;
`// Step 2: Training TSQA module`
Fine-tune module F_{TSQA} on W_{TSQA} with instructions;
`// Step 3: Training TEFV module`
for $j \leftarrow 1$ **to** m_f **do**
  for $i \leftarrow 1$ **to** n_j **do**
    $s\hat{a}_{j_i} \leftarrow F_{TSQA}(Prompt_Q(s\hat{q}_{j_i}, t_j))$;
  end for
end for
Define dataset $W_{TEFV} = \{t_j, s_j, \{s\hat{q}_{j_i}, s\hat{a}_{j_i}\}_{i=1}^{n_j}, v_j\}_{j=1}^{m_f}$;
Fine-tune module F_{TEFV} on W_{TEFV} with instructions;

3.3 Module Training

We utilize the annotated datasets in instruction-following format to fine-tune the LLM-based modules of LEAP with next-word prediction object. We only consider the target tokens like "{label}" in Fig. 3 during the calculation of cross-entropy loss. Besides, given the long input derived from tabular data, we adapt parameter-efficient fine-tuning strategy LoRA [10] in all training steps to reduce the training cost. The complete training process of LEAP is shown in Algorithm 1.

4 Experimental Setup

4.1 Datasets and Metrics

We train and evaluate our LEAP framework on four TFV datasets: **1) Tabfact** [3], a binary classification benchmark with 16K web tables from Wikipedia and 118K human-crafted statements, offering both simple and complex test subsets based on reasoning difficulty, plus a small subset for human evaluation. **2) Semtabfact** [23], with 2K scientific tables and 4K annotated statements, presents more challenges due to specialized vocabulary and hierarchical structures. **3) Infotabs** [8] is a 3-class dataset focused on horizontal tables, which incorporates neutral statements as a distinct category. It offers three distinct test sets—test-α_1, test-α_2, and test-α_3—each designed for evaluation in traditional, adversarial, and cross-domain scenarios, respectively. **4) Scitab** [17] consists of 1K scientific claims requiring compositional reasoning on tables from science papers. We use it as an out-of-domain test set for cross-domain evaluation.

To train the TSQA module in LEAP, we use six existing TQA datasets from diverse sources, including WikiTQ [20], WikiSQL [35], TabMCQ [11], TabMWP

[15], AIT-QA [12], and TAT-QA [37]. WikiTQ and WikiSQL focus on structured tables from Wikipedia and typically require concise text fragment answers. TabMWP contains table-based mathematical word problems, demanding step-by-step derivations for accurate solutions. TAT-QA, derived from corporate financial reports, necessitates multi-hop reasoning by integrating table data with contextual information. AIT-QA, sourced from airline annual reports, exclusively addresses questions involving numerical data. Finally, TabMCQ features science exam multiple-choice questions based on hand-crafted tables.

For the evaluation of TFV tasks, we use accuracy for Tabfact and Infotabs, Micro-F1 for Semtabfact, and Macro-F1 for Scitab. To assess the performance of intermediate steps, we employ LLM evaluators as detailed in Sect. 5.3.

4.2 Implementation Details

Models and Devices. We use the same LLM structure for all modules in our experiments. LLaMA3-8B and InternLM2-7B are chosen as our backbone models. For label distilling, we use the open-source model Qwen2-72B as the teacher model, which is deployed on a remote platform. Except for label distilling, all other fine-tuning and inference steps are implemented on 2 Nvidia-A800 GPUs with 80G memory.
Training Parameters. During training, each module undergoes 2 epochs with a learning rate of 1e−4, utilizing a cosine scheduler that includes a 10% warm-up phase. To balance training speed and memory usage, the batch size is set to 8 for the SD module and 4 for the TSQA and TEFV modules. All modules have their maximum input length capped at 8,192 tokens, while the maximum generation length during inference is set to 1,024 tokens. For other generation parameters, we use a temperature of 0.95, a top-p value of 0.7, and a top-k value of 50. For LoRA configuration, a default rank of 8 and a scale factor of 16 are used.

4.3 Baselines

In our work, we compare LEAP with advancing TFV baselines, which can be categorized into the following groups:
1) SPLM-based methods. SPLM-based methods typically follow pre-training and fine-tuning paradigm, effectively generating a unified representation of the table and statement, which is then fed into the classification layer. We choose TAPAS [9], TAPEX [14], LKA [34], PASTA [7] as our baselines, which either utilize table-based pre-training or incorporate table-aware components. Some methods augmenting the unified representation from logical forms, such as LFC [36] and ProgVGAT [28], are also considered for Tabfact. All these methods have been fine-tuned on the TFV training sets.
2) LLM-based prompting methods. LLM-based prompting methods aim to generate the answer label induced by task instruction without extra training. We select open-source LLMs including Qwen2-72B, LLaMA3-8B and InternLM2-7B under zero-shot setting as our baselines. Additionally, we compare our approach with some advanced prompting frameworks on Tabfact, such as Dater [29] and

Chain-of-table [24]. To align with the high-security scenarios, we adapt these frameworks by replacing their original closed-source LLMs with open-source LLMs as their backbone models.
3) LLM-based fine-tuning methods. LLM-based fine-tuning methods utilize instruction tuning to adapt LLMs for table-related tasks. In this study, we use four fine-tuning baselines: two generalist models for tables, TableLlama [31] and StructLM [38], and two fine-tuned general-purpose models, LLaMA3-8B and InternLM2-7B, which are adapted through instruction tuning on TFV datasets. Since TableLlama and StructLM have been trained on various table-related tasks, we directly prompt them to perform TFV task.

Table 1. The results of different methods on Tabfact. For Dater and Chain-of-table, we report the results of our reproduction based on open-source models without employing voting strategy out of respect for computational efficiency.

Methods	Tabfact(Accuracy)			
	test	simple test	complex test	small test
SPLM-based				
LFC	71.7	85.4	65.1	74.3
ProgVGAT	74.4	88.3	67.6	76.2
TAPAS	81.0	92.3	75.6	83.9
TAPEX	84.2	93.9	79.6	85.9
PASTA	**89.3**	**96.7**	85.6	90.6
LKA	84.87	94.06	80.31	87.40
LLM-based prompting				
InternLM2-7B	71.88	78.25	68.78	71.77
Dater(InternLM2-7B)	–	–	–	70.45
Chain-of-table(InternLM2-7B)	–	–	–	73.27
LLaMA3-8B	77.86	85.9	73.97	79.38
Dater(LLaMA3-8B)	–	–	–	77.12
Chain-of-table(LLaMA3-8B)	–	–	–	76.33
Qwen2-72B	82.09	90.6	77.96	84.78
Dater(Qwen2-72B)	–	–	–	83.15
Chain-of-table(Qwen2-72B)	–	–	–	87.65
LLM-based fine-tuning				
TableLlama	82.29	90.82	78.16	83.53
StructLM	80.19	91.94	74.49	83.68
InternLM2-7B	80.34	91.46	74.95	83.28
LLaMA3-8B	87.99	94.97	84.61	89.69
LEAP(InternLM2-7B)	81.89	91.97	77.01	84.08
LEAP(LLaMA3-8B)	88.97	95.61	**85.75**	**90.74**
Human	–	–	–	92.1

5 Result Analysis

5.1 Main Result

We compare our proposed method LEAP with different kinds of baselines on three in-domain datasets Tabfact, Infotabs, Semtabfact and one out-domain dataset Scitab. We train LEAP on a combination of Tabfact, Semtabfact, and Infotabs, which is noted as Mixed dataset. Table 1 presents the results on Tabfact. We find that LEAP surpasses all LLM-based approaches and achieves SoTA performance. Notably, it outperforms the table-specific pre-trained model PASTA on both complex and small test subsets. With the same backbone model, LEAP improves the accuracy over direct fine-tuning by 0.51%–3.06%, while it significantly outperforms prompting methods that combine LLM reasoning with programmatic execution. Compared with prompted Qwen2-72B, fine-tuned LLMs with 7B-8B parameters achieve comparable or even superior results, demonstrating that instruction tuning effectively enhances the verification performance. We also observe that LEAP exhibits greater improvement on complex examples than on simpler ones, highlighting its superiority in complex reasoning tasks.

Table 2. The results of different methods on Infotabs, Semtabfact and Scitab. Methods with "†" mark means they are not trained on the target dataset directly. Results marked with "*" are from models fine-tuned on the Mixed dataset.

Methods	Infotabs (Accuracy)			Semtabfact (Micro-F1)	Scitab (Macro-F1)
	test-α_1	test-α_2	test-α_3	test	test
SPLM-based					
TAPAS	73.22	61.83	60.88	75.33	50.30
TAPEX	76.50	67.55	66.38	75.47	56.06
LKA	82.05	74.94	73.55	78.54	–
PASTA	–	–	–	84.10	–
LLM-based prompting					
InternLM2-7B	68.06	66.06	68.28	74.52	62.24
LLaMA3-8B	64.89	60.67	63.39	79.31	63.05
Qwen2-72B	75.39	74.94	73.55	78.54	76.29
GPT4	–	–	–	–	**78.22**
LLM-based fine-tuning					
TableLlama†	52.83	52.89	52.83	73.18	59.22
StructLM†	59.06	54.50	56.94	76.25	59.89
InternLM2-7B	75.33	68.67	68.67	75.10	60.06*
LLaMA3-8B	84.72	78.61	78.05	82.76	61.15*
LEAP(InternLM2-7B)	76.94	70.39	69.72	80.46	**65.04***
LEAP(LLaMA3-8B)	**86.06**	**80.89**	**79.28**	**87.74**	63.52*
Human	84.04	83.88	79.33	–	–

Instead of relational tables in Tabfact, the tables in Infotabs and Semtabfact contain multi-domain and heterogeneous information. From Table 2, we find that our proposed LEAP method achieves SoTA results on these datasets with non-standard tables. Specifically, LEAP based on LLaMA3-8B outperforms direct fine-tuning methods with a nearly 5% higher accuracy on Semtabfact and outperforms PASTA, the previous SoTA model, by a margin of 3.54%. On the standard test set (test-α_1) of Infotabs, LEAP even surpasses human performance by 2.02%. These results highlight LEAP's adaptability to diverse table formats.

Furthermore, LEAP demonstrates robust performance on the adversarial set (test-α_2) and cross-domain set (test-α_3) of Infotabs, underscoring its excellent robustness and transferability across varying data distribution.

On the out-domain dataset Scitab, LEAP also demonstrates significant advantages among fine-tuning methods. LEAP improves accuracy by 4.98% and 2.37% over direct fine-tuning InternLM2-7B and LLaMA3-8B, respectively. This indicates that LEAP's task decomposition and sub-task solving capabilities extend beyond in-domain data. Surprisingly, fine-tuning on the Mixed dataset underperforms compared to zero-shot prompting on both two backbone models. We hypothesize that this phenomenon arises from the differences of data distribution between Scitab and in-domain datasets, while LEAP effectively bridges this gap by reducing task complexity. Although LEAP, based on open-source models with 7B or 8B parameters, still lags behind closed-source models such as GPT-4, it achieves satisfactory accuracy and serves as a viable alternative in high-security scenarios.

5.2 Ablation Study

Datasets for Fine-Tuning. We investigate the impact of different TFV training datasets on model performance. Based on two backbone models, we employ LEAP and the end-to-end method, comparing instruction tuning on the Mixed dataset to tuning on only the target dataset Tabfact, as well as directly prompting without any training.

Table 3. Accuracy results on Tabfact for different methods with different instruction tuning datasets. Given the output of LEAP is not stable with zero-shot backbone models, we don't report the results of LEAP without fine-tuning.

Methods	Setting	Accuracy			
		test	simple test	complex test	small test
InternLM2-7B					
End-to-end	w/o Fine-tuning	71.88	78.25	68.78	71.77
	w/ Tabfact	80.34	91.46	74.95	83.28
	w/ Mixed	80.79	91.87	75.42	83.78
LEAP	w/ Tabfact	75.70	87.12	70.17	78.28
	w/ Mixed	**81.89**	**91.97**	**77.01**	**84.08**
LLaMA3-8B					
End-to-end	w/o Fine-tuning	77.86	85.90	73.97	79.38
	w/ Tabfact	87.99	94.97	84.61	89.69
	w/ Mixed	88.45	95.01	85.27	89.94
LEAP	w/ Tabfact	88.69	95.59	85.35	90.64
	w/ Mixed	**88.97**	**95.61**	**85.75**	**90.74**

The experimental results on Tabfact are presented in Table 3. They indicate that all methods fine-tuned on the Mixed dataset outperform those fine-tuned solely on Tabfact, demonstrating that utilizing more diverse and enriched datasets for fine-tuning can significantly enhance the model's accuracy and generalization ability for specific task. Moreover, further analysis reveals that the performance of different dataset choices varies across the two backbone models. When fine-tuning InternLM2-7B model exclusively on Tabfact, LEAP shows a 5% accuracy drop compared to the end-to-end method on Tabfact small test; however, replacing the backbone model with LLaMA3-8B reverses this trend: LEAP outperforms the end-to-end method by approximately 1% (90.64% vs 89.69%). We hypothesize that the LLaMA3-8B model grasps more intrinsic knowledge, enabling it to learn effective evidence-augmented verification patterns even with fewer data. In contrast, LLMs with less intrinsic knowledge require more extensive datasets to fully exploit the advantages of LEAP.

Intermediate Modules. To assess the contributions of individual modules, we conduct ablation experiments with two configurations: (1) w/o TSQA: The TSQA module is removed, and the generated sub-questions are directly used as evidence for TFV. (2) w/o SD+TSQA: Both the SD and TSQA modules are removed, adopting end-to-end method for TFV.

Table 4. Ablation analysis of LEAP based on LLaMA3-8B

Methods	Tabfact (Accuracy)	Infotabs (Accuracy)	Semtabfact (Micro-F1)	Scitab (Macro-F1)	Avg.
LEAP	**88.97**	**86.06**	87.74	**63.52**	**81.57**
w/o TSQA	88.24	85.06	**87.93**	60.90	80.53
w/o SD+TSQA	88.45	**86.06**	87.36	61.15	80.76

We conduct experiments using LLaMA3-8B as the backbone model across four TFV datasets, with the results summarized in Table 4. The results indicate that removing TSQA module or both SD and TSQA modules leads to varying degrees of performance degradation on most test sets with average drops of 1.04% and 0.81%, respectively. Furthermore, the performance decline is notably more remarkable on complex datasets. For example, removing TSQA results in a 2.62% drop on the Scitab test set. Interestingly, jointly removing SD and TSQA modules (w/o SD+TSQA) sometimes slightly outperforms removing TSQA module alone (w/o TSQA). This suggests that the effectiveness of SD module may be influenced by the collaboration with TSQA modules, while the complete combination of steps can handle complex tasks more efficiently. In a few cases, performance slightly improves after module ablation, possibly due to LEAP's sensitivity to dataset biases or error propagation among modules.

5.3 Evaluation for Sub-modules

To evaluate the capability of the SD and TSQA module, we compare our fine-tuned modules against the vanilla backbone on TFV datasets. Given the lack of golden references, we prompt a large-scale LLM as an evaluator [6] to rate the sub-questions and their answers on 0–5 scales based on strict evaluation criteria, considering completeness, correctness, coherence, and faithfulness. Specifically, we present the LLM-based evaluator with comparative outputs from both methods rather than individual assessments, which could improve alignment between LLM judgments and human annotations. We employ LLaMA3-8B and Qwen2-72B as the vanilla backbone and the evaluator, respectively.

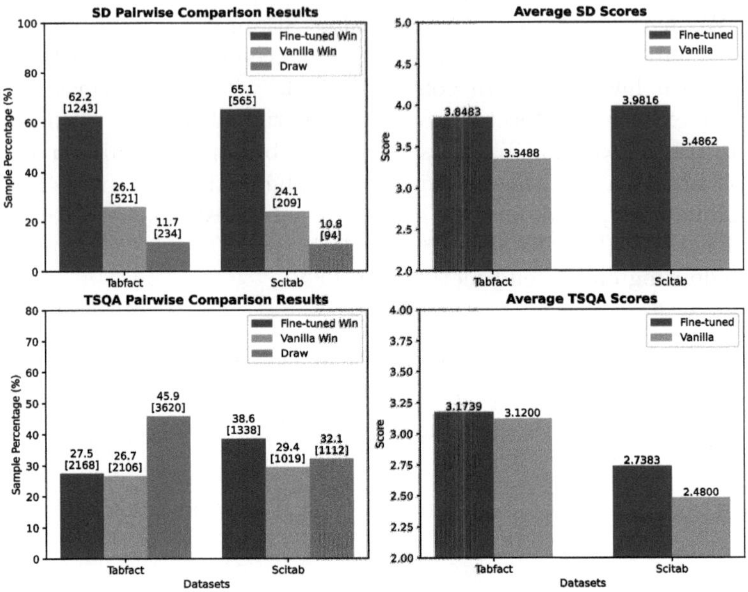

Fig. 4. The comparison between Fine-tuned SD/TSQA Module (on the Mixed dataset) and Vanilla backbone (LLaMA3-8B). The left two sub-figures show the percentage and number of samples with different battle results, while the right ones display the average score of samples on full datasets.

The pairwise comparison results and average scores on Tabfact (small-test) and Scitab are presented in Fig. 4. The fine-tuned SD module consistently outperforms the vanilla LLM, winning over 60% of samples and achieving an average score improvement of approximately 0.5 points, validating the effectiveness of instruction tuning on distilled datasets for sentence decomposition. For TSQA module, we evaluate sub-question answering using questions generated by the fine-tuned SD module. The fine-tuned TSQA module shows a slight gain (+0.05) on Tabfact, while it performs markedly better on Scitab, winning 38.6% of samples and improving the average score by 0.26 points. These results demonstrate that instruction tuning on diverse TQA datasets significantly boosts the LLM's sub-question answering capabilities.

6 Conclusion

In this paper, we introduce LEAP, a pipeline framework by augmenting the evidence of verification to improve the TFV ability. By instruction tuning open-source LLMs, it is especially well-suited for the scenarios demanding high levels of data privacy. LEAP is comprised of a sentence decomposer to decompose the original statements, a table-based sub-question answerer to solve the sub-questions, and a table-based evidence-augmented fact verifier to integrate the evidence and determine the veracity. To train these sub-modules, we construct task-specific instruction datasets and acquire training labels by employing data distillation and pseudo-labeling strategy. Experimental results show that LEAP significantly improves the accuracy and achieves SOTA performance on four TFV datasets, even surpassing human performance on Infotabs. Further analysis on sub-modules reveals that LEAP benefits from instruction tuning on diverse and enriched datasets, which equips it with robust capabilities for decomposing complex sentences and answering table-related questions. We also notice that the generated evidence can enhance the explainability of TFV tasks. For future work, we will explore the solutions to develop a more intelligent TFV agent in high-security scenarios, which is capable of planning the free-form procedures and utilizing offline external tools.

Acknowledgments. This work was supported by the National Natural Science Foundation of China (Nos. 62472419, 62472420).

References

1. Bian, J., et al.: HeLM: highlighted evidence augmented language model for enhanced table-to-text generation. arXiv preprint arXiv:2311.08896 (2023)
2. Chen, W.: Large language models are few(1)-shot table reasoners. In: Findings of ACL: EACL, pp. 1090–1100 (2023)
3. Chen, W., et al.: TabFact: a large-scale dataset for table-based fact verification. In: ICLR (2020)
4. Chen, Y., et al.: LongLoRA: efficient fine-tuning of long-context large language models. In: ICLR (2024)
5. Cheng, Z., et al.: Binding language models in symbolic languages. In: ICLR (2023)
6. Fu, J., Ng, S., Jiang, Z., Liu, P.: GPTScore: evaluate as you desire. In: NAACL, pp. 6556–6576 (2024)
7. Gu, Z., Fan, J., Tang, N., Nakov, P., Zhao, X., Du, X.: PASTA: table-operations aware fact verification via sentence-table cloze pre-training. In: EMNLP (2022)
8. Gupta, V., Mehta, M., Nokhiz, P., Srikumar, V.: INFOTABS: inference on tables as semi-structured data. In: ACL, pp. 2309–2324 (2020)
9. Herzig, J., Nowak, P.K., Müller, T., Piccinno, F., Eisenschlos, J.M.: Tapas: weakly supervised table parsing via pre-training. In: ACL (2020)
10. Hu, E.J., et al.: LoRA: low-rank adaptation of large language models. arXiv preprint arXiv:2106.09685 (2021)
11. Jauhar, S.K., Turney, P.D., Hovy, E.H.: TabMCQ: a dataset of general knowledge tables and multiple-choice questions. CoRR **abs/1602.03960** (2016)

12. Katsis, Y., et al.: AIT-QA: question answering dataset over complex tables in the airline industry. arXiv preprint arXiv:2106.12944 (2021)
13. Lee, D.H., et al.: Pseudo-label: the simple and efficient semi-supervised learning method for deep neural networks. In: Workshop on Challenges in Representation Learning, Atlanta, vol. 3, p. 896. ICML (2013)
14. Liu, Q., et al.: TAPEX: table pre-training via learning a neural SQL executor. In: ICLR (2022)
15. Lu, P., et al.: Dynamic prompt learning via policy gradient for semi-structured mathematical reasoning. In: ICLR (2023)
16. Lu, W., Zhang, J., Fan, J., Fu, Z., Chen, Y., Du, X.: Large language model for table processing: a survey. Frontiers Comput. Sci. **19**(2), 192350 (2025)
17. Lu, X., Pan, L., Liu, Q., Nakov, P., Kan, M.: SCITAB: a challenging benchmark for compositional reasoning and claim verification on scientific tables. In: EMNLP, pp. 7787–7813 (2023)
18. Nahid, M.M.H., Rafiei, D.: TabSQLify: enhancing reasoning capabilities of LLMs through table decomposition. In: NAACL, pp. 5725–5737 (2024)
19. Nan, L., et al.: Enhancing text-to-SQL capabilities of large language models: a study on prompt design strategies. In: Findings of ACL: EMNLP, pp. 14935–14956 (2023)
20. Pasupat, P., Liang, P.: Compositional semantic parsing on semi-structured tables. In: ACL-IJCNLP, pp. 1470–1480 (2015)
21. Shi, Q., Zhang, Y., Yin, Q., Liu, T.: Logic-level evidence retrieval and graph-based verification network for table-based fact verification. In: EMNLP, pp. 175–184 (2021)
22. Taori, R., et al.: Stanford alpaca: an instruction-following LLaMA model (2023). https://github.com/tatsu-lab/stanford_alpaca
23. Wang, N.X.R., Mahajan, D., Danilevsky, M., Rosenthal, S.: SemEval-2021 task 9: fact verification and evidence finding for tabular data in scientific documents (SEM-TAB-FACTS). In: Proceedings of the 15th International Workshop on Semantic Evaluation, SemEval@ACL/IJCNLP 2021, pp. 317–326 (2021)
24. Wang, Z., et al.: Chain-of-table: evolving tables in the reasoning chain for table understanding. In: ICLR (2024)
25. Wei, J., et al.: Finetuned language models are zero-shot learners. In: ICLR (2022)
26. Wei, J., et al.: Emergent abilities of large language models. Trans. Mach. Learn. Res. **2022** (2022)
27. Wu, Z., et al.: Divide-or-conquer? Which part should you distill your LLM? In: Findings of ACL: EMNLP, pp. 2572–2585 (2024)
28. Yang, X., Nie, F., Feng, Y., Liu, Q., Chen, Z., Zhu, X.: Program enhanced fact verification with verbalization and graph attention network. In: EMNLP, pp. 7810–7825 (2020)
29. Ye, Y., Hui, B., Yang, M., Li, B., Huang, F., Li, Y.: Large language models are versatile decomposers: decomposing evidence and questions for table-based reasoning. In: SIGIR (2023)
30. Zhang, H., Si, Q., Fu, P., Lin, Z., Wang, W.: Are large language models table-based fact-checkers? In: CSCWD, pp. 3086–3091 (2024)
31. Zhang, T., Yue, X., Li, Y., Sun, H.: TableLlama: towards open large generalist models for tables. In: NAACL-HLT (2024)
32. Zhang, X., Wang, D., Dou, L., Zhu, Q., Che, W.: A survey of table reasoning with large language models. Frontiers Comput. Sci. **19**(9), 199348 (2025)

33. Zhang, Y., Henkel, J., Floratou, A., Cahoon, J., Deep, S., Patel, J.M.: Reactable: enhancing react for table question answering. Proc. VLDB Endow. **17**(8), 1981–1994 (2024)
34. Zhao, G., Yang, P.: Table-based fact verification with self-labeled keypoint alignment. In: COLING, pp. 1401–1411 (2022)
35. Zhong, V., Xiong, C., Socher, R.: Seq2SQL: generating structured queries from natural language using reinforcement learning. CoRR **abs/1709.00103** (2017)
36. Zhong, W., et al.: LogicalFactChecker: leveraging logical operations for fact checking with graph module network. In: ACL, pp. 6053–6065 (2020)
37. Zhu, F., et al.: TAT-QA: a question answering benchmark on a hybrid of tabular and textual content in finance. In: ACL-IJCNLP, pp. 3277–3287 (2021)
38. Zhuang, A., et al.: StructLM: towards building generalist models for structured knowledge grounding. CoRR **abs/2402.16671** (2024)

Strategic Reading Skills Work: Perceiving Locally and then Reasoning Globally Improves Emotion Recognition

Chuwen Wang and Cheng Wang(✉)

School of Computer Science and Technology, Tongji University, Shanghai 201804, China
{2331924,cwang}@tongji.edu.cn

Abstract. Multimodal Emotion Recognition in Conversations (ERC) is crucial for understanding human communication, aiming to infer speakers' emotional states through verbal and nonverbal cues. While current graph-based models utilize Graph Neural Networks (GNNs) in conversation graphs, they often focus on developing complex neural network architectures and multimodal fusion processes, neglecting the optimization of the underlying graph structure. The graph structures where nodes are linked to a quantity of other nodes are of high complexity in computation but low efficiency in expressing temporal and local information for emotion inference. Inspired by strategic skills in reading comprehension, we propose a novel approach that transforms multimodal conversations into heterogeneous graphs with reduced edge density, which significantly decreases the computational cost of GNN operations. Accordingly, we complement this with our proposed Multimodal Graph Reasoning Network (MMGRN) to effectively process the optimized graph structure of rich temporal and local information. Our method outperforms the best graph model without recurrent neural network in two benchmark datasets, achieving 4.2% and 0.4% improvements in two typical multimodal ERC datasets MELD and IEMOCAP, respectively, while significantly increasing processing speed. This work demonstrates that improving the foundational graph structure, combined with efficient reasoning mechanisms, can lead to substantial advancements in multimodal emotion recognition tasks.

Keywords: Emotion Recognition · Graph Neural Network · Multi modality

1 Introduction

There is a prevalent phenomenon in our daily lives that text-only communication is more likely to be misinterpreted than video calls. One of the important factors is the lack of an accurate understanding of the other party's emotions. It highlights the significant role that emotions play in effective communication, emphasizing the need for Emotion Recognition in Conversation (ERC), and further multimodal ERC. Multimodal ERC enables real-time recognition of the speaker's emotions through various modalities of information, which can be a helpful technology to improve the experience of interpersonal or human-computer communication on devices [1].

M. Yoshikawa et al. (Eds.): ADMA 2025, LNAI 16197, pp. 222–236, 2026.
https://doi.org/10.1007/978-981-95-3453-1_15

Early research on ERC primarily focused on Recurrent Neural Network (RNN) models, such as the bc-LSTM model [2] and ICON [3]. RNN-based methods analyze conversations sequentially, processing utterances one by one. These approaches, while intuitive, lead to low computational efficiency, especially for long conversations or large datasets. The sequential nature of RNNs also limits their ability to capture long-range dependencies effectively.

To address the limitation in low efficiency of sequential processing, researchers turned to GNNs, which offer a more flexible framework for modeling conversational dynamics. DialogueGCN [6] pioneered this approach by introducing the concept of a conversation graph. In this paradigm, contextual information can be directly propagated among utterances through graph edges, while speaker information is easily incorporated into node representations. Following DialogueGCN, the use of conversation graphs and information extraction through Message Passing Neural Networks (MPNNs) [7–11] has become a prominent approach in multimodal ERC. Most existing GNN methods employ a fully connected graph structure. While this ensures comprehensive information flow, it also introduces significant computational overhead. The time and space complexity of these models grows quadratically with the number of utterances, making them still inefficient for long conversations. Moreover, the fully connected structure, while comprehensive, can lead to noise in emotion recognition for individual nodes (utterances). This is because each node receives information from all other nodes, including those that may not be directly relevant to its emotional context, and this structure also diminishes the sequential relationship of nodes. This indiscriminate information aggregation can dilute the signal of truly relevant emotional cues.

Optimizing graph structures and developing corresponding multimodal fusion methods presents a promising solution to the aforementioned challenges. In this paper, we draw inspiration from strategies employed in reading comprehension tests. Extracting related information from local contexts and reasoning based on global emotional foundations can effectively mitigate the challenges of efficiency and noise in GNN-based ERC models. In standardized reading comprehension assessments, a common approach involves scrutinizing details for each question within a passage rather than perusing the entire text. Typically, readers focus on a few sentences in the paragraph directly related to the query, subsequently retrieving supplementary information, such as main ideas, from the broader context. Emulating this cognitive process, we propose a novel "global-to-local" conversation graph construction paradigm to gather local information. By exclusively connecting to neighbor nodes in graph and limited MPNN layers, it preserves temporal information and captures speaker-specific data through a more localized and efficient structure among utterances in comparison with approaches based on fully connected graphs. We propose a novel plug-and-play Global Reasoning Module (GRM) to conduct ERC corresponding to our graph construction paradigm, which jointly forms the Multimodal Graph Reasoning Network (MMGRN). MMGRN is specifically designed to address the multimodal ERC task. It leverages a streamlined GATv2 [12] architecture, serving dual roles as both a contextual information extractor and a multimodal fusion module. GRM performs global reasoning on MMGRN-processed multimodal features and global graph memory which are both converted from processed graph node embeddings.

To evaluate the efficacy of MMGRN on our conversation graph, we conduct comprehensive experiments using two widely-recognized ERC datasets: IEMOCAP [13] and MELD [14]. The results unequivocally demonstrate that our approach surpasses other GNN-based models, achieving the best performance among GNN-based models as measured by the weighted-average F1 score on both datasets, while possessing outstanding efficiency.

In summary, the primary contributions of this work are as follows:

- We propose a novel methodology for transforming multimodal conversations into heterogeneous graphs for ERC tasks. It has outstanding efficiency so that it is promising to be implemented on devices with relatively limited computational capability.
- We introduce MMGRN, a network architecture tailored to our conversation graph representation, which outperforms existing GNN-based models in ERC tasks.
- We present a plug-and-play GRM that significantly enhances the performance of non-fully connected MPNNs, offering potential applications on GNNs beyond the scope of this study.

2 Related Work

Recent advances in ERC have primarily evolved along three methodological paradigms: GNN-based approaches, RNN-based approaches, and Large Language Model (LLM) augmented approaches. These approaches have demonstrated varying degrees of effectiveness in capturing conversational dynamics and emotional contexts.

2.1 GNN-Based Models

The GNN-based approaches has demonstrated significant potential in modeling conversational structures and emotional dynamics. DialogueGCN [6] pioneered the application of graph convolution networks for encoding speaker-level contextual information. MMGCN [8] advanced this approach by incorporating speaker embeddings and a sophisticated graph structure encompassing both intra-modal and inter-modal relationships. MM-DFN [9] further refined this architecture by introducing dynamic multimodal graph fusion. The current state-of-the-art GNN model, GraphCFC [10], introduced a Pair-wise Cross-modal Complementation (PairCC) mechanism, decomposing conversation graphs into modality-specific subgraphs for enhanced feature extraction. GraphMFT [11] extended this paradigm by implementing an enhanced Graph Attention Network architecture, incorporating skip-connections and GATv2 [12] modifications for robust feature extraction. These graph-based architectures typically adopt one of two fundamental graph construction paradigms: either a fully weighted connectivity scheme based on embedding similarities, exemplified by MMGCN, or a selective connectivity pattern linking utterances within specified temporal windows, as demonstrated in GraphCFC. However, both approaches necessitate preliminary temporal feature extraction through LSTM or similar mechanisms, potentially introducing computational inefficiencies and redundant information processing.

2.2 RNN-Based Models

The RNN-based approaches has emphasized the capture of temporal dynamics in emotional expressions. M2R2 [15] proposed RNN-based party attentive network to track all speakers' states and context. EmoCaps [16] introduced the innovative Emoformer architecture, incorporating multimodal emotion vectors with sentence embeddings through emotion capsules, while utilizing LSTM for temporal context modeling. MALN [17] enhanced multimodal fusion efficiency through adversarial learning techniques, explicitly modeling both commonalities and distinctions across modalities. CM-RoBERTa [18] implemented a sophisticated parallel self-attention and cross-attention mechanism for dynamic feature extraction across both inter-modal and intra-modal dimensions. CBERL [19] designed a multimodal generative adversarial network and proposed an RNN-based deep joint variational autoencoder to refine and fuse multimodal semantic information. The performance of RNN-based models have currently surpassed that of the GNN-based models with its chronological inference which is more fine-grained and consistent with the nature of conversation. However, its linear increase in time consumption limits its application in real-time ERC, especially for long conversations.

2.3 LLM-Augmented Models

The emergence of LLM-augmented approaches has introduced novel perspectives in emotional recognition. BiosERC [20] leverages LLMs for extracting and incorporating speaker biographical information, enhancing the understanding of individual emotional characteristics. SpeechCueLLM [21] developed an elegant approach for translating acoustic features into natural language descriptions, effectively utilizing LLMs' inherent language understanding capabilities. CKERC [22] advanced this paradigm by generating and incorporating interlocutors' commonsense knowledge through LLMs, enabling more nuanced emotional analysis in conversational contexts. LLM-augmented models leverage stored knowledge from LLMs' pretraining on large-scale corpus. These additional training data contributes to the enhancement in performance. However, its larger scale also leads to the limited application prospect in light devices and scenarios.

3 Methodology

The multimodal ERC task operates on a conversation comprising N utterances U

$$U = [u_1, \cdots, u_N], \tag{1}$$

$$u_i = \{u_i^a, u_i^v, u_i^t\}, \tag{2}$$

where each utterance u_i encompasses acoustic (u_i^a), visual (u_i^v), and textual (u_i^t) modalities. The conversation involves M speakers, and the speakers set P is represented as (Fig. 1):

$$P = \{p_1, \ldots, p_M\}(M \geq 1), \tag{3}$$

where speaker $p_{\phi(u_i)}$ articulates utterance u_i. The mapping function ϕ associates each utterance with its corresponding speaker. For speaker p_λ, the set of their utterances is denoted as U_λ:

$$U_\lambda = \{u_i | u_i \in U \wedge p_{\phi(u_i)} = p_\lambda, \forall i \in [1, N]\}, \lambda \in [1, M]. \tag{4}$$

The objective is to classify each utterance u_i into an emotion label y_i from a predefined emotion set $\mathcal{Y}$.

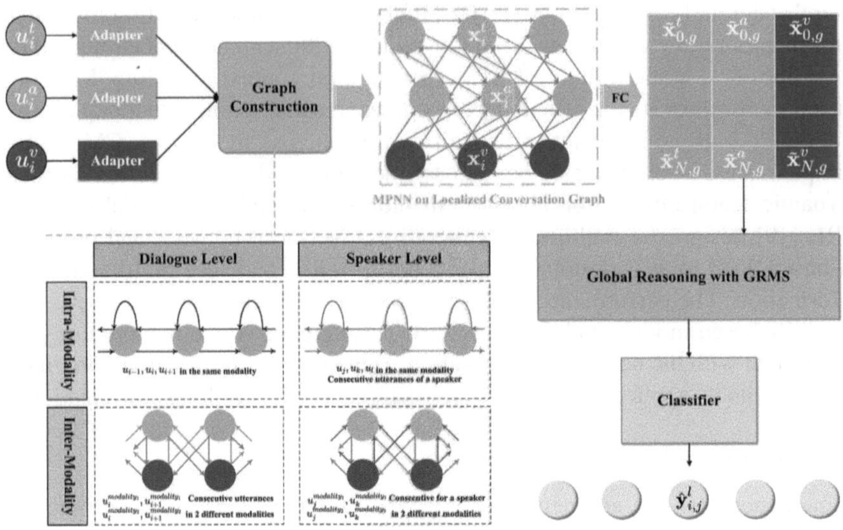

Fig. 1. Architecture of the Multimodal Graph Reasoning Network (MMGRN). The pipeline encompasses: (a) utterance embedding with adapters, (b) heterogeneous graph construction with localized intra-modal and inter-modal connections at both dialogue and speaker levels, (c) contextual information extraction and multimodal fusion through MPNN, (d) iterative global reasoning via cascaded GRMs based on global memory and reasoning clues, and (e) emotion classification based on refined representations.

3.1 Modality-Specific Data Pre-processing

The initial phase involves embedding the multimodal data into distinct vector spaces through pre-trained models augmented with Adapter layers, following the approach proposed in [23]. For the textual modality, we utilize BERT [24] as the pre-trained model, which directly converts natural language utterances in English into tensors of fixed feature size. For the acoustic modality, we utilize wav2vec2 [25] as the pre-trained model, which can directly convert audio files represented in matrix form into tensors of fixed feature size. For the visual modality, we follow the work of MM-DFN [9] to pre-process visual data. The pre-processed utterance i 's embeddings of textual, acoustic, and visual modality are respectively represented as $\mathbf{x}_i^t, \mathbf{x}_i^a, \mathbf{x}_i^v$

3.2 Local Message Passing on Multimodal Heterogeneous Graph

Localized Graph Construction. The proposed framework initiates the emotion recognition process by constructing a localized directed heterogeneous graph $\mathcal{G} = (\mathcal{V}, \mathcal{E})$ that captures contextual dependencies among adjacent utterances in multimodal conversations. The graph structure facilitates both contextual learning and multimodal fusion processes. The vertex set $\mathcal{V}$ for a conversation $U = [u_1, \cdots, u_N]$ is formally defined as:

$$\mathcal{V} = \{\mathbf{x}_i^\tau \mid i \in \{1, \cdots, N\}, \tau \in \{a, v, t\}\}. \tag{5}$$

The edge set $\mathcal{E}$ encompasses 18 distinct edge types, corresponding to the interactions between three modalities. Each vertex is connected to adjacent utterance vertices and itself in every modality. In a triple-modal graph, there are totally $(3N-2)$ edges for each edge type. In addition, to contain the speaker information, every vertex of utterances someone makes should be connected to vertices of the adjacent utterance he makes and the vertex itself. The graph incorporates speaker-specific connections, introducing $(3N - 2M)$ edges per type. The formal definition of $\mathcal{E}$ is expressed as:

$$\begin{cases} \mathcal{E} = \bigcup_{i,j}^{i,j\in\{a,v,t\}} \mathcal{E}_{i,j}^{\tau}, \tau \in \{C, S\}, \\ \mathcal{E}_{i,j}^{C} = \{(\mathbf{x}_m^i, \mathbf{x}_n^j) | m \in \{n-1, n, n+1\} \\ \qquad \wedge m, n \in \{1, \cdots, N\}\}, i, j \in \{a, v, t\}, \\ \mathcal{E}_{i,j}^{S} = \{(\mathbf{x}_m^i, \mathbf{x}_n^j) | (m = n \vee m \ adjcent \ to \ n) \\ \qquad \wedge m, n \in U_\lambda \wedge \lambda \in \{1, \cdots, M\}\}, i, j \in \{a, v, t\}, \end{cases} \tag{6}$$

where $(\mathbf{x}_1, \mathbf{x}_2)$ denotes a directed edge with source vertex $\mathbf{x}_1$ and target vertex $\mathbf{x}_2$.

Contextual Learning and Graph Multimodal Fusion. Based on the localized graph, the utterance information needs to be first spread among the local context through message passing, which imitates checking the adjacent sentences of the questioned sentence in reading test. We employ GATv2 [12] as the MPNN-based model for simultaneous multimodal fusion and contextual information extraction. This selection is motivated by GATv2's capability to perform isolated message passing across different edge types, enabling efficient multimodal fusion and contextual learning. The message passing process on graph $\mathcal{G}$ is formalized as following sequential equations:

$$\mathbf{x}_{i,e}^{t+1} = \text{GATv2}(\mathbf{x}_{i,e}^t, \mathcal{N}_e^t(i)) \tag{7}$$

$$\mathcal{N}_e^t(i) = \{j | (\mathbf{x}_j^t, \mathbf{x}_i^t) \in \mathcal{E}_e\}, \tag{8}$$

where $\mathbf{x}_i^t$ represents the vertex features whose initial features is $\mathbf{x}_i$ at the t-th GATv2 layer, e denotes one of 18 different types of edges, and $\mathbf{x}_{i,e}^{t+1}$ represents the features obtained through edges of type e after the t-th layer of GATv2.

3.3 Global Graph Reasoning

After the localized multimodal information fusion, it is necessary to further process the information of each utterance and then recognize emotion based on the global emotional foundations. We separate this process into three specific phases: 1) Obtaining global memory and reasoning clues. 2) Reasoning on global memory and reasoning clues through global reasoning module. 3) Conducting emotion recognition through classifier whose input is the reasoning result (Fig. 2).

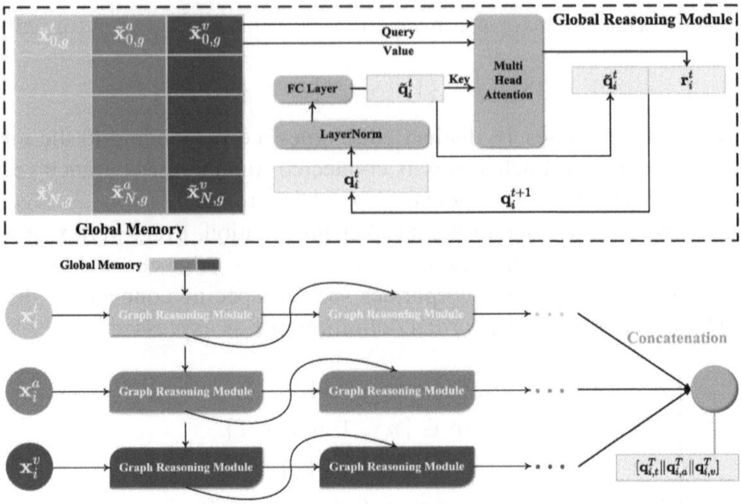

Fig. 2. Architecture of the Global Reasoning Module (GRM). The module implements multi-head attention mechanisms where global memory vectors are inputted as both queries and values. Another inputs are representations $\mathbf{x}_i^\tau$ which undergoes transformation into query vectors $\mathbf{q}_i^\tau$ to facilitate following iterative global reasoning operations.

Obtaining Global Memory and Reasoning Clues. Since local information has been processed in previous MPNN, the following step is refining the interpretation from a global perspective, which imitates understanding and then utilizing an article's main idea to answer some reading questions. The global reasoning process begins with the extraction of global memory and reasoning clues from the processed graph structure, like understanding the main idea. Given the post-MPNN vertex set $\mathcal{V} = \{\mathbf{x}_{i,g}^\tau | i \in \{1, \cdots, N\}, \tau \in \{a, v, t\}\}$ in the graph $\mathcal{G} = (\mathcal{V}, \mathcal{E})$, the global memory $\mathbf{g}$ and reasoning clues $\mathbf{q}$ are computed as following sequential equations:

$$\tilde{\mathbf{x}}_{i,g}^\tau = \mathbf{W}_g^\tau \mathbf{x}_{i,g}^\tau + \mathbf{b}_g^\tau, \tau \in \{a, v, t\}, \tag{9}$$

$$\mathbf{g}_i = [\tilde{\mathbf{x}}_{i,g}^a \| \tilde{\mathbf{x}}_{i,g}^v \| \tilde{\mathbf{x}}_{i,g}^t], \tag{10}$$

$$\mathbf{q}_i^\tau = \mathbf{W}_q^\tau \mathbf{x}_{i,g}^\tau + \mathbf{b}_q^\tau, \tau \in \{a, v, t\}. \tag{11}$$

Supposing $\mathbf{x}_{i,g}^\tau \in \mathbb{R}^{d_x}$, $\mathbf{W}_g^\tau \in \mathbb{R}^{d_x \times d_x}$, $\mathbf{W}_q^\tau \in \mathbb{R}^{d_x \times 6d_x}$, $\mathbf{b}_g^\tau \in \mathbb{R}^{d_x}$, $\mathbf{b}_q^\tau \in \mathbb{R}^{6d_x}$ are learnable parameters. After linear transformation and concatenation, we obtain multi-modal global memory $\mathbf{g}_i \in \mathbb{R}^{3d_x}$ and reasoning clues of each modality $\mathbf{q}_i^\tau \in \mathbb{R}^{6d_x}$.

Global Reasoning Module. GRM introduces attention mechanism to adaptively focus global reasoning of each utterance on different keypoints of the main idea. For the t-th reasoning iteration with input reasoning clues $\mathbf{q}_t$ and global clues $\mathbf{g}$, the comprehensive reasoning process is formalized as following sequential equations:

$$\tilde{\mathbf{q}}_i^t = \mathbf{W}_r \cdot \text{LayerNorm}(\mathbf{q}_i^t) + \mathbf{b}_r, \tag{12}$$

$$\mathbf{r}_i^t = \text{Attention}(\mathbf{g}_j, \tilde{\mathbf{q}}_i^t, \mathbf{g}_j), \tag{13}$$

$$\mathbf{q}_i^{t+1} = [\tilde{\mathbf{q}}_i^t \| \mathbf{r}_i^t], \tag{14}$$

where $\mathbf{W}_r \in \mathbb{R}^{3d_x \times 6d_x}$, $\mathbf{b}_r \in \mathbb{R}^{3d_x}$ represent the learnable transformation parameters. The architectural design of this reasoning module maintains dimensional consistency between input and output representations, facilitating arbitrary iterations of the reasoning process. Through multiple attention-based reasoning cycles, the module systematically accumulates and refines utterance-specific emotional information, ultimately generating classification-ready feature representations.

Emotion Recognition Through Classifier. The emotion recognition process leverages the multi-modal reasoning results from GRM, specifically the terminal classification features $\mathbf{q}_{i,a}^T, \mathbf{q}_{i,v}^T, \mathbf{q}_{i,t}^T$ corresponding to each original utterance u_i. The emotion classification architecture implements a fully connected neural network followed by softmax normalization, formalized as:

$$\hat{\mathbf{y}}_i = \text{Softmax}(\mathbf{W}_o[\mathbf{q}_{i,t}^T \| \mathbf{q}_{i,a}^T \| \mathbf{q}_{i,v}^T] + \mathbf{b}_o), \tag{15}$$

where the learnable parameters $\mathbf{W}_o \in \mathbb{R}^{6d_x \times n_{class}}$, $\mathbf{b}_o \in \mathbb{R}^{n_{class}}$ facilitate the transformation to the emotion class space, with n_{class} representing the size of the emotion class set. The optimization framework employs a cross-entropy loss function augmented with L2 regularization, mathematically expressed as:

$$\mathcal{L} = -\frac{1}{\sum_{l=1}^{L} \tau(l)} \sum_{i=1}^{L} \sum_{j=1}^{\tau(i)} \mathbf{y}_{i,j}^l \log(\hat{\mathbf{y}}_{i,j}^l) + \eta \|\Theta\|_2. \tag{16}$$

In this formulation, L denotes the number of conversations in the training conversation set, while $\tau(i)$ represents the number of utterances in the i-th conversation. The variables $\mathbf{y}_{i,j}^l$ and $\hat{\mathbf{y}}_{i,j}^l$ correspond to the ground truth one-hot vector and the predicted probability distribution for the j-th class of utterance i in conversation l, respectively. The regularization term incorporates all learnable parameters Θ with a L2-regularization weight η controlling the regularization strength. This comprehensive loss function effectively guides the learning process while preventing overfitting through appropriate regularization.

3.4 Fully Connected Versus Non-fully Connected MPNN

The construction of conversation graphs has employed two primary approaches for connecting utterances. The first approach implements a fully connected graph with weighted edges determined by embedding similarity [8], while the second establishes selective connections linking each utterance to its preceding j and subsequent k utterances [10]. Our model introduces a distinct methodology by establishing connections exclusively between adjacent utterances and self-loops. These architectural variations manifest significant differences in contextual information propagation dynamics. In fully connected architectures, utterance information propagates directly to all nodes within a single MPNN layer. Non-fully connected structures restrict information flow to predetermined local neighborhoods, necessitating multiple MPNN layers for comprehensive information dissemination. These approaches, lacking dedicated global information processing mechanisms, require complete sequential message propagation throughout the conversation.

Actually, we don't need to pass information to all nodes in the graph. It is possible to obtaining global information while focusing message passing on local. Following the strategic reading pattern we propose, the information only needs to be passed to a few local nodes, and then reasoning on global. In our model, it is unnecessary to employ as many rounds of MPNN as the length of the conversation in order to effectively pass contextual information from the first utterance to the last. We only implement few MPNN layers (three to five in our experiments) to pass information in local context, which reduce computational cost meanwhile. Next, The global retrieval and reasoning capability our GRM possessed will provide previously inaccessible global information.

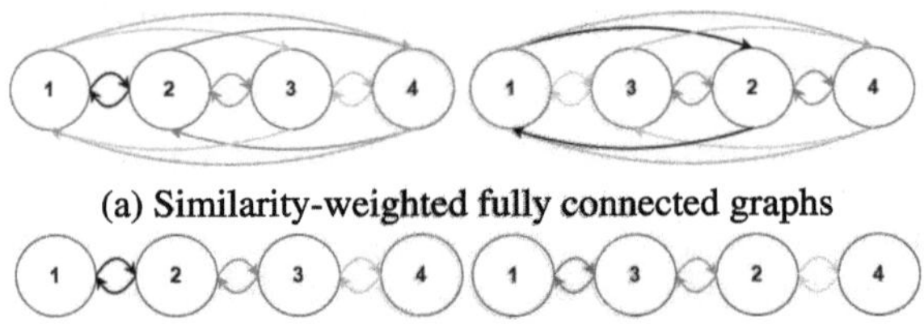

(a) Similarity-weighted fully connected graphs

Adjacent-only connected graphs

Fig. 3. In (a), position exchange between utterances produces equivalent graph structures, indicating failure to preserve temporal relationships. Edge weights remain consistent across edges of identical color, and each node is connected to all other nodes. The message passing process is unchanged after nodes switch. **In (b)**, position exchange between utterances generates distinct graph structures, demonstrating successful temporal relationship preservation. Node 1 is no longer connected to node 2, and node 4 no longer connected to node 3. Therefore, the message passing process is changed after nodes switch. The contextual relationship is preserved.

Compared to fully connected graphs with weights and non-fully connected graphs with a certain number of utterances, our connection methodology demonstrates superior temporal information preservation. As depicted in Fig. 3a, fully connected architectures inherently diminish temporal relationships among utterances. While position embeddings represent a conventional solution to temporal encoding, their application to conversations of variable and unbounded length necessitates interpolation. Figure 3b illustrates how our graph construction preserves temporal information through localized connectivity patterns, while avoiding the position embedding. The elimination of redundant edges reduces both spatial and temporal complexity of matrix operations from $O(N^2)$ to $O(N)$, where N represents the conversation length.

4 Experiments

4.1 Datasets

The **IEMOCAP** (Interactive Emotional Dyadic Motion Capture) dataset [13] represents a comprehensive multimodal emotional conversation corpus. The dataset encompasses English dialogues performed by a balanced gender distribution of ten professional actors (five males and five females). The conversational structure exhibits consistent patterns with approximately 49 turns per dialogue, averaging 11.4 words and 4.5 s per utterance. The multimodal nature of the dataset incorporates facial motion capture data, audio recordings, video sequences, head motion trajectories, angular measurements, conversation transcripts, and multi-level linguistic alignments (word, syllable, and phoneme).

The **MELD** (Multimodal EmotionLines Dataset) [14] presents a naturalistic multimodal emotional conversation corpus extracted from the television series *Friends*. The dataset comprises 1,433 conversations totaling 13,708 utterances, featuring dynamic multi-party interactions involving one to eight participants per conversation.

4.2 Baseline Models

To evaluate the effectiveness of our proposed MMGRN and GRM architectures, we establish a comprehensive comparison framework incorporating both classical and state-of-the-art GNN-based models. The baseline models encompass the foundational LSTM-based model and advanced graph-based models.

- **bc-LSTM** [2]: utilizes LSTM to extract contextual information and classify the emotion after a simple multimodal concatenation.
- **DialogueRNN** [28]: introduces speaker information through speaker embedding.
- **DialogueGCN** [6]: introduces conversation graph.
- **MMGCN** [8]: utilizes GCN to fuse multimodal features.
- **MM-DFN** [9]: introduces graph-based dynamic fusion models.
- **GraphMFT** [11]: introduces a new conversation graph structure and Improved GAT containing residual connection.
- **DialogueCRN** [27] introduces reasoning modules to refine the contextual clue.
- **GraphCFC** [10]: introduces a complex multimodal fusion structure.

4.3 Results

Table 1 presents performance evaluation across the IEMOCAP and MELD datasets, comparing our MMGRN architecture with eight baseline models through accuracy and weighted-average F1 score metrics across various emotion categories.

On IEMOCAP, MMGRN achieves state-of-the-art performance with a weighted-average F1 score of 69.36% and accuracy of 69.07%, marginally surpassing the previous benchmark set by GraphCFC (68.91% wa-f1, 69.13% acc). MMGRN demonstrates superior performance in recognizing "happiness" (53.91%), "neutral" (70.40%), and "frustration" (67.80%) emotions, while GraphCFC excels in detecting "sadness" (84.99%), "anger" (71.35%), and "excitement" (78.86%). As illustrated in Fig. 4, MMGRN exhibits more balanced performance across the six emotion categories compared to GraphCFC.

Results on MELD reveal more substantial improvements, with MMGRN achieving 64.64% accuracy and 63.03% weighted-average F1 score, significantly outperforming the previous leading model, MM-DFN (62.49% acc, 59.46% wa-f1). MMGRN particularly excels in "neutral" (78.44%) and "anger" (51.69%) recognition, while MM-DFN maintains superiority in "surprise" (50.69%) and "happiness" (54.78%) detection.

The comparative analysis reveals that while earlier models like bc-LSTM and DialogueRNN show decent performance, more recent architectures incorporating sophisticated graph-based mechanisms consistently achieve better results. With simpler and lighter graph structure, our MMRGN can achieve and surpass those models. This trend

Table 1. Performance of full version models under multimodal setting. Acc and wa-f1 refer to the overall accuracy and the weighted-average F1 score. Best results are highlighted in **BOLD**.

Methods	IEMOCAP		MELD	
	acc	wa-f1	acc	wa-f1
bc-LSTM	60.51	60.42	59.62	57.29
DialogueRNN	63.52	62.89	60.31	57.66
DialogueCRN	67.16	67.21	61.11	58.67
DialogueGCN	63.22	62.89	58.62	56.36
MMGCN	66.36	66.26	60.42	58.31
MM-DFN	68.21	68.18	62.49	59.46
GraphMFT	67.90	68.07	61.30	58.37
GraphCFC	**69.13**	68.91	61.42	58.86
MMGRN	69.07	**69.36**	**64.64**	**63.03**

suggests that effectively modeling the inter-modal and inter-utterance relationships through graph neural networks is crucial for emotion recognition in conversations. Our localized graph construction pattern and global reasoning model is one of the effective designs.

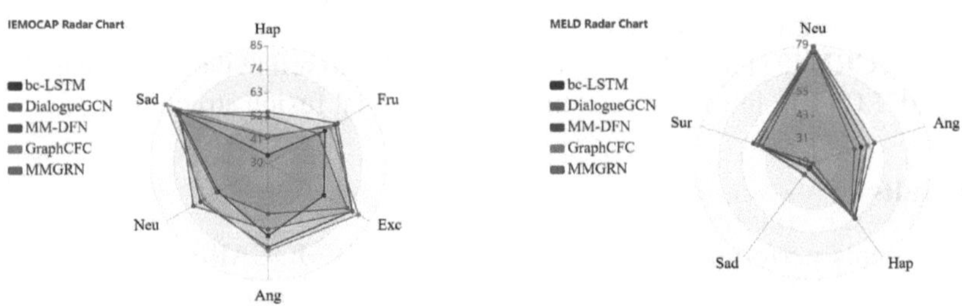

(a) IEMOCAP (b) MELD

Fig. 4. Performance distribution across emotion categories visualized through radar plots: (a) IEMOCAP, (b): MELD, demonstrating the comparative emotional recognition capabilities of the evaluated models.

The runtime of models with better performance is shown in Fig. 5. Our MMGRN significantly outperforms the current best GNN-based model, GraphCFC, in terms of running time on two datasets. Meanwhile, compared to two simpler models with fully connected graph structures that exhibit poorer performance, our MMGRN achieves the fastest inference speed on IEMOCAP, but not on MELD. The significant runtime advantage of MMGRN over GraphCFC stems from two architectural differences: (1) GraphCFC employs LSTM for preliminary feature extraction before graph processing and a more complex network with many layers, and (2) our localized graph construction reduces edge density from $O(N^2)$ to $O(N)$, dramatically decreasing GNN computational complexity. The LSTM preprocessing in GraphCFC becomes particularly expensive for longer conversations, explaining the larger speedup on IEMOCAP (49 turns average) compared to MELD (8 turns average). Therefore, on IEMOCAP, our model

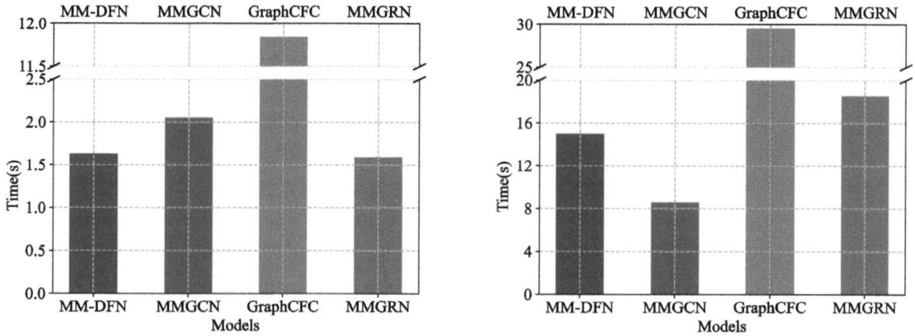

(a) IEMOCAP (b) MELD

Fig. 5. Runtime of four better models on test datasets when batch size is set 1. (a) IEMOCAP contains 151 conversations with an average length of 49 turns, while MELD contains 1433 conversations of an average length of 8 turns.

not only improves performance but also achieves a more significant improvement in inference speed compared to models based on a fully connected graph structure, while not making significant improvements in conversations with very few utterances. In this case, the improvement in inference performance brought by our GRM comes at the cost of inference speed.

4.4 Ablation Studies

To verify the effectiveness of both multimodal fusion and GRM in MMGRN, we conduct ablation experiments on IEMOCAP using MMGRN with different modality settings and on both datasets using MMGRN with or without GRM.

Table 2 presents the modality ablation results on IEMOCAP, revealing textual features as the strongest individual modality (wa-f1: 56.37%), followed by acoustic features (wa-f1: 55.44%), while visual features demonstrate comparatively lower performance (wa-f1: 35.59%). The acoustic-textual combination (A+T) emerges as the most effective dual-modality configuration (wa-f1: 64.35%), with the complete tri-modal system (A+V+T) achieving optimal performance (wa-f1: 69.36%), demonstrating that despite the relatively weak performance of visual features alone, they contribute meaningful complementary information when combined with other modalities.

Table 2. Performance of MMGRN in different modality settings on IEMOCAP, where A, V and T refer to acoustics, video and text. Best results are highlighted in **BOLD**.

Methods	acc	wa-f1	Methods	acc	wa-f1	Methods	acc	wa-f1
MMGRN (A)	57.55	55.44	A+V	61.49	60.91	**A+V+T**	**69.07**	**69.36**
MMGRN (V)	37.89	35.59	V+T	58.16	58.15			
MMGRN (T)	58.72	56.37	A+T	65.43	64.35			

Table 3. Performance of MMGRN with/without GRM. Best results are highlighted in **BOLD**.

Methods	IEMOCAP		MELD	
	acc	wa-f1	acc	wa-f1
MMGRN -w/o GRM	64.02	64.41	63.32	61.84
MMGRN -w GRM	**69.07**	**69.36**	**64.64**	**63.03**

The second ablation study examines the importance of GRM on IEMOCAP and MELD. As shown in Table 3, On IEMOCAP and MELD, removing GRM leads to significant performance drops IEMOCAP (wa-f1: −4.95%) and MELD (wa-f1: −1.19%). These consistent performance gaps across both datasets validate the effectiveness of GRM in global reasoning in local graph information.

4.5 Discussion

The superior performance of MMGRN over existing models, particularly its achievement of 69.36% wa-f1 on IEMOCAP and 63.03% wa-f1 on MELD, strongly validates our local-to-global ERC approach. This two-phase processing strategy demonstrates particular efficacy in handling complex emotional dynamics in conversations. The ablation studies further substantiate this design philosophy inspired from reading comprehension test strategy, particularly in the textual modality, where MMGRN achieves remarkable performance (58.72% accuracy with text alone, A+T achieving 65.43% accuracy). Just as human readers first process local sentences before constructing complete semantic understanding, our model's local-to-global architecture particularly excels in textual emotion recognition. This enhancement can be attributed to its ability to first capture fine-grained emotional cues at the local level before scrutinizing them into a global contextual understanding.

The comparison between MMGRN with and without GRM, provide compelling evidence for the critical role of global knowledge integration in ERC. The significant performance degradation observed when removing GRM (5.95% decrease in accuracy on IEMOCAP and 1.32% on MELD) underscores the importance of comprehensive contextual understanding. Simultaneously, MMGRN's graph-base architecture, a demonstrated by GRM ablation study, achieves superior performance while enabling parallel processing of information. The experimental results suggest that our approach successfully balances the need for comprehensive global context understanding with computational efficiency, offering a more practical solution for real-world applications while maintaining high performance.

5 Conclusion

This study introduces a novel methodology for conversation graph construction, integrating it with MMGRN. The synergistic combination demonstrates superior performance in multimodal ERC tasks compared to existing message-passing graph neural networks. Our architecture addresses fundamental challenges in multimodal ERC with GNN model through an optimal balance between information completeness and computational efficiency. The proposed graph structure enables efficient extraction of contextual information, speaker dynamics, and multimodal fusion within conversations through any MPNN. GRM enhances the graph-based model's capability to access previously unreachable global information and facilitates sophisticated reasoning about inter-utterance associations, thereby refining emotional recognition cues. The strategic implementation of graph construction and global reasoning reduces both temporal and spatial computational complexity.

References

1. Kumar, S., Dudeja, S., Akhtar, M.S., Chakraborty, T.: Emotion flip reasoning in multiparty conversations. IEEE Trans. Artif. Intell. **5**(3), 1339–1348 (2024)
2. Poria, S., Cambria, E., Hazarika, D., et al.: Context-dependent sentiment analysis in user-generated videos. In: Proceedings of the 55th Annual Meeting of the Association for Computational Linguistics, Vancouver, Canada, vol. 1, pp. 873–883. Association for Computational Linguistics (2017)
3. Hazarika, D., Poria, S., Mihalcea, R., et al.: ICON: interactive conversational memory network for multimodal emotion detection. In: Proceedings of the 2018 Conference on Empirical Methods in Natural Language Processing, Brussels, Belgium, October–November 2018, pp. 2594–2604. Association for Computational Linguistics
4. Fu, C., Qian, F., Su, K., et al.: HiMul-LGG: a hierarchical decision fusion-based local-global graph neural network for multimodal emotion recognition in conversation. Neural Netw. **181**, 106764 (2025)
5. Zhang, Y., Jia, A., Wang, B., et al.: M3GAT: a multi-modal, multi-task interactive graph attention network for conversational sentiment analysis and emotion recognition, vol. 42, no. 1, (2023)
6. Ghosal, D., Majumder, N., Poria, S., et al.: DialogueGCN: a graph convolutional neural network for emotion recognition in conversation. In: Proceedings of the 2019 Conference on Empirical Methods in Natural Language Processing and the 9th International Joint Conference on Natural Language Processing (EMNLP-IJCNLP), Hong Kong, China, November 2019, pp. 154–164. Association for Computational Linguistics (2019)
7. Zhang, B., Luo, S., Wang, L., He, D.: Rethinking the expressive power of GNNs via graph biconnectivity. In: The Eleventh International Conference on Learning Representations, ICLR 2023, Kigali, Rwanda, 1–5 May 2023 (2013)
8. Hu, J., Liu, Y., Zhao, J., Jin, Q.: MMGCN: multimodal fusion via deep graph convolution network for emotion recognition in conversation. In: Proceedings of the 59th Annual Meeting of the Association for Computational Linguistics and the 11th International Joint Conference on Natural Language Processing, August 2021, vol. 1, pp. 5666–5675. Association for Computational Linguistics (2021)
9. Hu, D., Hou, X., Wei, L., et al.: MM-DFN: multimodal dynamic fusion network for emotion recognition in conversations. In: ICASSP 2022 - 2022 IEEE International Conference on Acoustics, Speech and Signal Processing (ICASSP), pp. 7037–7041 (2022)
10. Li, J., Wang, X., Lv, G., Zeng, Z.: GraphCFC: a directed graph based cross-modal feature complementation approach for multimodal conversational emotion recognition. IEEE Trans. Multimedia **26**, 77–89 (2024)
11. Li, J., Wang, X., Lv, G., Zeng, Z.: GraphMFT: a graph network based multimodal fusion technique for emotion recognition in conversation. In: Neurocomputing, vol. 550, p. 126427 (2023)
12. Brody, S., Alon, U., Yahav, E.: How attentive are graph attention networks? In: The Tenth International Conference on Learning Representations, ICLR 2022, Virtual Event (2022)

13. Busso, C., Bulut, M., Lee, C., et al.: IEMOCAP: interactive emotional dyadic motion capture database. Lang. Resour. Evaluation **42**(4), 335–359 (2008)
14. Poria, S., Hazarika, D., Majumder, N., et al.: MELD: a multimodal multi-party dataset for emotion recognition in conversations. In: Proceedings of the 57th Annual Meeting of the Association for Computational Linguistics. Florence, Italy, July 2019, pp. 527–536. Association for Computational Linguistics (2019)
15. Wang, N., Cao, H., Zhao, J., et al.: M2R2: missing-modality robust emotion recognition framework with iterative data augmentation. IEEE Trans. Artif. Intell. **4**(5), 1305–1316 (2023)
16. Li, Z., Tang, F., Zhao, M., Zhu, Y.: EmoCaps: emotion capsule based model for conversational emotion recognition. In: Findings of the Association for Computational Linguistics: ACL 2022, Dublin, Ireland, pp. 1610–1618. Association for Computational Linguistics (2022)
17. Ren, M., Huang, X., Liu, J., et al.: MALN: multimodal adversarial learning network for conversational emotion recognition. IEEE Trans. Circuits Syst. Video Technol. **33**(11), 6965–6980 (2023)
18. Luo, J., Phan, H., Reiss, J.: Cross-modal fusion techniques for utterance-level emotion recognition from text and speech. In: ICASSP 2023 - 2023 IEEE International Conference on Acoustics, Speech and Signal Processing (ICASSP), pp. 1–5 (2023)
19. Meng, T., Shou, Y., Ai, W., et al.: Deep imbalanced learning for multimodal emotion recognition in conversations. IEEE Trans. Artif. Intell. **5**(12), 6472–6487 (2024)
20. Xue, J., Nguyen, M.-P., Matheny, B., Nguyen, L.M.: BiosERC: integrating biography speakers supported by LLMs for ERC tasks. In: Artificial Neural Networks and Machine Learning – ICANN 2024, pp. 277–292. Springer (2024)
21. Wu, Z., Gong, Z., Ai, L., et al.: Beyond silent letters: amplifying LLMs in emotion recognition with vocal nuances, *CoRR*, vol. abs/2407.21315 (2024)
22. Fu, Y.: CKERC: joint large language models with commonsense knowledge for emotion recognition in conversation, *CoRR*, vol. abs/2403.07260 (2024)
23. Houlsby, N., Giurgiu, A., Jastrzebski, S., et al.: Parameter-efficient transfer learning for NLP. In: Proceedings of the 36th International Conference on Machine Learning, ICML 2019, Long Beach, California, USA, 9–15 June 2019, Proceedings of Machine Learning Research, pp. 2790–2799 (2019)
24. Devlin, J., Chang, M.-W., Lee, K., Toutanova, K.: BERT: pre-training of deep bidirectional transformers for language understanding. In: Proceedings of the 2019 Conference of the North American Chapter of the Association for Computational Linguistics, Minneapolis, Minnesota. Association for Computational Linguistics, vol. 1, pp. 4171–4186 (2019)
25. Yang, S., Chi, P., Chuang, Y., et al.: SUPERB: speech processing universal performance benchmark. In: 22nd Annual Conference of the International Speech Communication Association, Interspeech 2021, Brno, Czechia, August 30–September 3, pp. 1194–1198 (2021)
26. Velickovic, P., Cucurull, G., Casanova, A., Romero, A., Liò, P., Bengio, Y.: Graph attention networks. In: 6th International Conference on Learning Representations, ICLR 2018, Vancouver, BC, Canada, April 30–May 3, 2018, Conference Track Proceedings
27. Hu, D., Wei, L., Huai, X.: DialogueCRN: contextual reasoning networks for emotion recognition in conversations. In: Proceedings of the 59th Annual Meeting of the Association for Computational Linguistics. Association for Computational Linguistics, vol. 1, pp. 7042–7052 (2021)
28. Majumder, N., Poria, S., Hazarika, D., et al.: DialogueRNN: an attentive RNN for emotion detection in conversations. In: The Thirty-Third AAAI Conference on Artificial Intelligence, AAAI 2019, Honolulu, Hawaii, USA, January 27–February 1 2019, pp. 6818–6825. AAAI Press (2019)

Curvature-Based Knee Detection for Robust and Non-robust Features

Xuanyu Li and Weitong Chen(✉)

School of Computer and Mathematical Sciences, The University of Adelaide, Adelaide, Australia
weitong.chen@adelaide.edu.au

Abstract. Despite their strong performance, deep networks remain fragile to adversarial perturbations, often because they rely on imperceptible, non-robust features. Yet there is no principled method to localize the boundary between non-robust and robust feature regimes. We introduce a curvature-based framework that identifies this transition via a continuous pixel-budget attack. By perturbing inputs in saliency-ranked order and analyzing the second derivative of adversarial-accuracy curves, we define a formal "knee" that marks the exhaustion of non-robust features. Across five ImageNet models, the "knee" consistently aligns with a shift in transferability dynamics: models with earlier knees produce more universal attacks, whereas models with later knees better preserve accuracy. Our method establishes a reproducible boundary between brittle and stable representations, offering a new axis for evaluating and comparing adversarial robustness across architectures.

Keywords: Adversarial robustness · Feature regimes · "knee" detection · Curvature analysis

1 Introduction

Adversarial examples have revealed a fundamental vulnerability in modern deep networks: they routinely exploit imperceptible, high-frequency patterns that bear little relation to human-interpretable features [7,19]. Early work focused on dense, norm-bounded perturbations, most notably FGSM and its iterative variants (e.g. PGD), which modify every pixel according to gradient sign information [29]. More recent sparse attacks, such as the one-pixel attack [8], relax this constraint by targeting only a handful of coordinates, yet still treat pixels as independent entities rather than manifestations of underlying feature representations. As a result, the vast literature on adversarial robustness has overlooked a deeper question: when, as we gradually exhaust low-level, non-robust cues, do networks transition to relying on genuinely robust features?

In this work, we introduce a curvature-based "knee" detection framework that fills this gap by supplying a clear, model-agnostic criterion for identifying when non-robust features are exhausted. We apply pixel-budget attacks in saliency-ranked order, perturbing the most important pixels first, and record the resulting

M. Yoshikawa et al. (Eds.): ADMA 2025, LNAI 16197, pp. 237–251, 2026.
https://doi.org/10.1007/978-981-95-3453-1_16

adversarial accuracy as a function of the pixel budget. By fitting a single global polynomial to the accuracy vs. budget curve, we compute its second derivative and define the "knee" point as the first local minimum of that derivative. This formulation yields an exact budget N^* at which the rate of accuracy decay transitions from rapid (dominated by non-robust features) to linear (dominated by inherently robust features).

Our extensive experiments on five ImageNet models , ResNet50, ResNet101, VGG19, InceptionV3, and MobileNetV2, demonstrate remarkable consistency in "knee" locations despite vast architectural differences. Perturbations applied before N^* transfer effectively across models, achieving cross-model accuracies within a few points of each network's self-attack baseline, whereas post-"knee" perturbations result in steep transfer losses. Furthermore, once the "knee" is reached, each model's accuracy curve settles into a stable plateau, approximately 40% for ResNet50, 35–37% for ResNet101 and InceptionV3, and roughly 20% for VGG19 and MobileNetV2, indicating that only sparse, semantically meaningful features remain vulnerable. These findings validate the "knee" as a universal marker of feature-regime change.

Theoretical analyses have begun to address the feature-learning dynamics underlying this transition. Li and Li [16] prove that standard training amplifies non-robust features, rendering models provably fragile under structured perturbations, but their framework stops short of offering a practical mechanism to detect the boundary between non-robust and robust regimes. Attribution methods and information-bottleneck studies provide post-hoc saliency maps that highlight important pixels or channels, yet none yield a quantifiable change-point separating brittle, spurious patterns from semantically meaningful representations. Consequently, robustness evaluations remain tied to arbitrarily chosen budgets or norm thresholds, hindering reproducibility and cross-architecture comparisons.

The contributions of this paper are,

- First, we propose a novel curvature - based method for detecting the precise moment when a model shifts from non-robust to robust feature usage.
- Second, we provide empirical evidence that this "knee" point is architecture-agnostic and correlates tightly with transferability and post-"knee" accuracy floors.
- Third, we offer a new perspective on adversarial robustness—one that emphasizes not only the magnitude of perturbations but also the phase at which the model's reliance on different feature types fundamentally changes.

This paper is organized as follows: Sect. 2 reviews key background on white-box adversarial attacks, with a focus on gradient- and saliency-based techniques. Section 3 outlines our proposed method, including saliency-driven pixel selection, curvature-based "knee" detection, and transferability analysis. Section 4 describes the experimental setup and evaluation protocol. Section 5 presents and analyzes the results across multiple models. Sections 6 and 7 discuss related work and conclude with a summary and future directions.

2 Preliminaries and Backgrounds (White-Box Attack)

The Fast Gradient Sign Method (FGSM) [7,22,29] perturbs inputs by a single gradient step:

$$\epsilon = \sigma \operatorname{sign}\big(\nabla_x J(\theta, x, y)\big), \tag{1}$$

where $\sigma > 0$ bounds the perturbation magnitude. Adding ϵ to x typically suffices to mislead the classifier.

The Jacobian-based Saliency Map Attack (JSMA) [11,15] targets specific classes by first computing

$$J_F(\mathbf{x}) \;=\; \left[\tfrac{\partial F_j(\mathbf{x})}{\partial x_i}\right]_{i=1..n}^{j=1..m}, \tag{2}$$

then deriving a saliency map

$$S(\mathbf{x}, t)_i = \begin{cases} 0, & J_F(\mathbf{x})_{t,i} < 0 \;\vee\; \sum_{j \neq t} J_F(\mathbf{x})_{j,i} > 0, \\ J_F(\mathbf{x})_{t,i} \left|\sum_{j \neq t} J_F(\mathbf{x})_{j,i}\right|, & \text{otherwise,} \end{cases} \tag{3}$$

and perturbing the highest-scoring pixels to achieve a targeted misclassification.

$$J_F(\mathbf{x}) = \left[\frac{\partial F_j(\mathbf{x})}{\partial x_i}\right]_{\substack{i=1,\dots,n\\ j=1,\dots,m}} \tag{4}$$

$J_F(\mathbf{x})$ is the Jacobian matrix and F represents the general function of the model. One can use this matrix to compute the saliency map $\mathbf{S}(\boldsymbol{x}, t)$ as:

$$\mathbf{S}(\mathbf{x}, t)_i = \begin{cases} 0, & \text{if } J_F(\mathbf{x})_{t,i} < 0 \;\vee\; \sum\limits_{j \neq t} J_F(\mathbf{x})_{j,i} > 0, \\ J_F(\mathbf{x})_{t,i} \left|\sum\limits_{j \neq t} J_F(\mathbf{x})_{j,i}\right|, & \text{otherwise.} \end{cases} \tag{5}$$

3 Methodology

Top Pixels Selection. Since the goal is to discover the relationship between adversarial accuracies and the corresponding number of modified pixels, the process of gradually modifying more pixels should obey certain rules. Combining the concepts of robust and non-robust features with saliency maps, the map can reveal a "ranking system" for all pixels according to the significance of each pixel in terms of misclassification. Before picking and perturbing the pixels that are most important for the model to correctly classify images, the saliency map should be determined for consideration. First, the model is defined as a function F. The image input is defined as $x \in \mathbb{R}^{C \times H \times W}$, where C is the number of channels (3 here due to RGB images), H is the height of the image, and W is the width of the image. The number of output units is also defined as K so that $F : x \in \mathbb{R}^{C \times H \times W} \to \mathbb{R}^K$. Then one can define a scalar function G as the sum of all output components and compute the gradient of G with respect to the input x as:

$$G(x) = \sum_{k=1}^{K} F_k(x) \tag{6}$$

$$\nabla_x G(x) \in \mathbb{R}^{C \times H \times W}. \tag{7}$$

Afterwards, by using the concept of calculating the Jacobian matrix according to formula (3), the saliency map S can be obtained by taking the absolute value of these gradients and summing over the channel dimension for each spatial location (h, w) as in formula (7); eventually, the pixel ranking system can be established via formula (8):

$$S(x)_{h,w} = \sum_{c=1}^{C} \left| \frac{\partial G(x)}{\partial x_{c,h,w}} \right| \tag{8}$$

$$P = \{(h, w) \in \{1, \ldots, H\} \times \{1, \ldots, W\} : S(x)_{h,w} \text{ is among the top } k \text{ values}\}. \tag{9}$$

Formula (8), with S defined as the saliency score within the saliency map S, defines the set P as the collection of pixel coordinates (h, w) within an image of height H and width W that have the highest saliency scores. From that step, we can then apply perturbations using one of the most effective methods, FGSM, via:

$$\tilde{I}(c, h, w) = \begin{cases} \min\{1, \ \max\{0, \ I(c, h, w) + \sigma \times \text{sign}(\frac{\partial J(\theta, x, y)}{\partial x_{c,h,w}})\}\}, & (h, w) \in P, \\ I(c, h, w), & (h, w) \notin P, \end{cases} \tag{10}$$

where $I(c, h, w)$ is the original pixel value at channel c and spatial position (h, w), σ is a small scalar perturbation magnitude, and $\text{sign}(\frac{\partial J(\theta, x, y)}{\partial x_{c,h,w}})$ is the sign of the standard loss at pixel (h, w) that indicates the direction (positive or negative) for the perturbation. The inner max ensures that the modified pixel value does not fall below 0, while the outer min ensures it does not exceed 1. This "clamping" keeps the pixel values within the valid range from 0 to 1 inclusively. For pixels not in the set P, the pixel value remains unchanged. Figure 1 visualises a simple implementation of this testing method. Detailed experimental settings will be mentioned in Section "Experimental Implementation".

Relationship Research. Besides obtaining the ranking system for gradually increasing the number of perturbed pixels, a formula for determining adversarial accuracies is also required:

$$A_{\text{self}}(N) = \frac{1}{T} \sum_{i=1}^{T} \mathbf{1}\{\hat{y}_i(N) = y_i\} \times 100. \tag{11}$$

Here, $A_{\text{self}}(N)$ denotes the adversarial accuracy (as a percentage) when N pixels are perturbed; $\hat{y}_i(N)$ is the predicted label for the ith image after applying the

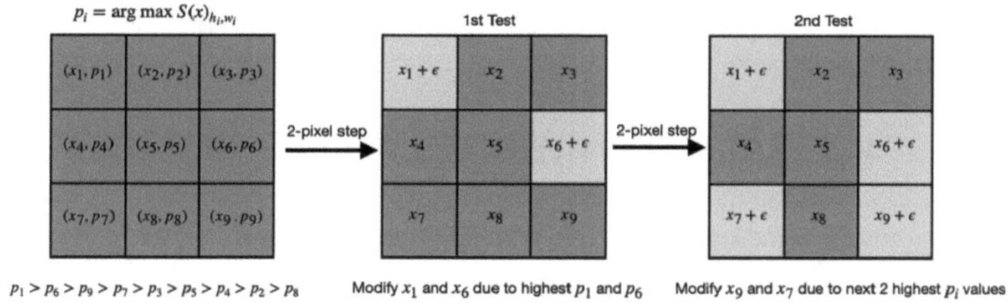

Fig. 1. A simple example of a testing adversarial accuracies only twice using 2-pixel-step top pixel selection process. ϵ has direction determined from $\text{sign}(\nabla_x J(\theta, x, y))$ (i.e. $\epsilon = \sigma \times \textbf{sign}(\frac{\partial J(\theta,x,y)}{\partial x_{c,h,w}})$) where as S computed from $G(\mathbf{x})$.

adversarial perturbation that modifies N pixels; y_i is the ground-truth label for the ith image; and $\mathbf{1}\{\hat{y}_i(N) = y_i\}$ is an indicator function that equals 1 if the perturbed image is still correctly classified and 0 otherwise.

After computing each adversarial accuracy with its corresponding number of perturbed pixels, a set of points with coordinates (A, N) can be plotted. Under the assumption that the points follow a continuous pattern, one can attempt a regression fitting for these points so that

$$A_{\text{self}}(N) = f(N) + e, \tag{12}$$

where $f(N)$ fits the relationship between N and A, with e representing an error term that accounts for noise or deviations between the fitted function $f(N)$ and the actual measured adversarial accuracies. From there, the first derivative $f'(N)$ and second derivative $f''(N)$ of the fitted function are particularly helpful in explaining the relationships and variations. After fitting $A_{\text{self}}(N)$, we then take N^* (the first local minimum of f''; see Sect. 5.1) as our "knee".

Further Testing and Cross-Model Validation. To verify that the curvature-based "knee" points correspond to an intrinsic transition between non-robust and robust feature regimes, rather than idiosyncrasies of any single architecture, we subject each "knee"-derived perturbation mask to transfer attacks on all other models. Concretely, for each source model s and target model t, we:

1. Compute the "knee" point

$$N^* = \min\{\, N > 0 \mid f''(N) \text{ is a local minimum} \,\}.$$

2. Generate adversarial examples for each test input x_i:

$$\tilde{x}_i = x_i + \epsilon\, M^{(s)}_{N^*}(x_i) \odot \text{sign}\big(\nabla_x J(\theta^{(s)}, x_i, y_i)\big),$$

where $M^{(s)}_{N^*}$ is the saliency mask from the source model s.

3. Evaluate the transfer accuracy on the target model t and compare it to the self-attack accuracy:

$$A^{s\to t}_{\text{transfer}}(N^*) = \frac{1}{T}\sum_{i=1}^{T} \mathbf{1}\{\hat{y}_i^{(t)}(\tilde{x}_i) = y_i\} \times 100, \qquad A_{\text{self}}(N^*).$$

By measuring $A^{s\to t}_{\text{transfer}}(N^*)$ for every source–target pair, we test whether the "knee" truly demarcates a universal shift from non-robust to robust features. Consistent patterns—such as high transfer gains for early-"knee" networks and losses for late-"knee" ones—confirm that our curvature-based "knee" detection captures a fundamental feature-regime boundary shared across architectures. Significant discrepancies would instead highlight model-specific artifacts and suggest directions for refining the evaluation pipeline.

4 Experimental Implementation

4.1 Testing Setups and Equipment

The adversarial testing was implemented using ResNet50 and ResNet101 [5], which are among the most typical classification models, including VGG19 [2], InceptionV3 [3], and MobileNetV2 [4]. Given the training and testing datasets used for these models, the ImageNet dataset [23] was selected for adding perturbations and testing. However, due to the large number of images, to reduce computational cost, a selection of 1,000 images from 1,000 different classes (one image per class) was used; these images are considered relatively more robust than other images in their respective classes according to a previous paper [17].

The experiments were run on a remote server equipped with an NVIDIA RTX 4090 GPU (24.0 GB) and an AMD EPYC 7282 16-core processor.

4.2 Procedures and Metrics

The perturbation added to the image pixels was set to 0.1 for all models. The saliency map was calculated before measurements. The values of adversarial accuracies, with the corresponding number of perturbed pixels, were recorded each time additional pixels were modified. To thoroughly evaluate the adversarial-accuracy variations from 0 modified pixels to nearly all pixels, the testing range was set from 0 to 40,000 (i.e., the size of the images in the ImageNet selection is 224×224), with a step of 100 between measurement points. Thus, after one complete test of a model, 401 points (including the point with no modified pixels) were obtained for regression fitting. Polynomial regressions were found to be the most accurate and informative. To assess the quality of the fittings and ensure no overfitting, a combination of evaluation metrics with assigned weights was utilised.

1. $\text{CVMSE} = \frac{1}{K}\sum_{k=1}^{K} \text{MSE}_k$, where MSE_k is the mean squared error from the kth fold [25].

2. $R^2_{\text{adj}} = 1 - \frac{(1-R^2)(n-1)}{n-p}$. For model selection, we use $1 - R^2_{\text{adj}}$ so that lower values indicate a better fit.
3. $\text{AIC} = n\ln\left(\frac{\text{SS}_{\text{res}}}{n}\right) + 2p$, where SS_{res} is the residual sum of squares [26].
4. $\text{BIC} = n\ln\left(\frac{\text{SS}_{\text{res}}}{n}\right) + p\ln(n)$ [27].
5. $\text{norm}(m) = \frac{m-\min\{m\}}{\max\{m\}-\min\{m\}}$, where the minimum and maximum are taken over all candidate models, to normalise the metrics as they differ in scale.

Finally, the combined composite score for analysing the performance of the regression is:

$$\begin{aligned}\text{Composite Score} = {} & w_{\text{cv}} \cdot \text{norm}(\text{CVMSE}) \\ & + w_{R^2} \cdot \text{norm}(1 - R^2_{\text{adj}}) \\ & + w_{\text{AIC}} \cdot \text{norm}(\text{AIC}) \\ & + w_{\text{BIC}} \cdot \text{norm}(\text{BIC}).\end{aligned} \tag{13}$$

Weights were assigned to each metric according to their importance: $w_{\text{cv}} = 0.4$, $w_{R^2} = 0.3$, $w_{\text{AIC}} = 0.15$, and $w_{\text{BIC}} = 0.15$. Each metric m is normalised using the min–max rule so that the best value becomes 0 and the worst becomes 1, placing all metrics on the same scale. The final score is a weighted sum of these normalised metrics, so a score of 0 indicates an optimal model (best in every metric). After the best-fit lines were determined, plots of the first and second derivatives were generated. "Knee" points were calculated and marked on the graphs and then used for the transfer-attack process. This is because, in some cases, no "knee" points occurred for either the original regression functions or the first-derivative functions (i.e., gradients). As a result, the concavity (second derivative) was used instead. Here, concavity represents the acceleration of the accuracy drop, and its "knees" indicate the points where the increase or decrease of that acceleration stops. Finally, comparisons were made between the testing results of self-attacked adversarial accuracies and transfer-attacked adversarial accuracies.

5 Results

5.1 Polynomial Regression

Before discussing the fitted curves, we compute the second derivative $f''(N)$ of each polynomial and define $N^* = \min\{N > 0 \mid f''(N) \text{ is a local minimum}\}$, i.e. the "knee" point. This N^* marks the point where the concavity of the accuracy curve first changes sign.

Despite methods like splines [28] may better capture local behaviour, we treat the adversarial robustness curve much like a physical quantity, where not only the value but the full hierarchy of its rates of change (velocity, acceleration, jerk, etc.) can carry meaningful information. We require a model whose derivatives of all orders are well defined and smooth. Cubic (or higher-order) splines guarantee continuity only up to a finite derivative (e.g. C^2 for cubic splines), whereas a single global polynomial is infinitely differentiable. Thus, to rigorously locate and compare extrema in the second derivative (and potentially higher derivatives) across the entire domain, we prefer the polynomial representation. The power of the fitted regression function was tested from 1 to 20, and the one with the lowest composite score was chosen as the final fit for each model. The knee points were also very similar in both polynomial and spline cases. As a result, polynomial regressions were preferred. Figure 2. illustrates images of the fitted functions and Fig 3. illustrates first and second derivatives with maximum and minimum points marked as triangles. In comparing polynomial fits on randomly selected pixel subsets (Fig. 4) versus our rank-ordered pixel perturbations (Fig. 2), a few key differences emerge. First, the random-selection curves consistently exhibit larger out-of-sample errors (CV MSEs often > 0.9) and higher information-criteria values (AIC/BIC significantly above those for the ranked data), indicating overfitting and poorer generalization. Second, high-degree polynomials on random data produce multiple spurious inflection points (two minima and a maximum in the 6th-degree case), whereas the ranked-pixel fits yield a single, well-defined "knee" with minimal noise. Finally, the residual-vs-fitted plots and Q–Q diagnostics for the ranked data show more homoscedasticity and normality, while the random-case diagnostics display heteroscedastic spreads and heavier-tail outliers. Together, these results demonstrate that ordering pixels by their saliency produces more stable, interpretable regression models than sampling pixel budgets at random.

As shown in Table 1, ResNet101, VGG19, InceptionV3, and MobileNetV2 reach their first local minimum of $f''(N)$ near 5000 pixels, while ResNet50's "knee" is delayed at 15673 pixels. Beyond these "knee"s, $f''(N) \approx 0$ and accuracy decays linearly, indicating that non-robust features have been exhausted. ResNet50 maintains $\sim 40\%$ accuracy, whereas VGG19 and MobileNetV2 drop to $\sim 20\%$. Minor oscillations from high-degree polynomial fits appear at large N, but they do not obscure the clear weak-to-robust transition. On CIFAR-100 (pretrained RN56, VGG19, RN44, RepVGG), the estimated non-robust feature fractions are 32.6%, 32.6%, 31.6%, and 28.0%, respectively (exact: 32.617188%,

Table 1. Polynomial fitting parameters and post-"knee" plateau accuracies. N^* denotes the "knee" point, first local minimum of the second derivative (weak-to-robust transition), with 5000 pixels as reference, using pixel ranking selections

Model	Degree	N^* ("knee" Point)	Final Acc. (%)	Interpretation
ResNet50	9	15673	~40	Strongly delayed vs. 5000, highest robustness floor
ResNet101	8	6429	~37	Slightly delayed vs. 5000, solid post-"knee" decay
VGG19	10	5794	~20	Near reference (slightly delayed), sharp drop to low floor
InceptionV3	10	4921	~36	Near reference (slightly early), moderate robustness plateau
MobileNetV2	12	4921	~20	Near reference (slightly early), lightweight low robustness

32.617188%, 31.640625%, 28.027344%), aligning with the ImageNet trends and supporting the method's generality.

5.2 Knee Validity Evaluation

Since the former research [24] already investigated and discovered that transferabilities of adversarial examples across models exist [21,24], to test whether this phenomenon behaves similarly at these "knee" points, the transfer-attacks were applied against all tested models with the adversarial examples generated from each model. Table 2 contains all relevant parameters and tested accuracies for all models.

Table 2. Transfer-attack performance across different original and target models. "Pixels Perturbed" is the turning point from the second-derivative analysis; "Params (M)" is the total trainable parameters in each target model. Bold rows indicate original = target.

Orig. Model	Target Model	Pixels Perturbed	Params (M)	Transfer Acc. (%)	Original Acc. (%)	GFLOPS
ResNet50	**ResNet50**	15634	25.6	62.3	62.3	8.17
	ResNet101		44.5	71.7	72.4	15.61
	VGG19		143.7	53.2	40.3	39.17
	MobileNetV2		3.4	54.3	43.8	0.60
	InceptionV3		23.9	74.8	66.2	5.67
ResNet101	ResNet50	6429	25.6	76.2	75.1	8.17
	ResNet101		44.5	82.4	82.4	15.61
	VGG19		143.7	68.2	58.0	39.17
	MobileNetV2		3.4	69.7	60.8	0.60
	InceptionV3		23.9	85.3	78.7	5.67
VGG19	ResNet50	5794	25.6	75.1	76.5	8.17
	ResNet101		44.5	81.1	84.2	15.61
	VGG19		143.7	60.0	60.0	39.17
	MobileNetV2		3.4	66.5	62.2	0.60
	InceptionV3		23.9	83.0	80.3	5.67
InceptionV3	ResNet50	4921	25.6	78.2	78.2	8.17
	ResNet101		44.5	83.0	85.2	15.61
	VGG19		143.7	67.0	62.8	39.17
	MobileNetV2		3.4	70.1	65.8	0.60
	InceptionV3		23.9	81.8	81.8	5.67
MobileNetV2	ResNet50	4921	25.6	75.4	78.2	8.17
	ResNet101		44.5	82.3	85.2	15.61
	VGG19		143.7	62.6	62.8	39.17
	MobileNetV2		3.4	65.8	65.8	0.60
	InceptionV3		23.9	85.4	81.8	5.67

Table 3. Transfer-attack summary at the first minimum of the second derivative. "Variation (%)" is defined as Avg. Transfer Acc. minus Original Acc.

Original Model	Pixels Perturbed	Avg. Transfer Acc. (%)	Variation (%)
ResNet50	15634	65.3	+3.0
ResNet101	6429	76.0	–6.4
VGG19	5794	76.9	+16.9
InceptionV3	4921	75.8	–6.0
MobileNetV2	4921	77.1	+11.3

From Fig. 3, once each model surpasses its "knee" point N^*, the accuracy curve $f(N)$ flattens ($f'' \approx 0$), indicating that dense non-robust features have been exhausted and only sparse, semantically meaningful robust features remain susceptible. Although ResNet50's curvature briefly approaches zero near $N \approx 5000$, it fails to form a formal trough and instead yields its first "knee" at $N = 15634$. This consistent "knee" behavior across all five architectures validates our curvature-based criterion for delineating non-robust and robust feature regimes.

Table 2 reports both self-attack (diagonal) and cross-model transfer accuracies at these "knee" points, while Table 3 summarises average transfer gains and deviations from each model's original self-attack baseline. Early-"knee" networks (VGG19 at 5794 pixels, MobileNetV2 at 4921 pixels) achieve the largest transfer improvements (+16.9%, +11.3%), reflecting reliance on widely shared non-robust cues, whereas late-"knee" models (ResNet101 at 6429, InceptionV3 at 4921) incur negative variations (–6.4%, –6.0%), indicating more idiosyncratic vulnerabilities. ResNet50's delayed "knee" yields a modest +3.0% gain. Notably, VGG19 and MobileNetV2 also exhibit the lowest post-"knee" accuracy floors ($\approx$ 20%) and the greatest inter-model variability (up to ±10%), underscoring the instability of their non-robust regimes.

Across the four early-"knee" networks, the detected knee points fall within a tight 800-pixel band (4921–5794), yielding a standard deviation of only 350 pixels, far smaller than the full 40000-pixel budget, underscoring the robustness of our curvature-based detection across architectures. Moreover, the mean knee among all five models is 7620 pixels, and the median knee for the early group is 5358, suggesting that the budget required to exhaust non-robust features is largely independent of network depth or total parameter count. Crucially, this knee location aligns with the onset of the linear decay regime in the first derivative (see Fig. 2), confirming that our second-derivative criterion captures the true transition between feature regimes rather than spurious inflection points from high-degree fits. Together, these results demonstrate that the curvature-based knee is a stable, architecture-agnostic marker of the shift from shared, brittle feature usage to sparse, robust representation exploitation.

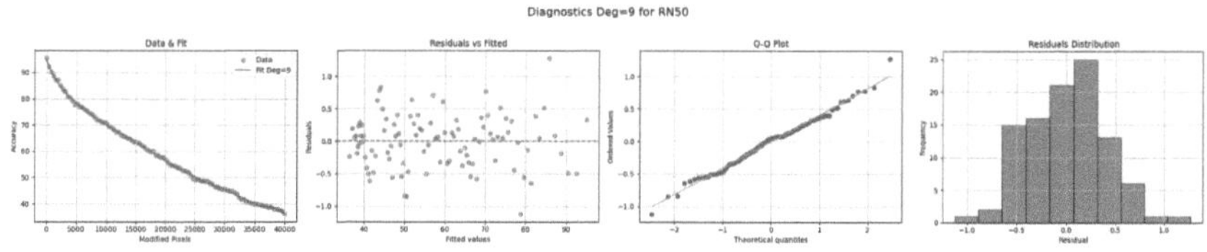

(a) ResNet50 diagnostics

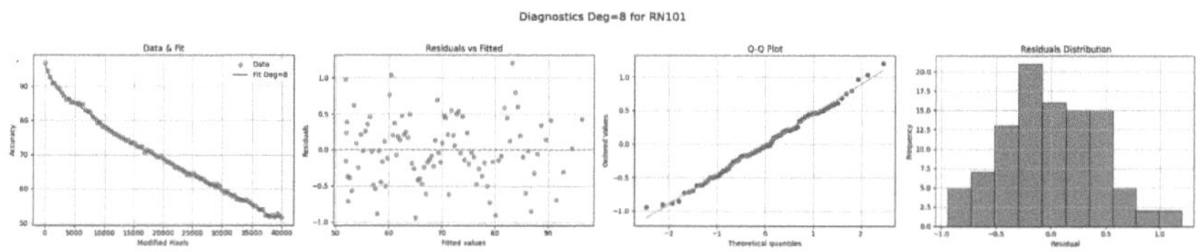

(b) ResNet101 diagnostics

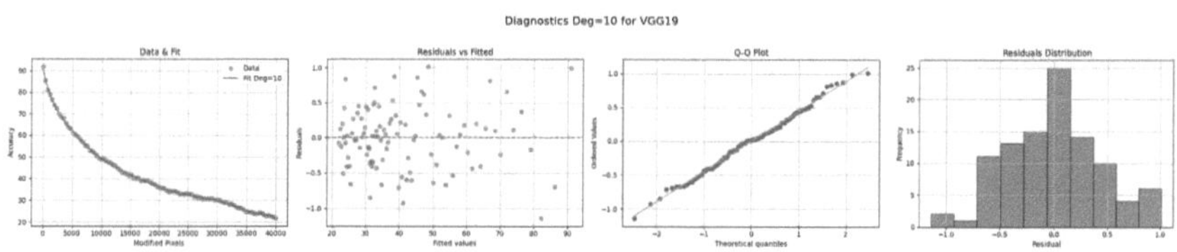

(c) VGG19 diagnostics

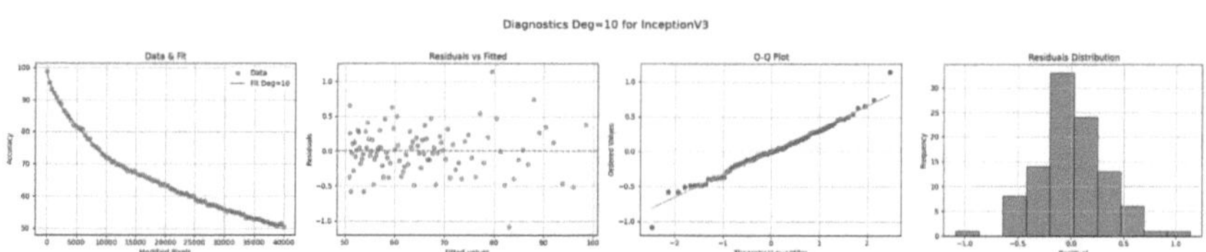

(d) InceptionV3 diagnostics

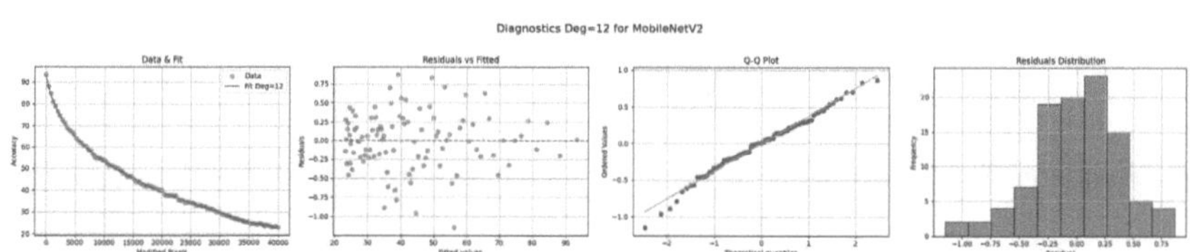

(e) MobileNetV2 diagnostics

Fig. 2. Diagnostics of polynomial-regression fittings for five backbone models.

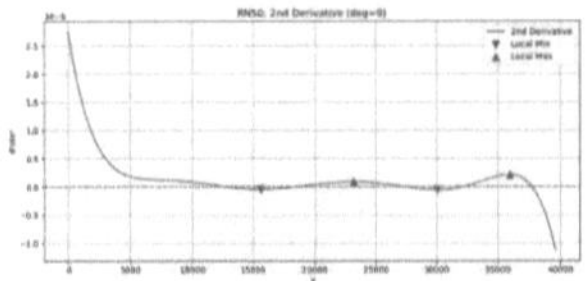

(a) ResNet50 2nd Deriv.

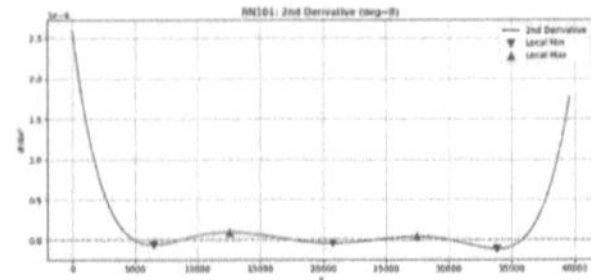

(b) ResNet101 2nd Deriv.

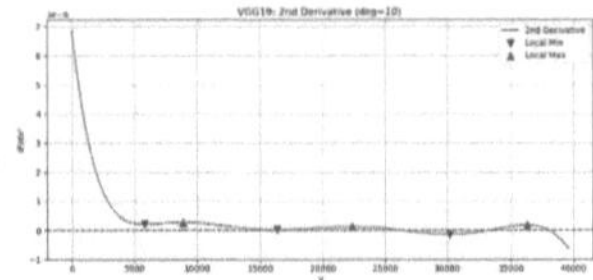

(c) VGG19 2nd Deriv.

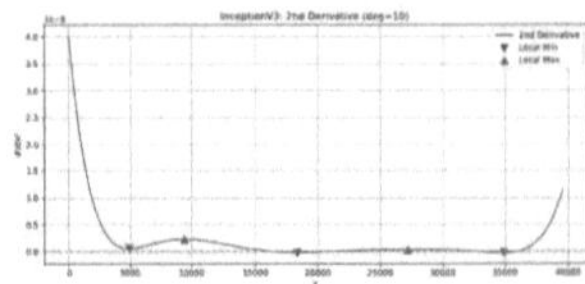

(d) InceptionV3 2nd Deriv.

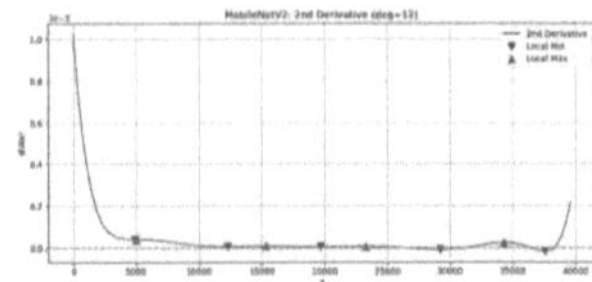

(e) MobileNetV2 2nd Deriv.

Fig. 3. Visualisations of polynomial-regression second derivatives for five backbone models with pixel ranking selections.

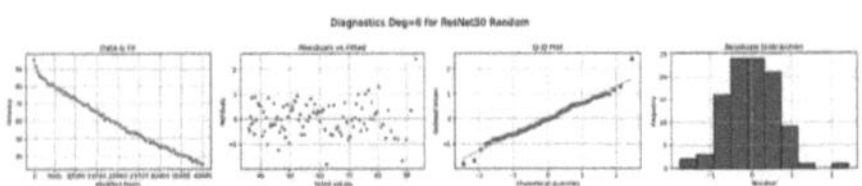

(a) ResNet50 diagnostics (random)

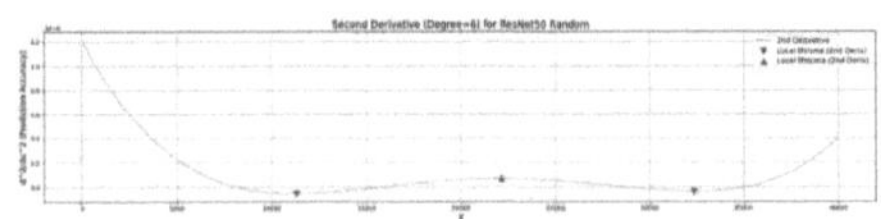

(b) ResNet50 Derivatives (random)

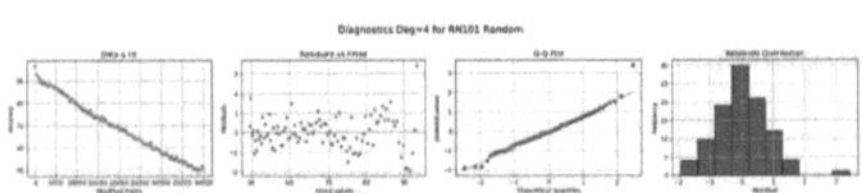

(c) ResNet101 diagnostics (random)

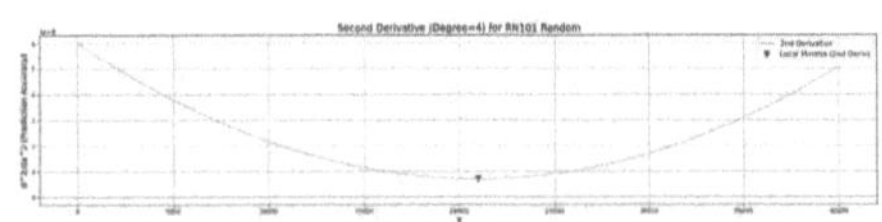

(d) ResNet101 Derivatives (random)

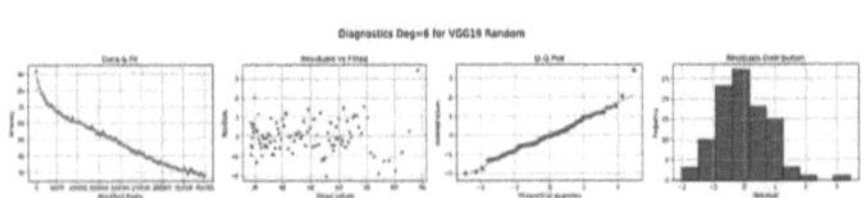

(e) VGG19 diagnostics (random)

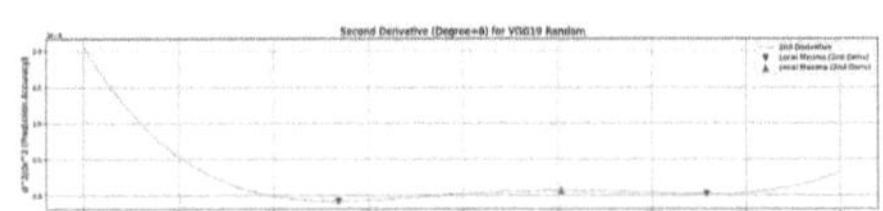

(f) VGG19 Derivatives (random)

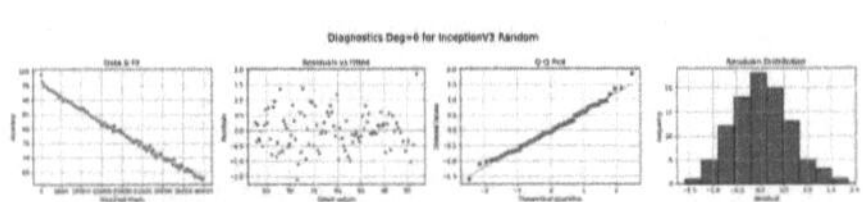

(g) InceptionV3 diagnostics (random)

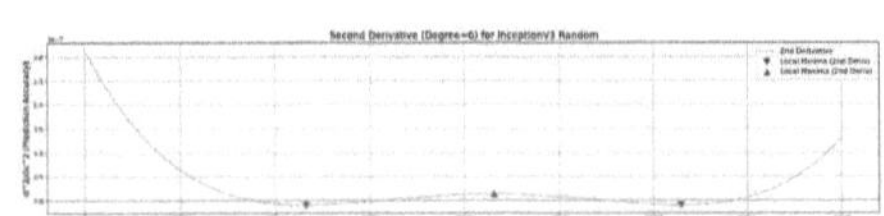

(h) InceptionV3 Derivatives (random)

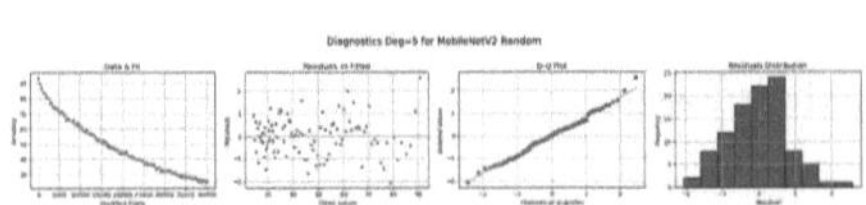

(i) MobileNetV2 diagnostics (random)

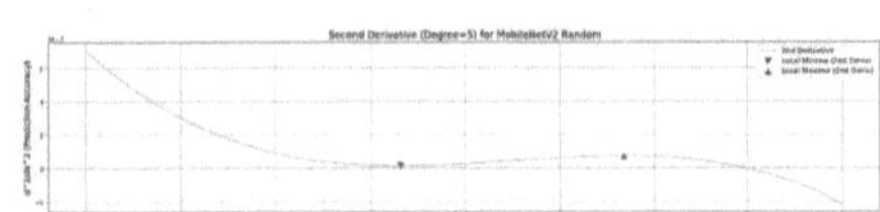

(j) MobileNetV2 Derivatives (random)

Fig. 4. Diagnostics and second-derivative visualisations for five backbone models with random selections.

6 Related Work

The bulk of adversarial-attack research has concentrated on crafting perturbations that modify all or nearly all pixels in an image, beginning with single-step methods such as FGSM [7] and its iterative extensions (e.g. PGD [29]), or on dense, norm-bounded noise (e.g. C&W [10]). Even sparse approaches like the one-pixel attack [8] relax the full-image constraint by targeting only a handful of pixel coordinates, yet they still do not account for the underlying feature semantics or distinguish fragile, non-robust patterns from semantically meaningful, robust ones.

A notable theoretical advance [16] rigorously characterizes how standard empirical risk-minimization procedures encourage neural networks to latch onto so-called "non-robust" features, subtle, high-frequency patterns that correlate strongly with class labels in the training data but lack semantic coherence from a human perspective. Under reasonable assumptions about high-dimensional data distributions, they decompose the learned representation into two orthogonal subspaces: one subspace captures robust, semantically meaningful features that correspond closely to human-interpretable visual concepts, while the other encodes brittle, non-robust cues that arise from spurious correlations within the training set.

Through a careful perturbation analysis of gradient-descent dynamics, they demonstrate that, during training, gradient-based optimization preferentially amplifies these non-robust components. Specifically, since non-robust features can often achieve low classification error on clean data more quickly than robust features, the optimization process gravitates toward them, causing the resulting classifier to rely disproportionately on high-frequency, dataset-specific artefacts. They then prove that any structured perturbation aligned with the non-robust subspace, however small under an ℓ_p norm will, with high probability, cause misclassification. In contrast, they derive lower bounds on the distortion required to attack solely through the robust subspace, showing that adversarial perturbations constrained to robust directions must be significantly larger in magnitude.

Although they outline potential spectral or frequency-domain filtering strategies to separate these subspaces since non-robust features often manifest as localized, high-frequency components, they stop short of providing a concrete, model-agnostic algorithm for inferring the precise boundary between robust and non-robust features at inference time. As a result, their framework illuminates the geometric underpinnings of adversarial fragility and explains why adversarial examples transfer across architectures, but it leaves open the practical challenge of isolating non-robust subspaces in a real-world pre-trained network. Meanwhile, parallel investigations using saliency map interpretation and information-bottleneck principles highlight heterogeneous feature importance across spatial locations but similarly fail to define a formal change-point separating stable, semantically grounded representations from brittle, non-robust ones.

7 Conclusion

We have presented a unified, continuous pixel-budget framework for systematically distinguishing non-robust from robust features in deep convolutional networks. By fitting global polynomials to accuracy-vs.-budget curves and detecting the first local minimum of the second derivative, we locate a formal "knee" that marks the transition from brittle, non-robust feature exploitation to steady degradation of robust features. Cross-model transfer experiments validate this "knee" as a model-agnostic regime boundary: early-"knee" architectures but also exhibit the largest transfer variability, reflecting the fragility of their non-robust features, while robust-feature-dominated models (ResNet101, InceptionV3) incur consistent transfer losses, and ResNet50 yields an intermediate gain. A brief CIFAR-100 check mirrors the ImageNet pattern, further supporting generality. These findings corroborate the weak-vs-robust feature hypothesis and provide practical benchmarks for crafting both universal and model-specific adversarial evaluations.

In future work, we will explore structured and non-gradient perturbation schemes, extend curvature-based "knee" detection to certified defense frameworks and randomized smoothing, and apply our methodology to other modalities and downstream tasks.

References

1. Hendrycks, D., Mazeika, M., Dietterich, T.: ImageNet-A: 1000 natural adversarial examples. arXiv preprint arXiv:2002.02133 (2020)
2. Simonyan, K., Zisserman, A.: Very deep convolutional networks for large-scale image recognition. In: International Conference on Learning Representations (2015). https://arxiv.org/abs/1409.1556v6
3. Szegedy, C., et al.: Going deeper with convolutions. arXiv preprint arXiv:1409.4842v1 (2014)
4. Howard, A.G., et al.: MobileNets: efficient convolutional neural networks for mobile vision applications. arXiv preprint arXiv:1704.04861v1 (2017)
5. He, K., Zhang, X., Ren, S., Sun, J.: Deep residual learning for image recognition. arXiv preprint arXiv:1512.03385v1 (2016)
6. Das, S., Suganthan, P.N.: Differential evolution: a survey of the state-of-the-art. IEEE Trans. Evol. Comput. **15**(1), 4–31 (2011)
7. Goodfellow, I.J., Shlens, J. and Szegedy, C.: Explaining and harnessing adversarial examples. arXiv preprint arXiv:1412.6572 (2015)
8. Su, J., Vargas, D.V., Sakurai, K.: One pixel attack for fooling deep neural networks. IEEE Trans. Evol. Comput. **23**(5), 828–841 (2019)
9. Yang, X., Wang, X.: Kolmogorov–Arnold Transformer (n.d.). https://github.com/Adamdad/kat
10. Carlini, N., Wagner, D.: Towards evaluating the robustness of neural networks. In: 2017 IEEE Symposium on Security and Privacy (SP), pp. 39–57 (2017)
11. Papernot, N., McDaniel, P., Jha, S., Fredrikson, M., Celik, Z.B., Swami, A.: The limitations of deep learning in adversarial settings. In: Proceedings of the 2016 IEEE European Symposium on Security and Privacy, pp. 372–387 (2016). https://ieeexplore.ieee.org/document/7546546

12. Papernot, N., McDaniel, P., Goodfellow, I., Jha, S., Celik, Z.B., Swami, A.: Practical black-box attacks against machine learning. In: Proceedings of the 2017 ACM Asia Conference on Computer and Communications Security (ASIACCS 2017), pp. 506–519 (2017). https://arxiv.org/abs/1602.02697
13. Bojarski, M., et al.: End to end learning for self-driving cars. NVIDIA (2016)
14. Chen, C., Seff, A., Kornhauser, A., Xiao, J.: DeepDriving: learning affordance for direct perception in autonomous driving. In: Proceedings of the IEEE International Conference on Computer Vision (ICCV), pp. 2722–2730 (2015)
15. Jacobi, C.G.J.: De determinantibus functionalibus. J. Reine Angew. Math. **11**, 262–293 (1841)
16. Li, B., Li, Y.: Adversarial training can provably improve robustness: theoretical analysis of feature learning process under structured data. J. Adversarial Mach. Learn. (2025, Forthcoming)
17. Dan, X., et al.: Robust image selection for benchmarking deep learning models. arXiv preprint arXiv:2101.12345 (2021)
18. Gu, T., Dolan-Gavitt, B., Garg, S.: BadNets: identifying vulnerabilities in the machine learning supply chain. arXiv preprint arXiv:1708.06733 (2017)
19. Ilyas, A., Engstrom, L., Athalye, A., Lin, J.: adversarial examples are not bugs, they are features. In: Advances in Neural Information Processing Systems, vol. 32, pp. 125–136 (2019)
20. Moosavi-Dezfooli, S.-M., Fawzi, A., Frossard, P.: DeepFool: a simple and accurate method to fool deep neural networks. In: Proceedings of the IEEE Conference on Computer Vision and Pattern Recognition, pp. 2574–2582 (2016)
21. Liu, Y., Chen, X., Liu, C., Song, D.: Delving into transferable adversarial examples and black-box attacks. In: Proceedings of the 34th International Conference on Machine Learning, pp. 3548–3557 (2017)
22. Kurakin, A., Goodfellow, I., Bengio, S.: Adversarial examples in the physical world. In: Proceedings of the International Conference on Learning Representations (ICLR) Workshop, pp. 39–46 (2017)
23. Deng, J., Dong, W., Socher, R., Li, L.-J., Li, K., Fei-Fei, L.: ImageNet: a large-scale hierarchical image database. In: Proceedings of the IEEE Conference on Computer Vision and Pattern Recognition (CVPR), pp. 248–255 (2009)
24. Wu, D., Wang, Y., Xia, S.-T., Bailey, J. Ma, X.: Skip connections matter: on the transferability of adversarial examples generated with ResNets. In: Proceedings of the International Conference on Learning Representations (ICLR) (2020). https://arxiv.org/abs/2002.05990
25. Stone, M.: Cross-validatory choice and assessment of statistical predictions. J. Roy. Stat. Soc. B **36**(2), 111–147 (1974)
26. Akaike, H.: A new look at the statistical model identification. IEEE Trans. Autom. Control **19**(6), 716–723 (1974)
27. Schwarz, G.: Estimating the dimension of a model. Ann. Stat. **6**(2), 461–464 (1978)
28. Hastie, T., Tibshirani, R., Friedman, J.: The Elements of Statistical Learning, 2nd edn. Springer, New York (2009). https://doi.org/10.1007/978-0-387-84858-7
29. Madry, A., Makelov, A., Schmidt, L., Tsipras, D., Vladu, A.: Towards deep learning models resistant to adversarial attacks. arXiv preprint arXiv:1706.06083 (2017)
30. Cohen, J., Rosenfeld, E., Kolter, J.Z.: Certified adversarial robustness via randomized smoothing. arXiv preprint arXiv:1902.02918 (2019)
31. Brown, T.B., et al.: Language models are few-shot learners. arXiv preprint arXiv:2005.14165 (2020)
32. Chen, M., et al.: Evaluating large language models trained on code. arXiv preprint arXiv:2107.03374 (2021)

Graph-Oriented Cross-Modality Diffusion for Multimedia Recommendation

Jiamin Chen[1], Tanzheng Jiang[1], Zhenghong Lin[1], Guofang Ma[2], and Yanchao Tan[1(✉)]

[1] College of Computer and Data Science, Fuzhou University, Fuzhou 350116, China
yctan@fzu.edu.cn

[2] School of Computer Science and Technology, Zhejiang Gongshang University, Zhejiang 310018, China
maguofang1991@zju.edu.cn

Abstract. Multimedia recommender systems have gained significant attention with the proliferation of multimedia-sharing platforms. While existing approaches primarily focus on modeling user-item bipartite graphs enhanced with multimodal features, they often overlook the rich structural information embedded in the cross-modality item-item graph. In this paper, we introduce the Graph-oriented cross-modality diffusion for multimedia Recommendation (GoodRec), a novel framework that excavates high-order relations between the cross-modality item-item graph for multimedia recommendations. Specifically, we first conceptualize a unified multi-modality item-item graph as a multivariate heat diffusion system, with an effective energy function to guide both intra-modality and inter-modality diffusion toward consistent representation learning. Then, we develop an enhanced multimedia recommendation module that constructs modality-specific graphs from diffusion-refined representations and employs adversarial mechanisms to strengthen user-item interactions. Extensive experiments on three real-world multimedia datasets demonstrate that GoodRec consistently outperforms state-of-the-art baselines, confirming the effectiveness of excavating high-order relations between cross-modality graph structure via diffusion.

Keywords: Graph diffusion · Multimedia recommendation · Graph convolutional network

1 Introduction

The rapid proliferation of multimedia platforms (e.g., Twitter and TikTok) has led to an explosive growth in user-generated content [8,14,27], making multimedia recommender systems (MRS) indispensable for guiding users to relevant items [3,20,26,28]. By integrating multimodal data, such as visual and textual modalities, MRS can capture more nuanced user preferences than conventional collaborative filtering approaches [9,13,17,25]. Existing research has thus focused on modeling the userâĂŞitem interaction graph augmented with

M. Yoshikawa et al. (Eds.): ADMA 2025, LNAI 16197, pp. 252–266, 2026.
https://doi.org/10.1007/978-981-95-3453-1_17

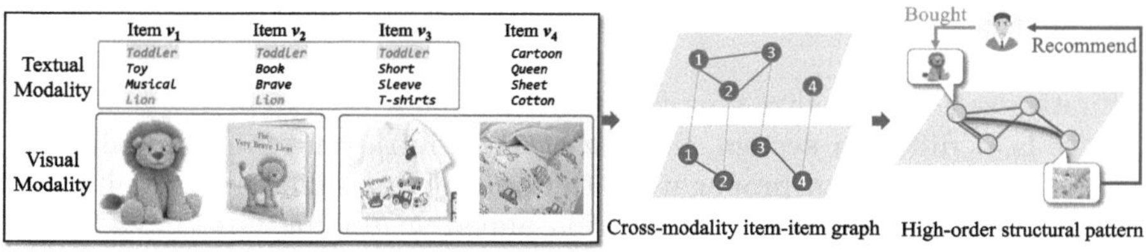

Fig. 1. Illustration of cross-modality item-item graph structure-enhanced multimedia recommendation.

modality-specific features. For example, MMGCN [23] captures users' modality-specific preferences for micro videos by constructing user-item bipartite graphs on each modality and utilizing the information transmission mechanism of graph neural networks. GRCN [22] improves the performance of graph convolutional network-based recommendation systems by adaptively refining the user-item interaction graph structure, identifying and trimming potential false-positive edges. These approaches, which integrate user interactions with modality-specific features, have to some extent improved the accuracy of recommendations.

However, beyond the user-item interaction graph lies a richer signal: the high-order structural information embedded in the cross-modality item-item graph. As illustrated in Fig. 1, items v_1, v_2, and v_3 all share a textual theme (toddler), while v_1 and v_2 are visually similar (both lion motifs), and v_3 and v_4 share another visual similarity (cartoon cars). A user purchasing a lion-shaped toy v_1 could receive recommendations for a car-patterned sheet v_4 via the cross-modality path $v_1 \xrightarrow{\text{textual}} v_3 \xrightarrow{\text{visual}} v_4$, which conventional models based solely on user-item interaction graphs may fail to capture. Given the potential of a cross-modality item-item graph structure to enhance multimedia recommendation, a natural question is:

How can we extract a fine cross-modality item-item graph structure and integrate it with the user-item interaction graph to enable recommendation?

To address this question, we draw inspiration from heat diffusion processes on graphs [16], which have proven effective at capturing nuanced relationships in complex networks. In graph heat diffusion theory, nodes correspond to heat reservoirs and edges carry energy values, where higher energy indicates weaker or lower-quality connections. The diffusion in heat systems can be seen as the energy reduction process (from high energy to low energy) by removing the low-quality edges. By treating items' visual and textual modalities as separate "heat reservoirs" and initializing fully connected intra- and inter-modality edges, we can frame the cross-modality itemâĂŞitem graph as a heat system. We then define an energy function to quantify the quality of each cross-modality edge, iteratively performing diffusion to prune low-quality associations and yield a refined graph structure that highlights high-order interactions.

Building on this intuition, we put forward **G**raph-**o**riented cr**o**ss-modality **d**iffusion for multimedia **Rec**ommendation (GoodRec), a unified framework that treats multimedia recommendation as a quality-driven diffusion process. The key steps are as follows:

Firstly, we conceptualize a unified multi-modality item-item graph as a multivariate heat diffusion system, initializing both intra- and inter-modality connections where heat can propagate within and across different modalities. We then propose an effective energy function aimed at guiding inter-modality and intra-modality diffusion towards a consistent direction for multi-modality representation learning.

Furthermore, in order to enhance the multimedia recommendation based on the representations optimized by the above diffusion step, we design a tailored Enhanced Multimedia Recommendation Module. This module constructs modality-specific graphs from post-diffusion representations and employs adversarial mechanisms to strengthen user-item interactions. Subsequently, multi-modality features are aggregated through graph message passing to obtain final representations for recommendation.

Finally, we conduct evaluations of GoodRec on three real-world multimedia datasets, Amazon-Baby, Amazon-Sports, and Allrecipes, comparing with 11 state-of-the-art baselines. Extensive experimental results demonstrate that GoodRec consistently achieves superior performance (e.g., achieving up to 33.91% improvement on Recall@10 over the best baseline on the Allrecipes dataset).

Contribution: (1) *Theoretical Formulation:* To the best of our knowledge, we are the first to model cross-modality itemâĂŞitem associations as a multivariate heat diffusion system, introducing an energy function to evaluate and refine cross-modality graph quality; (2) *Methodological Innovation:* GoodRec integrates heat-diffusion principles into multimedia recommendation, enabling the discovery of high-order structural information across modalities and improving representation learning through energy minimization; (3) *Empirical Validation:* We provide the convergence analysis about our diffusion process and theoretically prove that our energy equation has a definite upper bound, where experimental results demonstrate that our model achieves state-of-the-art performance in multimedia recommendation tasks.

2 Related Work

2.1 Multimedia Recommendation

The multimedia recommendation model recommends items that meet personal preferences to users by analyzing their historical interaction information on the multimedia platform and the multimodal features of items. For example, MKGAT [18] could introduce the multi-modal knowledge graph to the recommendation system innovatively. NCL [12] incorporated the potential neighbors into contrastive pairs for the preference of users over items. Mandari [15] was able to effectively construct the implicit associations among multimodal data and user choices. However, all the above methods still rely primarily on the user-item interaction graph and ignore the high-order structural information in multimedia recommendation data. In our work, we propose GoodRec to capture the high-order structural information.

2.2 Graph Neural Diffusion

Graph neural diffusion is a diffusion process that is led by partial differential equations (PDEs). Inspired by the numerical solution of partial differential equations on manifolds, PDE-GCN [2] was proposed to alleviate the oversmoothing phenomenon observed in graph convolutional networks. ADC [30] could learn the optimal neighborhood size from the data automatically. NIGCN [7] was applied to extract the unique features of a specific node in the process of diffusion, promoting the creation of top-notch node representations. MGDCF [6] was a method to weigh the two types of distances based on Markov processes. Although graph neural diffusion models have made ahead in multiple territories, their extension to complex multimodal learning remains unsolved due to the complex relationships between the different modalities.

3 Method

3.1 Preliminary and Problem Statement

Preliminary of Graph Diffusion Process. Let $\mathcal{G} = (\boldsymbol{G}, \boldsymbol{X})$ represents the graph, where $\boldsymbol{G} \in \mathbb{R}^{N\times N}$ and $\boldsymbol{X} = [\boldsymbol{x}_1; \cdots ; \boldsymbol{x}_N] \in \mathbb{R}^{N\times D}$. Here, N corresponds to the number of samples and D denotes the number of dimensions. Drawing inspiration from thermal diffusion on Riemannian manifolds, all instances are treated as a cohesive entity and propagated as a continuous flow of features. The smoothness of feature propagation between two instances is directly proportional to the disparity in their respective feature sets. Mathematically, the diffusion process of the heat system can be formally described as follows:

$$\frac{\partial \boldsymbol{F}(t)}{\partial t} = div\big(\boldsymbol{G} \odot \nabla \boldsymbol{F}(t)\big), \quad \boldsymbol{F}(0) = \boldsymbol{X}, \tag{1}$$

where $\boldsymbol{F}(t)$ represents the feature representation at time t. The symbol $\odot$ denotes Hadamard product, $\boldsymbol{G}_{ij}$ denotes the smoothness of feature propagation between instances i and j, ∇ indicates the difference between instances, and $div(\cdot)$ represents the cumulative feature flow. More specifically, for the i-th instance, the heat flow per unit of time into its interior corresponds to the summation of the heat changes over its space. Equation (1) can be written explicitly as:

$$\frac{\partial \boldsymbol{f}_i(t)}{\partial t} = \sum_{j=1}^{N} \boldsymbol{G}_{ij}\big(\boldsymbol{f}_j(t) - \boldsymbol{f}_i(t)\big). \tag{2}$$

Since Eq. (2) represents continuous dynamics, practical implementation requires the utilization of numerical methods for its solution. We extend this process to encompass more intricate multi-modality scenarios.

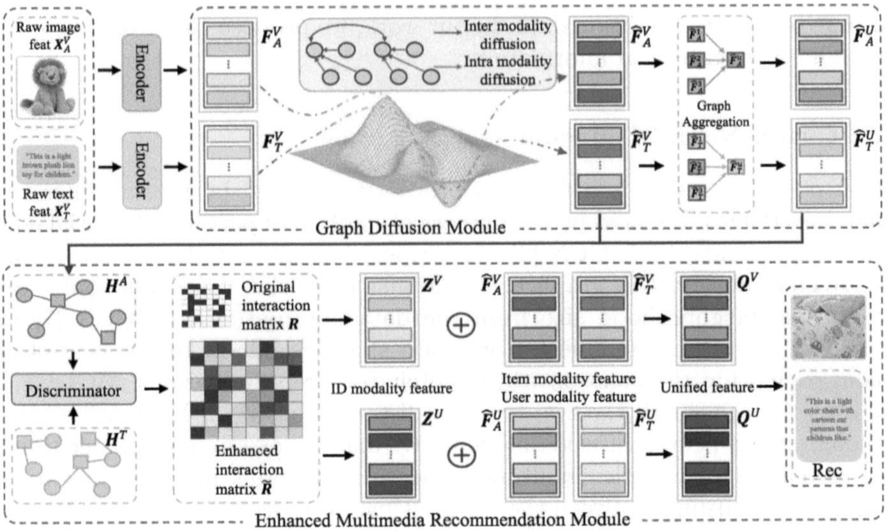

Fig. 2. The overall framework about GoodRec. In the upper part, the graph diffusion network is employed to capture the high-order relations among multiple modalities. In the under part, the enhancement mechanism based on a discriminator is designed to enhance multimedia recommendation.

Problem Statement. GoodRec aims to effectively capture the high-order relations present in multimedia recommendation data through a graph diffusion structure. The inputs of our proposed GoodRec contain a user-item interaction matrix $\boldsymbol{R} \in \mathbb{R}^{N_U \times N_V}$ and the raw item modality features $\{\boldsymbol{X}_1^V, ..., \boldsymbol{X}_m^V, ..., \boldsymbol{X}_M^V\}$, where N_U, N_V are the number of users and items and M is the number of item modalities (e.g., M is set to 2 with visual and textual modalities). The output is a probability matrix $\hat{\boldsymbol{R}} \in \mathbb{R}^{N_U \times N_V}$ for multimedia recommendation, where the value of $\hat{\boldsymbol{R}}_{ij}$ represents the probability that the j-th item is recommended to the i-th user. We summarize the key components of GoodRec framework in Fig. 2 and provide an overview.

3.2 Graph Diffusion Module

Cross-Modality Relation Capturing. Our proposed model begins with a diffusion process that considers the multi-modality data as a unified entity, facilitating diffusion within and among modalities via the flow of features. We denote $\boldsymbol{X}_A^V \in \mathbb{R}^{N_V \times d_A}$ as raw visual item features with latent dimension d_A and denote $\boldsymbol{X}_T^V \in \mathbb{R}^{N_V \times d_T}$ as raw textual item features with latent dimension d_T. Given that raw item features in the multi-modality data exhibit distinct dimensions, potentially impeding feature flow, we address this by independently mapping the features of each modality into a shared space. The specifics are outlined below: $\boldsymbol{F}_m^V = \boldsymbol{X}_m^V \boldsymbol{W}_m + \boldsymbol{b}_m$. Where $\boldsymbol{X}_m^V \in \mathbb{R}^{N_V \times d_m}$, $\boldsymbol{W}_m \in \mathbb{R}^{d_m \times d}$ and $\boldsymbol{b}_m \in \mathbb{R}^d$

are trainable weight matrices and bias in m-th modality. $\boldsymbol{F}_m^V \in \mathbb{R}^{N_V \times d}$ is m-th modality features which is mapped into a shared space.

After undergoing the aforementioned process, we extend the diffusion process into multimodal scenarios to capture cross-modality high-order relations. We rewrite Eq. (2) as follows:

$$\frac{\partial \boldsymbol{F}_i^m(t)}{\partial t} = \sum_{j=1}^{N_V} \boldsymbol{S}_{ij}^m(t)\Big(\boldsymbol{F}_j^m(t) - \boldsymbol{F}_i^m(t)\Big) + \sum_{v=1}^{M} \boldsymbol{P}_{mv}(t)\Big(\boldsymbol{F}_i^v(t) - \boldsymbol{F}_i^m(t)\Big), \quad (3)$$

where $\boldsymbol{S}_{ij}^m(t)$ denotes the intra-modality similarity between item i and item j within the m-th modality at time t, while $\boldsymbol{P}_{mv}(t)$ denotes the inter-modality similarity between the m-th modality and the v-th modality at item i at time t. $\boldsymbol{S}_{ij}^m(0)$ and $\boldsymbol{P}_{mv}(0)$ are Xavier-initialized embeddings. The interpretation of Eq. (3) is as follows: it represents that the representation update of item i is simultaneously affected by the intra-modality relations and inter-modality relations. According to previous theories [16], we can use explicit Euler method with step size α to approximate Eq. (3) as:

$$\begin{aligned} &\boldsymbol{F}^{(m,k+1)} = \Big[(1-2\alpha)\boldsymbol{I} + \alpha \boldsymbol{S}^{(m,k)}\Big] \boldsymbol{F}^{(m,k)} + \alpha \sum_{v=1}^{M} \boldsymbol{P}_{mv}^{(k)} \boldsymbol{F}^{(v,k)}, \\ &\text{s.t. } \boldsymbol{S}^{(m,k)}1 = 1, \boldsymbol{P}^{(k)}1 = 1, \boldsymbol{P}^{(k)} = \boldsymbol{P}^{(k)T}, \end{aligned} \quad (4)$$

where $\boldsymbol{S}^{(m,k)}$ denotes the intra-modality similarity matrix within the m-th modality at step k, while $\boldsymbol{P}_{mv}^{(k)}$ denotes the inter-modality similarity matrix between the m-th modality and the v-th modality at step k. $\boldsymbol{I}$ is the identity matrix. When the matrices $\boldsymbol{S}^{(m,k)}$ and the matrices $\boldsymbol{P}_{mv}^{(k)}$ satisfy the constraint conditions, if $0 < \alpha < 1$, Eq. (4) can converge iteratively. The detailed convergence process is shown as:

Theorem 1. *The diffusion process of Eq. (4) can converge iteratively when* $0 < \alpha < 1$.

Proof. The updating rules defined in Eq. (4) can be rewritten as:

$$\boldsymbol{F}^{(m,k+1)} = \boldsymbol{Q}^{(m,k)} \boldsymbol{F}^{(m,k)} + \alpha \sum_{v=1}^{M} \boldsymbol{P}_{mv}^{(k)} \boldsymbol{F}^{(v,k)}, \quad (5)$$

where $\boldsymbol{Q}^{(m,k)}$ and $\boldsymbol{P}^{(k)}$ are the intra-modality and inter-modality similarity matrix, respectively. To ensure convergence, we let $\boldsymbol{Q}^{(m,k)}$, with $\lambda_{\max}(\boldsymbol{S}^{(m,k)}) = 1$, where $\lambda_{\max}$ satisfies the following condition:

$$|(1-2\alpha) + \alpha\lambda_{\max}| < 1 \Rightarrow 0 < \alpha < 1. \quad (6)$$

This ensures a contraction in the vector space. Given $\boldsymbol{P}^{(k)}$ is doubly stochastic with $\lambda_{\max}(\boldsymbol{P}^{(k)}) = 1$, it maintains the norm of $\boldsymbol{F}^{(m,k)}$, thus reinforcing the constraint on α to keep ρ of the overall update matrix below 1.

Calibrated Energy Constraint. Since heat diffusion needs a reasonable energy function to constrain it, we construct our reliable energy function tailored for multimodal scenarios as:

$$E(\{\boldsymbol{F}^m\}_{m=1}^M) = \underbrace{\frac{1}{2}\sum_{m=1}^{M}\sum_{i,j}^{N_V}\eta(\|\boldsymbol{F}_i^m - \boldsymbol{F}_j^m\|_2^2)}_{intra-modality} + \underbrace{\frac{1}{4}\sum_{m,n}^{M}\delta(\|\boldsymbol{F}^m - \boldsymbol{F}^n\|_F^2)}_{inter-modality}, \tag{7}$$

where $\eta(\cdot)$ and $\delta(\cdot)$ denote the monotonically increasing concave functions. The first term quantifies the variance among items within the same modality, where a minimal variance correlates with reduced energy levels. Similarly, the second term quantifies the disparity between items across different modalities, with the principle that a lesser disparity also results in decreased energy. This dual-term representation underlines the system's equilibrium, emphasizing the importance of minimizing both intra- and inter-modality discrepancies to achieve optimal energy efficiency. Then we take the energy function into the diffusion process, Eq. (4) can be written in the form of Eq. (8) :

$$\begin{aligned} \boldsymbol{F}^{(m,k+1)} &= \left[(1-2\alpha)\boldsymbol{I} + \alpha\boldsymbol{S}^{(m,k)}\right]\boldsymbol{F}^{(m,k)} + \alpha\sum_{v=1}^{M}\boldsymbol{P}_{mv}^{(k)}\boldsymbol{F}^{(v,k)}. \\ \text{s.t. } &\boldsymbol{S}^{(m,k)}1 = 1, \boldsymbol{P}^{(k)}1 = 1, \boldsymbol{P}^{(k)} = \boldsymbol{P}^{(k)^T}, \\ &E(\{\boldsymbol{F}^{(m,k+1)}\}_{m=1}^M) < E(\{\boldsymbol{F}^{(m,k)}\}_{m=1}^M). \end{aligned} \tag{8}$$

Addressing the equation presented necessitates navigating through an extensive array of constraints, posing significant challenges in deriving appropriate values for $\boldsymbol{S}^{(m,k)}$ and $\boldsymbol{P}^{(k)}$. To illustrate the effectiveness of our proposed calibrated energy constraint in Eq. (8), we first analyze the upper bound of the calibrated energy constraint system and give the optimization process of Eq. (8). The analysis of the upper bound is from Eq. (9) to Eq. (10), and the optimization process is from Eq. (11) to Eq. (16). According to [24], we can learn that the proposed energy function upper bound is:

$$\begin{aligned} \tilde{E}(\{\boldsymbol{F}^m\}_{m=1}^M) =& \frac{1}{2}\sum_{m=1}^{M}\sum_{i,j}^{N_V}\left[\boldsymbol{S}_{ij}^m\|\boldsymbol{F}_i^m - \boldsymbol{F}_j^m\|_2^2 - \tilde{\eta}(\boldsymbol{S}_{ij}^m)\right] \\ &+\frac{1}{4}\sum_{m,n}^{M}\left[\boldsymbol{P}_{mn}\|\boldsymbol{F}^m - \boldsymbol{F}^n\|_F^2 - \tilde{\delta}(\boldsymbol{P}_{mn})\right], \end{aligned} \tag{9}$$

where $\tilde{\eta}(x)$ and $\tilde{\delta}(x)$ correspond to the conjugate functions of $\eta(x)$ and $\delta(x)$. The upper bound in Eq. (9) is realized if and only if the conditions meet the following equation:

$$\tilde{\boldsymbol{S}}_{ij}^{(m)} = \left.\frac{\partial\eta(\boldsymbol{L}^2)}{\partial\boldsymbol{L}^2}\right|_{\boldsymbol{L}^2=\|\boldsymbol{F}_i^{(m)}-\boldsymbol{F}_j^{(m)}\|_2^2}, \quad \tilde{\boldsymbol{P}}_{mn} = \left.\frac{\partial\delta(\boldsymbol{G}^2)}{\partial\boldsymbol{G}^2}\right|_{\boldsymbol{G}^2=\|\boldsymbol{F}^{(m)}-\boldsymbol{F}^{(n)}\|_F^2}, \tag{10}$$

where $\boldsymbol{S}_{ij}^{(m)} = \frac{\tilde{\boldsymbol{S}}_{ij}^{(m)}}{\sum_{j=1}^{N} \tilde{\boldsymbol{S}}_{ij}^{(m)}}$. In this paper, we specify the function $\eta(x) = \delta(x) = x - 2\log(e^{\frac{x}{2}-1} + 1)$, and then Eq. (10) can be rewrite as:

$$\tilde{\boldsymbol{S}}_{ij}^{(m)} = \frac{1}{1 + e^{-f(\boldsymbol{F}_i^{(m)}, \boldsymbol{F}_j^{(m)})}}, \quad \tilde{\boldsymbol{P}}_{mn} = \frac{1}{1 + e^{-g(\boldsymbol{F}^{(m)}, \boldsymbol{F}^{(n)})}}, \tag{11}$$

where $f : \mathbb{R}^d \times \mathbb{R}^d \rightarrow \mathbb{R}$, $g : \mathbb{R}^{N_V \times d} \times \mathbb{R}^{N_V \times d} \rightarrow \mathbb{R}$. To ensure the inter-modality similarity matrix $\boldsymbol{P}$ retains symmetry and bi-randomness throughout the computation, we employ the differentiable projection algorithm as proposed by previous work [1]:

$$\mathcal{J}_0(\tilde{\boldsymbol{P}}) = \frac{\text{softmax}_{dim=0}(\tilde{\boldsymbol{P}}) + \text{softmax}_{dim=1}(\tilde{\boldsymbol{P}})}{2}, \tag{12}$$

$$\mathcal{J}_1(\tilde{\boldsymbol{P}}) = \text{ReLU}(\tilde{\boldsymbol{P}}), \tag{13}$$

$$\mathcal{J}_2(\tilde{\boldsymbol{P}}) = \tilde{\boldsymbol{P}} - \frac{1}{M}(\tilde{\boldsymbol{P}}\mathbf{1} - \mathbf{1})\mathbf{1}^\top, \tag{14}$$

$$\mathcal{J}_3(\tilde{\boldsymbol{P}}) = \hat{\boldsymbol{P}} - \frac{1}{M}\mathbf{1}(\mathbf{1}^\top \tilde{\boldsymbol{P}} - \mathbf{1}^\top), \tag{15}$$

where $\mathbf{1}$ denotes all 1 vector. Equations (13)–(15) undergoes iterative computations, necessitating a large number of iterations to satisfy the desired conditions. To speed up the convergence of this process, Eq. (12) utilizes an initialization to approximate the constraint-satisfying matrix $\hat{\boldsymbol{P}}$. Ultimately, we obtain $\boldsymbol{P} = \mathcal{J}_3(\mathcal{J}_2(\mathcal{J}_1(\mathcal{J}_0(\tilde{\boldsymbol{P}}))))$. By selecting both intra-modality and inter-modality diffusion functions, the energy function reaches its upper bound. We compute the energy function $E(\{\boldsymbol{F}^{(m)}\}_{m=1}^{M})$ partial derivative with respect to $\boldsymbol{F}^{(m)}$. Following this computation, we implement a gradient descent algorithm in a step-wise manner, adopting a step size denoted by γ, as described below:

$$\begin{aligned} \boldsymbol{F}^{(i,k+1)} &= \boldsymbol{F}^{(i,k)} - \gamma \frac{\partial E(\{\boldsymbol{F}^{(m,k)}\}_{m=1}^{M})}{\partial} \\ &= \left[(1-2\gamma)\boldsymbol{I} + \gamma \boldsymbol{S}^{(i,k)}\right] \boldsymbol{F}^{(i,k)} + \gamma \sum_{v=1}^{M} \boldsymbol{P}_{iv}^{(k)} \boldsymbol{F}^{(v,k)}. \end{aligned} \tag{16}$$

Equation (16) reveals its structural similarity to Eq. (4), indicating that executing a single iteration of the update process effectively corresponds to a reduction in the overall system energy.

After performing K iterations, we obtain the final potential representation for item embeddings of each modality. We denote $\hat{\boldsymbol{F}}_m^V \in \mathbb{R}^{N_V \times d}$ as item embeddings of each modality after the graph diffusion operation. In order to obtain user embeddings of each modality, we aggregate the item embeddings through graph message aggregation as: $\hat{\boldsymbol{F}}_m^U = {\boldsymbol{D}^U}^{-1} \boldsymbol{R} \hat{\boldsymbol{F}}_m^V$, where $\boldsymbol{D}^U \in \mathbb{R}^{N_U \times N_U}$ is the diagonal degree matrix of user-item interaction matrix $\boldsymbol{R} \in \mathbb{R}^{N_U \times N_V}$.

3.3 Enhanced Multimedia Recommendation Module

Adversarial Enhanced Part. In order to enhance the multimedia recommendation based on the above graph diffusion process, we propose an enhancement mechanism based on adversarial learning. Firstly, we construct a unique graph structure under each modality by calculating the maximum probability likelihood of the user and the item modality embeddings in the m-th modality, and the calculation process can be formalized as:

$$pro(\boldsymbol{H}^m|\hat{\boldsymbol{F}}_m^V, \hat{\boldsymbol{F}}_m^U) = pro(\boldsymbol{H}^m[i,k] = 1|\hat{\boldsymbol{f}}_m^{V,i}, \hat{\boldsymbol{f}}_m^{U,k}), \tag{17}$$

where $\boldsymbol{H}^m \in \mathbb{R}^{N_U \times N_V}$ is a learnable probability matrix, $\hat{\boldsymbol{f}}_m^{V,i}$ and $\hat{\boldsymbol{f}}_m^{U,k}$ are the i-th item and k-th user representation under m-th modality of item and user representation set $\hat{\boldsymbol{F}}_m^V$ and $\hat{\boldsymbol{F}}_m^U$, respectively. Then, we can obtain $\boldsymbol{H}^m[i,k]$ through the cosine similarity:

$$\boldsymbol{H}^m[i,k] = \hat{\boldsymbol{f}}_m^{V,i} \cdot \hat{\boldsymbol{f}}_m^{U,k} / (||\hat{\boldsymbol{f}}_m^{V,i}||_2 \cdot ||\hat{\boldsymbol{f}}_m^{U,k}||_2), \tag{18}$$

where $\boldsymbol{H}^m[i,k]$ is the probability of assigning the i-th item to the k-th user; therefore, $\boldsymbol{H}^m$ can be regarded as the user's preference matrix for the item in the m-th modality.

In order to further enhance multimedia recommendation, we feed $\boldsymbol{H}^m$ and $\boldsymbol{R}$ into the discriminator to learn an enhanced interaction matrix that contains multimodal information: $\tilde{\boldsymbol{R}} = \boldsymbol{R} + \sum_{m=1}^{M} \lambda_m \boldsymbol{H}^m$, where λ_m is a hyperparameter under each modality graph structure. Based on the enhanced interaction matrix $\tilde{\boldsymbol{R}}$, We define our adversarial self-supervised learning (SSL) loss to optimize our relation generator $\mathcal{G}(\cdot)$ and discriminator $\mathcal{D}(\cdot)$ as follows:

$$\min_{\mathcal{G}} \max_{\mathcal{D}} \mathbb{E}_{\tilde{\boldsymbol{R}} \sim \mathbb{P}_r}[\mathcal{D}(\tilde{\boldsymbol{R}})] - \mathbb{E}_{\boldsymbol{H}^m \sim \mathbb{P}_f}[\mathcal{D}(\mathcal{G}(\hat{\boldsymbol{f}}_m^V))]. \tag{19}$$

Specifically, we use a commonly used adversarial (SSL) loss [21] to train the model as shown in Eq. (20):

$$\begin{cases} \mathcal{L}_{\mathcal{G}} = -\mathbb{E}_{\boldsymbol{H}^m}[\mathcal{D}(\boldsymbol{H}^m)], \\ \mathcal{L}_{\mathcal{D}} = -\mathbb{E}_{\tilde{\boldsymbol{R}}}[\mathcal{D}(\tilde{\boldsymbol{R}})] + \mathbb{E}_{\boldsymbol{H}^m}[\mathcal{D}(\boldsymbol{H}^m)] + \lambda_A \mathbb{E}_{\ddot{\boldsymbol{A}}}[(||\nabla_{\mathcal{D}(\ddot{\boldsymbol{A}})}|| - 1)^2], \end{cases} \tag{20}$$

where the optimized objectives $\mathcal{L}_{\mathcal{G}}$ and $\mathcal{L}_{\mathcal{D}}$ corresponding to the generator $\mathcal{G}$ and discriminator $\mathcal{D}$, respectively. λ_A denotes the balance weight, which is set to 1 according to previous works [11,21], $\ddot{\boldsymbol{A}}$ denotes the interpolation of matrix $\boldsymbol{H}^m$ and matrix $\tilde{\boldsymbol{R}}$.

Multimedia Recommendation. In order to introduce the collaborative multimedia recommendation signals, we first initialize the ID embeddings $\boldsymbol{Z}^V$ and $\boldsymbol{Z}^U$ of items and users, respectively. Then, we perform the ID-corresponding aggregation over the neighbors of users and items based on the enhanced interaction matrix $\tilde{\boldsymbol{R}}$.

$$\begin{aligned}\boldsymbol{Z}^{V,(l+1)} &= \sigma(\boldsymbol{D}^{V^{-1}}\tilde{\boldsymbol{R}}^{T}\boldsymbol{Z}^{U,(l)}\boldsymbol{W}^{V,(l)}),\\ \boldsymbol{Z}^{U,(l+1)} &= \sigma(\boldsymbol{D}^{U^{-1}}\tilde{\boldsymbol{R}}\boldsymbol{Z}^{V,(l)}\boldsymbol{W}^{U,(l)}),\end{aligned} \tag{21}$$

where $\boldsymbol{Z}^{V,(l)} \in \mathbb{R}^{N_V \times d}$, $\boldsymbol{Z}^{U,(l)} \in \mathbb{R}^{N_U \times d}$ as the ID-corresponding embedding of users and items in the l-th layer of graph neural networks, where the zero-layer embeddings $\boldsymbol{Z}^{V,(0)}$ and $\boldsymbol{Z}^{U,(0)}$ are initialized from a trainable lookup table. $\sigma(\cdot)$ is the activation function to introduce the nonlinear factors. $\boldsymbol{W}^{V,(l)}$ and $\boldsymbol{W}^{U,(l)}$ are trainable item and user weights of the l-layer for GNNs.

To generate our final item and user representations for predictions, we aggregate ID embeddings and multi-modal embeddings as follows:

$$\boldsymbol{Q}^{V} = \boldsymbol{Z}^{V} + \omega \sum_{m \in M}^{|M|} \frac{\hat{\boldsymbol{F}}_{m}^{V}}{||\hat{\boldsymbol{F}}_{m}^{V}||_2}, \boldsymbol{Q}^{U} = \boldsymbol{Z}^{U} + \omega \sum_{m \in M}^{|M|} \frac{\hat{\boldsymbol{F}}_{m}^{U}}{||\hat{\boldsymbol{F}}_{m}^{U}||_2}, \tag{22}$$

where ω is the aggregation weight. Then, we make predictions by using the probability matrix $\hat{\boldsymbol{R}}_{ij} = \boldsymbol{Q}^{U}\boldsymbol{Q}^{V^{T}}$.

In order to enhance multimodal recommendation, we adopt the BPR loss, which is a common loss function in recommendation tasks:

$$\mathcal{L}_{BPR} = \sum_{(i,j_p,j_n)}^{|\mathcal{E}|} -log(sigm(\hat{\boldsymbol{r}}_{ij_p} - \hat{\boldsymbol{r}}_{ij_n})), \tag{23}$$

where j_p and j_n denotes the positive and negative samples for user i. Finally, the combination loss of GoodRec is as follows:

$$\mathcal{L} = \mathcal{L}_{BPR} + \lambda_G \mathcal{L}_{\mathcal{G}}, \tag{24}$$

where λ_G is the hyperparameter to control the strength of adversarial networks, and λ_G is set to 1 according to the previous works [11].

3.4 Complexity Analysis

Our proposed GoodRec achieves promising results in multimedia recommendation scenarios. However, the trade-off between model performance and running complexity is essential during the deployment process. Therefore, in this section, we will provide our discussion about the complexity analysis of our GoodRec.

For our GoodRec framework, the running time is limited by two steps: graph diffusion and multimedia recommendation. We exploit the gradient descent algorithm in Eq. (16) with $O(2{N_V}^2 d + M{N_V}^2 d)$, therefore the complexity of our graph diffusion is $O((2 + M){N_V}^2 d)$. Here, M is the number of item modalities, N_V is the number of items, and d is the dimension of latent embeddings. The multimedia recommendation contains the adversarial learning module with $O(max(MN_U N_V d, MN_U N_V)) = O(MN_U N_V d)$ and the recommendation module with $O(N_U N_V d)$, where N_U is the number of users. Therefore, the total complexity of our GoodRec is $O((2+M){N_V}^2 d + MN_U N_V d + N_U N_V d)$.

4 Experiment

In this section, we conduct experiments to verify the validity of our GoodRec framework by answering the following research questions: **RQ1:** How does the GoodRec work compared to other state-of-the-art models for recommendation? **RQ2:** How does each module of the GoodRec assists to improve the performance? **RQ3:** What impact do the hyperparameters have on prediction performance, and how to determine the optimal values?

4.1 Experimental Settings

Dataset Descriptions. We apply three multimedia datasets to finish experiments, which are respectively Amazon-Baby, Amazon-Sports, and Allrecipes. These publicly available datasets provide both product descriptions and images, and have variable sizes across product categories. The initial data within each dataset goes through pre-processing using a 5-core configuration for both items and users, following the processing like [11,31]. The results of the 5-core filtering are delivered in Table 1.

Table 1. Statistics of the experimental datasets.

Datasets	Users	Items	Interactions	Sparsity
Baby	19,445	7,050	160,792	99.88%
Sports	35,598	18,357	296,337	99.95%
Allrecipes	19,805	10067	58322	99.97%

Evaluation Metrics. Following the related works [11,31], we choose two widespread metrics NDCG@K (N@K) and Recall@K (R@K) in recommender systems to test the model performance.

Baseline Models. To verify the effectiveness of our proposed framework in capturing cross-modality high-order correlations, we adopt the following representative state-of-the-art baselines for comparison, which can be divided into: (1) Single-modal recommendation models that rely only on implicit feedback (i.e., user-item interaction), including graph-based methods BPR [17], LightGCN [5], BUIR [9] and hypergraph-based models HCCF [25]. (2) Multimodal recommendation models that use implicit feedback and multimodal features include VBPR [4], MMGCN [23], GRCN [22], DualGNN [19], LATTICE [29], BM3 [31], and AD-DRL [10].

Experimental Setup Details. To ensure fair comparison, we have carefully tuned all hyperparameters for all baselines through cross-validation as suggested in their original papers to achieve their best performance. We implement our proposed framework GoodRec with Pytorch. The batch size is fixed as 1024 and the learning rate is initialized to $3e^{-4}$. In our graph diffusion module, we specify the following hyperparameters: step size $\alpha = 0.5$ and the number of

Table 2. Overall performance on Baby, Sports, and Allrecipes regarding Recall and NDCG metrics. The best performance is highlighted with bold and the second-best performance is underlined respectively.

Metrics	Single Modality				Multiple Modality							
	BPR	LightGCN	BUIR	HCCF	VBPR	MMGCN	GRCN	DualGNN	LATTICE	BM3	<u>AD-DRL</u>	**Our**
Baby												
R@10	0.0357	0.0479	0.0506	0.0521	0.0423	0.0378	0.0532	0.0448	0.0544	0.0564	<u>0.0579</u>	**0.0611**
R@20	0.0575	0.0754	0.0788	0.0822	0.0663	0.0615	0.0824	0.0716	0.0848	0.0883	<u>0.0893</u>	**0.0895**
N@10	0.0192	0.0257	0.0269	0.0280	0.0223	0.0200	0.0282	0.0240	0.0291	0.0301	<u>0.0305</u>	**0.0327**
N@20	0.0249	0.0328	0.0342	0.0357	0.0284	0.0261	0.0358	0.0309	0.0369	0.0383	<u>0.0397</u>	**0.0417**
Sports												
R@10	0.0432	0.0569	0.0467	0.0620	0.0558	0.0370	0.0559	0.0568	0.0618	0.0656	<u>0.0657</u>	**0.0671**
R@20	0.0653	0.0864	0.0733	0.0941	0.0856	0.0605	0.0877	0.0859	0.0947	0.0980	<u>0.0985</u>	**0.0999**
N@10	0.0241	0.0311	0.0260	0.0339	0.0307	0.0193	0.0306	0.0310	0.0337	0.0355	<u>0.0359</u>	**0.0378**
N@20	0.0298	0.0387	0.0329	0.0421	0.0384	0.0254	0.0389	0.0385	0.0422	0.0438	<u>0.0447</u>	**0.0469**
Allrecipes												
R@10	0.0250	0.0259	0.0261	0.0280	0.0259	0.0202	0.0279	0.0277	0.0304	0.0329	<u>0.0345</u>	**0.0462**
R@20	0.0440	0.0455	0.0459	0.0468	0.0447	0.0412	0.0459	0.0457	0.0491	0.0509	<u>0.0526</u>	**0.0686**
N@10	0.0176	0.0194	0.0189	0.0196	0.0190	0.0175	0.0196	0.0192	0.0239	0.0255	<u>0.0263</u>	**0.0329**
N@20	0.0258	0.0260	0.0262	0.0295	0.0262	0.0250	0.0268	0.0270	0.0312	0.0324	<u>0.0332</u>	**0.0411**

graph diffusion iterations $K = 2$. In our enhanced multimedia recommendation, we specify the following hyperparameters: the graph convolution layers $l = 2$ and the aggregation weight $\omega = 0.1$.

4.2 Overall Performance Comparison (RQ1)

In order to evaluate our model, we take some advanced baselines and let them work on the same dataset. From Table 2, we have the following observations.

Generally speaking, GoodRec outperforms all 11 baselines across all evaluation metrics on the three multimedia datasets. This answers RQ1 and demonstrates the effectiveness of the graph diffusion structure, which can capture the high-order correlations between different models. Compared with the second-best performance, GoodRec performance improvements in Recall and NDCG ranged from small (0.22% achieved with Recall@20 on Baby) to significantly large (33.91% achieved with Recall@10 on Allrecipes).

In addition, the single-modality models (e.g., BPR, BUIR and LightGCN) perform significantly worse than multimodal baselines (i.e., BM3, LATTICE and AD-DRL), which is consistent with the results of previous work.

4.3 Ablation Studies (RQ2)

To thoroughly evaluate our proposed methods, we conduct ablation studies for our framework, integrating each component incrementally. We choose LightGCN [5], which is a widely adopted general recommendation model, as our base. Then we gradually introduce the graph diffusion (+DIF) and discriminator (+DIS). By observing Table 3, we can obtain the following conclusions:

Table 3. Ablation study results (%) on Baby, Sports, and Allrecipes datasets.

Datasets	Baby		Sport		Allrecipes	
Metrics	R@10	N@10	R@10	N@10	R@10	N@10
Base	0.0479	0.0257	0.0569	0.0311	0.0259	0.0194
Base+DIF	0.0513	0.0289	0.0596	0.0334	0.0407	0.0254
Base+DIF+DIS (GoodRec)	0.0611	0.0327	0.0671	0.0378	0.0462	0.0329

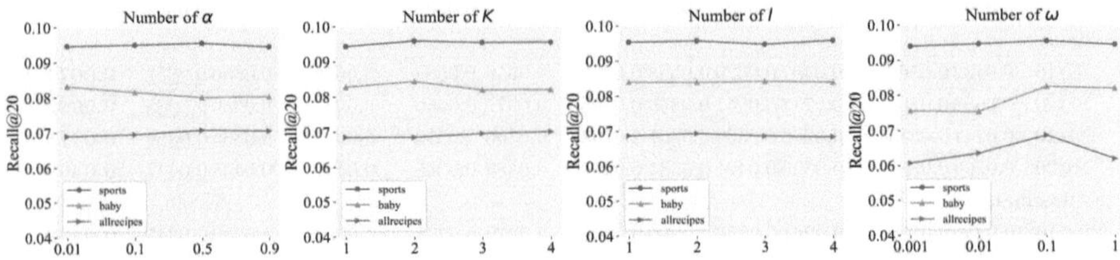

Fig. 3. Performance of hyperparameter study regarding Recall@20 of the GoodRec framework with varying hyperparameters on Baby, Sports, and Allrecipes datasets.

Base+DIF leads to performance gains ranging from 4.75% (achieved in Recall@10 on Sports) to 57.14% (achieved in Recall@10 on Allrecipes) compared to the base model. Base+DIF demonstrates the effectiveness of the graph diffusion module in capturing high-order cross-modality correlations.

Moreover, the performance gains of Base+DIF+DIS (i.e., our GoodRec), surpass Base+DIF ranges from 12.58% (achieved in Recall@10 on Sports) to 19.10% (achieved in Recall@10 on Baby). Although Base+Diffusion has already captured the diverse high-order relations present in multimedia recommendation data, GoodRec is able to improve the accuracy of multimedia recommendations with our enhancement mechanism.

4.4 Hyperparameter Study (RQ3)

Our proposed GoodRec framework involves four main hyperparameters, which are α, K, l and ω. From Fig. 3, we can observe the following information: (1) *alpha* is the graph diffusion step size, where the optimal values are about 0.01 and 0.5. The setting $alpha = 0.5$ seems to be the rule-of-thumb. (2) K is the number of graph diffusion iterations and the optimal values are about 2 and 4. In practice, $K = 2$ seems to be the rule-of-thumb. (3) l is the graph convolution layer and the optimal values are about 2 and 4. In practice, $l = 2$ seems to be the rule-of-thumb. (4) ω is the aggregation weight and we found that the optimal aggregation weight ω is 0.1. As shown in Fig. 3, the model is sensitive to changes in ω, and optimal performance can be achieved through slight tuning.

5 Conclusion

In this work, we propose GoodRec, a Graph-oriented Cross-modality Diffusion for Multimedia Recommendation, which can simultaneously capture the high-order structural information embedded in the cross-modality item-item graph. Our approach encompasses two key modules: the graph diffusion module and the enhanced multimedia recommendation module. The former captures high-order cross-modality correlations with both intra- and inter-modality information through a graph-oriented diffusion network, while the latter enhances the multimedia recommendation based on an adversarial enhancement mechanism. The experimental results confirm the efficacy of GoodRec, with consistent improvements over the state-of-the-art baselines, paving the way for more accurate and tailored approaches for MRS.

Acknowledgments. This work was supported in part by the National Natural Science Foundation of China under Grants (No.62302098), Fujian Provincial Natural Science Foundation of China under Grants (2025J01540), Zhejiang Provincial Natural Science Foundation of China under Grants (LQ23F020007), Zhejiang Provincial Department of Agriculture and Rural Affairs Project under Grants (2024SNJF044), and Fundamental Research Funds for the Provincial Universities of Zhejiang (FR25008Q), Fuzhou University Fund for Overseas Academic Visits of Outstanding Students.

References

1. Chen, Z., Wu, Z., Wang, S., Guo, W.: Dual low-rank graph autoencoder for semantic and topological networks. In: AAAI, vol. 37, pp. 4191–4198 (2023)
2. Eliasof, M., Haber, E., Treister, E.: Pde-gcn: novel architectures for graph neural networks motivated by partial differential equations. NeurIPS **34**, 3836–3849 (2021)
3. Fan, Q., Yu, P., Tan, Z., Bao, B.K., Lu, G.: Befa: a general behavior-driven feature adapter for multimedia recommendation. In: Proceedings of the AAAI Conference on Artificial Intelligence, vol. 39, pp. 11634–11644 (2025)
4. He, R., McAuley, J.: Vbpr: visual bayesian personalized ranking from implicit feedback. In: AAAI, vol. 30 (2016)
5. He, X., Deng, K., Wang, X., Li, Y., Zhang, Y., Wang, M.: Lightgcn: simplifying and powering graph convolution network for recommendation. In: SIGIR, pp. 639–648 (2020)
6. Hu, J., Hooi, B., Qian, S., Fang, Q., Xu, C.: Mgdcf: distance learning via markov graph diffusion for neural collaborative filtering. TKDE (2024)
7. Huang, K., Tang, J., Liu, J., Yang, R., Xiao, X.: Node-wise diffusion for scalable graph learning. In: WWW. pp. 1723–1733 (2023)
8. Jiang, Y., Xia, L., Wei, W., Luo, D., Lin, K., Huang, C.: Diffmm: multi-modal diffusion model for recommendation. In: Proceedings of the 32nd ACM International Conference on Multimedia, pp. 7591–7599 (2024)
9. Lee, D., Kang, S., Ju, H., Park, C., Yu, H.: Bootstrapping user and item representations for one-class collaborative filtering. In: SIGIR, pp. 317–326 (2021)
10. Li, Z., Liu, F., Wei, Y., Cheng, Z., Nie, L., Kankanhalli, M.: Attribute-driven disentangled representation learning for multimodal recommendation. In: MM, pp. 9660–9669 (2024)

11. Lin, Z., et al.: Contrastive intra-and inter-modality generation for enhancing incomplete multimedia recommendation. In: MM, pp. 6234–6242 (2023)
12. Lin, Z., Tian, C., Hou, Y., Zhao, W.X.: Improving graph collaborative filtering with neighborhood-enriched contrastive learning. In: WWW, pp. 2320–2329 (2022)
13. Liu, C., Yuan, H., Xu, Y., Wang, Z., Sun, Z.: A collaborative filtering recommendation method with integrated user profiles. In: ADMA, pp. 196–207. Springer (2022). https://doi.org/10.1007/978-3-031-22137-8_15
14. Liu, Q., et al.: Diffusion augmentation for sequential recommendation. In: CIKM, pp. 1576–1586 (2023)
15. Liu, X., Li, X., Cao, Y., Zhang, F., Jin, X., Chen, J.: Mandari: multi-modal temporal knowledge graph-aware sub-graph embedding for next-poi recommendation. In: ICME, pp. 1529–1534 (2023)
16. Lu, J., Wu, Z., Chen, Z., Cai, Z., Wang, S.: Towards multi-view consistent graph diffusion. In: MM, pp. 186–195 (2024)
17. Rendle, S., Freudenthaler, C., Gantner, Z., Schmidt-Thieme, L.: Bpr: Bayesian personalized ranking from implicit feedback. arXiv preprint arXiv:1205.2618 (2012)
18. Sun, R., et al.: Multi-modal knowledge graphs for recommender systems. In: CIKM, pp. 1405–1414 (2020)
19. Wang, Q., Wei, Y., Yin, J., Wu, J., Song, X., Nie, L.: Dualgnn: dual graph neural network for multimedia recommendation. TMM **25**, 1074–1084 (2021)
20. Wang, Z., Feng, Y., Zhang, X., Yang, R., Du, B.: Multi-modal correction network for recommendation. IEEE Trans. Knowl. Data Eng. (2024)
21. Wei, W., Huang, C., Xia, L., Zhang, C.: Multi-modal self-supervised learning for recommendation. In: Proceedings of the ACM Web Conference 2023, pp. 790–800 (2023)
22. Wei, Y., Wang, X., Nie, L., He, X., Chua, T.S.: Graph-refined convolutional network for multimedia recommendation with implicit feedback. In: MM, pp. 3541–3549 (2020)
23. Wei, Y., Wang, X., Nie, L., He, X., Hong, R., Chua, T.S.: Mmgcn: multi-modal graph convolution network for personalized recommendation of micro-video. In: MM, pp. 1437–1445 (2019)
24. Wu, Q., Yang, C., Zhao, W., He, Y., Wipf, D., Yan, J.: Difformer: scalable (graph) transformers induced by energy constrained diffusion. arXiv preprint arXiv:2301.09474 (2023)
25. Xia, L., Huang, C., Xu, Y., Zhao, J., Yin, D., Huang, J.: Hypergraph contrastive collaborative filtering. In: SIGIR, pp. 70–79 (2022)
26. Xu, J., Chen, Z., Yang, S., Li, J., Wang, H., Ngai, E.C.: Mentor: multi-level self-supervised learning for multimodal recommendation. In: Proceedings of the AAAI Conference on Artificial Intelligence, vol. 39, pp. 12908–12917 (2025)
27. Yang, C., Yuan, W., Qu, L., Nguyen, T.T.: Pdc-frs: privacy-preserving data contribution for federated recommender system. In: ADMA. pp. 65–79. Springer (2024). https://doi.org/10.1007/978-981-96-0850-8_5
28. Yu, P., Tan, Z., Lu, G., Bao, B.K.: Mind individual information! principal graph learning for multimedia recommendation. In: Proceedings of the AAAI Conference on Artificial Intelligence, vol. 39, pp. 13096–13105 (2025)
29. Zhang, J., Zhu, Y., Liu, Q., Wu, S., Wang, S., Wang, L.: Mining latent structures for multimedia recommendation. In: MM, pp. 3872–3880 (2021)
30. Zhao, J., Dong, Y., Ding, M., Kharlamov, E., Tang, J.: Adaptive diffusion in graph neural networks. NeurIPS **34**, 23321–23333 (2021)
31. Zhou, X., et al.: Bootstrap latent representations for multi-modal recommendation. In: WWW, pp. 845–854 (2023)

Zero-Shot Character Recognition Method of Korean Ancient Documents Based on the Chinese and Korean Characters Unified IDS Encoding

Mengling Zhao, Xiaofeng Jin(✉), Guirong Wang(✉), and Yankai Zhao

Department of Computer Science and Technology, Yanbian University, Yanji, China
{xfjin,0000004356}@ybu.edu.cn

Abstract. To address the recognition challenges in mixed-script Korean classical texts containing Chinese and Korean characters, this study proposes a unified character recognition method for ancient Korean documents named CKR-UniIDS based on unified encoding of Chinese and Korean characters. Firstly, considering the multi-script characteristics of classical Korean texts, we implement unified encoding for Chinese radicals, Korean letters, and 12 ideographic description characters based on structural similarities between Chinese and Korean characters. Secondly, to resolve the information loss caused by low-resolution text images in classical Korean documents, we employ multi-layer Spatial-Depth convolutional blocks (SPDConv) to mitigate severe information degradation during downsampling. To accommodate the rich stroke diversity of Chinese and Korean characters, we introduce linear deformable convolution (LDConv), whose adaptable kernel size and shape enable extraction of fine-grained stroke features within smaller receptive fields, thereby improving robustness to stroke variability. Finally, to address the limited samples and class imbalance in Korean ancient datasets, we implement a transfer learning strategy combining pre-training with Unicode font images and fine-tuning on ancient document datasets. Experimental results demonstrate that our method significantly outperforms existing approaches in zero-shot recognition of Korean classical texts, achieving a maximum accuracy of 44.94% on Korean classical datasets,28.98% higher than Liu's method. After fine-tuning, our model reaches 73.19% accuracy with notable improvements under limited training data conditions. Additionally, the CKR-UniIDS model comprehensively surpasses radical/stroke-based methods and their combinations in recognizing artistic Chinese typefaces under zero-shot scenarios.

Keywords: Korean ancient scripts recognition · Unified IDS coding · Multi-scale features · Transfer learning

1 Introduction

One of the most notable features of ideographic ancient Korean texts is their multilingual mixed layout, most commonly involving Chinese and Korean characters, which greatly increases the difficulty of text recognition.

M. Yoshikawa et al. (Eds.): ADMA 2025, LNAI 16197, pp. 267–280, 2026.
https://doi.org/10.1007/978-981-95-3453-1_18

Research on Chinese character recognition (CCR) has focused primarily on three hierarchical levels: whole characters, radicals, and strokes [1–10]. Because Chinese characters are thousands and have complex internal structures, zero-shot scenarios present a major challenge in practice. As a result, CCR methods based on the radical and stroke levels have attracted increasing attention. Chen [11] decomposed Chinese characters at the stroke level into combinations of five basic stroke types. Wang [12] proposed DenseRAN, which recognizes radicals and analyzes the spatial structure of Chinese characters: it treats each character as a two-dimensional structure combined with its radicals and achieves zero-shot recognition via an encoder-decoder architecture. Cao [13]introduced Hierarchical Decomposition Embedding (HDE), which uses radical-composition information to design an embedding vector for each character; by converting complex classes into basic primitive combinations and learning a compatibility function between samples and labels, HDE classifies zero-shot characters. Yu [14] combined stroke-level and radical-level information, proposing Radicals-Structured Stroke Tree (RSST), which represents each character as a stroke tree organized according to its radical structure, thereby leveraging the advantages of both hierarchies.

Because Korean script is constructed from a finite set of letters according to specific orthographic rules [15], and since those letters play a role analogous to Chinese radicals, they too can be decomposed by letter or stroke. Kim [16] decomposed Korean syllables into letters, reframing the recognition task from syllables to letters. This decomposition significantly alleviates class-imbalance issues by recognizing unseen syllables through learned consonant–vowel combinations, thus enabling full Korean-script recognition. Furthermore, Yeongseo [17] went a step further in decomposing diphthongs {ㅘ, ㅙ, ㅚ, ㅝ, ㅞ, ㅟ, ㅢ} into their more basic jamo components: { ㅡ, ㅣ, ㅏ, ㅗ, ㅐ, ㅜ, ㅓ, ㅔ }.

Addressing the multilingual mixed layout in ancient Korean texts, Liu [18] proposed a stroke-level recognition network that decomposes both Chinese and Korean characters into combinations of five stroke types and incorporates a coordinate attention mechanism to accurately localize stroke regions and capture structural features. This improves the model's ability to predict stroke encoding and distinguish confusable codes. However, because this method relies solely on Chinese stroke types and decomposition rules, clustering only those Korean letters that visually resemble Chinese strokes, it produces many confusing encodings.

To tackle the challenges of multilingual mixed layout, sparse datasets, and class imbalance in ancient Korean texts, this paper proposes a zero-shot recognition method based on unified Chinese and Korean coding. Our main contributions are as follows:

(1) Unified Chinese-Korean IDS Encoding (UCK-IDS): We introduce a UCK-IDS scheme that, drawing on Chinese character decomposition rules, assigns a unified code to decomposed Korean letters, Chinese radicals, and ideographic descriptive characters. This enables our CKR-UniIDS model to recognize both Chinese and Korean scripts simultaneously while reducing model complexity.

(2) Multi-Layer Spatial-Depth Convolution and Linear Deformable Convolution: We employ multi-layer spatial-to-depth convolution to transfer information lost by spatial downsampling into the channel dimension, preserving fine-grained stroke details. We also leverage the dynamic kernel-size and shape adjustment of linear deformable convolution to better adapt to stroke-diversity variations in both scripts.

(3) Pretraining and Fine-Tuning Strategy: To mitigate dataset sparsity and class imbalance, we first pretrain on extensive printed-font libraries of Chinese and Korean characters and then fine-tune the model using a limited ancient-text dataset.

2 Related Work

The internal structure of Chinese characters is complex, but they can be decomposed into specific radical sequences according to predetermined rules. The first-level commonly used characters [19] consist of 514 radicals and 12 Ideographic Description Characters (IDCs). The Unicode 3.0 Ideographic Description Sequence (IDS) specification provides a recursively defined ideographic description algorithm. Figure 1 illustrates the 12 IDC along with examples, while Fig. 2 demonstrates the decomposition process of the Chinese character "锈" and its corresponding IDS sequence.

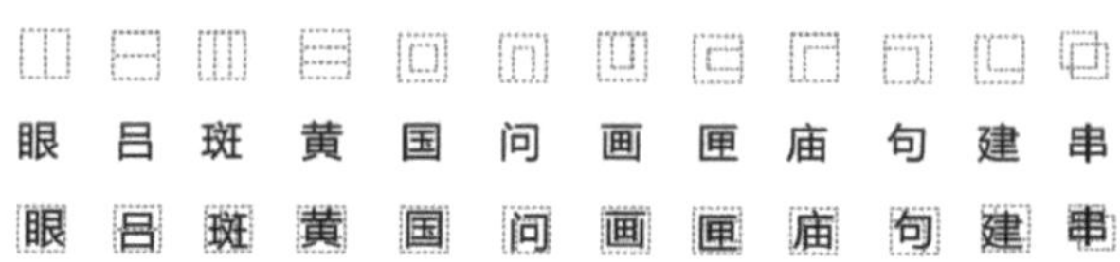

Fig. 1. 12 ideographic descriptive characters and their examples.

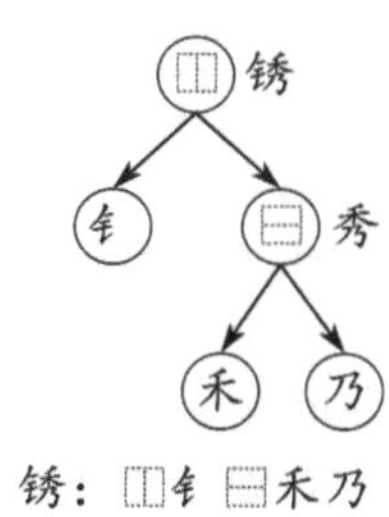

Fig. 2. The decomposition process of the Chinese character "锈" and the corresponding IDS sequence.

Korean characters are formed by combining 2 or 3 letters in specific configurations: either "vowel + consonant" or "vowel + consonant + final consonant", as illustrated in Fig. 3. With 21 vowels, 19 consonants, and 27 final consonants, the Korean script can theoretically generate 11,172 distinct characters.

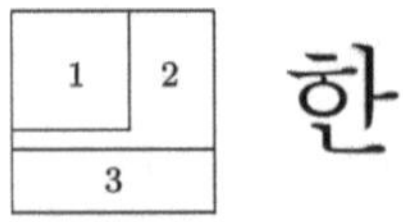

1. ㄱ ㄲ ㄴ ㄷ ㄸ ㄹ ㅁ ㅂ ㅃ ㅅ ㅆ ㅇ ㅈ ㅉ ㅊ ㅋ ㅌ ㅍ ㅎ
2. ㅏ ㅐ ㅑ ㅒ ㅓ ㅔ ㅕ ㅗ ㅘ ㅙ ㅚ ㅛ ㅜ ㅝ ㅞ ㅟ ㅠ ㅡ ㅢ ㅣ
3. (empty) ㄱ ㄲ ㄳ ㄴ ㄵ ㄶ ㄷ ㄹ ㄺ ㄻ ㄼ ㄽ ㄾ ㄿ ㅀ ㅁ ㅂ ㅄ ㅅ ㅆ ㅇ ㅈ ㅊ ㅋ ㅌ ㅍ ㅎ

Fig. 3. The structure of the Korean script, where 1 represents a consonant, 2 represents a vowel, and 3 represents a final-consonants.

Korean and Chinese characters share structural similarities, as both are block-shaped scripts composed of basic components. In Korean, these components are alphabetic letters, while in Chinese, they are called radicals. Both scripts can be decomposed into sequences of radicals (or letters) and strokes according to writing order and rules. Moreover, after decomposition, Korean characters have only three basic structures, corresponding to the IDC descriptors left-right, topbottom, and top-middle-bottom. Additionally, the IDC of Korean characters is completely included in Chinese characters. This structural parallelism provides a foundation for researching a unified Ideographic Description Sequence (IDS) encoding scheme for Chinese and Korean.

3 Methodology

3.1 Unified Chinese-Korean IDS Encoding

When decomposing characters in the Korean language,to reduce the number of basic Korean character components, the double final consonants {ㄳ, ㄵ, ㄶ,ㄺ, ㄻ, ㄼ, ㄽ, ㄾ, ㄿ, ㅀ, ㅄ} are further decomposed into {ㄱ, ㅅ, ㄴ, ㅈ, ㅎ,ㄹ, ㅁ, ㅂ, ㅌ, ㅍ}.
Likewise, the vowels {ㅘ, ㅙ, ㅚ, ㅝ, ㅞ, ㅟ, ㅢ} are further decomposed into {ㅡ, ㅣ, ㅏ, ㅗ, ㅐ, ㅜ, ㅓ, ㅔ}.

Using this decomposition method, a total of 35 basic components and 3 types of IDC are obtained. Thus, Korean characters can likewise be represented by IDS, as illustrated in Fig. 4.

The proposed Unified Chinese-Korean IDS Encoding Scheme (hereinafter referred to as UCK-IDS) follows the encoding rules outlined below: (1) Each Chinese character, Korean basic component, and IDC is assigned a unique code. (2) Chinese characters and Korean components that share identical or similar shapes are merged and assigned the same code. For example, Chinese and Korean

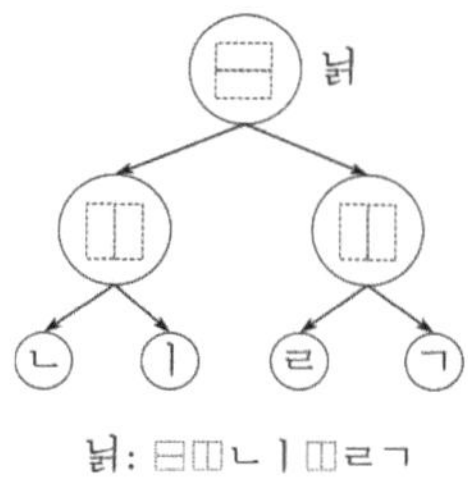

Fig. 4. An example of the breakdown of the Korean character 닑.

components like {一, ㅡ}, {人, ㅅ}, and {口, ㅁ}, which exhibit identical or similar shapes, are encoded uniformly.

The UCK-IDS encoding scheme contains a total of 515 codes. Among these, 29 codes correspond to multiple basic components (with 10 cases where Chinese and Korean basic components share the same code), and 486 codes correspond to only one basic component.

The UCK-IDS encoding scheme derived from the above rules significantly reduces the number of IDS codes required for both Chinese and Korean scripts. More importantly, UCK-IDS lays the foundation for a unified model framework capable of recognizing both Chinese and Korean characters simultaneously, thereby lowering model complexity and enabling zero-shot recognition of Korean ancient texts.

3.2 Model Structure

The method proposed in this paper for recognizing Korean ancient texts (hereafter referred to as CKR-UniIDS) has the model architecture shown in Fig. 5. It builds on the CCR-CLIP [20] framework and comprises two core modules: an image encoder and a text encoder. The CCR-CLIP model can exclusively recognize Chinese characters, and its radical encoding scheme is specifically designed for Chinese character radicals, with each radical corresponding to a single code. In contrast, CKR-UniIDS incorporates the UCK-IDS encoding scheme, enabling the model to simultaneously recognize both Chinese characters and Korean script. Furthermore, the image encoder has been enhanced to better extract fine-grained stroke features from text images as well as the diverse stroke characteristics of Chinese and Korean characters. The improved image encoder is abbreviated as S2D-LDConv (Space to Depth and Linear Deformable Convolutional encoder).

During training, the S2D-LDConv encoder is responsible for extracting visual features from input Chinese and Korean character images, while the text encoder extracts component sequence features of the characters. The entire training process is supervised using a contrastive loss to ensure that the visual features of Chinese and Korean character images are effectively aligned with the textual features of their component sequences. During testing, the visual features extracted

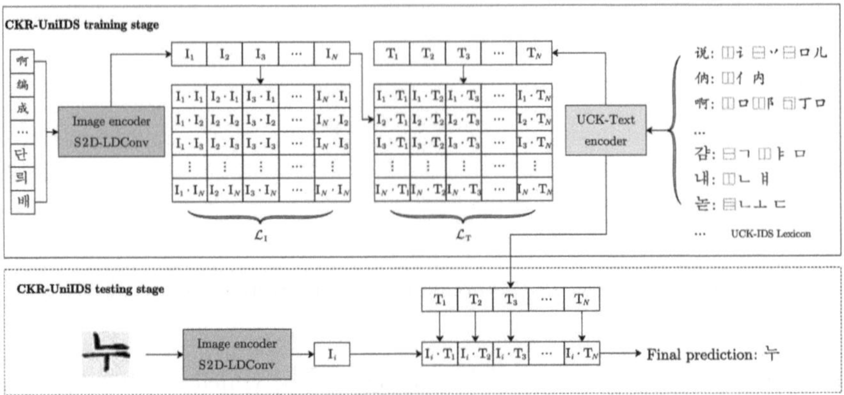

Fig. 5. Structure of the CKR-UniIDS model.

by the S2D-LDConv encoder for each query image are matched against the component sequence features produced by the text encoder, and the character with the highest similarity score is selected as the final prediction.

3.3 S2D-LDConv Image Encoder

The primary purpose of the image encoder is to extract visual features from character images and produce a corresponding visual embedding. Its effectiveness in capturing these features directly impacts the recognition accuracy of the model. To better mitigate the effects of low resolution and varying scales in text image samples, we adopt a multi-scale analysis approach based on Space to Depth convolution [21], enhancing the encoder's ability to model fine-grained stroke features. Additionally, the encoder incorporates Linear Deformable Convolution [22], whose dynamically adjustable kernel size and shape bolster the model's capacity to adapt to and represent the diverse stroke variations found in Chinese and Korean characters. The resulting improved image encoder is termed S2D-LDConv, and its architecture is illustrated in Fig. 6.

When traditional multi-scale analysis is applied to low-resolution text images, downsampling severely degrades the accurate extraction of fine stroke details. To address this information loss, we employ the Space to Depth convolution (SPD-Conv) [21]. SPD-Conv consists of a Space to Depth (SPD) layer and a non-strided convolution layer. The SPD layer transfers the spatial information that would be lost during dimensionality reduction into the channel dimension, thus preserving fine-grained details. The SPD-Conv network architecture is shown in Fig. 7.

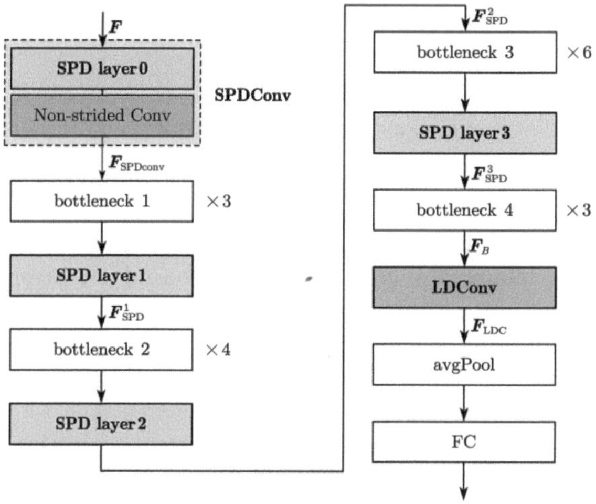

Fig. 6. Model structure of S2D-LDConv image encoder.

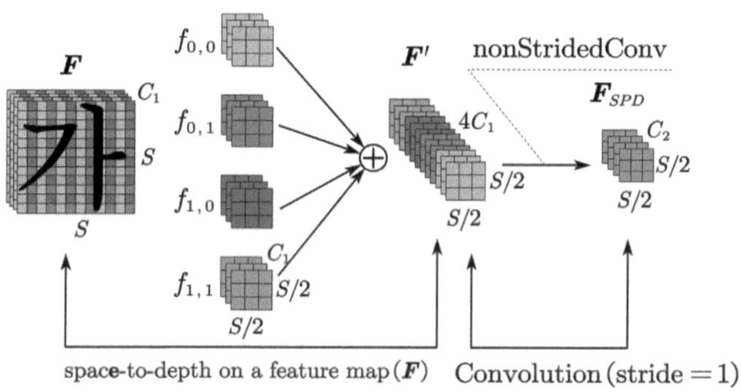

Fig. 7. Schematic diagram of the SPD-Conv network structure.

The input feature map F is partitioned according to Eq. (1) into four sub-maps $f_{0,0}, f_{0,1}, f_{1,0}, f_{1,1} \in \mathbb{R}^{S/2 \times S/2 \times C_1}$

$$\begin{cases} f_{0,0} = F\left[i : S, j : S\right] \\ f_{0,1} = F\left[i : S, m : S\right] \\ f_{1,0} = F\left[n : S, j : S\right] \\ f_{1,1} = F\left[n : S, m : S\right] \end{cases} \tag{1}$$

where i, j are even indices, m, n are odd indices.

These sub-maps are then concatenated along the channel dimension to form an intermediate feature map F'

$$F' = \text{connect}\left(f_{0,0}, f_{0,1}, f_{1,0}, f_{1,1}\right) \tag{2}$$

where, $F' \in \mathbb{R}^{S/2 \times S/2 \times 4C_1}$ Under the condition $C_2 < 4C_1$, F' is fed into a non-strided convolution layer (stride = 1), yielding the final SPD-Conv output

$F_{SPD} \in \mathbb{R}^{S/2 \times S/2 \times C_1}$. Using a stride = 1 preserves as many fine-grained stroke details as possible.

$$F_{SPDconv} = \text{nonStridedConv}\left(F'\right) \tag{3}$$

To capture the diverse stroke shape variations of Chinese and Korean characters, we further employ linear deformable convolution (LDConv) [22] to enhance the discriminative representation of stroke features. LDConv constructs a specific initial sampling shape for each kernel size based on prior knowledge and then, during adaptation to target variations, adjusts the irregular kernel sampling positions by applying learned offsets.

The feature extraction steps of the LDConv module within the S2D-LDConv encoder are as follows:

Step1: For the input feature F_B, first generate the initial sampling shape based on the initial kernel size N and compute the original coordinates.

Step2: Perform the convolution operation to compute the offset vectors for each kernel sampling point, then add these offsets to the original coordinates to obtain the new sampling locations. Figure 8 illustrates, for $N = 5$, how different shaped kernels are generated from the initial convolution kernel.

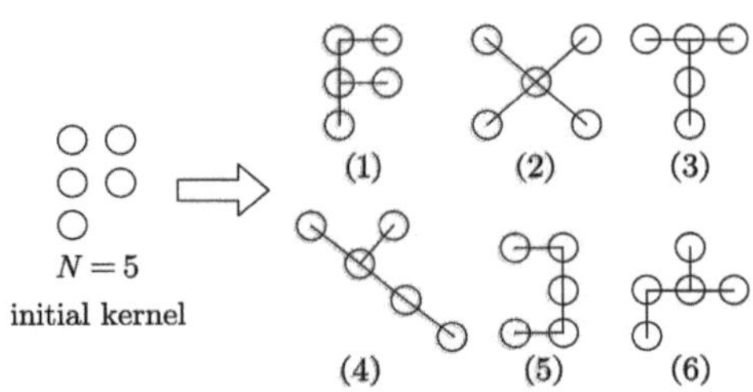

Fig. 8. Generating convolution kernels of different shapes from the initial convolution kernels.

Step3: The values at the new sampling locations are obtained via interpolation and resampling, yielding feature map Y, which is then reshaped.

$$Y' = \text{reshape}\left(Y\right) \tag{4}$$

Step4: Finally, Y' is processed through convolution, normalization, and activation layers to produce the LDConv output F_{LDC}.

$$F_{LDC} = \text{SiLU}\left(\text{norm}\left(\text{conv}\left(Y'\right)\right)\right) \tag{5}$$

The final visual feature embedding I is obtained by applying global average pooling to F_{LDC} and then projecting the pooled vector into the visual feature space.

$$I = \text{FC}\left(\text{avgpool}\left(F_{LDC}\right)\right) \tag{6}$$

3.4 UCK-Text Encoder

The UCK text encoder's primary function is to embed each Chinese and Korean character's UCK-IDS code from the vocabulary so that these embeddings can be compared with the visual embedding I produced by the image encoder. The UCK text encoder consists of an embedding layer followed by K transformer encoder layers.

Let the common Chinese and Korean character set be converted into a UCK-IDS code sequence $S = \{R_1, R_2, \cdots, R_N\}$, where $R_i = \{r_1, r_2, \cdots, r_L\}$ denotes the UCK-IDS code of the i-th character and r_L is the end of sequence token.

The sequence S is first passed through the embedding layer, mapping each discrete UCK-IDS code to a continuous vector. A positional encoding is added to preserve the order information. For code R_i, the embedding process is calculated by Eq. (7).

$$X_i^0 = \text{EmbedLayer}\,(R_i) \tag{7}$$

Secondly, text features are extracted through the K-layer Transformer.

$$X_i^k = \text{LayerNorm}\left(X_i^{k\text{-}1} + \text{MultiHead}\left(X_i^{k\text{-}1}\right)\right) \tag{8}$$

$$f_i = \text{LayerNorm}\left(X_i^K + \text{FFN}\left(X_i^K\right)\right) \tag{9}$$

where $k = 1, 2, \cdots, K, f_i \in \mathbb{R}^{1\times D}$.

In this way, the UCK-IDS encoding Eqs. (7) to (9) of all the characters in S yield $F = \{f_1, f_2, \cdots, f_N\}$.

Finally, project F onto T to obtain the output of the UCK-text encoder:

$$T = F \cdot W_{CK} \tag{10}$$

where W_{ck} is the projection matrix.

3.5 Loss Function

The CKR-UniIDS model employs a contrastive loss to align the visual embeddings of character images with their corresponding UCK-IDS code embeddings. For a batch of N image code pairs, the basic contrastive loss $\mathcal{L}_T$ is defined as

$$\mathcal{L}_T = -\sum_{j=1}^{N} \log \frac{\exp\left(I_j \cdot T_j\right)}{\sum\limits_{n=1}^{N} \exp\left(I_j \cdot T_n\right)} - \sum_{j=1}^{N} \log \frac{\exp\left(I_j \cdot T_j\right)}{\sum\limits_{n=1}^{N} \exp\left(I_n \cdot T_j\right)} \tag{11}$$

where, I_j and T_j are the visual and text embeddings of the j-th sample.

To reduce the prediction errors caused by different fonts or similar texts, in the CKR-UniIDS model, we also adopted the contrast loss between the visual features of text image samples with the same labels.

Suppose the given text image and the corresponding UCK-IDS code pair data $B = \{(C_1, R_1), ..., (C_N, R_N)\}$, where C_i and R_i respectively represent the i-th text image and its corresponding UCK-IDS code. If the S2D-LDConv encoder encodes C_i as the visual feature embedding representation I_i, the calculation of the loss function $\mathcal{L}_I$ is as follows:

$$\mathcal{L}_I = -\sum_{j=1}^{N} \log \frac{\sum_{I' \in \mathcal{U}_j} \exp(I_j \cdot I')}{\sum_{n=1}^{N} \exp(I_j \cdot I_n)} \tag{12}$$

where, $\mathcal{U}_j$ represents the set of R_j with the same UCK-IDS encoding, and I' is the visual feature embedding representation of any sample of $\mathcal{U}_j$.

The overall loss function of the CKR-UniIDS model is defined as

$$\mathcal{L}_{pre} = \mathcal{L}_T + \mathcal{L}_I \tag{13}$$

The model is supervised through the above losses to ensure that the model can effectively match the visual features of the character images of ancient Korean books with the encoded features of UCK-IDS.

4 Experiments

4.1 Experiment Setup

The Korean ancient data set was collected by scanning three classical works: "Similar Interpretation of the Same Text", "Explanation of Righteousness and Admonitions", and "Collection of Proverbs Explaining Pregnancy and Childbirth" then segmenting and manually annotating the text images. This yielded a total of 4,100 distinct Korean character classes (The ratio of Korean characters to Chinese characters is approximately 2:3). For our experiments, 1,000 of these classes were held out as a test set; from the remaining 3,100 classes, we formed five different training subsets of size 500, 1,000, 1,500, 2,000, and 3,100 classes (in alignment with Liu [18]). The character images range in size from $16 \times 16 \backsim 46 \times 46$ pixels, with the majority being smaller than 32×32, reflecting a low-resolution profile.

Printed artistic characters dataset It comprises the 3,755 Level-1 common Chinese characters, each rendered in 105 different artistic fonts. In our experiments, 1,000 classes were reserved for testing, and from the remaining 2,755 classes, we again created five training subsets of sizes 500, 1,000, 1,500, 2,000, and 2,755 classes. All images in this dataset are 32×32 pixels.

For pretraining the CKR-UniIDS model, we sampled fonts from both the Level-1 common Chinese character set (3,755 classes) and the common Korean character set (2,350 classes) that approximate the style of historical Korean texts, such as Song, Kai, Li, Batang, and Gungsuh. This resulted in a combined pretraining dataset of 6,105 classes, each with 150 character images.

During the training process, the Adam optimizer was used with a learning rate of 1e-4, trained for 100 epochs with a batch size of 32. The Transformer text encoder consisted of 12 layers.

4.2 Main Results

To evaluate the effectiveness of CKR-UniIDS, we conducted two sets of comparative zero-shot experiments: one on the Korean ancient dataset and one on the Chinese printed art font dataset.

Research on Korean historical document text recognition remains extremely scarce in both domestic and international literature. At the time of writing, only two relevant studies were identified: CCR-SLD [11] and the model from Reference [18]. Consequently, in the zero-shot experiment on the Korean ancient dataset, we compared our method against CCR-SLD and the model from Reference [18], both of which are zero-shot recognition approaches based on stroke decomposition. The results are shown in Table 1.

Table 1. Zero-shot comparative experiment results of Korean ancient document dataset. Both CCR-SLD and reference [18] are zero-shot recognition models based on the decomposition of character strokes

Method	recognition accuracy rates of training sets of different scales(%)				
	500	1000	1500	2000	3100
CCR-SLD [11]	9.45	23.71	29.56	57.94	65.59
Literature [18]	<u>11.79</u>	<u>27.98</u>	<u>33.41</u>	<u>60.59</u>	<u>68.37</u>
CKR-UniIDS	**30.11** (**+18.32**)	**41.92** (**+13.94**)	**46.86** (**+13.45**)	**64.24** (**+3.65**)	**73.19** (**+4.82**)

As seen in Table 1, CKR-UniIDS outperforms CCR-SLD and the model from Reference [18] across all training set sizes. Moreover, the accuracy gain over the second best method (the model from Reference [18]) is larger when the training set is smaller, demonstrating the superior zero-shot learning capability of our model. Additionally, as the size of the training set increases, the test accuracy of CKR-UniIDS improves, reaching 73.19% when trained on all 3,100 classes. This confirms that CKR-UniIDS is effective for recognizing Korean historical texts under challenging conditions such as mixed Chinese and Korean typesetting, sparse data, and class imbalance.

In the zero-shot experiment on the Chinese printed art font dataset, we held out 1,000 classes for testing and formed training sets of size 500, 1,000, 1,500, 2,000, and 2,755 classes from the remaining 2,755 classes. We compared CKR-UniIDS against DenseRAN, HDE, CCR-SLD, RSST, and CCR-CLIP. Except for RSST, which incorporates both radical and stroke knowledge, all other methods use either radicals or strokes alone. These methods are specifically designed for Chinese characters and cannot recognize Korean script. Thus, they are not applicable to recognizing text in Korean historical documents. The comparative results are presented in Table 2.

Table 2 shows that CKR-UniIDS consistently achieves higher recognition accuracy than the other methods on the art-font dataset. The improvement over

Table 2. Comparative experimental results of Chinese artistic printing fonts dataset.

Method	recognition accuracy rates of training sets of different scales(%)				
	500	1000	1500	2000	2755
DenseRAN [12]	0.2	2.26	7.89	10.86	24.8
HDE [13]	7.48	21.13	31.75	40.43	51.41
CCR-SLD [11]	7.03	26.22	48.42	54.86	65.44
CCR-CLIP [20]	15.7	37.16	52.62	60.25	71.51
RSST [14]	23.12	42.21	62.29	66.86	71.32
CKR-UniIDS	**30.5** (**+7.38**)	**52.19** (**+9.98**)	**62.44** (**+0.15**)	**67.73** (**+0.87**)	**76.24** (**+4.73**)

the next best method is particularly pronounced when the training set is small. As with the Korean dataset, our model's accuracy increases markedly with more training data, reaching 76.24% when trained on all 2,755 classes. These results demonstrate the strong cross-domain generalizability of CKR-UniIDS. Although the UCK-IDS encoding scheme unifies Chinese and Korean character IDS codes, it does not degrade performance on out-of-domain Chinese recognition tasks.

4.3 Ablation Study

To assess the individual contributions of the unified Chinese and Korean IDS encoding scheme, the enhanced image encoder, and transfer learning, we conducted an ablation study on the Korean ancient dataset. The results are summarized in Table 3.

Table 3. Experimental results of ablation study. Among them, FT stands for Fine-tune, SPD stands for spatial-depth convolution, and LD stands for linearly deformable convolution.

UCK-IDS	FT	SPD	LD	recognition accuracy rates of training sets of different scales (%)				
				500	1000	1500	2000	3100
w/	w/	w/	w/	**30.11**	**41.92**	**46.86**	**64.24**	**73.19**
w/o	w/o	w/o	w/o	5.47	7.83	9.45	13.28	15.96
w/	w/o	w/o	w/o	6.17	13.31	23.63	28.37	39.40
w/	w/o	w/o	w/	7.43	12.04	19.97	25.55	35.13
w/	w/o	w/	w/o	11.54	14.21	25.92	34.76	40.87
w/	w/o	w/	w/	12.85	17.80	29.89	38.84	44.94

When using SPD-Conv alone, the model's ability to learn fine-grained stroke features from low-resolution Korean ancient text images improved markedly, resulting in a significant increase in recognition accuracy. In contrast, employing

LDConv by itself led to a drop in accuracy, indicating that linear deformable convolution cannot, on its own, adapt effectively to the varied stroke shapes present in low-resolution images. Only when combined with SPD-Conv does LDConv deliver the intended benefit, an observation confirmed by the test results in the final row of Table 3.

From the first and last rows of Table 3, it is evident that pretraining on a mixed Chinese and Korean font dataset (comprising characters with similar glyph shapes) and then fine-tuning on the Korean ancient data further boosts recognition accuracy, particularly when the ancient training set is small. When the training set includes all 3,100 classes, the model achieves 73.19% accuracy. Moreover, since Chinese font image data are readily available, this pretraining strategy effectively mitigates the scarcity and class imbalance issues of the Korean ancient dataset.

5 Conclusion

This paper proposes CKR-UniIDS, a character recognition model for classical Korean texts. By implementing unified IDS (Ideographic Description Sequence) encoding for Chinese and Korean characters, the model addresses the recognition challenges posed by mixed Chinese-Korean script arrangements in classical Korean documents. To tackle the issues of low-resolution text images and stroke diversity in Chinese-Korean characters, an improved image encoder is introduced. Additionally, a pre-training strategy is adopted to effectively mitigate the problems of sparse annotated datasets and class imbalance in classical Korean texts. Experimental results demonstrate that the proposed CKR-UniIDS model outperforms previous stroke-based recognition methods in classical Korean text recognition tasks. Future research will focus on optimizing the unified IDS encoding scheme for Chinese and Korean characters, exploring stroke-level unified IDS encoding to compress the scale of the unified encoding framework while further enhancing the zero-shot recognition performance of classical Korean texts.

Acknowledgments. This work is supported by the Educational Commission of Jilin Province of China projects (No. JJKH20250415KJ).

References

1. Xiao, Y., Meng, D., Lu, C., Tang, C.K.: Template-instance loss for offline handwritten chinese character recognition. In: 2019 International Conference on Document Analysis and Recognition (ICDAR), pp. 315–322. IEEE (2019)
2. Gan, J., Chen, Y., Hu, B., Leng, J., Wang, W., Gao, X.: Characters as graphs: interpretable handwritten chinese character recognition via pyramid graph transformer. Pattern Recogn. **137**, 109317 (2023)
3. Hui, R., He, C., Zhang, W., Zhao, L., Li, M.: Handwritten chinese character text recognition based on convolutional recurrent neural networks. Science Technol. Eng. **25**(4) (2025)

4. Wang, T., Xie, Z., Li, Z., Jin, L., Chen, X.: Radical aggregation network for few-shot offline handwritten chinese character recognition. Pattern Recogn. Lett. **125**, 821–827 (2019)
5. Wang, S., Huang, G., Luo, X.: Hippocampus-heuristic character recognition network for zero-shot learning. arXiv preprint arXiv:2104.02236 (2021)
6. Luo, G.F., Wang, D.H., Du, X., Yin, H.Y., Zhang, X.Y., Zhu, S.: Self-information of radicals: a new clue for zero-shot Chinese character recognition. Pattern Recogn. **140**, 109598 (2023)
7. Chen, Z., Yang, W., Li, X.: Stroke-based autoencoders: self-supervised learners for efficient zero-shot Chinese character recognition. Appl. Sci. **13**(3), 1750 (2023)
8. Zeng, J., Xu, R., Wu, Y., Li, H., Lu, J.: Star: zero-shot chinese character recognition with stroke-and radical-level decompositions. arXiv preprint arXiv:2210.08490 (2022)
9. Zu, X., Yu, H., Li, B., Xue, X.: Chinese character recognition with augmented character profile matching. In: Proceedings of the 30th ACM International Conference on Multimedia, pp. 6094–6102 (2022)
10. Chen, C., Jiang, M., Zhang, M.: A Chinese character recognition scheme based on multi-dimensional representation. Softw. Eng. **27**(8), 24–29 (2024)
11. Chen, J., Li, B., Xue, X.: Zero-shot chinese character recognition with stroke-level decomposition. arXiv preprint arXiv:2106.11613 (2021)
12. Wang, W., Zhang, J., Du, J., Wang, Z.R., Zhu, Y.: Denseran for offline handwritten Chinese character recognition. In: 2018 16th International Conference on Frontiers in Handwriting Recognition (ICFHR), pp. 104–109. IEEE (2018)
13. Cao, Z., Lu, J., Cui, S., Zhang, C.: Zero-shot handwritten Chinese character recognition with hierarchical decomposition embedding. Pattern Recogn. **107**, 107488 (2020)
14. Yu, H., Chen, J., Li, B., Xue, X.: Chinese character recognition with radical-structured stroke trees. Mach. Learn. **113**(6), 3807–3827 (2024)
15. Choi, H.: Handwritten hangul recognition model using multi-label classification. J. Korean Soc. Indust. Appli. Mathem. **27**(2), 135–145 (2023)
16. Kim, G., Son, J., Lee, K., Min, J.: Character decomposition to resolve class imbalance problem in hangul ocr. arXiv preprint arXiv:2208.06079 (2022)
17. Ha, Y., Hwang, H., Kim, M., Lee, C., Shim, J.: A prior study on the improvement of the recognition rate of medieval Korean using clas compres ion and division in object detection. J. Korea Multimedia Soc. **26**(6), 795–803 (2023)
18. technology, C.: Research and application of Korean ancient books character recognition with stroke-level decomposition. Ph.D. thesis
19. Chen, J., et al.: Benchmarking Chinese text recognition: datasets, baselines, and an empirical study, vol. 3(4), p. 5. arxiv: 2112.15093 (2021)
20. Yu, H., Wang, X., Li, B., Xue, X.: Chinese text recognition with a pre-trained clip-like model through image-ids aligning. In: Proceedings of the IEEE/CVF International Conference on Computer Vision, pp. 11943–11952 (2023)
21. Sunkara, R., Luo, T.: No more strided convolutions or pooling: a new cnn building block for low-resolution images and small objects. In: Joint European conference on machine learning and knowledge discovery in databases. pp. 443–459. Springer (2022). https://doi.org/10.1007/978-3-031-26409-2_27
22. Zhang, X., et al.: Ldconv: linear deformable convolution for improving convolutional neural networks. Image Vis. Comput. **149**, 105190 (2024)

Noise-Robust Learning via Full Consistency

Zhen Wang[1], Xueying Chang[1], Wenxin Zhao[1], Wenlong Yu[2], Xiaohui Lei[3], and Yongfeng Dong[1(✉)]

[1] School of Computer Science and Engineering, Hebei University of Technology, Tianjin, China
{wangzhen,202432803017,202232805008,Dongyf}@hebut.edu.cn
[2] School of Computer Science and Technology, Tianjin University, Tianjin, China
[3] Renren Crowdsourcing (Tianjin) Technology Co., Ltd., Tianjin, China

Abstract. When dealing with label noise, sample selection is a commonly used method in which data with small losses are typically considered to be correctly labeled. However, this method may fail to identify challenging examples with large losses, which are crucial for improving the model's generalization performance. In this paper, we propose a novel selection strategy *Noise-robust Learning via Full Consistency (NLFC)* that measure the extent to which each example is forgotten during different training stages of the model. Based on this measure, we filter the examples and divide the data into clean and noisy sets. Furthermore, we address the issue of underutilization of labeled noisy examples by introducing consistency at the feature and image levels. Experimental results on benchmark-simulated noisy datasets and real-world noisy datasets demonstrate that our proposed method outperforms state-of-the-art methods.

Keywords: noisy labels · sample selection · consistency regularization · label noise learning · semi-supervised learning

1 Introduction

Deep Neural Networks (DNNs) have been remarkably successful in various tasks, but their success largely depends on high-quality annotated data [8,11,26,39] However, collecting accurately labeled large-scale data in many real-world scenarios can be challenging, and such data may inevitably contain noisy labels. Unfortunately, previous studies [42] have shown that DNNs can easily overfit random labels, resulting in poor generalization performance.

One common method in existing methods of Label Noise Learning (LNL) is to train the classifier using selected samples [9,12,21,23,32]. This method is based on the observation that Deep Neural Networks (DNNs) tend to first learn dominant patterns and then overfit to rare ones. To take advantage of this memorization effect and handle noisy data, a typical strategy is to start with

M. Yoshikawa et al. (Eds.): ADMA 2025, LNAI 16197, pp. 281–295, 2026.
https://doi.org/10.1007/978-981-95-3453-1_19

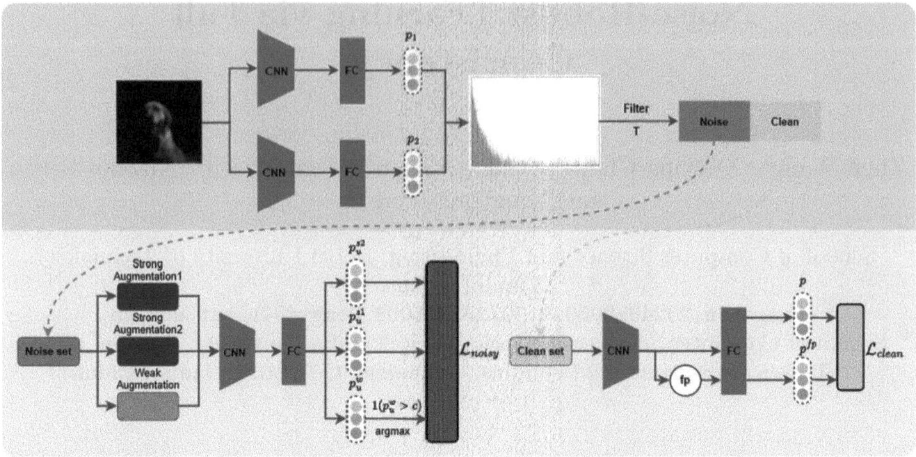

Fig. 1. The sample whose threshold is above a specific threshold would be in clean set in yellow, otherwise in noisy set in green. As shown, the backbone will first extract the features and the features are fed to decoder to get predictions. After that the sample will go through the trained network, and then be divided into clean sets and noise sets according to tolerance. Samples in different data sets are divided into different modules for processing. (Color figure online)

a small clean data set and gradually expand it, thus preventing DNNs from overfitting to noisy data. There are two primary methods to incorporate clean samples in the learning process. The first method involves identifying samples with clean high-probability labels and training the classifier on these samples, often referred to as the small-loss criterion [7,9,12,19,34]. The second method involves controlling the learning process of the classifier to prioritize learning clean high-probability samples from noisy datasets, commonly known as "early stopping" [3,10,17,24,30]. Although these methods have favorable properties for LNL, there is a drawback in selecting samples with small losses. These methods may discard boundary samples that have large losses but are often intertwined with noisy instances. In this paper, we develop a sample selection strategy called Noise-robust Learning via Full Consistency (NLFC) that filters out clean and noisy samples. We monitor the number of correct predictions for each sample in different training stages and determine the learning degree of each sample at these stages. By using this criterion, NLFC can identify and remove noisy samples while retaining boundary samples for optimization. Additionally, the number of correct predictions for all samples is dynamically updated at different learning stages as the learning progresses, and the samples are divided after each stage. Our proposed method introduces feature-level perturbation to clean samples, which has been shown to improve testing accuracy when training supervised models with perturbations. Unlike most current works that only utilize noisy samples, we add feature-level perturbations to clean sets for further learning. Additionally, we integrate the FixMatch [25] semi-supervised method into our framework to explore the useful information in the abandoned noisy samples, resulting in significant improvements to the model's noise-robust learning performance.

This paper makes the following key contributions:

1. We propose a novel sample selection strategy, Noise-robust Learning via Full Consistency (NLFC), which tracks the correct prediction count of each sample across training stages to dynamically identify clean, noisy, and boundary samples for robust learning.

2. We introduce feature-level perturbations to clean samples and integrate the FixMatch semi-supervised framework, jointly enhancing model generalization and enabling effective utilization of informative noisy samples.

3. Extensive experiments on benchmark noisy-label datasets demonstrate that NLFC achieves superior robustness and accuracy, outperforming state-of-the-art methods in label noise learning.

2 Related Work

2.1 Sample Selection Under Noisy Labels

Sample-selection methods aim to identify clean training examples in the presence of label noise. A common strategy is the "small-loss" criterion: deep networks tend to fit clean samples before noisy ones, so low-loss instances are treated as reliable.MentorNet [12] uses a pre-trained network to guide the selection of clean instances. Co-teaching [9] introduces a dual-network framework where each network selects small-loss samples to train the other. More recent approaches combine probabilistic models and semi-supervised learning. DivideMix [16] fits a Gaussian mixture model on per-sample losses to partition the data into a labeled-clean set and an unlabeled-noisy set, then applies MixMatch-style pseudo-labeling on both parts. These techniques significantly improve robustness by focusing training on high-confidence examples.

2.2 Consistency-Based Semi-supervised Learning

Consistency-based semi-supervised learning (SSL) methods have been adapted to noisy-label scenarios by treating dubious labels as unlabeled data. SSL algorithms such as MixMatch [5] combine pseudo-labeling with strong data augmentation. FixMatch [25] simplifies this paradigm by generating pseudo-labels from weakly-augmented images and enforcing consistency under strong augmentations only when confidence is high. Noisy Student Training [38] similarly uses iterative self-training with noise injection to refine labels. These ideas inform noisy-label learning: for instance, DivideMix [16] treats noisy examples as unlabeled and uses a co-training MixMatch pipeline, including label co-refinement, to exploit consistency. Overall, consistency-based SSL provides a powerful framework to leverage all data while mitigating noisy annotations.

2.3 Feature-Level Robustness

Another line of work focuses on learning robust representations at the feature level to counter-label noise. Contrastive and prototype-based methods are prominent here. MoPro [15] introduces *momentum prototypes* into a contrastive learning framework, enabling online correction of noisy labels and removal of outlier samples during training. In this way, class prototypes learned in feature space guide the model away from corrupted labels. These methods exploit the intrinsic structure of the feature space - often leveraging self-supervised pretraining or pairwise similarity - to improve noise robustness. Empirically, they yield more stable and transferable representations under label corruption.

3 Method

In this section, we propose a novel learning method with a noisy label to achieve robust learning. First, we introduce our pipeline and then describe the details of the proposed method: the sample selection module, the Clean Feature-level Consistency module, and the Noisy Image-level Consistency module. Finally, we introduce the loss function for learning.

3.1 Preliminary and Overview

Our proposed LNL method is presented in Fig. 1. The algorithm consists of two main modules. Firstly, the sample selection module divides the samples into clean and noisy samples. Secondly, different training methods are applied to the two types of samples divided by the sample selection method.

Supposing the training set $\mathcal{D} = \left\{\left(x^i, y^i\right)\right\}_{i=1}^{n} \in (x, y)$ with corrupted labels and $y^i \in [K] = \{1, 2, \cdots, k\}$, the output distribution of a classifier f_θ after t training epochs can be written as $P^{(t)} = f(x; \theta)$. Here, θ is the learnable parameters. The goal of learning with label noise to find the optimal parameter θ^* which can achieve admirable generalization performance on the clean testing set.

The training process of most methods in learning with label noise based on sample selecting methods includes the following processes: Firstly selecting the reliable set $\tilde{\mathcal{D}}$ from the polluted dataset $\mathcal{D}$ through a certain selection strategy. such as small-loss criterion, which select the top τ of samples with the smaller loss values in the current mini-batch as the clean samples, where τ is the noise ratio estimated by cross-validation. Secondly, training the classifier f on the selected set $\tilde{\mathcal{D}}$, and update the parameter as $\theta^{(t+1)} = \theta^t - \eta \bigtriangledown \left(\frac{1}{|\tilde{\mathcal{D}}|}\right) \sum_{(x,y)\in\tilde{\mathcal{D}}} \mathcal{L}\left(x, y; \theta^t\right)$, where η and $\mathcal{L}$ are the given learning rate and loss function, respectively. Then repeating the above processes until the optimal parameter θ^* is found and return the classifier f.

3.2 Sample Selection

Recent studies on the memorization dynamics of deep neural networks [1] suggest that clean-labeled samples are typically learned in the early stages of training, while noisy-labeled samples are often memorized later. Leveraging this insight, a number of noise-robust learning algorithms, such as Co-teaching and DivideMix, incorporate sample selection strategies to prioritize early-learned, presumably clean data.Our sample selection strategy is grounded in the memorization behavior observed in deep neural networks during training. Previous studies [2] [41] have demonstrated that different training examples contribute unequally to the learning dynamics of a model. While some instances are easily learned and remain correctly classified throughout training, others are more difficult and exhibit frequent misclassifications, particularly during the initial epochs. This observation suggests that removing well-memorized samples at an appropriate stage of training may enhance convergence speed without compromising generalization performance.

To further refine the selection process, we incorporate the concept of "forgetting events", which has proven effective in identifying noisy or ambiguous samples. Inspired by the work of Toneva et al. [31], we define a forgetting event as a transition where a sample, previously classified correctly at step t, becomes misclassified at step $t+1$. Formally, given a dataset $\mathcal{D} = \{(x_i, y_i)\}$, a forgetting event occurs when the predicted label $\hat{y}_i^t$ for sample x_i changes from correct to incorrect.

We classify training samples by introducing a threshold based on the number of times a sample is "forgetting events" during training. Inspired by studies on the memorization behavior of deep neural networks [1], we divide the training process into three stages. After these stages, all samples are categorized into a clean set $\mathcal{D}_{\text{clean}}$ and a noisy set $\mathcal{D}_{\text{noise}}$, based on their classification consistency. A tolerance threshold τ is introduced as a criterion to assess sample reliability. This threshold-based filtering strategy is consistent with recent findings, which suggest that samples frequently forgotten during training tend to be noisy or ambiguous [28,31]. The sample selection process can be summarized as the following phases. Phase1: In the easy phase, a sample that has been correctly classified more than τ is considered a clean sample, which can be divided into clean set $\mathcal{D}_{clean}$, and a noise sample when the times each sample is classified less than τ, which can be divided into noise set $\mathcal{D}_{noise}$. Then samples with correct Classification times greater than τ in the hard phase are divided into the clean set $\mathcal{D}_{clean}$ and are divided into $\mathcal{D}_{noise}$ with times less than τ. Lastly, the number of times each sample is classified correctly greater than τ are divided into clean set $\mathcal{D}_{clean}$, otherwise the samples will be divided into noise set $\mathcal{D}_{noise}$.

3.3 Clean Feature-Level Consistency

In a study conducted by Yann et al. [4,27], they discovered that perturbed samples exhibit highly clustered behavior during training, forming a centroid around the average embedding of perturbed samples sharing the same original image.

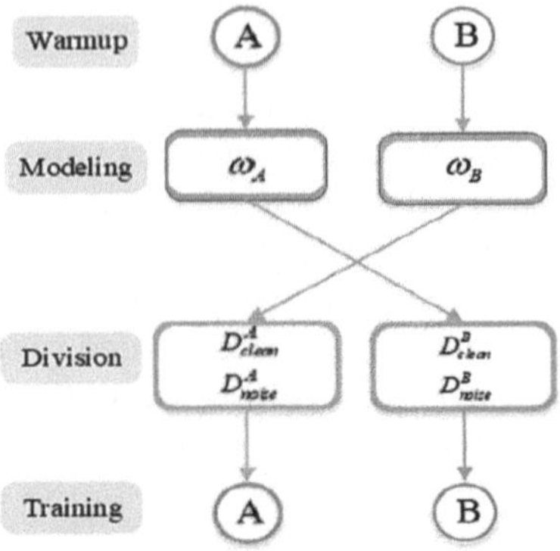

Fig. 2. Training pipeline of Noise-robust Learning via Full Consistency (NLFC). In brief, NLFC consists of two individual networks (A, B) which work in a co-teaching manner. More specifically, NLFC first warmups both A and B by using Eq. 2 for initialization. After that, at each epoch, the following procedure is performed. First, the network A/B models the number of per-sample correctly classified by trained network to estimate the correctly annotated threshold τ for each sample and then feed τ into B/A for further training. The next step will divide the data into two subsets, i,e., $\mathcal{D}_{noisy}$ and $\mathcal{D}_{clean}$.

However, existing label noise learning methods based on sample selection only utilize the selected samples themselves after sample selection, and perturbations on selected samples are not considered. Other related research has shown that training a supervised model with perturbations slightly decreases the training accuracy, yet significantly improves the test accuracy. Therefore, to improve the test accuracy of the model for samples in the clean set $\mathcal{D}_{clean}$, we adopt the Clean Feature-level Consistency method.

The sample selection strategy is based on memorization in deep networks, which can produce a clean set with a high pure ratio despite varying noise rates and types. To further utilize this clean set, we implement a feature-level consistency algorithm.

Specifically, on raw images, we predefined image-level strong perturbation, while on the extracted feature of images, we insert an embarrassingly simple channel dropout *FP* in Fig. 2. Dropout [27], originally proposed to reduce overfitting in deep neural networks, randomly deactivates a subset of feature channels during training. Introducing stochasticity into the feature space, it improves the robustness of learned representations, which is particularly beneficial in the presence of label noise or uncertain supervision. Formally, a label noise learning model f can be decomposed into an encoder g and a decoder h. In addition to acquiring feature p after the encoder, we also obtain p^{fp} from an auxiliary feature perturbation stream by

$$\begin{aligned} e &= g\left(x_{clean}\right) \\ p^{fp} &= h\left(\mathcal{P}\left(e\right)\right) \\ p &= h\left(e\right) \end{aligned} \tag{1}$$

where e is extracted features of x_{clean} which is in the clean set x_{clean}, and $\mathcal{P}$ denotes feature perturbations, *e.g.*, dropout or adding uniform noise.

By leveraging the proposed Feature-level consistency module, we can train the classification network with the selected clean samples. Typically, the supervised term $\mathcal{L}_{clean}$ is the cross-entropy loss between model predictions and the ground truth label including the samples as well as the sample under the feature-level perturbation.

$$\mathcal{L}_{clean} = \frac{1}{C}\sum H\left(p, y\right) + H\left(p^{fp}, y\right) \tag{2}$$

where C is batch size of selected samples and $H\left(p, y\right)$ minimizes the entropy between two probability distribution terms.

It is worth noting that we do not aim at proposing a novel feature perturbation method in this work. Actually, an extremely simple Channel Dropout is well-performed enough. We hope to add perturbations to clean samples so that the model can learn more clustering feature (Fig. 3).

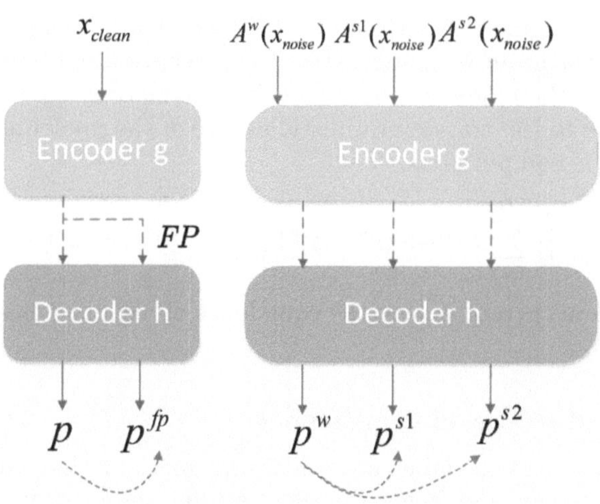

Fig. 3. Our Feature-level Consistency on the clean set $\mathcal{D}_{clean}$ and Our Image-level Consistency on the noise set $\mathcal{D}_{noise}$.

3.4 Noisy Image-Level Consistency

Assuming that the mislabeled samples in noise set $\mathcal{D}_{noise}$ are unlabeled samples. Several semi-supervised learning methods have been studied by leveraging the structure of unlabeled data including consistency regularization [37].

Furthermore, we aim to fully explore the significant benefits of image-level strong perturbations. Recent advancements in self-supervised learning and semi-supervised classification have inspired us to construct multiple views for mislabeled data as inputs, which can better exploit the perturbations. Specifically, each mislabeled sample is simultaneously perturbed by two operators, namely a weak perturbation $\mathcal{A}^w$ like cropping, and a strong perturbation $\mathcal{A}^s$ like adding Gaussian Noise.

And the unsupervised loss function for the noise set $\mathcal{L}_{noise}$ regularizes the prediction of the sample under strong perturbations to be the same as that under weak perturbations. FixMatch introduces Pseudo-Labeling techniques related to entropy minimization in the consistency regularization process. The improved consistency loss function can be formulated as:

$$\mathcal{L}_{noise} = \frac{1}{B}\sum_{i=1}^{B} 1\left(\max\left(p^w\right) \geq \tau\right) H\left(\hat{p^w}, p^s\right) \tag{3}$$

where p^w and p^s represent the prediction distribution of the weakly-perturbed version and strongly-perturbed version, respectively. $\hat{p^w} = argmax\left(p^w\right)$ is the hard target. τ is a confidence threshold and H represents the cross-entropy loss function.

Additionally, we have produced multiple strongly-perturbed versions for the model to learn, and we are curious if this simple idea can also benefit our label noise learning. In Figure 2, we attempt to independently yield dual-stream perturbations from the input by using a strong perturbation pool. Instead of feeding a single value into the model, we use two values, namely x^{s1}_{noise} and x^{s2}_{noise}, which are not equal due to the non-deterministic nature of the predefined pool $\mathcal{A}^s$. The final noise term is computed as:

$$\mathcal{L}_{noise} = \frac{1}{B}\sum_{i=1}^{B} 1\left(\max\left(p^w\right) \geq \tau\right)\left(H\left(\hat{p^w}, p^{s1}\right) + H\left(\hat{p^w}, p^{s2}\right)\right) \tag{4}$$

Following some current works, the confidence threshold τ is set as 0.95 in all our experiments.

3.5 Clean Feature-Level Consistency

We introduce a two-stage label noise learning method that effectively distinguishes between clean and noisy samples. Additionally, we propose two techniques to enhance the performance of clean and noisy samples separately: Clean Feature-level Consistency and Noisy Image-level Consistency. Our method maintained two auxiliary feed-forward streams, one for perturbation on features of x, and the other for multi-view learning of $\left(x^{s1}, x^{s2}\right)$. The overall loss function is computed as follows:

$$\mathcal{L} = \lambda_1 \mathcal{L}_{clean} + \lambda_2 \mathcal{L}_{noisy} \tag{5}$$

where λ_1 and λ_2 denote the weights of the $\mathcal{L}_{clean}$ and $\mathcal{L}_{noisy}$ and their value are equally set as 0.5.

4 Experiments

4.1 Experiment Setup

Dataset. We evaluate our method on four synthetic noisy datasets F-MNIST [35], SVHN [20], CIFAR-10 [13], CIFAR-100 [13], and two real-world datasets Clothing1M [36] and WebVision 1.0 [18].

Noise. We employ three types of noise (Symmetric, Asymmetric and Pairflip noises) with various noise rates (20% and 40%). Besides, we also evaluate our method on two types of long-tailed noise (exponential simulation (exp) and linear simulation (line)) with Asymmetric 20%, 30%, 40% and 45%.

Baselines. We compare our NLFC with the following state-of-the-art methods: CE [14], Co-teaching [9], Co-teaching-plus [40], JoCoR [32], Co-learning [29], CoDis [33] Bare [22] and LRA [6]. We implement all methods with default parameters using PyTorch.

Table 1. Mean and standard deviations of test accuracy (%) on four benchmark datasets with five noise types, which noise type has two noise rates (20%, 40%). The test accuracy is calculated by the last ten epochs. The best mean results are in **bold** and the second best mean results in <u>underlined</u>

	Method	Sym.		Asym.		Pair	
		20%	40%	20%	40%	20%	40%
CIFAR-10	CE	75.74±0.33	58.64±0.81	82.32±0.14	72.19±0.32	76.38±0.35	55.03±0.27
	Co-teaching	82.24±0.18	77.16±0.10	80.76±0.11	72.85±0.11	82.55±0.10	75.74±0.14
	Co-teaching-plus	81.96±0.12	71.49±0.33	79.68±0.13	70.96±0.69	79.71±0.14	58.39±0.76
	JoCoR	85.23±0.09	79.77±0.18	85.62±0.21	73.50±1.25	81.75±0.26	68.29±0.30
	Co-learning	88.82±0.23	85.99±0.22	88.58±0.28	82.66±0.44	89.13±0.11	79.68±0.27
	CoDis	82.36±0.24	77.04±0.09	84.58±0.05	75.30±0.32	82.53±0.23	70.86±0.22
	Bare	85.30±0.61	78.90±0.70	86.44±0.39	81.20±0.46	84.98±0.57	74.53±0.28
	LRA	<u>90.34±0.61</u>	**90.13±0.19**	<u>90.34±0.11</u>	**86.99±0.07**	<u>90.38±0.14</u>	<u>83.71±0.24</u>
	NLFC	**91.05±0.16**	<u>86.30±0.11</u>	**90.56±0.13**	<u>83.67±0.12</u>	**90.74±0.16**	**86.77±0.18**
CIFAR-100	CE	46.10±0.53	33.20±0.58	46.11±0.53	33.20±0.58	46.01±0.33	33.16±0.32
	Co-teaching	50.87±0.31	43.38±0.35	48.88±0.29	35.92±0.34	49.57±0.35	35.11±0.45
	Co-teaching+	51.72±0.33	44.31±0.67	51.48±0.28	34.20±0.64	50.71±0.77	34.29±0.27
	JoCoR	51.61±0.37	42.78±0.26	51.21±0.09	42.68±0.23	51.46±0.32	42.01±0.21
	Co-learning	62.43±0.31	<u>57.18±0.35</u>	<u>63.04±0.36</u>	<u>49.69±0.31</u>	<u>62.53±0.31</u>	**49.29±0.41**
	CoDis	50.65±0.35	43.44±0.27	50.14±0.38	35.43±0.38	50.80±0.41	35.15±0.29
	Bare	**63.32±0.32**	55.33±0.62	61.62±0.26	40.88±1.17	61.22±0.51	40.92±1.40
	LRA	56.84±1.57	54.25±1.22	56.54±0.58	44.57±1.33	56.25±1.58	46.03±1.07
	NLFC	<u>62.63±0.16</u>	**57.34±0.20**	**63.06±0.19**	**50.11±0.21**	**64.44±0.31**	<u>47.55±0.19</u>

4.2 Implementation Detail

We implement our method on F-MNIST, SVHN, CIFAR-10, and CIFAR-100 datasets using a 9-layer CNN network architecture. The optimizer settings are as follows: SGD with a momentum of 0.9, weight decay of $5e^{-4}$, batch size of 128, initial learning rate of 0.003, and a decay factor of 10 at 80 epochs. We train the network for 200 epochs for CIFAR-100, with a warming-up stage of 20 epochs and with a linear decay of the learning rate from 80 to 200 epochs.

For experiments on Clothing1M, we use a ResNet with 18 layers, an Adam optimizer with a momentum of 0.9, and a batch size of 64. During the training stage, we run a total of 15 epochs and set the learning rate to 8×10^{-4}, 5×10^{-4}, and 5×10^{-5} for 5 epochs each.

For experiments on WebVision 1.0, we utilized InceptionResNetV2 as the backbone for the network. The network was trained using an SGD optimizer with a momentum of 0.9 and a weight decay of $5e^{-5}$. We used a batch size of 32 and initially set the learning rate to 0.002 for 100 epochs, with a decrease by a factor of 10 at the 30th and 60th epochs.

Table 2. Test accuracy (%) on Clothing1M. The best results are in **bold.**

Method	Acc.
CE	67.22
Co-teaching	69.21
Co-teaching-plus	59.32
JoCoR	70.30
Co-learning	68.72
CoDis	70.48
Bare	70.32
NLFC	**71.22**

Table 3. Top-1 and Top-5 test accuracy (%) on the WebVision1.0 and ILSVRC12 datasets. The best results are in **bold.**

Method	WebVision		ILSVRC12	
	top1	top5	top1	top5
Co-teaching	63.58	85.20	61.48	84.70
Co-teaching-plus	68.56	86.64	65.60	86.60
JoCoR	61.84	83.72	59.16	84.16
CoDis	70.52	87.88	66.88	87.20
Bare	69.60	88.84	66.48	88.76
NLFC	**70.69**	**89.24**	**67.66**	**90.12**

4.3 Results on Simulated Noisy Datasets

We conducted experiments on two noise-simulated datasets under three types of noise with varying noise ratios to evaluate the performance of our proposed

Table 4. Test accuracy(%) on noisy long-tailed datasets. The best mean results are in **bold** and the second best mean results is underlined

	Method	Asym.20%	Asym.30%	Asym.40%	Asym.45%
	CE	85.77±0.17	80.99±0.45	74.90±0.25	70.90±0.37
	Co-teaching	89.41±0.13	86.83±0.16	84.21±0.12	82.27±0.21
	Co-teaching-plus	90.00±0.17	88.60±0.20	82.36±0.26	76.19±0.57
F-MNIST-exp	JoCoR	91.00±0.10	88.12±0.12	84.10±0.17	78.41±0.29
	Co-learning	86.70±0.25	86.95±0.29	86.78±0.41	83.02±0.66
	CoDis	89.11±0.17	87.11±0.16	83.84±0.26	80.88±0.13
	Bare	91.43±0.15	90.89±0.18	88.45±1.31	86.41±1.15
	NLFC	**92.21±0.11**	**91.07±0.08**	**90.17±0.07**	**88.88±0.06**
	CE	86.94±0.22	81.38±0.33	75.26±0.39	71.68±0.32
	Co-teaching	89.79±0.10	87.89±0.15	85.24±0.26	83.43±0.14
	Co-teaching-plus	90.61±0.14	87.79±0.20	82.83±0.35	77.28±0.58
F-MNIST-line	JoCoR	91.12±0.15	89.02±0.17	85.29±0.26	84.01±0.26
	Co-learning	87.84±0.23	86.22±0.41	85.66±0.43	83.09±1.04
	CoDis	90.13±0.10	87.79±0.16	84.77±0.26	81.46±0.13
	Bare	92.11±0.13	91.79±0.09	90.55±0.28	89.92±0.61
	NLFC	**92.63±0.08**	**92.12±0.07**	**91.17±0.05**	**90.69±0.10**
	CE	89.02±0.21	83.54±0.21	76.37±0.40	73.09±0.27
	Co-teaching	93.08±0.16	91.22±0.24	88.49±0.14	85.21±0.28
	Co-teaching-plus	93.10±0.13	89.57±0.19	83.58±0.43	73.28±0.69
SVHN-exp	JoCoR	95.01±0.08	94.62±0.08	92.65±0.11	89.95±0.07
	Co-learning	91.67±0.18	90.60±0.23	86.52±0.40	81.97±0.58
	CoDis	92.80±0.17	91.09±0.19	86.89±0.11	81.25±0.13
	Bare	95.42±0.14	92.51±0.23	89.48±0.33	83.31±0.65
	NLFC	**95.87±0.08**	**95.43±0.05**	**94.69±0.08**	**94.12±0.06**
	CE	88.90±0.15	83.33±0.20	76.92±0.29	72.57±0.24
	Co-teaching	92.92±0.10	91.17±0.10	88.33±0.25	85.77±0.26
	Co-teaching-plus	92.73±0.13	89.41±0.20	82.28±0.37	73.94±0.47
SVHN-line	JoCoR	95.34±0.04	94.88±0.07	93.56±0.06	91.00±0.07
	Co-learning	91.90±0.09	90.70±0.16	88.08±0.25	81.80±0.46
	CoDis	92.94±0.17	90.78±0.19	86.71±0.11	80.93±0.13
	Bare	95.40±0.11	93.38±0.22	89.62±0.16	86.35±0.72
	NLFC	**95.72±0.05**	**95.48±0.05**	**95.07±0.04**	**92.30±0.05**

method, as shown in Table 1. Our proposed method NLFC outperforms existing methods tremendously, as indicated in Table 1. Especaially on the 40% Pair noise in CIFAR-10 and the 20% Pair noise in CIFAR-100, our method is superior to previous best results by 3.06% and 1.93% respectively.

Since NLFC utilizes three views, one feature perturbation view, and dual image perturbation views, to train the model, the results also validate that constructing diverse perturbations is beneficial, and much better than blindly maintaining three parallel image perturbations.

4.4 Results on Noisy Long-Tailed Datasets

Two types of noisy long-tailed noise with various noise rates are employed to evaluated our method, the results is present in Table 4. Our method consistently outperformed other methods, demonstrating its effectiveness to long-tailed noises.

4.5 Results on Real-World Noisy Datasets

We also adopt real-world noisy datasets to validate the effectiveness and superiority of our method. Tables 2 and 3 present the results on Clothing1M and WebVision datasets. Our method outperforms state-of-the-art methods on both datasets, demonstrating its superior performance in handling real-world noisy datasets.

In the experiments conducted on WebVision 1.0 with real-world datasets, we utilized InceptionResNetV2 as the backbone for the network. The network was trained using an SGD optimizer with a momentum of 0.9 and a weight decay of $5e^{-5}$. We used a batch size of 32 and initially set the learning rate to 0.002 for 100 epochs, with a decrease by a factor of 10 at the 30th and 60th epochs. In the Clothing1M experiment, we opted for ResNet-18 as the backbone network and trained it with the Adam optimizer using a batch size of 64 in the Table 2 and Table 3. The network was trained for 20 epochs, starting with an initial learning rate of $8e^{-4}$.

Table 5. Effect of sample selection, clean feature-level consistency and noisy image-level consistency

	Acc.
Base	86.64±0.14
+ Sample selection	88.28±0.23
+ Clean Feature-level Consistency	89.20±0.15
+ Noisy Image-level	90.05±0.16

4.6 Ablation Study

To evaluate the effect of each component (sample selection, clean feature-level consistency and noisy image-level consistency) in the proposed method, we conducted ablation experiments on the CIFAR-10 dataset with Symmetry-0.2 noise, as shown in Table 5. We can observe that each module plays a non-trivial role in our method. The adoption of the sample selection, clean feature-level consistency and noisy image-level consistency can promote the performance by a large margin.

5 Conclusions

In this paper, we focus on addressing the challenge of existing sample selection methods being unable to effectively select clean hard examples. To overcome this challenge, we propose a novel method called *Noise-robust Learning via Full Consistency (NLFC)*. Unlike traditional sample selection methods, our method is rooted in memorization in deep networks that differentiate between mislabeled and clean examples based on the degree of forgetting of sample re-identification by the trained network. Experimental results on synthetic datasets demonstrate that our proposed method is both straightforward and effective in handling learning with noisy labels. In future work, we aim to enhance the sample selection criterion to more accurately capture the intrinsic robust learning ability of DNNs, which could potentially strengthen the effectiveness of our framework, especially under more complex noise conditions.

Acknowledgments. This work was supported in part by the National Natural Science Foundation of China (NSFC) under Grant Nos. 62306103 and 62376194, and in part by the Higher Education Science and Technology Research Project of Hebei Province under Grant No. QN2023262.

References

1. Arpit, D., et al.: A closer look at memorization in deep networks. arXiv preprint arXiv:1706.05394 (2017)
2. Bai, Y., Liu, T.: Me-momentum: extracting hard confident examples from noisily labeled data. In: Proceedings of the IEEE/CVF International Conference on Computer Vision, pp. 9312–9321 (2021)
3. Bai, Y., et al.: Understanding and improving early stopping for learning with noisy labels. Adv. Neural. Inf. Process. Syst. **34**, 24392–24403 (2021)
4. Ben-Shaul, I., Shwartz-Ziv, R., Galanti, T., Dekel, S., LeCun, Y.: Reverse engineering self-supervised learning. arXiv preprint arXiv:2305.15614 (2023)
5. Berthelot, D., Carlini, N., Goodfellow, I., Papernot, N., Oliver, A., Raffel, C.A.: Mixmatch: a holistic approach to semi-supervised learning. Adv. Neural Inform. Process. Syst. **32** (2019)
6. Chen, J., et al.: Label-retrieval-augmented diffusion models for learning from noisy labels. Adv. Neural Inform. Process. Syst. **36** (2024)

7. Han, B., et al.: Sigua: forgetting may make learning with noisy labels more robust. In: International Conference on Machine Learning, pp. 4006–4016. PMLR (2020)
8. Han, B., et al.: A survey of label-noise representation learning: Past, present and future. arXiv preprint arXiv:2011.04406 (2020)
9. Han, B., et al.: Co-teaching: robust training of deep neural networks with extremely noisy labels. Adv. Neural Inform. Process. Syst. **31** (2018)
10. Hu, W., Li, Z., Yu, D.: Simple and effective regularization methods for training on noisily labeled data with generalization guarantee. arXiv preprint arXiv:1905.11368 (2019)
11. Huang, Z., et al.: Harnessing out-of-distribution examples via augmenting content and style. arXiv preprint arXiv:2207.03162 (2022)
12. Jiang, L., Zhou, Z., Leung, T., Li, L.J., Fei-Fei, L.: Mentornet: learning data-driven curriculum for very deep neural networks on corrupted labels. In: International Conference on Machine Learning, pp. 2304–2313. PMLR (2018)
13. Krizhevsky, A., Hinton, G., et al.: Learning multiple layers of features from tiny images (2009)
14. Krizhevsky, A., Sutskever, I., Hinton, G.E.: Imagenet classification with deep convolutional neural networks. In: Advances in Neural Information Processing Systems (NeurIPS), vol. 25, pp. 1097–1105. Curran Associates, Inc. (2012)
15. Li, J., Hoi, S.C.H., Socher, R.: Towards noise-robust contrastive learning. In: Advances in Neural Information Processing Systems (NeurIPS) (2020)
16. Li, J., Socher, R., Hoi, S.C.: Dividemix: Learning with noisy labels as semi-supervised learning. arXiv preprint arXiv:2002.07394 (2020)
17. Li, M., Soltanolkotabi, M., Oymak, S.: Gradient descent with early stopping is provably robust to label noise for overparameterized neural networks. In: International Conference on Artificial Intelligence and Statistics, pp. 4313–4324. PMLR (2020)
18. Li, Y., Yang, J., Song, Y., Cao, L., Luo, J., Li, L.J.: Learning from noisy labels with distillation. In: Proceedings of the IEEE International Conference on Computer Vision, pp. 1910–1918 (2017)
19. Liu, Y., Guo, H.: Peer loss functions: learning from noisy labels without knowing noise rates. In: International Conference on Machine Learning, pp. 6226–6236. PMLR (2020)
20. Netzer, Y., Wang, T., Coates, A., Bissacco, A., Wu, B., Ng, A.Y.: Reading digits in natural images with unsupervised feature learning (2011)
21. Nguyen, D.T., Mummadi, C.K., Ngo, T.P.N., Nguyen, T.H.P., Beggel, L., Brox, T.: Self: learning to filter noisy labels with self-ensembling. arXiv preprint arXiv:1910.01842 (2019)
22. Patel, D., Sastry, P.: Adaptive sample selection for robust learning under label noise. In: Proceedings of the IEEE/CVF Winter Conference on Applications of Computer Vision, pp. 3932–3942 (2023)
23. Pleiss, G., Zhang, T., Elenberg, E., Weinberger, K.Q.: Identifying mislabeled data using the area under the margin ranking. Adv. Neural. Inf. Process. Syst. **33**, 17044–17056 (2020)
24. Rolnick, D., Veit, A., Belongie, S., Shavit, N.: Deep learning is robust to massive label noise. arXiv preprint arXiv:1705.10694 (2017)
25. Sohn, K., et al.: Fixmatch: simplifying semi-supervised learning with consistency and confidence. Adv. Neural. Inf. Process. Syst. **33**, 596–608 (2020)
26. Song, H., Kim, M., Park, D., Shin, Y., Lee, J.G.: Learning from noisy labels with deep neural networks: a survey. IEEE Trans. Neural Netw. Learn. Syst. (2022)

27. Srivastava, N., Hinton, G., Krizhevsky, A., Sutskever, I., Salakhutdinov, R.: Dropout: a simple way to prevent neural networks from overfitting. J. Mach. Learn. Res. **15**(1), 1929–1958 (2014)
28. Swayamdipta, S., et al.: Dataset cartography: Mapping and diagnosing datasets with training dynamics. arXiv preprint arXiv:2009.10795 (2020)
29. Tan, C., Xia, J., Wu, L., Li, S.Z.: Co-learning: learning from noisy labels with self-supervision. In: Proceedings of the 29th ACM International Conference on Multimedia, pp. 1405–1413 (2021)
30. Tanaka, D., Ikami, D., Yamasaki, T., Aizawa, K.: Joint optimization framework for learning with noisy labels. In: Proceedings of the IEEE Conference on Computer Vision and Pattern Recognition, pp. 5552–5560 (2018)
31. Toneva, M., Sordoni, A., Tsvetkov, Y., Jaakkola, T., Bengio, G.: An empirical study of example forgetting during deep neural network learning. In: International Conference on Learning Representations (ICLR) (2019)
32. Wei, H., Feng, L., Chen, X., An, B.: Combating noisy labels by agreement: a joint training method with co-regularization. In: Proceedings of the IEEE/CVF Conference on Computer Vision and Pattern Recognition, pp. 13726–13735 (2020)
33. Xia, X., Han, B., Zhan, Y., Yu, J., Gong, M., Gong, C., Liu, T.: Combating noisy labels with sample selection by mining high-discrepancy examples. In: Proceedings of the IEEE/CVF International Conference on Computer Vision, pp. 1833–1843 (2023)
34. Xia, X., et al.: Sample selection with uncertainty of losses for learning with noisy labels. arXiv preprint arXiv:2106.00445 (2021)
35. Xiao, H., Rasul, K., Vollgraf, R.: Fashion-mnist: a novel image dataset for benchmarking machine learning algorithms. arXiv preprint arXiv:1708.07747 (2017)
36. Xiao, T., Xia, T., Yang, Y., Huang, C., Wang, X.: Learning from massive noisy labeled data for image classification. In: Proceedings of the IEEE Conference on Computer Vision and Pattern Recognition, pp. 2691–2699 (2015)
37. Xie, Q., Dai, Z., Hovy, E., Luong, T., Le, Q.: Unsupervised data augmentation for consistency training. Adv. Neural. Inf. Process. Syst. **33**, 6256–6268 (2020)
38. Xie, Q., Luong, M.T., Hovy, E., Le, Q.V.: Self-training with noisy student improves imagenet classification. In: Proceedings of the IEEE/CVF Conference on Computer Vision and Pattern Recognition (CVPR) (2020)
39. Yang, X., Song, Z., King, I., Xu, Z.: A survey on deep semi-supervised learning. IEEE Trans. Knowl. Data Eng. (2022)
40. Yu, X., Han, B., Yao, J., Niu, G., Tsang, I., Sugiyama, M.: How does disagreement help generalization against label corruption? In: International Conference on Machine Learning, pp. 7164–7173. PMLR (2019)
41. Zhang, C., Bengio, S., Hardt, M., Recht, B., Vinyals, O.: Understanding deep learning requires rethinking generalization. In: International Conference on Learning Representations (ICLR) (2017)
42. Zhang, C., Bengio, S., Hardt, M., Recht, B., Vinyals, O.: Understanding deep learning (still) requires rethinking generalization. Commun. ACM **64**(3), 107–115 (2021)

Emotional Earth Mover's Distance for Fine-Grained Hierarchical Emotion Analysis

Hai-Tao Yu[1(✉)], Dawei Li[1], and Xin Kang[2]

[1] University of Tsukuba, Tsukuba, Ibaraki, Japan
yuhaitao@slis.tsukuba.ac.jp
[2] Tokushima University, Tokushima, Japan
kang-xin@is.tokushima-u.ac.jp

Abstract. Effective emotion understanding has become increasingly important in developing human-centered interactive systems, such as chatbots and companion/service robots. Despite the remarkable successes made by previous studies, we found that *the hierarchical information among emotion labels has been ignored when either quantifying the optimization objective or evaluating the performance, hindering the achievement of fine-grained hierarchical emotion analysis.* To bridge this gap, we propose *Emotional Earth Mover's Distance* (EEMD)[1], a novel framework that extends the *Earth Mover's Distance* (EMD) to emotion analysis by explicitly encoding the hierarchical structure of emotion labels. This hierarchical distance is integrated into both the training loss and evaluation metric, allowing EEMD to effectively capture the hierarchical emotion nuances throughout the entire cycle of emotion analysis. To demonstrate the effectiveness of the proposed approach, we conduct a series of experiments on the widely used GoEmotions dataset. In addition to comparing our approach with representative traditional methods based on the pre-trained language model (e.g., BERT), we also compare it with different types of few-shot prompting methods based on *large language models* (LLMs). Furthermore, since traditional metrics for emotion analysis such as F1 score and subset accuracy do not effectively reflect a model's ability to perform fine-grained hierarchical emotion analysis, we propose using EMD as a hierarchy-aware evaluation metric that captures the severity of misclassifications based on label structure. Our extensive empirical experiments reveal that: (1) Benefiting from the integration of hierarchical information among labels during the training process, EEMD outperforms other methods by a large margin. (2) Because LLMs (such as Llama3 and GPT) are general-purpose base models without task-specific fine-tuning, they show poor performance in hierarchical emotion analysis, especially given a large number of labels (e.g., 28 labels in GoEmotions). (3) EMD serves as an effective evaluation metric for evaluating a model's ability to handle fine-grained hierarchical emotion analysis. (The source code: https://github.com/Shinnaaa/Hierarchical_Emotions.

Keywords: Hierarchical emotion analysis · Earth mover's distance

M. Yoshikawa et al. (Eds.): ADMA 2025, LNAI 16197, pp. 296–310, 2026.
https://doi.org/10.1007/978-981-95-3453-1_20

1 Introduction

Over the past decade, the field of *affective computing* has made significant progress in developing effective models for emotion understanding. The prior studies typically treat emotion labels as independent categories in a flat space [18], such as the two-group categorization (positive and negative) or using Ekman's six basic emotions. As a result, capturing the subtleties of human emotions in the text remains a challenging task, particularly in effectively capturing the structured relationships among emotion labels. Taking the hierarchical emotion labels shown in Fig. 1(a) as an example, "*Joy*" is a top-level label that expresses happiness at a coarse granularity, which can be further subdivided into finer-grained emotions, such as "*Love*", "*Optimism*", "*Amusement*", and others. Consider a scenario where the ground-truth label for a given text is "*Optimism*". When a model incorrectly classifies it as either "*Love*" or "*Fear*", traditional methods would typically assign the same penalty to both errors, which is counterintuitive. Confusing "*Optimism*" with "*Fear*" that belongs to a different top-level category reflects a more serious misunderstanding than misclassifying it as "*Love*" that resides under the same high-level category of "*Joy*".

Motivated by the above observations, this work proposes a novel framework for effective fine-grained hierarchical emotion analysis, which is referred to as *Emotional Earth Mover's Distance* (EEMD). The key idea is to tailor the *Earth Mover's Distance* (EMD) to quantify both the training loss and the model performance, which enables us to effectively capture hierarchical emotion nuances throughout the cycle of emotion analysis. To demonstrate the effectiveness of the proposed approach, we conduct a series of experiments based on the benchmark dataset GoEmotions. Besides comparing with representative traditional methods based on the pre-trained language model (e.g., BERT), we also compare with different types of few-shot prompting methods based on *large language models* (LLMs). However, evaluating performance in hierarchical emotion analysis remains challenging, as traditional evaluation metrics such as F1 score and subset accuracy treat all misclassifications equally. To address this limitation, we propose EMD as a hierarchy-aware evaluation metric that penalizes misclassifications in proportion to the distance between predicted and true labels within the hierarchical label structure.

The remainder of this work is structured as follows. In Sect. 2, we briefly survey the related work. In Sect. 3, we provide the general problem formulation for emotion analysis. In Sect. 4, we detail the proposed framework of EEMD. In Sects. 5 and 6, we introduce the experimental setup, and analyze the results, respectively. Finally, we conclude the paper in Sect. 7.

2 Related Work

In this section, we first provide a brief description of EMD and its typical applications in different fields. Then, we concisely survey representative studies in emotion analysis. Finally, we introduce the recent studies on LLM-based emotion analysis.

2.1 EMD and Its Wide Applications

In the following, we use $\mathbf{1}_n$ to denote the n-dimensional vector with all ones, and use $\Delta_n = \{\mathbf{o} \in \mathbb{R}_+^n : \mathbf{1}_n^\intercal \mathbf{o} = 1\}$ to denote the probability simplex, whose elements are called probability vectors, or equivalently histograms. Given two probability vectors $p \in \Delta_n$ and $q \in \Delta_n$, we use $\Pi(p, q) = \{\pi \in \mathbb{R}_+^{n \times n} : \pi^\intercal \mathbf{1}_n = p, \pi \mathbf{1}_n = q\}$ to denote the set of coupling matrices. In the context of optimal transport, p and q are viewed as mass distributions. An element $\pi \in \Pi(p, q)$ is referred to as a transport plan. An entry π_{ij} indicates how much of the mass from p_j is transported to q_i. Given the cost matrix $C \in \mathbb{R}_+^{n \times n}$, where C_{ij} indicates the corresponding transport cost per unit mass from p_j to q_i, the optimal transport problem is defined as $\mathcal{W}(p, q) = \min_{\pi \in \Pi(p,q)} < C, \pi >_F$, where $< \cdot, \cdot >_F$ denotes the Frobenius inner product of two matrices. $\mathcal{W}(p, q)$ is called *the earth mover's distance* (EMD) (also known as the Wasserstein-1 distance), which indicates the minimum transportation cost. Computationally, it is quite expensive to obtain the optimal solution. For a general cost matrix C, the worst-case complexity of achieving the optimum scales in $\mathcal{O}(n^3 \log n)$ [7]. Cuturi [7] proposed regularizing the optimal transport problem with an entropy term, which achieves a near-$\mathcal{O}(n^2)$ complexity.

Thanks to the work by Cuturi [7], the theory of optimal transport has been revitalized, establishing it as a prevalent framework in numerous machine learning domains. For instance, recent applications [1,27] in generative model learning have demonstrated the potential of optimal transport. Rolet et al. [36] and Huang et al. [14] utilized the Wasserstein distance for dictionary learning. Yu et al. [44] explored how to use EMD to quantify the discrepancy between the predicted ranking and that generated based on the ground truth. Regarding the application of EMD for emotion analysis, the most related studies to our work are Park et al. [28] and Mitsios et al. [26]. They delved into the continuous dimensions of *Valence, Arousal, and Dominance* using corpora with categorical emotion annotations. Though using EMD to quantify the loss, they did not take into account the structural information among labels when either quantifying the optimization objective or evaluating the performance. Yet, to the best of our knowledge, we are the first to adapt EMD for fine-grained hierarchical emotion analysis through loss quantification and performance evaluation.

2.2 Traditional Emotion Analysis

Over the past decade, emotion analysis has evolved from basic sentiment analysis to fine-grained emotion analysis [30]. The traditional methods categorize emotions into coarse groups, such as positive, negative, or neutral [2,11,50], Ekman's six basic emotions [9] and Plutchik's wheel of emotions [31]. With the boom of deep learning, significant advancements have been made in applying deep learning techniques to fine-grained emotion analysis [51]. Among these advancements, transformer-based pre-trained models [41,47] (e.g., BERT and RoBERTa) have achieved the state-of-the-art performance. Recently, Zhang et al. [46] and Pontiki et al. [32] explored how to identify and analyze emotions related to specific aspects within the text. Poria et al. [34] and Zadeh et al. [45] explored multimodal emotion analysis, showing improved performance in capturing the full spectrum of emotional expression. Additionally, emotion recognition in conversations has advanced through the works of Hazarika et al. [12] and Poria et al. [33], with models incorporating contextual and interactive elements to better understand emotions in dialogues. Emotion-cause detection tasks have also remained a popular topic. For example, Gui et al. [10] and Xia et al. [43] explored how to identify the reasons behind specific emotions. Furthermore, for fine-grained emotion analysis, Liu et al. [22] and Chen et al. [4] have explored utilizing context to enhance the model's performance. Singh et al. [37] utilized emotion definition modeling and multiple annotations, analyzed model performance on fused emotion categories and sub-sampled training data, and leveraged correlations between emotion categories to achieve superior results. The hierarchical structure of emotions is explored in several studies [15,21]. However, most fine-grained emotion analysis methods still treat emotion labels in a flat space, and the structural information is often ignored [39], leaving a research gap in leveraging the structural properties of emotional labels [13,24].

2.3 LLM-Based Emotion Analysis

Coming after the aforementioned traditional methods is the *LLM revolution* for emotion analysis. Given that LLMs, such as GPT [35] and Llama [38], have shown remarkable progress in various NLP tasks [25,29], many efforts [19,20,42,49] have been made to explore the effectiveness of LLMs in emotion analysis. Specifically, Wang et al. [42] evaluated the emotional intelligence of LLMs, demonstrating their ability to understand and simulate complex emotions through psychological assessments. Similarly, Li et al. [20] investigated the enhancement of LLMs' performance using emotional stimuli, finding that appropriate emotional cues can significantly improve the models' understanding and response to emotional contexts. In the realm of conversational emotion recognition, Zhang et al. [49] introduced DialogueLLM, which tunes LLMs with contextual and emotional knowledge, resulting in improved accuracy and robustness in conversation-based emotion analysis. Batbaatar et al. [3] conducted a comparative study of LLMs, such as GPT-3 and IBM Watson, highlighting the superior performance of LLMs in emotion and sentiment analysis tasks compared to traditional models. Chun et al. [6] presented a novel framework for using GPT-4 in diachronic sentiment analysis, which involves tracking and analyzing the sentiment evolution in narratives over time, providing explanations for sentiment shifts and their underlying causes. Deng et al. [48] provided a comprehensive evaluation of LLMs in various sentiment analysis tasks, underscoring the models' strengths and identifying areas for future improvement.

The above studies collectively underscore the potential of LLMs in advancing the field of emotion analysis, offering insights into their practical applications and highlighting the ongoing research efforts to optimize their performance [40]. However, a comprehensive and in-depth investigation of LLMs' effectiveness in fine-grained hierarchical emotion analysis is still lacking.

3 Problem Formulation

Let $\mathcal{X}$ and $\mathcal{Y}$ be the text space and the label space, respectively, for each sentence $t \in \mathcal{X}$, its emotion labels are denoted as y. In practice, we get *independently and identically distributed* (i.i.d) samples $\mathcal{S} = \{(t_i, y_i)\}_{i=1}^{n}$ from an unknown joint distribution $P(\cdot, \cdot)$ over $\mathcal{X} \times \mathcal{Y}$. We use f_θ parameterized by θ to denote the classification function. We measure the loss of emotion analysis for a text using f_θ with the loss function $\mathcal{L}(f_\theta(t), y)$. The goal is to learn the optimal classification function over a hypothesis space $\mathcal{F}$ of classification functions that can *minimize the expected risk* as defined below:

$$\min_{f_\theta \in \mathcal{F}} \Re(f_\theta) = \min_{f_\theta \in \mathcal{F}} \int_{\mathcal{X} \times \mathcal{Y}} \mathcal{L}(f_\theta(t), y) dP(t, y) \tag{1}$$

Typically, $\Re(f_\theta)$ is intractable to optimize directly and the joint distribution is unknown. Thus, we appeal to *empirical risk minimization* to approximate the expected risk, which is defined as follows:

$$\min_{f_\theta \in \mathcal{F}} \hat{\Re}(f_\theta; \mathcal{S}) = \min_{f_\theta \in \mathcal{F}} \frac{1}{n} \sum_{i=1}^{n} \mathcal{L}(f_\theta(t), y) \tag{2}$$

With this general emotion analysis framework, one can develop various methods by deploying different strategies (e.g., using neural networks or tree-based models) for implementing the classification function and designing different loss functions.

4 Emotional Earth Mover's Distance

In this section, we elaborate on how to perform fine-grained hierarchical emotion analysis. The advantage of using EMD described in Sect. 2.1 for quantifying the discrepancy between predicted labels and standard labels is that it provides us with the flexibility of designing different application-specific cost matrices, which determines the optimization direction during the training process. In the context of fine-grained hierarchical emotion analysis, we believe that the misclassification loss should be consistent with the hierarchical distance among the labels. As shown in Fig. 1b, let f denote the prediction function with θ as the parameters, R_h denote the ground-truth hierarchical labels, φ' and φ represent two specific normalizing functions, given the two probability distributions $\varphi'(f_\theta(t))$ and $\varphi(R_h)$ derived from the prediction and standard labels, respectively, our key idea is to quantify the discrepancy $\mathcal{W}(\varphi'(f_\theta(t)), \varphi(R_h))$ in a cost metric space that aligns well with the ground-truth hierarchical labels. Next, we show different ways for determining a hierarchy-aware cost matrix, and the detailed training process.

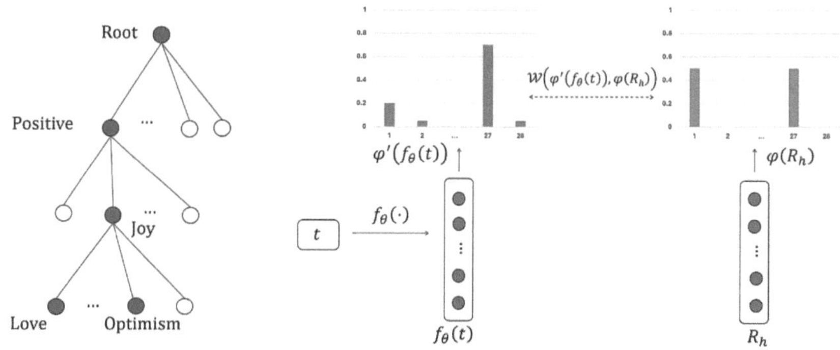

(a) The hierarchical labels (b) The EMD loss for quantifying emotional discrepancy

Fig. 1. An illustration of the proposed EEMD framework.

4.1 Hierarchy-Aware Cost Matrix

Tree-Based Cost Matrix. In Fig. 1a, we show the tree-structured hierarchical labels. The labels are first grouped into four main categories: positive, negative, ambiguous, and neutral. Each main category is further divided into more fine-grained categories. For the example text "*I like seeing this. There is hope. I want to hug this guy. He'll be alright*", its annotated labels are "*Love*" and "*Optimism*". Based on the tree-structured hierarchical labels, let i and j be the indices of emotion labels, we are able to define the tree distance $d(i, j)$ between two labels as the number of edges in the shortest path between the two label nodes, where each edge has a uniform weight of 1. For the tree-based cost matrix, we define the transportation cost of C_{ij} as the tree distance between the corresponding emotion labels within the hierarchical label tree. If i and j are the same, the cost is 0.

Embedding-Based Cost Matrix. A closer examination of the aforementioned tree-based cost matrix reveals that it fails to differentiate the transportation cost among sibling labels that are under the same parent label, since the tree distance between each pair of labels is identical. To cope with this shortcoming, we employ the embedding-based approach. Specifically, we embed each emotion label with a vector of 768 dimensions based on the pre-trained BERT model. For the embedding-based cost matrix, we define the transportation cost of C_{ij} as the Euclidean distance between the two vectors of the corresponding labels.

4.2 The Training Process

Algorithm 1 shows the detailed training steps of EEMD. Specifically, given a batch of text inputs, we first obtain their vector representations via a BERT-based encoder (i.e., Step-1). Next, the fully connected layer (FCL) maps these

Algorithm 1. The training process of EEMD

Input: The training data $\mathcal{D}$, multi-hot ground truth R_{h}, normalized multi-hot ground truth $\varphi(R_h)$, hierarchy-aware cost matrix $\mathbf{C}$, λ_1, λ_2
for each batch $\mathcal{B} \in \mathcal{D}$ **do**
  $\mathbf{H} \leftarrow \mathrm{BERT}(\mathbf{B})$ ▷ Step-1
  $\varphi'(f_\theta(t)) \leftarrow \mathrm{softmax}(\mathrm{FCL}(\mathbf{H}))$ ▷ Step-2
  $\mathcal{L}_B \leftarrow \mathrm{BCE}(\varphi'(f_\theta(t)), R_{\mathrm{h}})$ ▷ Step-3
  $\mathcal{L}_E \leftarrow \mathcal{W}(\varphi'(f_\theta(t)), \varphi(R_h))$ ▷ Step-4
  $(w_B, w_E) \leftarrow \mathrm{ALB}(\mathcal{L}_B, \mathcal{L}_E)$ ▷ Step-5
  Gradient backpropagation using the combined loss: $w_B \cdot \mathcal{L}_B + w_E \cdot \mathcal{L}_E$
end for

representations to logits. Unlike typical multi-label classification that uses a Sigmoid function for independent predictions, we intentionally apply a Softmax function to the logits (Step-2). This normalizes the outputs into a single probability distribution, compelling the model to learn the relative importance and relationships among all candidate emotions for a given input, rather than treating each as an independent binary decision. For Step-3 and Step-4, the binary cross-entropy (BCE) loss and the EMD loss are computed, respectively. In Step-5, a hybrid loss function by adaptively weighting BCE and EMD is deployed, which is referred to as an *adaptive loss balancer* (ALB) [17]. To dynamically balance the two distinct loss objectives, we adapt the uncertainty-based weighting method proposed by Kendall et al. [17]. This technique learns the relative weights of each loss term automatically during training by framing it as a multi-task learning problem where each loss corresponds to a task with its own homoscedastic uncertainty. This allows the model to down-weigh tasks with higher uncertainty, preventing a single, potentially noisy loss from dominating the training process and ensuring a stable balance between capturing binary classifications (BCE) and hierarchical relationships (EMD). This combination provides a balanced approach that enhances overall model performance.

The main reason for incorporating BCE is its effectiveness in handling independent label predictions. BCE directly optimizes the probability of each label, ensuring accurate predictions for individual emotions. This is crucial because it allows the model to address straightforward binary classifications efficiently. On the other hand, EMD is incorporated to capture the hierarchical structure and relationships between emotions, which cannot be captured by BCE alone.

5 Experimental Setup

In this section, we describe the experimental setup. First, we introduce the adopted dataset. Then, we detail the evaluation metrics. Finally, we describe the baseline methods, including their configurations and training procedures.

5.1 Dataset

This study utilizes the widely adopted benchmark dataset GoEmotions [8], which includes 58,000 English Reddit comments categorized into 28 emotion categories including *Neutral*. The dataset was carefully curated to mitigate biases and ensure quality, with comments ranging from 3 to 30 tokens in length and annotated through a rigorous multi-labeling process designed to balance emotion categories, rendering the dataset robust for nuanced emotion analysis. This dataset consists of texts with both single and multiple emotion labels, making it well-suited for fine-grained emotion analysis.

5.2 Evaluation Metrics

To provide an in-depth and comprehensive comparison, we use two groups of evaluation metrics.

Hierarchy-Agnostic Metrics. This group includes *F1 score*, *Subset accuracy* (SA) and *Hamming loss* (HL). The common characteristic of these metrics is that they treat each label independently and equally, regardless of whether some labels are hierarchically related. Specifically, **F1 score**, computed as the harmonic mean of precision and recall, balances these two aspects and provides a single measure of a model's performance, which is especially useful in scenarios with imbalanced datasets. In particular, we use three variants of F1: *Micro F1*, *Macro F1*, and *Weighted F1*. **Micro F1** calculates global metrics by aggregating the contributions of all classes. **Macro F1** computes the F1 for each class individually and then takes the unweighted average. **Weighted F1** computes the F1 for each class and then takes the average weighted by the number of true instances per class. **Subset accuracy** measures the proportion of exact matches between the predicted and true label sets, offering a strict metric that requires all predicted labels to match exactly with the true labels for a given instance. **Hamming loss** calculates the fraction of incorrect labels relative to the total number of labels, providing insight into the model's prediction error rate and accounting for false positives and false negatives in multi-label classification tasks.

Hierarchy-Aware Metric. In this work, we further propose using EMD as a hierarchy-aware evaluation metric to more realistically and meaningfully reflect prediction quality in the tree-structured label space. Specifically, we apply the hierarchy-aware cost matrix described in Sect. 4.1 when computing EMD over the two probability distributions derived from the prediction and standard labels.

5.3 Baseline Methods

In this study, we evaluate emotion analysis using both traditional methods based on pre-trained language models (e.g., BERT) and few-shot prompt-based methods using LLMs.

BERT with BCE Loss. The BERT-based model with a BCE loss function, as originally proposed with the GoEmotions [8] dataset, is adopted as a representative baseline. We employ the standard architecture without any modifications.

For model training, we use a batch size of 16, a learning rate of 5e-5, and no learning rate scheduler. For optimization, we employ the AdamW optimizer [23]. All experiments are conducted over 10 epochs.

LLM-oriented Methods. To investigate the effectiveness of prompt-based methods using LLMs for hierarchical emotional analysis, we evaluate several advanced LLMs, including GPT-3.5 Turbo, GPT-4 Turbo, and the Llama3 series. For Llama3, both the 8B and 70B versions are tested. We adopt two prompt design strategies: few-shot prompting with 2 examples and chain-of-thought (CoT) prompting. The few-shot strategy provides the LLM with instructions for the emotion analysis task and two carefully selected examples, one with a single-label annotation and one with multi-label annotations. These examples are chosen to cover a diverse range of emotions, ensuring that the model is exposed to varied scenarios. The CoT strategy breaks down the reasoning process into a series of intermediate steps, helping the model to follow a logical decision path. Additionally, to investigate the effect of incorporating hierarchical emotion information, each LLM is tested under two settings: (1) *Hierarchy-agnostic prompting*, where no hierarchy information is provided, serving as a control for assessing the capability of each model based on text only; and (2) *Hierarchy-aware prompting*, where the prompt includes hierarchical information about the relationships among emotion labels, allowing us to assess the influence of this additional context on predictive performance.

Figure 2 shows the prompt used in our experiments. In the prompt, *Emotion labels* represent the 28 emotion labels, *Hierarchical structure* provides the label hierarchy, and *Text* denotes the textual input. The inclusion of hierarchy information is controlled by a binary setting.

6 Results and Analysis

In this section, we report the experimental results and conduct a detailed analysis. In particular, we aim to demonstrate how effective EEMD is compared to the baseline methods and provide insights into why it achieves improved performance. In the following, we first describe the overall performance, and then explore how sensitive the performance of EEMD is to the combination of parameter settings and hierarchy information.

Table 1 shows the detailed results, where the best result for each metric is indicated in bold, and the second-best result is underlined. From Table 1, we can draw the following key insights:

The Superiority of EEMD: EEMD, which applies a novel hybrid loss function integrated with ALB, achieves state-of-the-art performance across multiple evaluation metrics, including traditional F1 scores and EMD. The improvements in terms of EMD score indicate that EEMD not only enhances performance but also enables the model to distinguish and quantify the severity of misclassifications, thereby producing predictions more closely aligned with the hierarchical ground-truth labels.

Analyze the following text using a "chain of thought" approach for multi-label emotion classification. Identify key phrases in the text that suggest emotional content. Then, explain how each phrase relates to specific emotions based on the categories provided below. If there is no specific emotion, return the label "Neutral".

Only use the following emotion categories: ***Emotion labels.txt***

Additionally, consider the hierarchical structure provided below to understand the relationships among the emotion categories: ***Hierarchical structure.json***

Hierarchy-aware setting

Few-shots:

Example 1:

Text: "I love you like a love song, baby. And if you know that song, it's now in your head."

Step 1: Identify key phrases - "love you", "love song".

Step 2: Relate phrases to emotions - "love you" and "love song" suggest strong positive emotions associated with Love.

Final Emotions: Love

Example 2:

Text: "That was hot!"

Step 1: Identify key phrases - "That was hot!".

Step 2: Relate phrases to emotions - "That was hot!" suggests Admiration and possibly Neutral.

Final Emotions: Admiration, Neutral

The text to be analyzed is: ***text.txt***

Fig. 2. The prompt used in our experiments.

The Comparison of Single Loss and Hybrid Loss: The models trained merely using the EMD loss function exhibit underperformance. Although those using only the BCE loss function achieve reasonable classification performance and form a strong baseline, they fail to consider the hierarchical structure among labels and thus cannot reflect the severity of misclassifications. The experimental results show that neither the BCE alone nor the EMD alone can yield the best performance. In contrast, EEMD combines the strong classification capability of BCE with the structure-awareness ability of tailored EMD. Compared to using a single loss function, incorporating the EMD loss allows the model to quantify misclassification severity more effectively.

The Comparison with LLMs: Advanced LLMs, including GPT-3.5 Turbo, GPT-4 Turbo, and Llama3 (8B and 70B), demonstrate inferior performance compared to traditional pre-trained language models optimized with the proposed methods. These LLMs commonly generate overly complex or irrelevant responses, which may lead to poor performance in fine-grained hierarchical emotion analysis. For example, given the text *"Ask him out for a drink,"*, of which the standard label is "Neutral", GPT-4 Turbo classifies it as "Desire" and "Optimism," clearly over-interpreting the intent. In another example, GPT-3.5 Turbo assigns labels not included in the 28 designated categories. The text *"I trust Safeway over Great Canadian Superstore for my meats. I know I'm not alone in that regard. Safeway has some pretty good cheaper meats."* was labeled by GPT-

Table 1. The performance of each method on GoEmotions. The best result in terms of each metric is indicated in bold, and the second-best is underlined.

Model	Hierarchy-aware	Micro F1	Macro F1	Weighted F1	SA	HL	EMD
BERT (BCE)	No	58.03%	46.32%	58.14%	44.93%	<u>0.034</u>	2.1133
GPT-3.5 Turbo	No	33.50%	30.37%	38.87%	15.88%	0.078	3.2179
GPT-4 Turbo	No	32.19%	31.92%	39.42%	11.76%	0.091	3.2794
Llama3-8B	No	26.03%	22.52%	30.60%	7.05%	0.091	3.4760
Llama3-70B	No	34.24%	34.59%	40.93%	9.41%	0.087	3.2569
EEMD (Embedding)	Yes	**61.18%**	**54.26%**	**60.12%**	**57.21%**	**0.032**	<u>2.0306</u>
EEMD (Hierarchical)	Yes	<u>59.09%</u>	<u>53.79%</u>	<u>58.80%</u>	<u>56.95%</u>	**0.032**	**2.0034**
EEMD (EMD)	Yes	30.75%	12.90%	21.59%	29.80%	0.058	2.7509
GPT-3.5 Turbo	Yes	37.66%	32.81%	42.90%	19.41%	0.068	2.8820
GPT-4 Turbo	Yes	32.96%	31.96%	38.89%	8.24%	0.089	3.2707
Llama3-8B	Yes	29.55%	24.51%	34.09%	8.24%	0.086	3.3643
Llama3-70B	Yes	36.27%	34.81%	39.95%	12.35%	0.079	3.1384

3.5 Turbo with the emotions of "Trust", "Approval", "Pride," and "Satisfaction". Unfortunately, "Trust" and "Satisfaction" do not belong to the standard label set. This type of behavior, commonly referred to as *over-interpretation* [5,40] or *hallucination* [16], is a known limitation of LLMs and contributes to their inferior performance in this task.

The Impact of Hierarchy-Awareness: In this work, we particularly evaluate the models under two settings: hierarchy-agnostic and hierarchy-aware. For the same model under the same parameter configuration, the hierarchy-aware version consistently outperforms the hierarchy-agnostic version across all evaluation metrics. This finding highlights the importance of incorporating structural information for fine-grained emotion analysis. On one hand, these results demonstrate the effectiveness of EEMD and highlight the benefits of leveraging the hierarchical structure in emotion analysis. On the other hand, they also indicate the limitations of LLMs when applied to tasks that require a deep understanding of structured label relationships. For instance, LLMs without structure awareness frequently misclassify similar emotions like "gratitude" and "realization," while incorporating hierarchy improves classification accuracy by better contextualizing the labels.

The Comparison of GPT and Llama3: The comparison among GPT-3.5 Turbo, GPT-4 Turbo, and Llama3 (8B and 70B) under identical conditions reveals some interesting patterns. The overall performance of GPT-4 Turbo is not better than that of GPT-3.5 Turbo, while the Llama3 70B version outperforms its 8B counterpart. The general expectation would be that a model with more parameters performs better. The Llama3 series follows this pattern, showing improved performance with increased model size. However, GPT-4 Turbo, despite presumably having more parameters and a more complex architecture, underperforms compared to GPT-3.5 Turbo in fine-grained emotion analysis.

This discrepancy implies that: (1) The model architecture and training data play crucial roles in final performance. GPT-4's architecture and training strategies may not be well suited to fine-grained emotion analysis. (2) While larger models can capture more complex patterns, they may also be prone to generating overly complicated responses if not properly tuned for the specific task. In contrast, the Llama3 series, particularly the 70B version, appears to benefit from their larger capacity in capturing subtle emotional cues. These observations also indicate the importance of task-specific tuning and optimization. Different models are suited to different tasks, and a one-size-fits-all approach may not yield optimal results. Therefore, it is essential to select and fine-tune models based on specific requirements. Moreover, balancing model complexity and task suitability is crucial for ensuring both effectiveness and efficiency.

7 Conclusion and Future Work

In this paper, we proposed a novel framework called EEMD to address the task of fine-grained hierarchical emotion analysis. The key idea is to utilize a tailored combination of EMD and BCE to measure the discrepancy between predicted labels and ground-truth labels. By imposing a hierarchy-aware cost matrix, we can incorporate the hierarchical relationships among emotion labels during both the training and evaluation processes, ensuring that misclassifications between hierarchically distant labels are penalized more heavily than those between adjacent ones. Experimental results demonstrate that EEMD achieves superior performance, significantly outperforming the baseline models. Moreover, our results show that LLM-based methods perform poorly on hierarchical emotion analysis, particularly when the label space is large (e.g., 28 labels in GoEmotions).

For future work, several practical directions are worth exploring. First, we plan to conduct additional experiments on more datasets and investigate the effectiveness of transfer learning in enhancing the robustness of EEMD. Second, we aim to further examine how to effectively integrate hierarchy information into LLM-oriented methods. A third direction is to explore non-uniform edge weights in the tree-based cost matrix, potentially learning the weights from data to better reflect semantic distances. Finally, we also intend to study the impact of contextual factors on fine-grained hierarchical emotion analysis, including persona information and dialogue context.

Acknowledgments. This research has been supported by the Project of Discretionary Budget of the Dean, Graduate School of Technology, Industrial and Social Sciences, Tokushima University.

References

1. Arjovsky, M., Chintala, S., Bottou, L.: Wasserstein generative adversarial networks. In: International Conference on Machine Learning, pp. 214–223. PMLR (2017)
2. Bostan, L., Kim, E., Klinger, R.: Goodnewseveryone: a corpus of news headlines annotated with emotions, semantic roles, and reader perception. In: Proceedings of the 12th Language Resources and Evaluation Conference, pp. 1554–1564 (2020)
3. Carneros-Prado, D., Villa, L., Johnson, E., Dobrescu, C.C., Barragán, A., García-Martínez, B.: Comparative study of large language models as emotion and sentiment analysis systems: a case-specific analysis of gpt vs. ibm watson. In: International Conference on Ubiquitous Computing and Ambient Intelligence, pp. 229–239. Springer (2023)
4. Chen, H., Liu, F., Yang, W.: Leveraging multi-task learning for fine-grained emotion classification. In: Proceedings of the 8th International Conference on Data Mining (ICDM), pp. 101–110 (2023)
5. Chen, M., Ma, Y., Song, K., Cao, Y., Zhang, Y., Li, D.: Learning to teach large language models logical reasoning. arXiv preprint arXiv:2310.09158 (2023)
6. Chun, J., Elkins, K.: explainable ai with gpt4 for story analysis and generation: a novel framework for diachronic sentiment analysis. Int. J. Digit. Humanities **5**(2), 507–532 (2023)
7. Cuturi, M.: Sinkhorn distances: lightspeed computation of optimal transport. In: Advances in Neural Information Processing Systems 26, pp. 2292–2300 (2013)
8. Demszky, D., Movshovitz-Attias, D., Ko, J., Cowen, A., Nemade, G., Ravi, S.: GoEmotions: a dataset of fine-grained emotions. In: 58th Annual Meeting of the Association for Computational Linguistics (ACL) (2020)
9. Ekman, P.: An argument for basic emotions. Cogn. Emotion **6**(3–4), 169–200 (1992)
10. Gui, L., Xu, R., Wu, D., Lu, Q., Zhou, Y.: Event-driven emotion cause extraction with corpus construction. In: Social Media Content Analysis: Natural Language Processing and Beyond, pp. 145–160. World Scientific (2018)
11. Guo, Y., Choi, J.: Enhancing cognitive models of emotions with representation learning. In: Proceedings of the Workshop on Cognitive Modeling and Computational Linguistics (CMCL), pp. 157–167 (2021)
12. Hazarika, D., Poria, S., Zimmermann, R., Mihalcea, R.: Conversational transfer learning for emotion recognition. Inf. Fusion **65**, 1–12 (2021)
13. Hu, H., Dey, D., Hebert, M., Bagnell, J.A.: Learning anytime predictions in neural networks via adaptive loss balancing. In: Proceedings of the 33rd AAAI Conference on Artificial Intelligence, pp. 3812–3821, July 2019. https://doi.org/10.1609/aaai.v33i01.33013812, https://ojs.aaai.org/index.php/AAAI/article/view/4268
14. Huang, G., Guo, C., Kusner, M.J., Sun, Y., Sha, F., Weinberger, K.Q.: Supervised word mover's distance. In: Advances in Neural Information Processing Systems, vol. 29 (2016)
15. Jin, H., Hou, L., Li, J., Dong, T.: Fine-grained entity typing via hierarchical multi-graph convolutional networks. In: Proceedings of the 2019 Conference on Empirical Methods in Natural Language Processing and the 9th International Joint Conference on Natural Language Processing (EMNLP-IJCNLP), pp. 4969–4978. Hong Kong, China (2019)
16. Jing, Z., et al.: When large language models meet vector databases: a survey. arXiv preprint arXiv:2402.01763 (2024)
17. Kendall, A., Gal, Y., Cipolla, R.: Multi-task learning using uncertainty to weigh losses for scene geometry and semantics. In: Proceedings of the IEEE Conference on Computer Vision and Pattern Recognition, pp. 7482–7491 (2018)

18. Kim, S., Park, J., Lee, S.: Hierarchical emotion classification using deep neural networks. In: Proceedings of the 10th International Conference on Artificial Intelligence (ICAI), pp. 567–576 (2021)
19. Kocoń, J., et al.: Chatgpt: jack of all trades, master of none. Inf. Fusion **99**, 101861 (2023)
20. Li, C., et al.: Large language models understand and can be enhanced by emotional stimuli. arXiv preprint arXiv:2307.11760 (2023)
21. Lian, R., Sethares, W.A., Hu, J.: Learning label hierarchy with supervised contrastive learning. In: Findings of the Association for Computational Linguistics: EACL 2024, pp. 1569–1581 (2024)
22. Liu, J., Wang, M., Zhou, L.: Contextualized fine-grained emotion recognition in social media texts. In: Proceedings of the 11th International Conference on Computational Linguistics (COLING), pp. 245–254 (2022)
23. Loshchilov, I., Hutter, F.: Decoupled weight decay regularization. arXiv preprint arXiv:1711.05101 (2017)
24. Lyu, Y., Tsang, I.W.: Curriculum loss: Robust learning and generalization against label corruption. In: Proceedings of the International Conference on Learning Representations (ICLR) (2020)
25. Minaee, S., et al.: Large language models: a survey. arXiv preprint arXiv:2402.06196 (2024)
26. Mitsios, M., et al.: Improved text emotion prediction using combined valence and arousal ordinal classification. In: Proceedings of the 2024 Conference of the North American Chapter of the Association for Computational Linguistics (NAACL), pp. 1234–1245 (2024)
27. Montavon, G., Müller, K.R., Cuturi, M.: Wasserstein training of restricted boltzmann machines. In: Advances in Neural Information Processing Systems, vol. 29 (2016)
28. Park, S., Kim, J., Ye, S., Jeon, J., Park, H.Y., Oh, A.: Dimensional emotion detection from categorical emotion. In: Proceedings of the 2021 Conference on Empirical Methods in Natural Language Processing (EMNLP), pp. 4367–4380. Online and Punta Cana, Dominican Republic (2021)
29. Patil, R., Gudivada, V.: A review of current trends, techniques, and challenges in large language models (llms). Appl. Sci. **14**(5), 2074 (2024)
30. Picard, R.W.: Affective Computing. MIT Press (2000)
31. Plutchik, R.: A general psychoevolutionary theory of emotion. In: Theories of Emotion, pp. 3–33. Elsevier (1980)
32. Pontiki, M., et al.: Semeval-2016 task 5: aspect based sentiment analysis. In: Proceedings of the 10th International Workshop on Semantic Evaluation (SemEval-2016), pp. 19–30. San Diego, California (2016)
33. Poria, S., Cambria, E., Hazarika, D., Majumder, N., Zadeh, A., Morency, L.P.: Context-dependent sentiment analysis in user-generated videos. In: Proceedings of the 55th Annual Meeting of the Association for Computational Linguistics, pp. 873–883 (2017)
34. Poria, S., Majumder, N., Hazarika, D., Cambria, E., Gelbukh, A., Hussain, A.: Multimodal sentiment analysis: addressing key issues and setting up the baselines. IEEE Intell. Syst. **33**(6), 17–25 (2018)
35. Radford, A., Narasimhan, K., Salimans, T., Sutskever, I.: Improving language understanding by generative pre-training. Technical Report, OpenAI (2018), technical Report
36. Rolet, A., Cuturi, M., Peyré, G.: Fast dictionary learning with a smoothed wasserstein loss. In: Artificial Intelligence and Statistics, pp. 630–638. PMLR (2016)

37. Singh, G., Brahma, D., Rai, P., Modi, A.: Text-based fine-grained emotion prediction. IEEE Trans. Affect. Comput. (2024)
38. Touvron, H., et al.: Llama: open and efficient foundation language models. arXiv preprint arXiv:2302.13971 (2023)
39. Wali, R.: Xtreme margin: a tunable loss function for binary classification problems. arXiv preprint arXiv:2211.00176 (2022)
40. Wang, K., Jing, Z., Su, Y., Han, Y.: Large language models on fine-grained emotion detection dataset with data augmentation and transfer learning. arXiv preprint arXiv:2403.06108 (2024)
41. Wang, L., Zhang, Y., Zhao, P.: Improving fine-grained emotion detection with attention mechanisms. In: Proceedings of the 9th International Conference on Machine Learning (ICML), pp. 789–798 (2021)
42. Wang, X., Li, X., Yin, Z., Wu, Y., Liu, J.: Emotional intelligence of large language models. J. Pac. Rim Psychol. **17**, 18344909231213960 (2023)
43. Xia, R., Ding, Z.: Emotion-cause pair extraction: a new task to emotion analysis in texts. arXiv preprint arXiv:1906.01267 (2019)
44. Yu, H.T., Jatowt, A., Joho, H., Jose, J.M., Yang, X., Chen, L.: Wassrank: listwise document ranking using optimal transport theory. In: Proceedings of the Twelfth ACM International Conference on Web Search and Data Mining, pp. 24–32 (2019)
45. Zadeh, A., Zellers, R., Pincus, E., Morency, L.P.: Multimodal sentiment intensity analysis in videos: facial gestures and verbal messages. IEEE Intell. Syst. **31**(6), 82–88 (2016)
46. Zhang, L., Wang, S., Liu, B.: Deep learning for sentiment analysis: a survey. Wiley Interdisc. Rev. Data Min. Knowl. Disc. **8**(4), e1253 (2018)
47. Zhang, W., Li, X., Chen, H.: Fine-grained emotion detection in online reviews using transformer models. In: Proceedings of the 12th International Conference on Natural Language Processing (ICON), pp. 123–132 (2023)
48. Zhang, W., Deng, Y., Liu, B., Pan, S.J., Bing, L.: Sentiment analysis in the era of large language models: a reality check. arXiv preprint arXiv:2305.15005 (2023)
49. Zhang, Y., Wang, M., Tiwari, P., Li, Q., Wang, B., Qin, J.: Dialoguellm: context and emotion knowledge-tuned llama models for emotion recognition in conversations. arXiv preprint arXiv:2310.11374 (2023)
50. Zhao, S., Jia, Z., Chen, H., Li, L., Ding, G., Keutzer, K.: Pdanet: polarity-consistent deep attention network for fine-grained visual emotion regression. In: Proceedings of the 27th ACM International Conference on Multimedia, pp. 192–201 (2019)
51. Zhou, H., Huang, M., Zhang, T., Zhu, X., Liu, B.: Emotional chatting machine: Emotional conversation generation with internal and external memory. In: Proceedings of the AAAI Conference on Artificial Intelligence, vol. 32 (2018)

Explain Before Classify: Contrastive Rationale Distillation for Academic Opinion Recognition

Mengting Zhang[1,2], Zhixiong Zhang[1,2](✉), Yajiao Wang[1,2], Yang Li[1,2], Xin Lin[1,2], and Meng Wang[1]

[1] National Science Library, Chinese Academy of Sciences, Beijing 100190, China
{zhangzhx,wangyajiao,liyang2022,linxin,wangmeng2022}@mail.las.ac.cn
[2] Department of Information Resources Management, School of Economics and Management, University of Chinese Academy of Sciences, Beijing 100190, China

Abstract. Academic opinion recognition aims to identify subjective expressions in scientific texts, particularly those conveyed through implicit reasoning or discourse-level cues. Existing approaches struggle with such subtle expressions, often relying on shallow lexical cues or flat classification, which neglect reasoning structures and limit interpretability. In contrast, we propose an explanation-driven distillation framework that captures academic subjectivity through linguistically grounded rationale templates. Drawing on linguistic theory, we summarize seven prototypical expression patterns and encode them into a three-step reasoning template, which guides large language models as teacher models to generate rationales. These rationales serve as the sole supervision signal for training a compact student model, from which labels are deterministically parsed. To enhance reasoning fidelity, we introduce a contrastive distillation strategy based on counterfactual rationales with inverted logic, encouraging the student model to distinguish valid from flawed explanations. Experiments demonstrate superior performance in opinion classification and explanation quality, with an F1-score of 89.10% and ROUGE-L of 74.25%. Ablation studies confirm that both rationale structure and the contrastive learning strategy are essential to achieve interpretability and robust opinion recognition.

Keywords: Academic Opinion Recognition · Rationale Distillation · Contrastive Learning

1 Introduction

Scientific literature conveys more than verifiable facts: it often reflects the author's evaluative stance on research topics, methods, or outcomes through implicit and inferential language. These expressions, which we refer to as academic opinions, are not casual impressions or strictly verifiable claims, but subjective judgments grounded in empirical contexts and shaped by theoretical

M. Yoshikawa et al. (Eds.): ADMA 2025, LNAI 16197, pp. 311–325, 2026.
https://doi.org/10.1007/978-981-95-3453-1_21

understanding or interpretive reasoning [1]. Identifying such expressions is crucial for understanding author perspective and supporting tasks like controversy detection, trend analysis, and scientific knowledge synthesis [2].

Among academic opinion, opinion-with-fact expressions are particularly subtle and challenging to detect. Though surface-level factual, they convey subjectivity through concessive constructions, contrastive logic, or discourse cues rather than overtly evaluative words. For example, "*While previous methods achieve stable performance on standard datasets, it remains unclear whether they generalize well to low-resource settings*" appears objective but implicitly questions the adequacy of prior work. Their factual surface often obscures subjectivity, blurring the boundary between opinion and fact [3]. Recent work in fact-checking, claim detection, and argument mining has begun to delineate such expressions to better support their own task objectives [4,5].

However, existing approaches to academic opinion recognition remain limited. Many rely on lexical features and lack systematic modeling of expression patterns, making it difficult to capture the diverse and implicit subjectivity in scholarly writing. Others treat it as sentence-level classification, focusing on label prediction while ignoring reasoning processes and offering no traceable decision basis. To compensate for the lack of reasoning supervision, recent studies have introduced LLM-generated rationales in tasks such as machine translation, summarization, and question answering. Although often fluent, these outputs suffer from semantic mismatch, incoherent logic, and weak structure. Their suitability for opinion recognition remains uncertain.

We propose an explanation-driven framework for academic opinion recognition that aims to identify implicit subjectivity in scholarly writing. Grounded in linguistic theory, we summarize seven prototypical expression patterns and encode them into a three-step reasoning template. Large language models serve as teacher models to generate structured rationales via this template, providing interpretable supervision for distilling reasoning ability into a compact student model. To improve reasoning fidelity, we introduce a contrastive distillation strategy with counterfactual rationales of similar structure but inverted logic, encouraging the model to distinguish between valid and invalid inference. Our contributions are threefold: (1) a structured rationale framework grounded in linguistic theory to model subjectivity in academic opinions; (2) a rationale-based distillation method that aligns predictions with structured reasoning steps from which labels are derived; and (3) a contrastive learning strategy that improves the model's ability to distinguish valid from flawed reasoning chains.

2 Related Work

2.1 Academic Opinion Recognition

Academic opinion recognition aims to identify subjective expressions in scientific literature, including evaluations of prior work, interpretations of current findings, and projections for future research. It is typically framed as a sentence-level binary classification task distinguishing subjective from objective expressions. Early work relied on lexicons and syntactic patterns: Wiebe et al. [6]

introduced a subjectivity annotation scheme based on private-state expressions (e.g., opinions, beliefs, emotions), while Xuan et al. [7] showed that syntax-based features improve classification. To overcome the limits of surface features, later approaches employed deep learning to capture contextual semantics. Satapathy et al. [8] proposed a BERT-based multitask framework jointly modeling subjectivity and polarity in sentiment analysis, using a neural tensor network for feature sharing. Biswas et al. [9] applied sentiment-oriented transformers to multilingual texts with preprocessing techniques such as POS tagging and attention masking. However, these models are designed for user-generated or sentiment-rich content, where subjectivity often aligns with polarity cues. Applied to scientific writing, they often fail to capture more implicit, fact-like opinions conveyed through concessive or contrastive structures. While recent work has begun to distinguish facts, factual claims, and opinions via verifiability and epistemic stance [3], most approaches still rely on shallow classification pipelines and lack interpretability.

2.2 Knowledge Distillation

Knowledge distillation transfers knowledge from large teacher models to smaller student models for efficient deployment under resource constraints [10]. With the rise of LLMs, recent work explores distilling their reasoning traces—rationales—as auxiliary supervision to improve interpretability and reasoning. This paradigm treats step-by-step explanations as process-level knowledge complementing traditional label supervision. Hsieh et al. [11] propose a multi-task model where the student jointly generates rationales and predicts labels, aligning explanation with prediction. Chen et al. [12] maximize mutual information between rationales and labels via a variational training objective. Wang et al. [13] use self-distillation with rationale replay to preserve generalization in machine translation. While rationale distillation is effective in QA and translation, it remains unexplored in academic opinion recognition, where modeling subjective reasoning is crucial.

3 Task Definition and Linguistic Grounding

3.1 Task Formulation

We formulate academic opinion recognition as an explanation-based classification task: instead of directly predicting the binary label, the model first generates a rationale that supports its decision. Formally, let $x_i \in \mathcal{X}$ denote a sentence from scientific texts, and let $y_i \in \mathcal{Y} = \{0, 1\}$ be its subjectivity label, where 1 denotes an *opinion sentence* and 0 denotes a *non-opinion sentence*. To enable explanation supervision, each instance is additionally paired with a natural language rationale r_i, generated by a teacher model to justify the label. This yields a rationale-augmented corpus $\mathcal{D}^{\text{rationale}} = \{(x_i, y_i, r_i)\}_{i=1}^{N}$, where N is the number of labeled training instances.

Rather than learning a direct mapping from $\mathcal{X}$ to $\mathcal{Y}$, we decompose the task into two stages: the model first generates a rationale that captures the underlying subjectivity, and then infers the classification label based on this explanation. This process can be described as a composition of two mappings:

$$\mathcal{X} \xrightarrow{f^{(1)}} \mathcal{R} \xrightarrow{f^{(2)}} \mathcal{Y} \tag{1}$$

3.2 Subjective Expression Patterns in Academic Writing

In scientific writing, subjective opinions are often conveyed through patterned linguistic strategies. To support interpretable academic opinion recognition, we summarize seven expression modes signaling subjectivity. Grounded in linguistic theory, these modes provide the semantic structure for our rationale framework.

(1) Explicit Authorial Stance. Typically conveys the author's stance using first-person pronouns and epistemic verbs. In Hyland's [14] stance framework, such markers show commitment and responsibility for the proposition. Phrases like "we argue", "we believe", or "we find that" foreground the author's presence and interpretive involvement.

(2) Epistemic Modality. Encodes epistemic stance via hedging that signals uncertainty without asserting factuality. In Palmer's [15] typology of *epistemic modality*, this includes modal verbs (e.g., "may", "might") and adverbials (e.g., "possibly", "likely") qualifying a proposition's truth status.

(3) Subjective Value Evaluation. Conveys value-based appraisal of a method, result, or concept in terms of quality, effectiveness, or significance. In Appraisal Theory [16], it aligns with *Appreciation*, which concerns evaluations of entities and processes. Terms such as "effective", "promising", or "insufficient", sometimes modified by intensifiers, to signal positive or critical stance.

(4) Subjective Comparison. Uses comparative constructions to imply preference or superiority. In Appraisal Theory [16], these fall under the Graduation subsystem, which captures how attitudinal meanings are adjusted in strength or intensity. Expressions like "outperforms" or "more efficient than' express a value judgment by ranking one entity above another.

(5) Implicit Critical Discourse. Conveys critique through concessive, restrictive, or negative discourse. As noted by Hunston and Thompson [17], concessive markers (e.g., "however", "although", "nevertheless") act as *counter-expectation* signals rejecting prior assumptions, while negative formulations (e.g., "fails to", "lack of", "does not consider") point to insufficiency or gaps.

(6) Subjective Causal Inference. Attributes causality based on interpretation rather than empirical verification. Sweetser's [18] framework distinguishes three causal domains: the *content domain* (objective world relations), the *epistemic domain* (speaker's reasoning), and the *speech-act domain* (justifying speech acts). In academic writing, phrases like "due to" or "attributed to" often function in the *epistemic domain*, signaling inference from evidence.

(7) Future-Oriented Projection. Extends beyond current findings to anticipate developments, applications, or open questions. Bybee et al. [19] distinguish *future certainty* and *future possibility* by confidence level. Phrases like "can be extended", "has the potential to", or "should be further explored" represent not temporal facts, but intentional projections.

4 Methodology

We propose a structured rationale distillation framework for academic opinion recognition, which extracts interpretable reasoning patterns from LLMs and distills them into student models. The overall architecture is illustrated in Fig. 1.

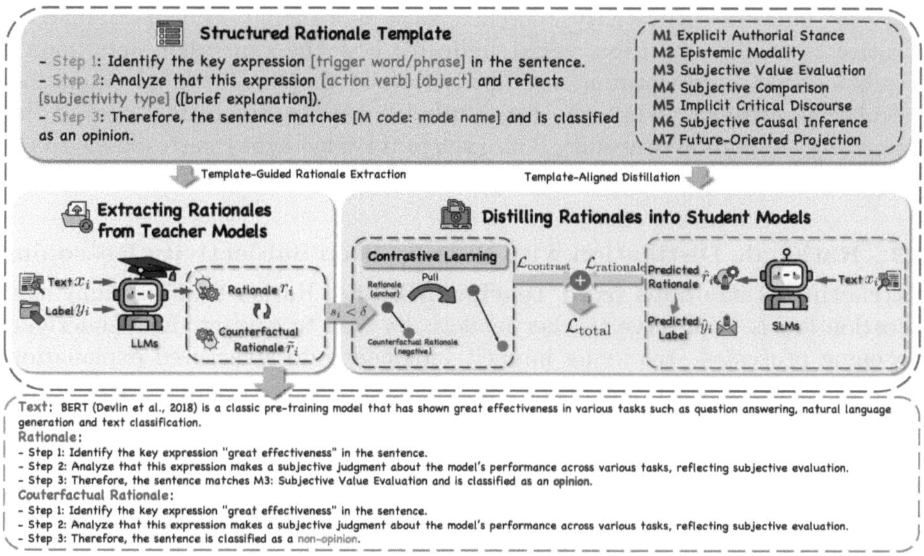

Fig. 1. The overall architecture of our method.

4.1 Modeling Subjectivity Through Structured Rationale Templates

To support interpretable academic opinion recognition, we model how linguistic subjectivity is encoded and linked to classification. The seven expression patterns in Sect. 3.2 are systematized into a set of subjectivity modes $\mathcal{M} = \{M_1, \ldots, M_7\}$, defining a semantic space for reasoning about how authors express stance. As a sentence may involve multiple strategies, our framework allows alignment with one or more modes in $\mathcal{M}$.

Based on this space, subjectivity reasoning is formulated as a constrained natural language explanation task, where classification is derived from rationale structures grounded in linguistic cues. Each rationale $r \in \mathcal{R}$ is formalized as a triplet $r = (s_1, s_2, s_3)$, representing a three-step interpretive chain: identifying a surface-level trigger expression (s_1); analyzing how this expression performs a specific subjective action upon a linguistic target and reflects a particular subjectivity strategy (s_2); and concluding with a classification decision that explicitly maps the sentence to one or more subjectivity modes in $\mathcal{M}$ (s_3). This triplet is operationalized via a structured template:

- **Step 1:** *Identify the key expression* `[trigger word/phrase]` *in the sentence.*

- **Step 2:** *Analyze that this expression* `[action verb] [object]` *and reflects* `[subjectivity type]` (*with brief explanation*).
- **Step 3:** *Therefore, the sentence matches* `[M code: mode name]` *and is classified as opinion/non-opinion.*

In this template, the `[trigger word/phrase]` refers to the surface expression that initiates subjectivity inference, such as a modal, evaluative, or comparative term. The `[action verb]` indicates how the expression acts upon a linguistic target by performing a subjectivity-related operation (e.g., evaluation, speculation, critique), while the `[object]` denotes the linguistic target of this action (e.g., a method or result). The `[subjectivity type]` corresponds to the modes in $\mathcal{M}$, serving as the semantic anchor for the final decision.

4.2 Rationale Distillation with Step-by-Step Subjectivity Reasoning

Extracting Rationales from Teacher Models. Rather than distilling classification labels from large teacher models, we seek to extract their underlying reasoning processes—rendering implicit judgment into structured explanation. The objective is not to imitate decisions, but to encode the interpretive path that leads to them. To this end, we prompt a teacher model to generate a structured rationale $r_i \in \mathcal{R}$ for each labeled academic sentence (x_i, y_i), where $x_i \in \mathcal{X}$ is the input sentence and $y_i \in \{0, 1\}$ is the manually assigned subjectivity label. The rationale r_i is generated under the condition of a fixed prompt template, given the input sentence and its gold label as content inputs. This template enforces consistency with the subjectivity reasoning schema defined in Sect. 4.1:

$$r_i = \texttt{TeacherGenerate}(x_i, y_i \mid \texttt{Prompt}_{\text{template}}) \tag{2}$$

For opinion sentences ($y_i = 1$), the generated rationale explicitly links a surface-level linguistic expression to one or more subjectivity modes $M_k \in \mathcal{M}$, as defined in Sect. 3.2. This alignment ensures that the explanation reflects not only the presence of subjectivity, but also its linguistic realization. For non-opinion sentences ($y_i = 0$), the rationale justifies the absence of such expressions under the same structured template, indicating the lack of subjective stance. In both cases, the output rationale conforms to a fixed explanatory structure, enabling interpretable supervision and facilitating the downstream model's acquisition of mode-aware classification behavior. Instead of relying solely on discrete labels, the model learns to associate subjectivity with identifiable linguistic cues.

Distilling Rationales into Student Models. To transfer the interpretive reasoning ability of large teacher models into smaller student models, we adopt a rationale-centered distillation framework. Instead of directly supervising the student with classification labels, we guide it to generate structured rationales $\hat{r}_i$ that follow the same three-step subjectivity reasoning template used by the teacher to explain its own predictions.

$$\hat{r}_i = \texttt{StudentGenerate}(x_i \mid \texttt{Prompt}_{\text{template}}) \tag{3}$$

The rationale $\hat{r}_i$ serves as an intermediate explanation from which the final label $\hat{y}_i$ is deterministically derived via a rule-based parser:

$$\hat{y}_i = \texttt{ParseLabel}(\hat{r}_i) \tag{4}$$

Counterfactual Contrastive Learning. We introduce a contrastive distillation objective to regularize the explanation space using *counterfactual rationales*—adversarial explanations constructed by inverting the final classification clause in the rationale template (e.g., from "*classified as an opinion*" to "*classified as a non-opinion*"), while preserving surface fluency. Unlike conventional counterfactuals that flip model outputs via input perturbations [20], our counterfactuals alter the reasoning outcome within a fixed explanation structure. These rationales are coherent but logically incorrect, serving as negative examples. Let r_i be a gold rationale and $\tilde{r}_i$ its counterfactual. We encode both using the decoder's hidden state at the first generation step:

$$\mathbf{z}_i = f_{\text{dec}}(r_i), \quad \tilde{\mathbf{z}}_i = f_{\text{dec}}(\tilde{r}_i) \tag{5}$$

where $f_{\text{dec}}(\cdot)$ denotes the decoder embedding function. We then compute the cosine similarity between the normalized embeddings:

$$s_i = \cos < \mathbf{z}_i, \tilde{\mathbf{z}}_i >= \frac{\mathbf{z}_i^\top \tilde{\mathbf{z}}_i}{||\mathbf{z}_i||\,||\tilde{\mathbf{z}}_i||} \tag{6}$$

To focus contrastive supervision on semantically ambiguous regions, we apply a similarity-based mask $\mathbb{I}_{[s_i<\delta]}$, where s_i denotes the cosine similarity between the gold rationale and its counterfactual, and δ is a predefined similarity threshold. The mask activates when $s_i < \delta$, retaining rationale pairs that are structurally similar yet semantically conflicting. This filtering excludes trivially dissimilar examples and ensures that contrastive learning emphasizes the most challenging distinctions. We compute a margin-based triplet loss over a batch of B samples, where the gold rationale serves as both anchor and positive, and its counterfactual acts as the negative.

$$\begin{aligned} \mathcal{L}_{\text{contrast}} &= \frac{1}{B}\sum_{i=1}^{B} \mathbb{I}_{[s_i<\delta]} \cdot \max\left(0, m + \cos < \mathbf{z}_i, \tilde{\mathbf{z}}_i > - \cos < \mathbf{z}_i, \mathbf{z}_i >\right) \\ &= \frac{1}{B}\sum_{i=1}^{B} \mathbb{I}_{[s_i<\delta]} \cdot \max\left(0, m + s_i - 1\right) \end{aligned} \tag{7}$$

The final training objective integrates contrastive regularization into explanation generation:

$$\mathcal{L}_{\text{total}} = \mathcal{L}_{\text{rationale}} + \gamma \cdot \mathcal{L}_{\text{contrast}} \tag{8}$$

where $\mathcal{L}_{\text{rationale}}$ supervises explanation generation based on teacher outputs, while $\mathcal{L}_{\text{contrast}}$ encourages semantic separation between valid and invalid reasoning. The hyperparameter γ controls the strength of contrastive regularization.

5 Experiments

5.1 Experimental Setup

Dataset. Existing public datasets for opinion mining are mostly derived from social media or restricted to abstracts, lacking coverage of subjective expression in full scientific texts. Therefore, we construct a new dataset of 10,000 sentences sampled from 157 full papers randomly selected from the ACL Anthology (2024). To mitigate label imbalance, we extract samples from sections that are typically rich in subjective content, including *Abstract*, *Introduction*, *Related Work*, *Discussion*, and *Conclusion*. Two PhD students with relevant domain expertise annotate each sentence following the seven subjectivity expression modes defined in Sect. 3.2, with final labels binarized as opinion (1) or non-opinion (0). Inter-annotator agreement achieves a Cohen's kappa of 0.901, and disagreements are resolved through adjudication. The resulting dataset contains 5,246 non-opinion and 4,754 opinion sentences, partitioned into training, validation, and test sets with an 8:1:1 ratio, as summarized in Table 1.

Table 1. Statistical Information of Our Dataset.

Label	Dataset			
	Train	Valid	Test	Total
Opinion	3,780	490	482	4,754
Non-Opinion	4,220	510	518	5,246
Total	8,000	1,000	1,000	10,000

To provide interpretive supervision for student models, we use GPT-4o as the teacher to generate structured rationales for each annotated sentence. A fixed prompt template is used across all examples, which includes an instruction, the input sentence with its gold label, and two illustrative exemplars: one opinion sentence and one non-opinion sentence, each paired with a rationale that follows our three-step reasoning template. GPT-4o is instructed to generate explanations consistent with this template, ensuring alignment with the subjectivity patterns defined in Sect. 3.2. The prompt is shown in Fig. 2.

Evaluation Metrics. To comprehensively assess both classification accuracy and explanation quality, we adopt a dual-metric strategy encompassing label-level and rationale-level evaluation. For label prediction, we report Accuracy, macro Precision, Recall, and F1-score, which reflect the model's ability to distinguish subjective from objective statements across categories. For rationale evaluation, we compute ROUGE scores (ROUGE-1, ROUGE-2, and ROUGE-L) by comparing model-generated explanations with gold rationales produced by the teacher model. ROUGE captures lexical and structural overlap at various n-gram and sequence levels, thus measuring the faithfulness and informativeness of the explanations. This combination of metrics provides a holistic evaluation of the model's performance in both prediction and explanation tasks.

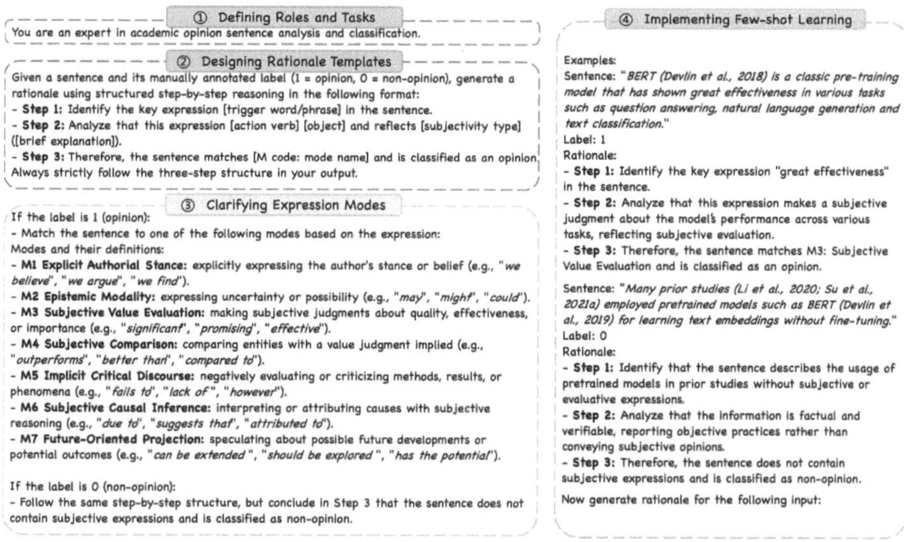

Fig. 2. Structure of the prompt used to elicit rationales from GPT-4o.

Implementation Details. We adopt the PyTorch framework (v 1.12.1) with the Transformers library (v 4.24.0) to develop our models, running on an A100-80 GB GPU with Python 3.8. We use the AdamW optimizer with an initial learning rate of 5e-5 and apply a linear learning rate decay schedule. The batch size is set to 8, and training lasts up to 50 epochs. For contrastive distillation, the margin is set to 0.5, the contrastive weight γ is fixed at 1, and the similarity threshold δ for identifying counterfactual rationales is set to 0.8, selected for the best performance on the validation set. For all baseline models, early stopping is triggered when the validation macro F1-score does not improve for 10 consecutive epochs. In contrast, our model adopts ROUGE-L on the validation set as the early stopping criterion to better align with rationale quality, since the final classification label is deterministically parsed from the generated rationale. This strategy helps mitigate instability due to minor decoding variations and ensures consistent supervision through explanation-oriented guidance.

5.2 Baseline Models

We compare our method against two representative baselines: standard supervised fine-tuning and rationale-based step-by-step distillation.

Supervised Fine-Tuning. We fine-tune a range of pretrained models with varying parameter scales, including BERT [21], SciBERT [22], RoBERTa [23], and the T5 family (small, base, large, xl) [24], on the binary opinion classification task. Encoder-based models use a classification head on top of the [CLS] representation, while T5-based models treat classification as text-to-text generation.

All models are trained solely on gold labels without any rationale supervision, providing performance baselines under purely label-driven learning.

Step-by-Step Distillation. We also implement the rationale-based distillation framework from Hsieh et al. [11], where a single model is prompted with different prefixes to generate rationales and classification labels. Their method uses a multi-task loss that jointly optimizes rationale generation (guided by teacher-provided explanations) and label prediction (based on gold annotations), with a weighting coefficient λ controlling the trade-off between the two. In our experiments, we adopt a variant of this formulation, where the loss is defined as $\mathcal{L} = \lambda\mathcal{L}_{\text{label}} + (1-\lambda)\mathcal{L}_{\text{rationale}}$. Our final model corresponds to the setting $\lambda = 0$, i.e., training is driven purely by rationale generation loss, and additionally incorporates a contrastive loss $\mathcal{L}_{\text{contrast}}$ to enhance the discriminative power of explanation representations. This separation encourages faithful reasoning and enhances interpretability by disentangling explanation from prediction.

5.3 Comparative Evaluation

Performance of Supervised Fine-Tuning. We evaluate standard fine-tuning across a range of pretrained language models. As shown in Fig. 3, T5-base achieves the best overall performance, with the highest macro F1-score (88.30%). Among encoder-based models, RoBERTa performs best (87.68%), slightly ahead of BERT (87.40%) and SciBERT (87.39%), which may be attributed to its dynamic masking and diverse pretraining data that improve generalization. T5-small underperforms (86.59%), likely due to its limited capacity. Interestingly, T5-large (87.19%) and T5-xl (86.39%) perform worse than T5-base, suggesting that scaling model size without appropriate regularization may lead to overfitting. Overall, T5-base offers a favorable balance between model capacity and robustness, and is therefore selected as the backbone for our distillation framework.

Performance of Step-by-Step Distillation. We further evaluate the step-by-step distillation framework under different values of the balancing coefficient λ, which controls the relative weight between label prediction and rationale generation. As shown in Table 2, classification performance remains relatively stable across a broad range of λ values, while the quality of generated explanations improves significantly as λ decreases.

In most λ settings, classification performance surpasses that of direct fine-tuning, suggesting that incorporating rationale supervision can improve predictive accuracy. At $\lambda = 0.5$, the model achieves the highest F1-score (89.00%) but produces uninformative rationales (ROUGE-2: 17.90%), indicating that excessive emphasis on label supervision suppresses explanation quality. As λ decreases, the model allocates more capacity to rationale modeling, leading to consistent gains in ROUGE scores. The best explanation quality is observed at $\lambda = 0.1$, with ROUGE-L reaching 72.93% while maintaining a competitive F1-score of

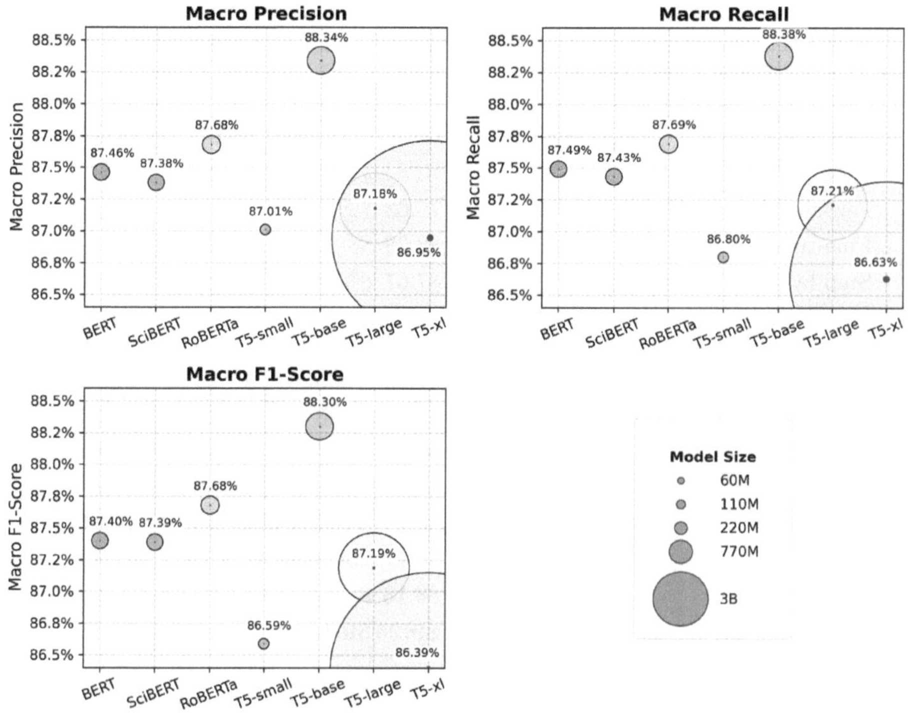

Fig. 3. Performance of supervised fine-tuning across different pretrained models. Bubble size indicates model scale (from 60M to 3B parameters).

88.00%, highlighting the potential of explanation-oriented supervision to balance interpretability and accuracy effectively.

Our method corresponds to the case where $\lambda = 0$, i.e., training is fully guided by rationale generation, with classification labels deterministically parsed. Moreover, we introduce contrastive learning to enhance the discriminability of explanation representations. As a result, our model achieves the highest F1-score (89.10%) and surpasses all baselines in rationale quality (ROUGE-L: 74.25%). These results demonstrate that treating explanation as the primary learning target—rather than a secondary constraint—not only improves interpretability, but also enhances generalization in opinion classification.

5.4 Ablation Study

We conduct ablation experiments to assess the effects of contrastive learning (**CL**) and structured rationale guidance (**SR**), where SR refers to the three-step template introduced in Sect. 4.1, designed based on observed patterns of academic opinion expression. Removing SR means the teacher model generates free-form rationales without being guided by our structured reasoning schema. This leads to a clear performance drop from 89.10% to 88.26% in F1-score,

Table 2. Comparison of classification and explanation performance across methods. We report Acc (Accuracy), P (Precision), R (Recall), F1 (F1-score), and ROUGE metrics (R-1, R-2, R-L).

Method	λ	Label				Rationale		
		Acc(%)	P(%)	R(%)	F1(%)	R-1(%)	R-2(%)	R-L(%)
Few-shot Learning	–	74.80	81.74	75.63	73.73	–	–	–
Fine-tuning	–	88.30	88.34	88.38	88.30	–	–	–
Step-by-step Distillation	0.9	87.70	87.69	87.67	87.68	47.14	6.91	36.64
	0.8	88.90	88.90	88.95	88.90	47.21	7.99	37.52
	0.7	88.70	88.68	88.72	88.69	47.75	10.45	39.76
	0.6	88.40	88.42	88.46	88.40	49.75	13.21	41.50
	0.5	<u>89.00</u>	<u>89.03</u>	<u>89.07</u>	<u>89.00</u>	53.87	17.90	47.82
	0.4	88.80	88.80	88.85	88.80	64.37	36.35	56.86
	0.3	85.50	86.30	85.77	85.47	76.07	60.18	70.81
	0.2	86.50	87.27	86.77	86.48	78.17	63.52	72.75
	0.1	88.00	88.33	88.18	88.00	<u>78.30</u>	<u>63.62</u>	<u>72.93</u>
Ours	–	**89.10**	**89.31**	**89.25**	**89.10**	**79.18**	**65.25**	**74.25**

highlighting the importance of rationale supervision grounded in subjectivity expression patterns. ROUGE scores are not reported in this setting (nor in w/o ALL), since the rationales used for supervision during training are no longer aligned with the gold three-step rationales, making direct comparison inappropriate. Removing CL results in a slight drop in F1-score but an increase in ROUGE scores, suggesting that contrastive supervision encourages the model to distinguish semantically inconsistent explanations, thereby improving prediction quality at the cost of surface-level similarity. When both SR and CL are removed, the F1-score further declines to 86.86%, confirming their complementary roles. Specifically, SR offers interpretable structural constraints grounded in reasoning patterns, whereas CL improves the model's ability to reject implausible rationales and focus on decision-consistent explanations (Table 3).

5.5 Case Study

To better understand model behavior, we present a representative case in Table 4. The input sentence describes an approach that "protects LLM from being misled by distracting information" and "avoids adding extra input tokens," implicitly highlighting the method's benefits. Our model correctly classifies the sentence as an opinion (*M3: Subjective Value Evaluation*) and generates a rationale that explicitly interprets the evaluative implication – e.g., *it implies a positive assessment of the approach's effectiveness.* This level of semantic reasoning even surpasses the gold rationale, which lists key phrases but lacks interpretive depth.

Interestingly, all ablated variants recognize the benefit-related content in their rationales. However, only the variant with contrastive learning (w/o SR) cor-

Table 3. Ablation study on the impact of Contrastive Learning (CL) and Structured Rationale (SR).

Method	Label				Rationale		
	Acc(%)	P(%)	R(%)	F1(%)	R-1(%)	R-2(%)	R-L(%)
Ours (full model)	**89.10**	**89.31**	**89.25**	**89.10**	79.18	65.25	74.25
w/o CL	89.00	89.22	89.15	89.00	**79.37**	**65.81**	**74.57**
w/o SR	88.30	88.41	88.20	88.26	–	–	–
w/o ALL	86.90	86.98	86.81	86.86	–	–	–

Table 4. Case studies comparing our model with ablated variants.

Case: *This approach protects LLM from being misled by distracting information, and avoids adding extra input tokens to LLM.*

Gold Rationale: - *Step 1: Identify the key expression "protects LLM from being misled by distracting information" in the sentence. - Step 2: Analyze that this expression makes a subjective judgment regarding the effectiveness of the approach in avoiding misleading information, reflecting subjective evaluation. - Step 3: Therefore, the sentence matches M3: Subjective Value Evaluation and is classified as an* ***opinion***.

Ours	(✓) - *Step 1*: Identify the key expression "protects LLM from being misled" in the sentence. - *Step 2*: Analyze that this expression makes a subjective judgment about the effectiveness of the approach in preventing misled LLMs, reflecting subjective evaluation (*it implies a positive assessment of the approach's effectiveness*). - *Step 3*: Therefore, the sentence matches *M3: Subjective Value Evaluation* and is classified as an ***opinion***.
w/o **SR**	(✓) The sentence describes the ***benefits of an approach*** by stating that it "protects LLM from being misled by distracting information" and "avoids adding extra input tokens to LLM." The use of terms like "protects" and "avoids" suggests a *positive evaluation* of the approach's effectiveness and efficiency. This sentence is classified as an ***opinion***.
w/o **CL**	(✗) - *Step 1*: Identify that the sentence describes the function and ***benefits of an approach*** without subjective or evaluative expressions. - *Step 2*: Analyze that the information is factual and verifiable, explaining the technical aspects of the approach rather than conveying subjective opinions. - *Step 3*: Therefore, the sentence does not contain subjective expressions and is classified as ***non-opinion***.
w/o **ALL**	(✗) The sentence provides a factual description of the ***benefits of the approach***, specifically its ability to protect LLM from misled by distracting information and avoid adding extra input tokens. It does not express any subjective judgment or evaluation. Therefore, this sentence is classified as ***non-opinion***.

rectly links this recognition to an opinion prediction. The others (w/o CL, w/o ALL) misclassify the sentence as non-opinion, indicating a disconnect between recognizing benefit and interpreting it as subjective. Although w/o SR makes the correct prediction, its rationale fails to capture the implicit subjectivity. This case highlights the role of contrastive learning in reinforcing the connection between subtle evaluative language and opinion classification, and demonstrates our model's superiority in both prediction accuracy and rationale quality.

6 Conclusion

In this work, we address the challenge of recognizing implicit subjective expressions in scientific texts with a structured, explanation-driven framework. Our approach models academic subjectivity through linguistically grounded rationale templates and distills step-by-step reasoning into a compact student model. To enhance reasoning fidelity, we introduce a contrastive learning objective based on counterfactual rationales, encouraging the model to distinguish coherent and flawed explanations. Experimental results show superior performance in both opinion classification and explanation quality, and ablation studies verify the complementary roles of structured rationales and the contrastive learning strategy. While the current evaluation is conducted on a single-domain dataset with one teacher model due to computational cost considerations, the results underscore the value of integrating explanation structure with reasoning discrimination, offering a robust and interpretable solution to academic opinion recognition.

Acknowledgments. This work is supported by the National Key Research and Development Program of China: "Key Technologies and Software for Deep Mining and Intelligent Analysis of Scientific Literature Content" (Grant No.2022YFF0711900).

Disclosure of Interests. The authors have no competing interests to declare that are relevant to the content of this article.

References

1. Govier, T.: A Practical Study of Argument, 7th edn. Wadsworth, Cengage Learning (2010)
2. Hariri, W.: Sentiment analysis of citations in scientific articles using ChatGPT: identifying potential biases and conflicts of interest. arXiv preprint arXiv:2404.01800 (2024)
3. Ni, J., Shi, M., Stammbach, D., et al.: AFaCTA: assisting the annotation of factual claim detection with reliable LLM annotators. arXiv preprint arXiv:2402.11073 (2024)
4. Reddy, R. G., Chetan, S., Wang, Z., et al: NewsClaims: a new benchmark for claim detection from news with attribute knowledge. arXiv preprint arXiv:2112.08544 (2021)
5. Lawrence, J., Reed, C.: Argument mining: a survey. Comput. Linguist. **45**(4), 765–818 (2020)

6. Wiebe, J., Wilson, T., Cardie, C.: Annotating expressions of opinions and emotions in language. Lang. Resour. Eval. **39**, 165–210 (2005)
7. Xuan, H.N.T., Le, A.C., Nguyen, L.M.: Linguistic features for subjectivity classification. In: Proceedings of the 2012 International Conference on Asian Language Processing, pp. 17–20. IEEE (2012)
8. Satapathy, R., Pardeshi, S.R., Cambria, E.: Polarity and subjectivity detection with multitask learning and BERT embedding. Future Internet **14**(7), 191 (2022)
9. Biswas, M.R., Abir, A.T., Zaghouani, W.: Nullpointer at CheckThat! 2024: identifying subjectivity from multilingual text sequence. arXiv preprint arXiv:2407.10252 (2024)
10. Hinton, G., Vinyals, O., Dean, J.: Distilling the knowledge in a neural network. arXiv preprint arXiv:1503.02531 (2015)
11. Hsieh, C.-Y., Li, C.-L., Yeh, C.-K., et al.: Distilling step-by-step! outperforming larger language models with less training data and smaller model sizes. arXiv:2305.02301 (2023)
12. Chen, X., Huang, H., Gao, Y., Wang, Y., Zhao, J., Ding, K.: Learning to maximize mutual information for chain-of-thought distillation. In: Proceedings of the 62nd Annual Meeting of the Association for Computational Linguistics (ACL), pp. 1245–1259 (2024)
13. Wang, J., Zhao, Y., Xu, Y., Liu, B., Zong, C.: Boosting LLM translation skills without general ability loss via rationale distillation. arXiv preprint arXiv:2410.13944 (2024)
14. Hyland, K., Tse, P.: Metadiscourse in academic writing: a reappraisal. Appl. Linguis. **25**(2), 156–177 (2004)
15. Palmer, F.R.: Mood and Modality. Cambridge University Press, Cambridge (2001)
16. Martin, J.R., White, P.R.: The Language of Evaluation, vol. 2. Springer, Berlin (2003)
17. Hunston, S., Thompson, G.: Evaluation in Text: Authorial Stance and the Construction of Discourse. Oxford University Press, Oxford (2000)
18. Sweetser, E.: From Etymology to Pragmatics: Metaphorical and Cultural Aspects of Semantic Structure, vol. 54. Cambridge University Press, Cambridge (1990)
19. Bybee, J.L., Perkins, R.D., Pagliuca, W.: The Evolution of Grammar: Tense, Aspect, and Modality in the Languages of the World, vol. 196. University of Chicago Press, Chicago (1994)
20. Wachter, S., Mittelstadt, B., Russell, C.: Counterfactual explanations without opening the black box: automated decisions and the gdpr. Harvard J. Law Technol. **31**, 841 (2017)
21. Devlin, J., Chang, M.-W., Lee, K., et al: BERT: pre-training of deep bidirectional transformers for language understanding. In: NAACL-HLT, pp. 4171–4186 (2019)
22. Beltagy, I., Lo, K., Cohan, A.: SciBERT: a pretrained language model for scientific text. In: EMNLP-IJCNLP, pp. 3615–3620 (2019)
23. Liu, Y., Ott, M., Goyal, N., et al.: RoBERTa: a robustly optimized BERT pretraining approach. arXiv:1907.11692 (2019)
24. Raffel, C., Shazeer, N., Roberts, A., et al.: Exploring the limits of transfer learning with a unified text-to-text transformer. J. Mach. Learn. Res. **21**(140), 1–67 (2020)

ZeroRFF: A Random Fourier Features and Analytic Learning Method for Generalized Class Incremental Learning

Duc-Hung Nguyen, Tri-Thanh Nguyen, Thanh-Hai Dang, and Quynh-Trang Pham Thi(✉)

Vietnam National University, University of Engineering and Technology, Hanoi, Vietnam
{22021109,ntthanh,hai.dang,trangptq}@vnu.edu.vn

Abstract. Continual learning aims to train models across multiple tasks, each containing different knowledge. The biggest challenge in continual learning is that models quickly lose previously learned knowledge when acquiring new tasks, as they lack access to their old knowledge. Generalized Class Incremental Learning (GCIL) poses the challenge that knowledge may not be uniform and may overlap between tasks. Some current methods store some old knowledge to prevent the model from forgetting, but this can violate data security. This research proposes the ZeroRFF (Zero-Memory Random Fourier Features) method, which solves the generalized class incremental learning problem without storing old knowledge (exemplar-free). This method uses analytic learning techniques; instead of relying on gradient descent, ZeroRFF finds optimal solutions by directly solving equations where the derivative equals zero. Additionally, ZeroRFF integrates Random Fourier Features (RFF) to transform linearly non-separable data into a new space where the data becomes linearly separable, enhancing the model's classification capability. Experimental results on the CIFAR100 dataset show that the ZeroRFF method achieves superior performance compared to other methods.

Keywords: continual learning · exemplar-free · analytic learning · Random Fourier Features

1 Introduction

Catastrophic forgetting is a major phenomenon in continual learning [1,2]. This phenomenon occurs when trained models lose knowledge about previous tasks or classes when learning new information. For example, if an image recognition model is trained to recognize new animal species, it may forget how to identify previously learned species. The main cause of this phenomenon is the fundamental nature of gradient-based training algorithms, which tend to focus

M. Yoshikawa et al. (Eds.): ADMA 2025, LNAI 16197, pp. 326–336, 2026.
https://doi.org/10.1007/978-981-95-3453-1_22

on and exhibit bias toward recent tasks (*task-recency bias*). Specifically, model predictions often prioritize classes from new tasks, reducing performance on old classes [3].

To illustrate, consider a student learning foreign languages. If the student only focuses on learning French after having learned English, without reviewing English, their English proficiency may significantly decline. Similarly, in continual learning, models require special mechanisms—this is precisely the goal of the continual learning problem: ensuring that models can still acquire new knowledge while simultaneously "reviewing" to avoid forgetting old knowledge.

In traditional Class-Incremental Learning (CIL) scenarios, tasks are assumed to have a fixed number of samples, and classes in each task are completely separate from previous tasks. However, this assumption does not reflect reality, where training data often includes both new classes and previously encountered classes, while the number of samples in each task may vary unevenly. A new scenario studied recently, which is closer to this reality, is called Generalized Class-Incremental Learning (GCIL) [4,5].

The uneven distribution of sample numbers and classes between tasks in GCIL can exacerbate the problem of *catastrophic forgetting*. In particular, samples belonging to minority classes in a data batch may be overlooked, leading to incomplete model learning during training. For example, in an e-commerce product recognition system, new data may include both popular products (such as smartphones) and less popular products (like specialised accessories). If the model prioritises popular classes due to their larger sample sizes, it may overlook or poorly represent less popular classes, thereby reducing overall performance.

To mitigate the phenomenon of *catastrophic forgetting* in continual learning, a simple yet effective method is to use memory storage mechanisms (replay-based methods). Memory storage methods [6] store a small number of samples from old classes, allowing the model to review previous knowledge while learning new information, thereby minimizing knowledge forgetting. However, this mechanism poses risks related to data privacy due to the storage of old samples.

This research focuses on solving the Generalized Continual Learning (GCIL) problem, a recently studied continual learning scenario where training data includes both new classes and previously encountered classes, with uneven sample distribution across tasks. To overcome challenges related to *catastrophic forgetting* and data privacy limitations, the method proposed in this study employs Analytic Continual Learning (ACL), a branch of Exemplar-Free Class-Incremental Learning (EFCIL) that does not require storing old data samples. This not only helps minimize privacy risks but also meets computational resource efficiency requirements. Furthermore, Analytic Continual Learning differs from many popular gradient descent-based training methods in that it does not seek minima by using iterative operations to gradually approach the target point, but instead finds optimal solutions by solving the equation where the derivative equals zero. To enhance the model's learning capability, the method integrates Random Fourier Features (RFF), a technique that transforms original feature vectors from a linearly non-separable space to a higher-dimensional space. In this new space, data becomes linearly separable, allowing the model to learn more effective representations and improve classification performance for complex classes. The effectiveness of the method is evaluated through several standard metrics in the continual learning field, including area under the accuracy curve, average accuracy, and final task accuracy. Experimental results demonstrate that the proposed method achieves superior performance compared to both replay-based methods and exemplar-free methods. This confirms the potential of combining ACL and RFF in building flexible continual learning systems suitable for real-world scenarios with complex and continuously changing data.

2 Preliminaries

2.1 Generalized Class Incremental Learning

Class-Incremental Learning (CIL) is a common scenario in continual learning, where the model learns new classes in stages, with the assumption that classes in each task are completely disjoint and the number of samples in each task is fixed, as illustrated in Fig. 1. However, this assumption often does not accurately reflect real-world scenarios, where task data may include both new classes and previously encountered classes, while the number of samples for each class is unevenly distributed across tasks.

Generalized Class-Incremental Learning (GCIL) is proposed as an extended scenario of CIL, addressing the limitations of CIL by allowing the model to handle data with characteristics more similar to reality. In GCIL, training data in a task may contain previously learned classes mixed with new classes, and the number of samples in each task is not fixed but changes flexibly. GCIL better reflects real-world scenarios, where data is not neatly organized into separate tasks but often appears in a mixed and unpredictable manner.

The main difference between GCIL and CIL lies in flexibility and the ability to handle complex data. While CIL requires clearly divided tasks, GCIL allows the model to learn under conditions closer to reality, where data continuously changes and does not follow fixed boundaries. However, the complexity of GCIL also increases the challenge of *catastrophic forgetting*, especially when minority classes may be overlooked during training due to uneven data distribution.

Specifically, the continual learning problem includes a sequence of continuous tasks $\mathfrak{D} = \{\mathcal{D}_1^{\text{train}}, \ldots, \mathcal{D}_T^{\text{train}}\}$, where $\mathcal{D}_t^{\text{train}} = \{x_i, y_i\}_{i=1}^{N_t}$ is the training dataset of the t-th task, and T is the total number of tasks.

The training dataset $\mathcal{D}_t^{\text{train}}$ consists of N_t data pairs, y_i is the label of data x_i and y_i is a one-hot vector of size $|\mathcal{C}|$, where $\mathcal{C}$ is the label set. x_i is an input image with dimensions $c \times w \times h$. The goal of GCIL in the k-th task is to train a model using the dataset $\mathcal{D}_k^{\text{train}}$ and evaluate the performance of that model on the test dataset $\mathcal{D}_{1:k}^{\text{test}}$. Here, $\mathcal{D}_{1:k}$ denotes the merged dataset including tasks from 1 to k.

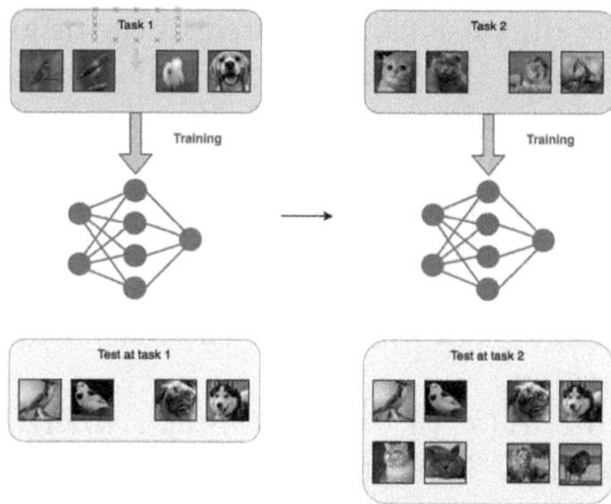

Fig. 1. Illustration of Class-Incremental Learning (CIL).

The Generalized Class-Incremental Learning (GCIL) scenario simulates real-world continual learning, where data distribution in terms of quantity and type can vary between tasks. In this scenario, the model is trained on a set of tasks, where each task may include classes belonging to the set $\mathcal{S}$, with a total number of classes being N. The number of samples of different classes in each task k is represented by a random vector $\mathbf{c}_k \in \mathbb{R}^N$. With $c_{k,i}$ being the i-th component of vector $\mathbf{c}_k$, representing the number of samples of class i in task k.

2.2 Analytic Continual Learning

Analytic Continual Learning (ACL) is an emerging branch in the field of Continual Learning, developed based on the foundation of Analytic Learning research [7]. Unlike traditional deep learning methods, which primarily rely on gradient descent algorithms to find optimal solutions for loss functions, ACL has a completely different approach. Specifically, instead of using iterative steps to adjust model parameters through gradients, ACL seeks to directly solve the equation where the derivative equals zero to determine the optimal solution. The ACIL method [8] transforms the continual learning problem into a least squares regression problem, eliminating the need for sample storage by maintaining correlation matrices. RanPAC [9] applies this technique to pre-trained models. GKEAL [10] focuses on continual learning scenarios with few samples (few-shot CIL) by using a Gaussian kernel projection. ACL is an emerging competitive branch in the field of continual learning.

2.3 Kernel Function

Kernel functions play an important role in solving nonlinear classification and regression problems. To handle data that is not linearly separable, we need a transformation to map the original data to a new space where the data becomes linearly separable. This transformation often creates data in a space with higher dimensions, even infinite dimensions, compared to the original data. Direct computation of these functions can cause memory and computational performance issues. Instead of directly computing each data point in the new space, an efficient approach is to use kernel functions to describe the relationship between any two data points in the new space. Some popular kernel functions widely used in machine learning applications include:

- **Linear Kernel**: $K(\mathbf{x}, \mathbf{y}) = \mathbf{x}^\top \mathbf{y}$. This is the basic case, suitable when data can be linearly separated.
- **Polynomial Kernel**: $K(\mathbf{x}, \mathbf{y}) = (\mathbf{x}^\top \mathbf{y} + c)^d$, where $c \geq 0$ and d is the degree of the polynomial. This function allows modeling nonlinear relationships between features.
- **Radial Basis Function (RBF) Kernel**: $K(\mathbf{x}, \mathbf{y}) = \exp(-\gamma \|\mathbf{x} - \mathbf{y}\|^2)$, where $\gamma > 0$. RBF is one of the most popular kernel functions, particularly effective in handling nonlinear data.
- **Sigmoid Kernel**: $K(\mathbf{x}, \mathbf{y}) = \tanh(\kappa \mathbf{x}^\top \mathbf{y} + c)$, with parameters κ and c chosen appropriately. This function originates from neural networks and can model nonlinear relationships.

3 Methodology

3.1 Overall Architecture

Figure 2 illustrates the architecture of our proposed method ZeroRFF, for Generalized Class-Incremental Learning (GCIL), which combines transformer-based feature extraction with analytic learning mechanisms to enable fast learning and uses a Fourier Features Module to enhance the model's classification capability for sequential task learning.

The process begins with the input of task k, represented as $\mathcal{D}_k^{\text{train}} = \{X_k^{\text{train}}, Y_k^{\text{train}}\}$, containing training samples and their corresponding labels. The methodology is organized into four distinct phases, each serving a specific purpose in the continual learning pipeline.

Feature Extraction Phase: The input data first passes through a pre-trained DeiT-S/16 backbone network with fixed parameters Θ_{backbone}. This feature extractor, which has been pre-trained on ImageNet, transforms the raw input images into high-level feature representations $X_k^{(E)} = f_{\text{backbone}}(X_k^{\text{train}})$. The backbone remains frozen throughout the training process to prevent catastrophic forgetting at the feature level and ensure consistent representations across different tasks.

Kernel Mapping Phase: The extracted features $X_k^{(E)}$ are then transformed using Random Fourier Features (RFF) to obtain $X_k^{(B)} = \text{RFF}(X_k^{(E)})$. This transformation maps the input features into a higher-dimensional space where the data becomes approximately linearly separable. The RFF transformation is defined as $z(x) = \sqrt{\frac{2}{D}} \cos(\omega^T x + \beta)$, where $\omega \sim \mathcal{N}(0, \sigma^{-2} I)$ and $\beta \sim$

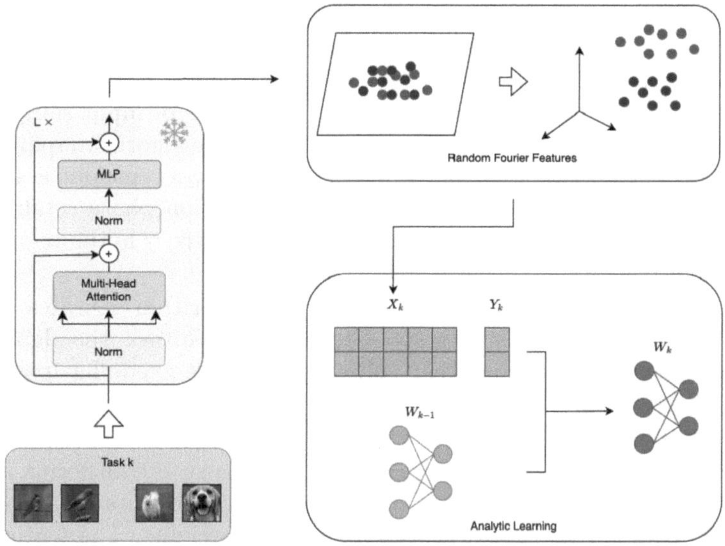

Fig. 2. Overall architecture of ZeroRFF.

Uniform$(0, 2\pi)$. With dimension $D = 5000$, this approximation effectively captures the behavior of the RBF kernel while avoiding the computational overhead of explicit kernel matrix calculations.

Exemplar-Free Learning Phase: The transformed features enter the analytic learning component, which represents the core innovation of our approach. Unlike traditional gradient-based optimization methods, this phase employs direct matrix solutions to find optimal parameters. The process maintains two key matrices: the correlation matrix R_k and the weight matrix $\hat{W}^{(k)}$. The correlation matrix is updated using the Sherman-Morrison formula for efficient recursive computation, while the weight matrix is updated through a direct analytical solution. Crucially, this phase requires no storage of previous task exemplars, addressing privacy concerns and reducing memory requirements.

Task Management Phase: After processing each task, the system checks whether all tasks have been completed. If additional tasks remain, the process increments the task counter ($k = k + 1$) and loads the next task dataset $\mathcal{D}_{k+1}^{\text{train}}$, creating a seamless loop for sequential task processing. Upon completion of all tasks, the methodology outputs the final model $\hat{W}_{\text{FCN}}^{(K)}$, which is ready for inference on the complete set of learned classes.

3.2 Algorithm Description

Algorithm 1 presents the complete ZeroRFF training procedure, which implements an exemplar-free approach to Generalized Class-Incremental Learning. The algorithm is designed to learn from a sequence of tasks without storing previous task data, addressing both privacy concerns and memory constraints inherent in traditional continual learning approaches.

Input and Initialization: The algorithm takes as input a sequence of training tasks $\mathcal{D}_1^{\text{train}}, \ldots, \mathcal{D}_K^{\text{train}}$, where each task $\mathcal{D}_k^{\text{train}}$ consists of input samples X_k^{train} and their corresponding labels Y_k^{train}. Additionally, the algorithm requires a pre-trained backbone model with fixed weights Θ_{backbone}, typically a DeiT-S/16 transformer pre-trained on ImageNet. The initialization phase establishes two critical matrices: the correlation matrix $R_0 = \gamma I$, where γ is the regularization parameter, and the weight matrix $\hat{W}_{\text{FCN}}^{(0)} = 0$.

Sequential Task Processing: The core of the algorithm operates through a sequential loop over all tasks from $k = 1$ to K. For each task, the algorithm performs four essential operations that collectively enable exemplar-free continual learning.

The first operation (Line 3) involves feature extraction using the pre-trained backbone: $X_k^{(E)} \leftarrow f_{\text{backbone}}(X_k^{\text{train}}, \Theta_{\text{backbone}})$. This step transforms raw input images into high-level feature representations while maintaining consistency across tasks through the use of frozen pre-trained weights. The feature extraction process ensures that the input data is represented in a meaningful semantic space suitable for subsequent kernel transformations.

The second operation (Line 4) applies the Random Fourier Features transformation: $X_k^{(B)} \leftarrow \text{rff}(X_k^{(E)})$. This transformation maps the extracted features into a higher-dimensional space where the data becomes approximately linearly separable. The RFF transformation is crucial for enabling the analytic learning approach, as it provides a finite-dimensional approximation to the infinite-dimensional feature space induced by the RBF kernel.

Analytic Learning Updates: The third and fourth operations (Lines 5–6) implement the core analytic learning mechanism through recursive matrix updates. The correlation matrix update (Line 5) employs the Sherman-Morrison formula:

$$R_k \leftarrow R_{k-1} - R_{k-1} X_k^{(B)T} (I + X_k^{(B)} R_{k-1} X_k^{(B)T})^{-1} X_k^{(B)} R_{k-1} \tag{1}$$

This update maintains the inverse of the accumulated covariance matrix without explicitly storing or accessing previous task data. The Sherman-Morrison formula enables efficient rank-one updates to matrix inverses, making the computation tractable even for high-dimensional feature spaces.

The weight matrix update (Line 6) computes the optimal parameters using a direct analytical solution:

$$\hat{W}_{\text{FCN}}^{(k)} \leftarrow \hat{W}_{\text{FCN}}^{(k-1)} - R_{k-1} X_k^{(B)T} X_k^{(B)} \hat{W}_{\text{FCN}}^{(k-1)} + R_k X_k^{(B)T} Y_k^{\text{train}} \tag{2}$$

This recursive formulation ensures that the weight matrix incorporates information from all previous tasks while being updated solely based on the current task data. The direct analytical nature of this update eliminates the need for iterative optimization procedures, resulting in faster convergence and more stable learning dynamics.

Computational Efficiency: The algorithm's computational complexity is dominated by matrix operations involving $D \times D$ matrices, where $D = 5000$ is the RFF dimension. The Sherman-Morrison update requires $O(D^2)$ operations per task, while the weight matrix update scales as $O(D \cdot C)$, where C is the number of classes. This complexity is significantly lower than approaches that require storing and processing exemplars from previous tasks.
Memory Efficiency: A key advantage of Algorithm 1 is its constant memory footprint with respect to the number of tasks. The algorithm maintains only the correlation matrix R_k and weight matrix $\hat{W}_{\text{FCN}}^{(k)}$, both of which have fixed sizes independent of the number of previously encountered tasks. This stands in stark contrast to replay-based methods, which require memory that grows linearly with the number of tasks.
Theoretical Foundation: The algorithm is grounded in the principle of analytic learning, which seeks to find closed-form solutions to optimization problems. By formulating the continual learning objective as a regularized least squares problem and employing the RFF approximation, the algorithm can compute exact solutions without iterative optimization. This theoretical foundation provides convergence guarantees and eliminates the hyperparameter tuning typically required for gradient-based methods.

The algorithm concludes by returning the final weight matrix $\hat{W}_{\text{FCN}}^{(K)}$, which encapsulates all learned knowledge from the sequence of tasks. This final model can then be used for inference on any class from any of the encountered tasks, effectively achieving the goal of continual learning without catastrophic forgetting for recursive updates throughout the learning process. The recursive nature of the updates, combined with the direct analytical solutions, enables the methodology to maintain performance across tasks without requiring access to historical data, making it particularly suitable for scenarios where data privacy and storage efficiency are paramount concerns.

Algorithm 1. ZeroRFF Training Algorithm

Require: Tasks $\mathcal{D}_1^{\text{train}}, \dots, \mathcal{D}_K^{\text{train}}$ with $\mathcal{D}_k^{\text{train}} \sim \{X_k^{\text{train}}, Y_k^{\text{train}}\}$; Pre-trained model with fixed weights Θ_{backbone}
Ensure: Weight matrix $\hat{W}_{\text{FCN}}^{(K)}$
1: Initialize: $R_0 \leftarrow \gamma I$, $\hat{W}_{\text{FCN}}^{(0)} \leftarrow 0$
2: **for** $k = 1$ to K **do**
3: $X_k^{(E)} \leftarrow f_{\text{backbone}}(X_k^{\text{train}}, \Theta_{\text{backbone}})$ ▷ Feature extraction
4: $X_k^{(B)} \leftarrow \text{rff}(X_k^{(E)})$ ▷ Apply Random Fourier Features
5: $R_k \leftarrow R_{k-1} - R_{k-1}X_k^{(B)T}(I + X_k^{(B)}R_{k-1}X_k^{(B)T})^{-1}X_k^{(B)}R_{k-1}$ ▷ Update correlation matrix
6: $\hat{W}_{\text{FCN}}^{(k)} \leftarrow [\hat{W}_{\text{FCN}}^{(k-1)} - R_{k-1}X_k^{(B)T}X_k^{(B)}\hat{W}_{\text{FCN}}^{(k-1)} + R_kX_k^{(B)T}Y_k^{\text{train}}]$ ▷ Update weight matrix
7: **end for**

4 Experiment and Results

4.1 Settings

We conducted experiments on a popular dataset in the fields of machine learning and computer vision: CIFAR-100 [11]. We evaluated under the Si-Blurry scenario [12], with 5 different seed values to ensure stability and reliability of the results. In the Si-Blurry setting, data is organized to reflect complex real-world scenarios where classes are not completely separated between tasks and there is label overlap. Specifically, the disjoint class ratio (*disjoint class ratio* (r_d)) was set to 50%, and the blurry sample ratio (*blurry sample ratio* (r_b)) was set at 10%.

4.2 Results

On the CIFAR-100 dataset, the proposed ZeroRFF method excels in the memory-free scenario, achieving $A_{\text{AUC}} = 60.48 \pm 4.33$, $A_{\text{AVG}} = 58.55 \pm 6.05$, và $A_{\text{Last}} = 72.44 \pm 0.13$. Compared to the second-best memory-free method, GACL ($A_{\text{AUC}} = 57.99 \pm 2.46$, $A_{\text{AVG}} = 56.24 \pm 3.12$, $A_{\text{Last}} = 70.31 \pm 0.06$), ZeroRFF improves by approximately 2% across all three criteria. Furthermore, the ZeroRFF method eliminates the dependence on memory storage while still outperforming most memory-based methods. Compared to MVP-R ($A_{\text{AUC}} = 60.62 \pm 1.03$ with memory size 2000), ZeroRFF is only slightly inferior in A_{AUC}, but superior in A_{AVG} and A_{Last}.

Table 1. Experimental results on CIFAR-100 (%).

Memory-size	Method	A_{AUC} (%)	A_{AVG} (%)	A_{Last} (%)
2000	EWC++ [13]	53.31 ± 1.70	50.95 ± 1.50	52.55 ± 0.71
	ER [14]	56.17 ± 1.84	53.80 ± 1.46	55.60 ± 0.69
	RM [15]	53.22 ± 1.82	52.99 ± 1.65	55.25 ± 0.61
	MVP-R [12]	60.62 ± 1.03	57.58 ± 0.56	64.30 ± 0.29
500	EWC++ [13]	48.31 ± 1.81	44.56 ± 0.40	40.52 ± 0.83
	ER [14]	51.59 ± 1.91	48.03 ± 0.80	44.09 ± 0.80
	RM [15]	41.07 ± 1.30	38.10 ± 0.59	32.66 ± 0.34
	MVP-R [12]	56.20 ± 1.44	53.61 ± 0.04	55.35 ± 0.43
0	LwF [16]	40.71 ± 1.21	38.49 ± 0.56	27.03 ± 2.92
	L2P [17]	42.68 ± 2.70	39.89 ± 0.45	28.59 ± 3.34
	DualPrompt [18]	41.34 ± 2.59	38.59 ± 0.82	22.74 ± 3.40
	MVP [12]	45.07 ± 1.24	44.93 ± 0.54	39.94 ± 0.47
	SLDA [19]	53.00 ± 3.85	50.09 ± 2.77	61.79 ± 3.81
	GACL [20]	57.99 ± 2.46	56.24 ± 3.12	70.31 ± 0.06
	ZeroRFF(Ours)	$\mathbf{60.48 \pm 4.33}$	$\mathbf{58.55 \pm 6.05}$	$\mathbf{72.44 \pm 0.13}$

5 Conclusion

This paper proposes ZeroRFF, an exemplar-free Generalized Class-Incremental Learning (GCIL) method aimed at addressing the catastrophic forgetting phenomenon and data privacy limitations. By integrating Analytic Continual Learning (ACL) techniques to directly solve for optimal solutions and Random Fourier Features (RFF) to transform data into a linearly separable space. Experimental results confirm that ZeroRFF outperforms other methods, providing an effective solution to address the continual learning problem.

References

1. French, R.M.: Catastrophic forgetting in connectionist networks. Trends Cogn. Sci. **3**(4), 128–135 (1999)
2. Goodfellow, I.J., Mirza, M., Xiao, D., Courville, A., Bengio, Y.: An empirical investigation of catastrophic forgetting in gradient-based neural networks. arXiv preprint arXiv:1312.6211 (2013)
3. Hou, S., Pan, X., Loy, C. C., Wang, Z., Lin, D.: Learning a unified classifier incrementally via rebalancing. In Proceedings of the IEEE/CVF Conference on Computer Vision and Pattern Recognition, pp. 831–839 (2019)
4. Aljundi, R., Lin, M., Goujaud, B., Bengio, Y.: Gradient based sample selection for online continual learning. In: Advances in Neural Information Processing Systems, vol. 32 (2019)
5. Mi, F., Kong, L., Lin, T., Yu, K., Faltings, B.: Generalized class incremental learning. In Proceedings of the IEEE/CVF Conference on Computer Vision and Pattern Recognition Workshops, pp. 240–241 (2020)
6. Rebuffi, S. A., Kolesnikov, A., Sperl, G., Lampert, C.H.: icarl: incremental classifier and representation learning. In: Proceedings of the IEEE conference on Computer Vision and Pattern Recognition, pp. 2001–2010 (2017)
7. Guo, P., Lyu, M.R.: A pseudoinverse learning algorithm for feedforward neural networks with stacked generalization applications to software reliability growth data. Neurocomputing **56**, 101–121 (2004)
8. Zhuang, H., Weng, Z., Wei, H., Xie, R., Toh, K.A., Lin, Z.: Acil: analytic class-incremental learning with absolute memorization and privacy protection. Adv. Neural. Inf. Process. Syst. **35**, 11602–11614 (2022)
9. McDonnell, M.D., Gong, D., Parvaneh, A., Abbasnejad, E., Hengel, A.: Ranpac: random projections and pre-trained models for continual learning. Adv. Neural. Inf. Process. Syst. **36**, 12022–12053 (2023)
10. Zhuang, H., Weng, Z., He, R., Lin, Z., Zeng, Z.: Gkeal: Gaussian kernel embedded analytic learning for few-shot class incremental task. In Proceedings of the IEEE/CVF Conference on Computer Vision and Pattern Recognition, pp. 7746–7755 (2023)
11. Krizhevsky, A., Hinton, G.: Learning multiple layers of features from tiny images (2009)
12. Moon, J.Y., Park, K.H., Kim, J.U., Park, G.M.: Online class incremental learning on stochastic blurry task boundary via mask and visual prompt tuning. In Proceedings of the IEEE/CVF International Conference on Computer Vision, pp. 11731–11741 (2023)

13. Kirkpatrick, J., et al.: Overcoming catastrophic forgetting in neural networks. Proc. Natl. Acad. Sci. **114**(13), 3521–3526 (2017)
14. Rolnick, D., Ahuja, A., Schwarz, J., Lillicrap, T., Wayne, G.: Experience replay for continual learning. In: Advances in Neural Information Processing Systems, vol. 32 (2019)
15. Bang, J., Kim, H., Yoo, Y., Ha, J. W., Choi, J.: Rainbow memory: continual learning with a memory of diverse samples. In Proceedings of the IEEE/CVF Conference on Computer Vision and Pattern Recognition, pp. 8218–8227 (2021)
16. Li, Z., Hoiem, D.: Learning without forgetting. IEEE Trans. Pattern Anal. Mach. Intell. **40**(12), 2935–2947 (2017)
17. Wang, Z., et al.: Learning to prompt for continual learning. In Proceedings of the IEEE/CVF Conference on Computer Vision and Pattern Recognition, pp. 139–149 (2022)
18. Wang, Z., et al.: Dualprompt: complementary prompting for rehearsal-free continual learning. In European Conference on Computer Vision, pp. 631–648. Springer, Cham, October 2022
19. Hayes, T.L., Kanan, C.: Lifelong machine learning with deep streaming linear discriminant analysis. In Proceedings of the IEEE/CVF Conference on Computer Vision and Pattern Recognition Workshops, pp. 220–221 (2020)
20. Zhuang, H., et al.: GACL: exemplar-free generalized analytic continual learning. Adv. Neural. Inf. Process. Syst. **37**, 83024–83047 (2024)

LegalDuet: Learning Fine-Grained Representations for Legal Judgment Prediction via a Dual-View Contrastive Learning

Buqiang Xu[1], Xin Dai[1], Zhenghao Liu[1(✉)], Huiyuan Xie[2], Xiaoyuan Yi[3], Shuo Wang[2], Yukun Yan[2], Liner Yang[4], Yu Gu[1], and Ge Yu[1]

[1] School of Computer Science and Engineering, Northeastern University, Shenyang, China
liuzhenghao@mail.neu.edu.cn
[2] Department of Computer Science and Technology, Institute for AI, Tsinghua University, Beijing, China
[3] Microsoft Research Asia, Beijing, China
[4] School of Information Science, Beijing Language and Culture University, Beijing, China

Abstract. Legal Judgment Prediction (LJP) is a fundamental task of legal artificial intelligence, aiming to automatically predict the judgment outcomes of legal cases. Existing LJP models primarily focus on identifying legal triggers within criminal fact descriptions by contrastively training language models. However, these LJP models overlook the importance of learning to effectively distinguish subtle differences among judgments, which is crucial for producing more accurate predictions. In this paper, we propose LegalDuet, which continuously pretrains language models to learn a more tailored embedding space for representing legal cases. Specifically, LegalDuet designs a dual-view mechanism to continuously pretrain language models: 1) **Law Case Clustering** retrieves similar cases as hard negatives and employs contrastive training to differentiate among confusing cases; 2) **Legal Decision Matching** aims to identify legal clues within criminal fact descriptions to align them with the chain of reasoning that contains the correct legal decision. Our experiments on the CAIL2018 dataset demonstrate the effectiveness of LegalDuet. Further analysis reveals that LegalDuet improves the ability of pretrained language models to distinguish confusing criminal charges by reducing prediction uncertainty and enhancing the separability of criminal charges. The experiments demonstrate that LegalDuet produces a more concentrated and distinguishable embedding space, effectively aligning criminal facts with corresponding legal decisions. The code is available at https://github.com/NEUIR/LegalDuet.

Keywords: Legal Judgment Prediction · Contrastive Learning · Legal Decision · Pretrained Language Models

B. Xu and X. Dai—Equal contribution.

M. Yoshikawa et al. (Eds.): ADMA 2025, LNAI 16197, pp. 337–352, 2026.
https://doi.org/10.1007/978-981-95-3453-1_23

1 Introduction

Legal Judgment Prediction (LJP) aims to predict judicial outcomes—including applicable law articles, charges, and imprisonment—based on descriptions of criminal facts. This task plays a vital role in assisting the legal judgment process, improving judicial efficiency, and raising public legal awareness [32,42]. Current LJP models predominantly leverage Pretrained Language Models (PLMs), such as BERT [4] and RoBERTa [20], to encode criminal fact descriptions and then predict legal judgment labels.

To improve the accuracy and reliability of LJP, recent research has focused on conducting more effective reasoning over criminal facts. Some studies primarily aim to learn fine-grained representations of criminal facts by leveraging handcrafted trigger words and legal attributes [13,22,24] or incorporating predictive information from related LJP tasks [30,37,41]. With the development of pretraining technologies, the focus has shifted from manually designed features to continuously pretrain language models on legal-domain corpora [3,6,25,31,43], enabling PLMs to capture more semantics from legal texts. However, these models typically employ masked language modeling [4] to help language models learn token-level semantics, rather than learning more tailored representations of entire criminal facts to make more accurate prediction results.

This paper introduces LegalDuet, a framework for continuously pretraining language models to learn more fine-grained representations of legal cases. As illustrated in Fig. 1, LegalDuet employs a dual-view contrastive learning mechanism, which integrates two key tasks: Law Case Clustering and Legal Decision Matching. The Law Case Clustering task leverages hard negatives [34] during contrastive learning to enable PLMs to capture subtle differences between similar legal cases. For instance, it helps models discern nuanced differences in the

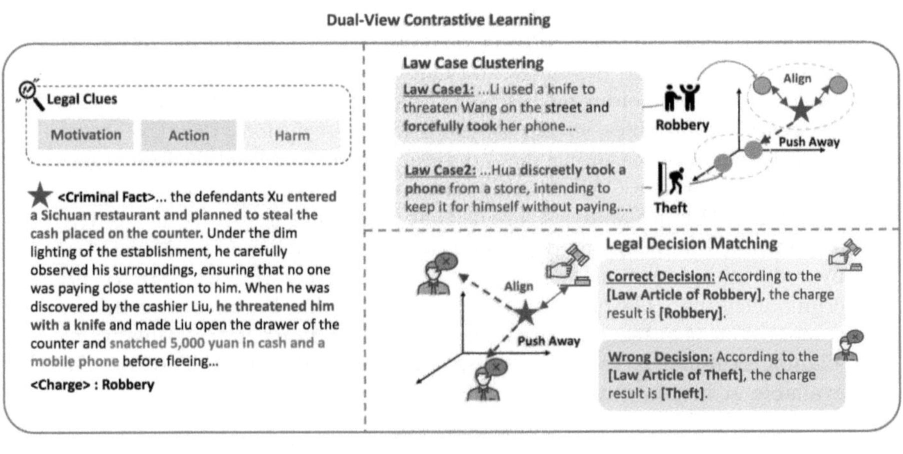

Fig. 1. An Example of Dual-View Contrastive Learning Mechanism in LegalDuet. LegalDuet incorporates both Law Case Clustering (LCC) and Legal Decision Matching (LDM) tasks for continuously pretraining language models.

description of legal cases, such as "forcefully took" versus "discreetly took", which are critical for distinguishing between charges like "robbery" and "theft".

Besides Law Case Clustering, the Legal Decision Matching task further enhances the representations of legal cases by aligning criminal facts with their corresponding legal decisions. Rather than directly predicting legal judgment labels [3,41], LegalDuet verbalizes, permutes and combines the labels of law articles and charges into reasoning chains to construct alternative legal decisions. PLMs are then tasked with identifying the correct matching between legal cases and decisions. This approach not only encourages PLMs to infer causal relationships between law articles and charges but also stimulates language models to pay more attention to these legal triggers, ultimately improving their ability to represent legal cases comprehensively.

Our experiments on the China Artificial Intelligence and Legal Challenges (CAIL2018) [33] dataset demonstrate the effectiveness of LegalDuet, yielding an improvement on both CAIL-small dataset and CAIL-big dataset compared to the SAILER [18] model. Notably, LegalDuet extends its effectiveness to other PLMs, such as BERT-xs [43] and BERT-Chinese [4], showcasing its strong generalization capability. Further analysis reveals that LegalDuet effectively constructs a tailored embedding space for fine-grained legal reasoning by capturing nuanced legal semantics from criminal case facts. LegalDuet produces more compact and coherent representations for legal cases within the same charge category while better separating cases with differing charges. These properties of the learned embeddings give LegalDuet the ability to effectively differentiate between similar criminal charges and address ambiguous points in complex criminal cases.

2 Related Work

Early Legal Judgment Prediction (LJP) models primarily rely on feature-based approaches. They aim to extract expert-defined legal cues, such as trigger words and templates, to support legal judgment predictions [24]. However, the performance of these models is constrained by the quality of these handcrafted features, making it difficult to distinguish between confusing criminal facts [7]. In contrast, more recent works utilize neural networks to automatically extract legal semantics and clues from criminal facts, such as LSTM [13,37], CNN [14], and Transformer [4,18]. After encoding legal cases with these neural models, LJP systems can independently predict judgment labels for subtasks like law articles, charges, and imprisonment. To address the gap between different subtasks, some research has focused on modeling task dependencies and employing multi-task learning, which has been shown to improve judgment accuracy [6,37].

Pretrained Language Models (PLMs), such as BERT [4], have demonstrated strong capabilities in both representing legal cases and making more accurate legal judgments [40]. To bridge the gap between general domain knowledge and the specialized legal domain [9], many studies have focused on continuously training PLMs using legal corpora. Techniques like masked language modeling have been employed to help models learn legal semantics more effectively [36]. Additionally, SAILER [18] pretrains language models to learn the intrinsic logical structures of legal case documents by decoding the reasoning and decisions of judges. However, these methods mainly focus on token-level clues, often overlooking the need for more nuanced and fine-grained crime representations.

To address the limitation of token-level focus, recent works have focused on two key areas: contrastive training and multi-case reasoning. Contrastive learning encourages building a more tailored embedding space for legal texts, bringing similar cases closer together [5,36]. It has been applied to LJP tasks to enhance the encoding of legal semantics [19,40]. For instance, Zhang et al. [40] propose a supervised contrastive learning approach for LJP, leveraging the structure of law articles to help models differentiate between laws and charges. Similarly, Liu et al. [19] incorporate both similar and dissimilar cases during prediction to improve judgment accuracy. Another promising direction is multi-case reasoning, where Graph Neural Networks (GNNs) [11] are used to propagate semantic information across criminal cases [6,38]. Researchers represent criminal facts as nodes and use predefined legal keywords to form edges, constructing hierarchical graphs to enrich case representations. Unlike these approaches, LegalDuet focuses on contrastively pretraining PLMs to learn fine-grained representations of legal cases without requiring additional LJP-specific architectures.

3 Methodology

In this section, we present our LegalDuet model. We begin by defining the task of Legal Judgment Prediction (LJP) (Sect. 3.1). Next, we describe our dual-view contrastive learning mechanism (Sect. 3.2) to continuously pretrain language models for producing fine-grained representations of criminal facts.

3.1 Preliminary of Legal Judgment Prediction

Given a criminal fact F, legal judgment prediction (LJP) models first encode its textual description X_F using pretrained language models such as BERT [4]:

$$h_F = \text{Linear}_F(\text{BERT}(X_F)), \tag{1}$$

where h_F is the encoded representation of the criminal fact F. The Linear_F layer reduces the dimensionality of the BERT-generated 768-dimensional embedding to a 256-dimensional space. This reduction helps decrease computational complexity and mitigates the risk of overfitting, consistent with prior studies [21,39].

For a given LJP task $\mathcal{T}$, the criminal fact representation h_F is then used to predict the label $y_\mathcal{T}$ for that task. LJP tasks aim to identify relevant law articles, charges, or imprisonment terms:

$$P(y_\mathcal{T}|F) = \text{softmax}_{y_\mathcal{T}}(\text{Linear}_\mathcal{T}(h_F)), \tag{2}$$

where $\text{Linear}_\mathcal{T}$ is used to project the hidden representation h_F and compute the logits for different classification labels in the LJP task.

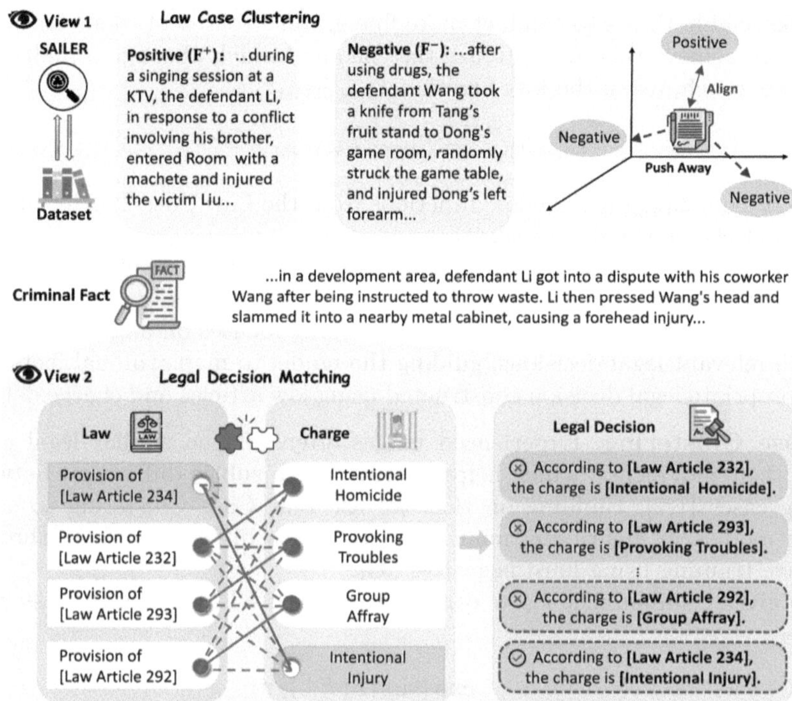

Fig. 2. Illustration of Our LegalDuet.

Finally, the LJP models can be trained by using the cross-entropy loss $\mathcal{L}_\mathcal{T}$ for the LJP task $\mathcal{T}$:

$$\mathcal{L}_\mathcal{T} = \text{CrossEntropy}(y^*_\mathcal{T}, P(y_\mathcal{T}|F)), \tag{3}$$

where $y^*_\mathcal{T}$ is the ground truth label for the given fact F.

To better adapt PLMs to legal scenarios, many models focus on continuously pretraining using mask language modeling [3,9]. Similarly, SAILER [18] also employs mask language modeling but proposes a structure-aware approach for legal case retrieval that leverages the inherent structure of legal documents. A complete legal document usually consists of five parts: procedure, criminal fact, reasoning, legal decision, and tail. Specifically, SAILER utilizes an asymmetric encoder-decoder architecture to seamlessly integrate multiple pretrained objectives. In LegalDuet, we utilize SAILER as the backbone to build LJP models.

3.2 Fine-Grained Legal Representation Learning Through the Dual-View Contrastive Learning

As shown in Fig. 2, LegalDuet leverages a dual-view contrastive learning mechanism to continually pretrain language models from two complementary perspectives: Law Case Clustering (LCC) and the Legal Decision Matching (LDM).

Both tasks enable PLMs to conduct more fine-grained representations for criminal facts. To achieve this, we optimize the encoder models through a multi-task learning approach, using the loss function $\mathcal{L}_{\text{LegalDuet}}$:

$$\mathcal{L}_{\text{LegalDuet}} = \mathcal{L}_{\text{LCC}} + \mathcal{L}_{\text{LDM}}, \tag{4}$$

where $\mathcal{L}_{\text{LCC}}$ and $\mathcal{L}_{\text{LDM}}$ are the loss functions from the Law Case Clustering task and Legal Decision Matching task.

Intuitively, Law Case Clustering helps PLMs to learn semantics from legal cases with the same charge and identify subtle distinctions among confusing cases. On the other hand, Legal Decision Matching focuses on aligning criminal facts with relevant legal decisions, guiding the model to map criminal facts to a more appropriate legal decision constructed using law articles and charge details.

Law Case Clustering. Experienced judges often rely on similar legal cases to guide their judgments, which helps to distinguish subtle differences between confusing cases [1,2]. Building on this, we use Law Case Clustering to assist language models in identifying nuanced distinctions in criminal facts through contrastive training using hard negatives [34].

For a given criminal fact F, we contrastively train PLMs using the loss function $\mathcal{L}_{\text{LCC}}$:

$$\mathcal{L}_{\text{LCC}} = -\log \frac{\exp\left(\text{sim}(F, F^{+})/\tau\right)}{\exp\left(\text{sim}(F, F^{+})/\tau\right) + \sum_{F^{-} \in \mathcal{N}_F(F)} \exp\left(\text{sim}(F, F^{-})/\tau\right)}, \tag{5}$$

where τ is the temperature used to control the sharpness of the softmax distribution. F^{+} is the criminal case that shares the same charge and law article labels with F, while F^{-} is the negative case that are annotated with different charge or law article labels from F. The similarity score between criminal facts F and $F^{+/-}$ is computed using cosine similarity:

$$\text{sim}(F, F^{+/-}) = \cos(h_F, h_{F^{+/-}}). \tag{6}$$

To construct the set $\mathcal{N}_F(F)$ of negative legal cases, we use SAILER to retrieve criminal facts that are related to F. We select the highest-ranked criminal case that shares the same label as F as the positive fact F^{+}. For negative samples, we sample from top-ranked criminal cases with different charges or legal articles to process these hard negatives for contrastive training.

Legal Decision Matching. Professional judges often look for triggers in legal facts to support their judgments [10,16,17,26]. The Legal Decision Matching mechanism builds a decision chain by associating law articles with criminal charges and verbalizing the decision chain. This enables language models to pretrain on identifying judgment triggers inherent in legal facts.

We first define the textual description X_D of legal decisions D, which consist of law articles and criminal charges with the following template:

$$X_D = \text{Template}(X_L, X_C). \tag{7}$$

where X_L and X_C are the descriptions of law article L and charge C.

For a given criminal fact F, we contrastively train the PLM to align criminal facts with corresponding legal decisions and distinguish these unmatched legal decisions:

$$\mathcal{L}_{\text{LDM}} = -\log \frac{\exp(\text{sim}(F, D^+)/\tau)}{\exp(\text{sim}(F, D^+)/\tau) + \sum_{D^- \in \mathcal{N}_D(F)} \exp(\text{sim}(F, D^-)/\tau)}, \tag{8}$$

where τ is the temperature hyperparameter. D^+ represents the legal decision that correctly matches the input criminal fact F according to the ground-truth label, while the negative legal decisions D^- are those that do not match the input fact F, either due to a different charge or law article within the decision. The legal decision is encoded using the same encoder as the criminal fact:

$$h_{D^{+/-}} = \text{Linear}_D(\text{BERT}(X_{D^{+/-}})), \tag{9}$$

where we use BERT model and Linear_D for encoding, which shares the same parameters with the encoder in the Eq. 1. The similarity of criminal facts and legal decisions can be calculated with cosine similarity:

$$\text{sim}(F, D^{+/-}) = \cos(h_F, h_{D^{+/-}}). \tag{10}$$

Rather than randomly selecting negative decisions, we specifically focus on using hard negatives of law articles and charges for contrastive training. Then we incorporate these hard negatives into the candidate pool alongside correct law article and charge. Then we use Eq. 7 to construct both negative and positive legal decisions. To construct the set of negative legal decisions $\mathcal{N}_D(F)$, we use the LJP models to predict the law article and charge labels according to the given fact F:

$$(X_{L^-}, X_{C^-}) = \text{Classifier}(F). \tag{11}$$

The Classifier is initialized with SAILER and then use the training set of CAIL-big to train the LJP models.

4 Experiment Methodology

In this section, we describe the datasets, baselines, evaluation metrics and implementation details.

Datasets. We utilize a Chinese LJP dataset CAIL2018 [33] in our experiments, which is the same as previous work [13,19,35,37,38,41]. It consists of three testing scenarios, CAIL-small, CAIL-big, and CAIL-rest. Specifically, we use CAIL-rest as the pretrained dataset, and both CAIL-small and CAIL-big are adopted during both finetuning and evaluation stages. For both finetuning and evaluation, we follow previous work [35] and keep the same experiment setting[1].

[1] https://github.com/prometheusXN/LADAN/tree/master/data_and_config.

Table 1. Data Statistics of CAIL2018.

Datasets	Finetuning		Pretraining
	Small	Big	Rest
Raw Dataset			
Case	142,238	1,774,122	701,999
Law Articles	103	118	59
Charges	119	130	62
Term of Penalty	11	11	11
Data Split			
Training	101,685	1,430,135	696,999
Development	13,787	158,759	5,000
Test	26,766	185,228	-

To guarantee the data quality, we follow Xu et al. [35] to process the dataset for pretraining and finetuning. The data statistics are shown in Table 1.

Evaluation Metrics. We use the official evaluation metrics of CAIL 2018 [33], including Accuracy (Acc), Macro-Precision (MP), Macro-Recall (MR), and Macro-F1 (F1). These metrics are provided in the official script[2]. We regard Macro-F1, which balances precision and recall, as our primary evaluation metric.

Baselines. We utilize several LJP baselines, including feature-based models, neural LJP models, and PLM-based models.

Feature-Based Models. We use `TF-IDF` [23] to extract legal features and employ `SVM` [27] to predict the corresponding legal judgments.

Neural LJP Models. We compare two convolutional neural network based models, `TextCNN` [15] and `DPCNN` [14], which encode criminal facts using convolutional neural networks. Additionally, six LSTM-based LJP models are evaluated. The `LSTM` [12] model constructs a text classification system using LSTM for predicting legal judgments. LJP models, such as `TopJudge` [41] and `MPBFN` [37], further model the dependencies between different subtasks of LJP. Besides task dependence, many LJP models also focus on fine-grained reasoning by capturing relations among criminal facts. For instance, `Few-Shot` [13] leverages annotated legal attributes to better represent criminal facts, significantly enhancing LJP performance in the few-shot scenario. Models like `LADAN` [35] and `CTM` [19] enhance LJP performance by incorporating case relations and frequencies. Specifically, `LADAN` uses graph neural networks to capture subtle differences among criminal facts, while `CTM` directly encodes similar and dissimilar facts for LJP.

PLM-Based Models. For PLM-based methods, we first compare with two BERT-based language models: BERT-Chinese and BERT-xs. These models finetune

[2] https://github.com/china-ai-law-challenge/cail2018.

BERT-Chinese [4] and BERT-xs [43] for LJP tasks. For multi-tasks based LJP modeling methods, NeurJudge$^+$ [38] separates fact descriptions into different circumstances using prediction results of other LJP tasks to leverage different LJP perspectives for fine-grained prediction. SAILER [18] is a structure-aware pretrained language model for legal case retrieval. It combines both approaches by continuously training BERT to fill in masked tokens and decode corresponding reasoning results and legal judgments.

Implementation Details. All experiments are implemented with PyTorch and start from the checkpoints of PLMs from Hugging Face Transformers [28].

Pretraining. We initialize the LJP models using the checkpoint of SAILER[3] and continuously train the language model using LegalDuet. During pretraining, we set the maximum number of epochs to 5, optimize the parameters using the AdamW optimizer, and configure the learning rate to 1e-5 with a batch size of 32. Following SimCSE [8], we adopt the optimal temperature hyperparameter setting ($\tau = 0.05$) to scale the similarity score.

Then, we present the experimental details for contrastive training. For Law Case Clustering, we construct a negative sample pool using retrieval results from SAILER and select the top-15 ranked criminal facts as hard negatives for each instance. For Legal Decision Matching, we employ a fine-tuned SAILER model trained on the CAIL-big dataset to classify each criminal fact and identify 3 negative law articles and 3 negative charges to construct the legal decisions.

Finetuning. During finetuning, we optimize both classification layer and BERT parameters. The training process spans 10 epochs, utilizing the AdamW optimizer with a learning rate of 5e-6 and a batch size of 64. For the Feature-based baseline, we restrict the number of extracted features to the top 2,000 terms. For the CNN-based baselines, we set a maximum document length of 512 and perform word segmentation using THULA[4]. For LSTM-based models, we retain the first 15 sentences in the criminal facts and cap each sentence at a maximum of 100 words. For PLM-based models, maximum sequence length is 512.

5 Evaluation Results

In this section, we evaluate the effectiveness of LegalDuet on LJP tasks. We first present its overall performance and conduct ablation studies to assess the contributions of each module of LegalDuet. Subsequently, we conduct detailed analyses to explore the characteristics of LegalDuet.

5.1 Overall Performance

The LJP performance of LegalDuet and baseline models is presented in Tables 2 and 3. LegalDuet outperforms all baseline models across various testing scenarios by learning more fine-grained representations of criminal facts.

[3] https://huggingface.co/CSHaitao/SAILER_zh.
[4] https://github.com/thunlp/THULAC-Python.

Table 2. LJP Performance on the CAIL-big Dataset. The best results are in **bold**, and the underlined scores indicate the second-best results.

Model	Law Articles				Charges				Imprisonment			
	Acc	MP	MR	F1	Acc	MP	MR	F1	Acc	MP	MR	F1
TF-IDF+SVM	94.83	83.99	68.21	72.23	94.60	86.71	71.18	75.36	51.63	39.62	29.41	30.28
TextCNN	93.53	74.97	59.40	62.37	93.10	78.44	62.12	65.41	51.15	39.76	28.23	28.07
DPCNN	93.81	76.53	65.91	68.57	93.58	78.57	68.83	71.40	51.65	40.56	30.81	33.57
LSTM	94.27	76.69	63.78	65.73	93.87	78.58	66.67	68.81	52.29	36.67	32.41	32.71
TopJudge	94.54	76.83	66.98	68.82	94.18	81.08	69.44	71.16	52.64	38.86	34.19	33.71
MPBFN	95.85	85.08	72.99	76.24	95.51	88.81	74.81	78.82	56.59	49.55	38.28	40.20
LADAN	96.34	85.75	78.45	80.88	96.22	87.64	81.60	83.74	58.60	50.39	43.21	44.67
CTM	97.24	86.56	81.45	82.11	97.00	87.45	83.53	83.73	56.70	46.81	39.81	41.65
Few-Shot	96.48	86.78	78.30	80.98	96.41	90.13	82.59	85.34	58.26	51.71	41.53	44.31
NeurJudge+	94.88	85.01	75.68	78.12	95.47	81.77	72.77	75.00	56.93	47.17	40.72	42.10
BERT-xs	96.20	82.30	75.29	77.29	96.09	87.15	79.64	81.72	58.06	48.07	43.47	44.71
w/ LegalDuet	96.76	85.55	79.31	80.94	96.70	89.52	83.35	85.42	58.89	49.00	45.31	46.24
BERT-Chinese	97.09	87.65	82.28	83.90	97.05	90.89	85.93	87.65	62.17	54.45	49.82	51.34
w/ LegalDuet	97.29	**88.76**	84.52	85.93	97.27	91.51	87.73	89.14	62.21	53.80	51.40	51.79
SAILER	97.35	88.69	84.29	85.89	97.34	**91.88**	87.94	89.49	62.75	54.15	**52.39**	52.77
w/ LegalDuet	**97.39**	88.73	**85.20**	**86.32**	**97.44**	91.86	**88.80**	**89.97**	**63.06**	**55.05**	52.34	**53.16**

As shown in Table 2, LegalDuet demonstrates its effectiveness by surpassing SAILER, particularly in the law articles prediction task. Furthermore, as presented in Table 3, the performance gains of LegalDuet become even more pronounced, highlighting its strong capability in few-shot learning scenarios. Notably, our approach eliminates the need for complex model architectures, such as explicitly modeling dependencies among different LJP tasks or extracting fine-grained legal cues. Despite leveraging only law articles and charges during continuous pretraining, LegalDuet achieves substantial improvements across other tasks, including imprisonment prediction, demonstrating its adaptability to various legal judgment prediction tasks. Additionally, LegalDuet exhibits strong generalization by extending its effectiveness to different PLMs, including BERT-xs and BERT-Chinese.

Table 3. LJP Performance on the CAIL-small Dataset. The best results are in **bold**, and the underlined scores indicate the second-best results.

Model	Law Articles				Charges				Imprisonment			
	Acc	MP	MR	F1	Acc	MP	MR	F1	Acc	MP	MR	F1
TF-IDF+SVM	77.76	77.89	72.84	72.70	80.31	82.74	77.56	78.30	35.33	29.24	26.74	26.13
TEXTCNN	75.81	71.33	70.40	68.37	76.90	74.80	74.95	73.05	34.31	32.84	30.10	28.93
DPCNN	76.10	70.93	70.46	68.89	79.95	76.01	77.75	76.23	35.03	31.31	29.51	29.78
LSTM	77.79	76.07	73.67	72.55	81.60	80.22	79.16	78.56	36.53	26.91	28.82	26.41
TopJudge	77.29	76.47	74.67	73.08	82.91	80.86	79.75	79.03	36.24	28.52	29.90	27.50
MPBFN	79.88	79.36	75.50	75.29	83.02	84.06	80.93	81.41	37.04	36.87	29.81	29.05
LADAN	80.92	77.74	78.79	77.14	84.97	83.34	83.67	83.03	37.70	37.04	33.48	34.94
CTM	83.10	79.12	81.81	79.23	87.92	86.31	86.08	85.71	37.81	34.66	30.32	30.90
Few-Shot	79.17	77.54	74.78	74.00	81.91	82.63	79.61	79.64	35.87	34.38	28.86	30.08
NeurJudge+	80.02	77.39	76.97	75.50	83.24	83.26	80.95	83.24	38.59	35.35	33.19	32.33
BERT-xs	80.99	79.23	76.62	76.03	85.08	83.68	82.45	82.28	38.59	32.77	33.15	31.45
w/ LegalDuet	83.20	80.54	80.23	79.21	87.48	85.47	83.60	83.85	39.20	35.97	33.72	32.65
BERT-Chinese	82.47	79.75	79.15	78.51	87.66	86.36	85.94	85.65	41.93	39.21	37.53	36.85
w/ LegalDuet	83.93	82.48	81.86	81.00	89.74	87.92	87.27	87.19	42.43	39.61	38.08	**37.91**
SAILER	84.23	81.67	82.59	81.00	89.75	88.12	87.65	87.59	42.76	38.87	38.00	37.15
w/ LegalDuet	**85.90**	**83.92**	**83.26**	**82.65**	**90.47**	**88.90**	**88.36**	**88.29**	**43.25**	**40.62**	**38.25**	37.85

Table 4. Performance of Ablation Models on LJP Tasks. All models are evaluated on the CAIL-small dataset. LDM and LCC indicate the Legal Decision Matching and Law Case Clustering tasks, respectively.

Model	Law Articles				Charges				Imprisonment			
	Acc	MP	MR	F1	Acc	MP	MR	F1	Acc	MP	MR	F1
BERT-xs	80.99	79.23	76.62	76.03	85.08	83.68	82.45	82.28	38.59	32.77	33.15	31.45
w/ LDM	81.96	79.76	79.30	78.10	86.49	84.73	82.86	82.99	**39.40**	35.13	**34.02**	**32.72**
w/ LCC	83.07	80.34	79.23	78.42	86.86	85.17	83.39	83.63	39.38	35.53	33.84	32.41
LegalDuet	**83.20**	**80.54**	**80.23**	**79.21**	**87.48**	**85.47**	**83.60**	**83.85**	39.20	**35.97**	33.72	32.65
BERT-Chinese	82.47	79.75	79.15	78.51	87.66	86.36	85.94	85.65	41.93	39.21	37.53	36.85
w/ LDM	**84.22**	81.69	81.95	81.00	89.17	87.27	86.42	86.47	41.93	**39.86**	**38.47**	37.66
w/ LCC	83.66	81.99	**82.00**	80.86	89.10	86.93	86.64	86.35	42.20	39.33	37.91	37.29
LegalDuet	83.93	**82.48**	81.86	**81.00**	**89.74**	**87.92**	**87.27**	**87.19**	**42.43**	39.61	38.08	**37.91**
SAILER	84.23	81.67	82.59	81.00	89.75	88.12	87.65	87.59	42.76	38.87	38.00	37.15
w/ LDM	85.68	83.32	**83.32**	82.45	89.86	88.56	**88.46**	88.16	**43.58**	**41.61**	37.44	37.63
w/ LCC	84.83	83.76	83.12	82.30	90.21	88.80	88.03	87.97	42.95	39.95	37.76	37.46
LegalDuet	**85.90**	**83.92**	83.26	**82.65**	**90.47**	**88.90**	88.36	**88.29**	43.25	40.62	**38.25**	**37.85**

5.2 Ablation Study

Then, we conduct ablation studies, as presented in Table 4, to evaluate the contributions of different modules in LegalDuet across Law Articles, Charges, and Imprisonment prediction tasks.

Compared to vanilla PLMs, the LCC module significantly enhances LJP performance across various tasks, by extracting case-specific legal cues—critical actions, motivations, and contextual details—essential for fine-grained legal reasoning. Similarly, the LDM module also surpasses vanilla PLMs, with its impact being particularly evident in tasks requiring precise alignment, such as Law Articles and Imprisonment prediction. This underscores the effectiveness of the Legal Decision Matching module in aligning criminal facts with corresponding legal decisions, ensuring that the model integrates legal knowledge into its reasoning process. When both modules are incorporated, LegalDuet consistently outperforms the baseline across all tasks, emphasizing their complementary roles. Law Case Clustering aids in distinguishing between criminal facts linked to different charges by extracting case-specific clues, while Legal Decision Matching ensures alignment between criminal facts and legal decisions. Together, these modules empower LegalDuet to capture fine-grained semantics and extract legally relevant information, thereby enhancing the model's overall LJP capabilities.

5.3 Learned Embeddings of Criminal Facts Optimized by LegalDuet

In this subsection, we evaluate the effectiveness of LegalDuet in learning fine-grained representations of criminal facts through two sets of experiments. The first experiment examines model uncertainty in legal judgment prediction (LJP) across different models, while the second investigates the learned embedding space to assess the impact of LegalDuet based pretraining.

Prediction Confidence of Different LJP Models. We begin by quantitatively analyzing model uncertainty in LJP by presenting cross entropy scores, as illustrated in Fig. 3. A higher cross-entropy score indicates lower model confidence in predicting the ground truth.

As shown in the evaluation results, LegalDuet generally reduces prediction entropy scores compared to SAILER, enabling LJP models to make more confident and precise predictions. Across different testing scenarios, LegalDuet is

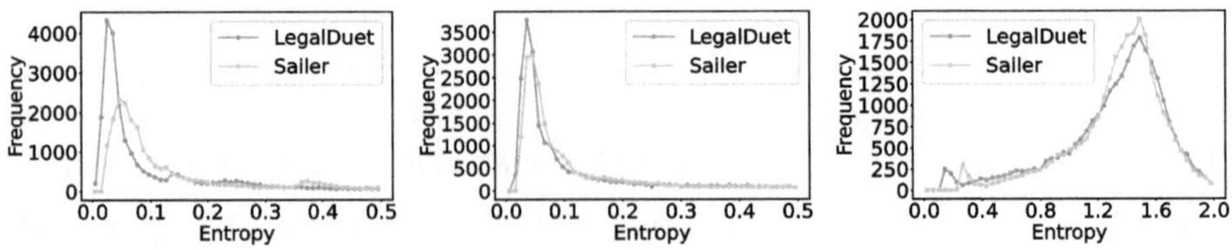

(a) Entropy Distribution of Charge Predictions. (b) Entropy Distribution of Law Articles Predictions. (c) Entropy Distribution of Imprisonment Predictions.

Fig. 3. Entropy Distributions of Legal Judgment Predictions.

particularly effective in lowering entropy scores for the charge prediction task, showing better semantic alignment between criminal facts and charges. This improvement may stem from LegalDuet's approach of verbalizing charge decisions and leveraging contrastive training to distinguish them, thereby enhancing charge comprehension and representation learning ability of PLMs.

Characteristics of the Learned Embedding Space by LegalDuet. The entropy-based analysis highlights how LegalDuet enhances prediction confidence, particularly in ambiguous cases. To further investigate its capability in fine-grained differentiation, we examine how it encodes legal cases into the embedding space, as shown in Fig. 4.

Quantity Analyses of the Learned Embedding Space. We first compute the Davies-Bouldin Index (DBI), a metric that quantifies cluster separability, where lower values indicate more compact and well-separated clusters. Specifically, we select six ambiguous criminal charges (Provoking Troubles, Robbery, Fraud, Intentional Homicide, Theft, and Intentional Injury) and extract all corresponding cases from the CAIL-small dataset for DBI computation. Figure 4a illustrates the DBI reduction values when comparing the embeddings learned by SAILER and LegalDuet. The consistently positive reduction values demonstrate that LegalDuet can effectively map more legal cases into an embedding space by improving the quality of legal case clustering. Notably, LegalDuet demonstrates superior effectiveness in the tasks of Provoking Troubles, Theft, and Intentional Injury.

Visualization of Learned Embedding Space. To provide a more intuitive view of how LegalDuet organizes criminal facts in the embedding space, we employ t-SNE to visualize the distribution of embeddings. Compared to SAILER (Fig. 4), LegalDuet (Fig. 4c) produces a more distinct cluster behavior with clearer boundaries between similar charges, aligning with the principles of contrastive training [29]. This suggests that LegalDuet not only learns more structured representations but also better preserves subtle semantic distinctions between criminal facts, ultimately leading to improved classification of ambiguous charges.

(a) DBI Reduction Values. (b) SAILER. (c) LegalDuet.

Fig. 4. DBI Reduction Values and Embedding Visualizations of SAILER and LegalDuet. The embeddings of criminal facts, **Provoking Troubles**, **Robbery**, **Fraud**, **Intentional Homicide**, **Theft**, **Intentional Injury**, are annotated.

6 Conclusion

In this paper, we introduce LegalDuet, a continuous pretrained method designed to enhance language models' ability to learn more fine-grained representations for criminal facts. LegalDuet emulates the reasoning processes of judges and incorporates a dual-view legal contrastive learning mechanism. Specifically, it comprises Law Case Clustering and Legal Decision Matching, which helps PLMs to better cluster criminal facts and align the semantics between criminal facts with corresponding legal decisions. Our experiments demonstrate that LegalDuet outperforms the baseline across most evaluation metrics, instead of using more sophisticated reasoning architectures. Furthermore, its effectiveness can be generalized to various legal judgment prediction tasks and different PLMs.

Acknowledgments. This work is partly supported by the Natural Science Foundation of China (No. 62206042), and the Fundamental Research Funds for the Central Universities (No. N25ZLL045).

Disclosure of Interests. The authors have no competing interests to declare that are relevant to the content of this article.

References

1. Ashley, K.D.: Modeling legal argument - reasoning with cases and hypotheticals. Artifi. Intell. Legal Reasoning (1991)
2. Atkinson, K., Bench-Capon, T.: Legal case-based reasoning as practical reasoning. Artifi. Intell. Law (2005)
3. Chalkidis, I., Fergadiotis, M., Malakasiotis, P., Aletras, N., Androutsopoulos, I.: LEGAL-BERT: the muppets straight out of law school. In: Findings of the Association for Computational Linguistics: EMNLP 2020, pp. 2898–2904 (2020)
4. Devlin, J., Chang, M.W., Lee, K., Toutanova, K.: BERT: pre-training of deep bidirectional transformers for language understanding. In: Proceedings of NAACL-HLT, pp. 4171–4186 (2019)
5. Fang, H., Wang, S., Zhou, M., Ding, J., Xie, P.: Cert: Contrastive self-supervised learning for language understanding. ArXiv preprint (2020)
6. Feng, Y., Li, C., Ng, V.: Legal judgment prediction: a survey of the state of the art. In: Proceedings of IJCAI, pp. 5461–5469 (2022)
7. Gan, L., Kuang, K., Yang, Y., Wu, F.: Judgment prediction via injecting legal knowledge into neural networks. In: Proceedings of AAAI, pp. 12866–12874 (2021)
8. Gao, T., Yao, X., Chen, D.: SimCSE: simple contrastive learning of sentence embeddings. In: Proceedings of EMNLP, pp. 6894–6910 (2021)
9. Gururangan, S., et al.: Don't stop pretraining: adapt language models to domains and tasks. In: Proceedings of ACL, pp. 8342–8360 (2020)
10. Haar, C.M., Sawyer, J.P., Cummings, S.J.: Computer power and legal reasoning: a case study of judicial decision prediction in zoning amendment cases. Am. Bar Foundation Res. J. (1977)
11. Hamilton, W.L., Ying, Z., Leskovec, J.: Inductive representation learning on large graphs. In: Proceedings of NeurIPS.,pp. 1024–1034 (2017)
12. Hochreiter, S., Schmidhuber, J.: Long short-term memory. Neural Comput. (1997)

13. Hu, Z., Li, X., Tu, C., Liu, Z., Sun, M.: Few-shot charge prediction with discriminative legal attributes. In: Proceedings of COLING, pp. 487–498 (2018)
14. Johnson, R., Zhang, T.: Deep pyramid convolutional neural networks for text categorization. In: Proceedings of ACL, pp. 562–570 (2017)
15. Kim, Y.: Convolutional neural networks for sentence classification. In: Proceedings of EMNLP, pp. 1746–1751 (2014)
16. Lauderdale, B.E., Clark, T.S.: The supreme court's many median justices. Am. Political Sci. Rev. **106**(4), 847–866 (2012)
17. Levi, E.H.: An introduction to legal reasoning. University of Chicago Press (2013)
18. Li, H., et al.: Sailer: structure-aware pre-trained language model for legal case retrieval. In: Proceedings of SIGIR. p, 1035–1044 (2023)
19. Liu, D., Du, W., Li, L., Pan, W., Ming, Z.: Augmenting legal judgment prediction with contrastive case relations. In: Proceedings of COLING, pp. 2658–2667 (2022)
20. Liu, Y., et al.: Roberta: A robustly optimized bert pretraining approach. ArXiv preprint (2019)
21. Liu, Z., Zhang, H., Xiong, C., Liu, Z., Gu, Y., Li, X.: Dimension reduction for efficient dense retrieval via conditional autoencoder. In: Proceedings of EMNLP, pp. 5692–5698 (2022)
22. Liu, Z., Chen, H.: A predictive performance comparison of machine learning models for judicial cases. In: 2017 IEEE Symposium Series on Computational Intelligence (SSCI), pp. 1–6. IEEE (2017)
23. Salton, G., Buckley, C.: Term-weighting approaches in automatic text retrieval. Inform. Process. Manag. **24**(5), 513–523 (1988)
24. Saravanan, M., Ravindran, B., Raman, S.: Improving legal information retrieval using an ontological framework. Artifi. Intell. Law **17**, 101–124 (2009)
25. Shao, Y., et al.: BERT-PLI: modeling paragraph-level interactions for legal case retrieval. In: Proceedings of IJCAI (2020)
26. Sun, Z., Xu, J., Zhang, X., Dong, Z., Wen, J.R.: Law article-enhanced legal case matching: a causal learning approach. In: Proceedings of SIGIR, pp. 1549–1558 (2023)
27. Suykens, J.A., Vandewalle, J.: Least squares support vector machine classifiers. Neural Process. Lett. **9**, 293–300 (1999)
28. Vaswani, A., et al.: Attention is all you need. In: Proceedings of NeurIPS, pp. 5998–6008 (2017)
29. Wang, T., Isola, P.: Understanding contrastive representation learning through alignment and uniformity on the hypersphere. In: Proceedings of ICML (2020)
30. Wu, Y., et al.: Towards interactivity and interpretability: a rationale-based legal judgment prediction framework. In: Proceedings of EMNLP. pp. 4787–4799 (2022)
31. Xiao, C., Hu, X., Liu, Z., Tu, C., Sun, M.: Lawformer: a pre-trained language model for chinese legal long documents. AI Open **2**, 79–84 (2021)
32. Xiao, C., Liu, Z., Lin, Y., Sun, M.: Legal knowledge representation learning. In: Representation Learning for Natural Language Processing, pp. 401–432. Springer Nature Singapore Singapore (2023)
33. Xiao, C., et al.: Cail2018: a large-scale legal dataset for judgment prediction. ArXiv preprint (2018)
34. Xiong, L., et al.: Approximate nearest neighbor negative contrastive learning for dense text retrieval. In: Proceedings of ICLR (2021)
35. Xu, N., Wang, P., Chen, L., Pan, L., Wang, X., Zhao, J.: Distinguish confusing law articles for legal judgment prediction. In: Proceedings of ACL, pp. 3086–3095 (2020)

36. Yan, Y., Li, R., Wang, S., Zhang, F., Wu, W., Xu, W.: ConSERT: a contrastive framework for self-supervised sentence representation transfer. In: Proceedings of ACL, pp. 5065–5075 (2021)
37. Yang, W., Jia, W., Zhou, X., Luo, Y.: Legal judgment prediction via multi-perspective bi-feedback network. In: Proceedings of IJCAI, pp. 4085–4091 (2019)
38. Yue, L., et al.: Neurjudge: a circumstance-aware neural framework for legal judgment prediction. In: Proceedings of SIGIR, pp. 973–982 (2021)
39. Zbontar, J., Jing, L., Misra, I., LeCun, Y., Deny, S.: Barlow twins: self-supervised learning via redundancy reduction. In: Proceedings of ICML, pp. 12310–12320 (2021)
40. Zhang, H., Dou, Z., Zhu, Y., Wen, J.R.: Contrastive learning for legal judgment prediction. ACM Trans. Inform. Syst. (2023)
41. Zhong, H., Guo, Z., Tu, C., Xiao, C., Liu, Z., Sun, M.: Legal judgment prediction via topological learning. In: Proceedings of EMNLP, pp. 3540–3549 (2018)
42. Zhong, H., Xiao, C., Tu, C., Zhang, T., Liu, Z., Sun, M.: How does NLP benefit legal system: a summary of legal artificial intelligence. In: Proceedings of ACL, pp. 5218–5230 (2020)
43. Zhong, H., Zhang, Z., Liu, Z., Sun, M.: Open chinese language pre-trained model zoo. Tech. rep. (2019)

MalHdb:Malware Detection Based on Heterogeneous Dual-Branch Neural Networks

Yiming Li[1,2], Meichen Liu[3], Nan Li[1,2](✉), Meimei Li[1,2], and Chao Liu[1,2]

[1] Institute of Information Engineering, Chinese Academy of Sciences, Beijing, China
liyiming@iie.ac.cn
[2] School of Cyber Security, University of Chinese Academy of Sciences, Beijing, China
[3] Cncert, Beijing, China

Abstract. With the rapid iteration and mutation of malware, traditional detection methods face great challenges, especially the limitation of single-feature detection leads to insufficient accuracy. In this paper, we propose a malware detection method based on Heterogeneous Dual Branch Neural Network (MalHdb), by converting malware binaries into grayscale graphs and extracting static API sequences. It incorporates global features and behavioral features for comprehensive analysis. The grayscale graph can captures the overall structural features of the malware, and its structure remains largely unchanged even with minor code modifications, thereby effectively detect the malware and its variants. Meanwhile, the static API sequences can reflect the potential behavioral intent of the malware at the code level by extracting the list of API functions it may invoke. Experimental results indicate that the method significantly improves the detection accuracy and the F1 score across various datasets, effectively tackling the detection challenges of malware variants. In addition, this paper establishes the optimal ratio of the dual branch output dimensions via grid search and verifies the effectiveness of the model through ablation experiments. Future work will concentrate on refining the detection of obfuscation techniques and further bolstering the robustness of the model to counteract increasingly sophisticated malware attacks.

Keywords: Heterogeneous dual branch neural networks · malware classification · grayscale graphs · static API sequences

1 Introduction

Malicious software, commonly known as malware [1], is typically designed to damage, tamper with, steal, or otherwise cause harm to a computer system, server, client, smartphone, or other devices. It includes viruses, worms, Trojan horses, ransomware, spyware, adware, and other malicious programs whose purpose is generally to gain illegal benefits or to carry out cyber attacks. Meanwhile, as malware continues to update and iterate, it poses a serious security threat to users and organizations. Data from Orange Cyberdefense's Cybersecurity Evolution Guide report Security Navigator 2024

M. Yoshikawa et al. (Eds.): ADMA 2025, LNAI 16197, pp. 353–367, 2026.
https://doi.org/10.1007/978-981-95-3453-1_24

[2], reveals that hacking continues to be the most prominent attack category today, accounting for nearly one-third (30.32%) of confirmed security incidents; followed by privilege abuse (16.61%) and malware (12.98%). Therefore, malware detection remains an area for continued research.

In recent years, with the continuous development of deep learning, neural networks have been utilized in various fields, including malware detection [3]. Deep learning models can automatically extract high-level representations of malware by learning features and patterns from a wide range of datasets. Consequently, a practice has emerged where malware source is converted into a grayscale image, and then neural networks are used to extract image features to distinguish between different malware families [4]. However, its accuracy is not as good as some of the current mainstream detection methods. Therefore, to enhance the accuracy of neural networks in malware detection, we propose a novel approach that employs a heterogeneous two branch neural network structure. In this structure, one branch processes the grayscale maps converted by the malware, while the other branch is augmented by using the static API sequences of the malware. Subsequently, the features of these two branches were merged for malware classification.

But in the MalHdb model, the features extracted from the two branches may exhibit distributional differences or information redundancy. To ensure that the fused features can fully represent the diversity of the input data without introducing redundancy or noise, it is necessary to determine the optimal ratio of the output dimensions for each branch. To address this problem, we employ a grid search method to explore various combinations of output dimensions for each branch. We experiment with multiple ratios and refine our search grid to find the optimal ratio of the dimensions of each branch based on the contribution of the features extracted from each branch to the final classification task, among other methods.

The contributions of our work are summarized below:

- The approach we propose involves converting malware into grayscale graphs and extracting static API sequences from malware disassembly files using dictionaries. This aids our model in effectively managing large-scale and novel malware detection.
- The MalHdb model we constructed can fully leverage the multi-dimensional features of malware and determine the optimal ratio of the output dimensions for each branch, pertaining to image and API sequences, through experiments.
- Experimental results show that our proposed model achieves outstanding performance on multiple malware classification datasets and further improves the accuracy compared to traditional methods in malware detection.

The rest of the paper is organized as follows. Section 2 expounds related work on malware detection focusing on visualization and API sequencing. Section 3 provides a detail explanation of the model structure and methodology employed by MalHdb. Section 4 presents the experimental results. Finally, Sect. 5 discusses the limitations of our approach and concludes the paper.

2 Related Work

2.1 State of the Art in Image-Based Malware Detection Research

Image based malware detection is currently an important research direction in the field of network security. It transforms the characteristics, behaviors, and network activities of malware into graphics, helping security experts identify and analyze potential threats more intuitively.

In 2011, Natarj et al. [4] pioneered an innovative approach by visualizing and automatically classifying malware. They converted malware binaries into grayscale images and utilized image processing techniques. In their study, the researchers discovered that malware images of the same family exhibit a high degree of similarity in texture and layout. Consequently, classification can be achieved using standard image features without the need for code or execution, and the method has a low computational resource consumption. The method achieves 98% accuracy on the dataset; On the other hand, it provides a new and effective approach for automatic malware analysis, which is still in use today. Kalash et al. [6] and Chu et al. [7] achieved malware detection by converting malware binaries into grayscale maps and then constructing classifiers using convolutional neural networks. Additionally, Chen Xiaohan et al. [8] detected by converting opcode sequences into grayscale maps, and Qiao Yanchen et al. [9] detected based on converting assembly instruction word vectors into grayscale maps. Han et al. [10] and Zhao et al. [11] performed texture segmentation and texture extraction after converting binary files into grayscale images, and their method also achieved good results.

2.2 State of the Art of API Sequence-Based Malware Detection Research

Malicious software detection based on API call sequences is also an important branch of current research in the field of network security. It distinguishes malicious software from normal software by analyzing and identifying the API call patterns during program execution.

Iwamoto et al. [12] proposed using the similarity between the presence or absence of specific API function call pairs in the extracted API sequences and the identified malware features in the executable code as a basis for malware discrimination. Zahra et al. [13] used the combination of API names and parameters in the call sequences of API to determine malware. Amer et al. [14, 15] employed word embedding technique to comprehend the contextual relationship between API functions in the call sequences of malware and benign software. API functions with similar contextual features were clustered into groups, and then the behavioral patterns of API call sequences were analyzed by constructing a Markov chain model to differentiate between malware and benign software. Mathew et al. [16] used N-gram model for feature extraction from API call sequences and used the TF-IDF technique for feature selection. They then trained an Long Short-Term Memory (LSTM) network to detect malware. Li et al. [17] applied deep learning to extract features from API call sequences. Converting each API into a vector through embedding layer and using a convolutional layer to extract the API phrase features. Li et al. [18] eliminated consecutive repetitions of the API calls and subsequences of API call sequences as input for malware classification using RNN model.

2.3 Status of the Multimodal Based Malware Detection Research

Meanwhile, many authors use multimodal methods to detect malicious software, among which Xuan et al. [19] use binary file Opcode, combine operand and API call information with virtual address encoding, and generate RGB images to extract static and structural features. Xue et al. [20] proposed a malware classification system called Malscore based on probability scoring and machine learning. The system combines static analysis of grayscale images with dynamic analysis of native API call sequences generated in a sandbox environment, and combines the random forest algorithm to improve the accuracy and efficiency of classification. Alsumaidaee and others [21] respectively use static CNN-LSTM and dynamic CNN1D-LSTM to detect and classify malicious software based on its static features (such as binary file patterns or opcode sequences) and dynamic behavioral features (such as API call sequences). However, in the above-mentioned multi-mode malware detection methods, the fusion of too much information may introduce more useless information, and the generation of color images and dynamic API sequences requires more computing resources, so there are still many areas that need improvement.

3 Methodology

In this section, we detail the model and methodology for malware classification, including the data preprocessing component. Figure 1 illustrates an overview of our approach.

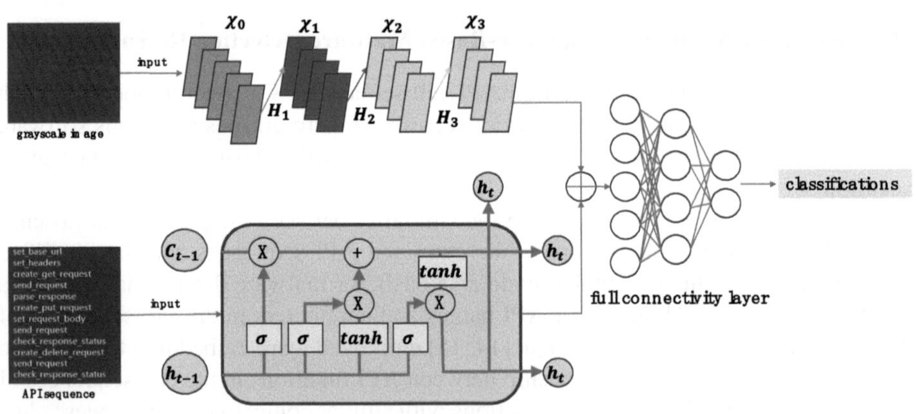

Fig. 1. MalHdb model architecture.

In the above figure, we first divide the input into two parts, namely the grayscale image and the API function name sequence. Then, we perform feature extraction through different neural networks, fuse the features extracted from the two branches, and finally output the classification result through a fully connected layer. Below, we will offer a detailed explanation of how the inputs for each branch of the model are processed and the construction of the model itself.

3.1 Malicious Software Grayscale Images Converting

First, the image width is dynamically determined according to the file size. This is specifically realized through a nonlinear mapping function as in:

$$W = f(S) \tag{1}$$

where the specific mapping of $f(\cdot)$ is shown in Table 1 [4].

Table 1. Corresponding widths of converted images for malicious files of different sizes

File Size	Image Width	File Size	Image Width
<10 kB	32	100 kB-200 kB	384
10 kB-30 kB	64	200 kB-500 kB	512
30 kB-60 kB	128	500 kB-1000 kB	768
60 kB-100 kB	256	>1000 kB	1024

Next, The hexadecimal data of the file is then read as a byte array $B = [b_1, b_2, \ldots, b_N]$, where each byte $b_i \in [0.255]$. Then, the data is populated with the image height calculated using equation $H = \frac{N}{W}$, and ensuring that the tensor can be completely reshaped into a 2D image. Lastly, the tensor is normalized to the range [0, 1] and reshaped into a single-channel grayscale image I, As shown in (2).

$$I = reshape(\frac{P_{padded}}{255}, (H, W)) \tag{2}$$

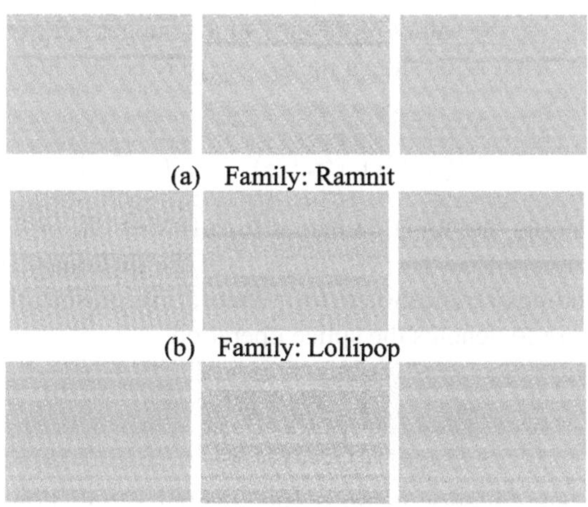

(a) Family: Ramnit

(b) Family: Lollipop

(c) Family: Tracur

Fig. 2. Gray-scale diagram of malware visualization

Upon examining the images generated by malware, we observe that the same families exhibit visually similar characteristics, while there are significant differences between different families. For easier observation, we have color reversed and locally enlarged the grayscale images generated by some families of malicious software, as shown in Fig. 2. And most of the new malware variants are generated by attackers making minor code updates, changes that do not substantially alter the overall structure of the gray-scale image [4]. Therefore, this visualization technique can detect malware and its variants more effectively by converting malware into images.

3.2 API Sequences Extracting

In Our model, we use the static API function sequence. The following will introduce the extraction and generation methods of the sequence. The malware sample is disassembled by IDA Pro (Interactive Disassembler Professional) to generate an.asm file, where function calls are usually introduced with an "extrn" instruction, as indicated by the blue label in Fig. 3.

```
; DWORD __stdcall GetTickCount()
        extrn GetTickCount:dword
                                          ; CODE XREF: sub_10001018+51□p
                                          ; sub_100044B0+61□p
                                          ; DATA XREF: ...
; BOOL __stdcall VirtualFree(LPVOID lpAddress, SIZE_T dwSize, DWORD dwFreeType)
        extrn VirtualFree:dword           ; CODE XREF: sub_10003B60+52□p
                                          ; DATA XREF: sub_10003B60+15□r
; BOOL __stdcall TlsSetValue(DWORD dwTlsIndex, LPVOID lpTlsValue)
        extrn TlsSetValue:dword           ; CODE XREF: sub 10004BC0+23C□p
```

Fig. 3. API function names extraction

We extract the part after "extrn" from the.asm file and clean up the format to obtain the function names $\mathcal{F} = \{f_1, f_2, \ldots, f_M\}$, where f_i denotes the i-th function name. Next, these function names are constructed as a dictionary $\mathcal{D}$, where the keys represent strings of API function names f_i and the values represent unique integer indexes $\sqsubseteq_i$ corresponding to the function names. That is, as shown in (3), where K is the total number of function names in the dictionary.

$$\mathcal{D} = \{(f_1, \sqsubseteq_1), (f_2, \sqsubseteq_2), \ldots, (f_k, \sqsubseteq_k)\} \tag{3}$$

For each malware sample, the function names it calls $S = [s_1, s_2, \ldots, s_L]$ are mapped to a sequence of numbers $\mathcal{V} = [v_1, v_2, \ldots, v_L]$ via the dictionary. With fixed lengths achieved through padding or truncation to form a uniform representation of the sequence, As shown in (4), where $\oplus$ denotes the splice operation.

$$\mathcal{V}_{fixed} = \begin{cases} V[:T] if L > T \\ V \oplus 0_{T-L} if L < T \end{cases} \tag{4}$$

Finally, when a new function name is encountered, the dictionary is dynamically updated and saved to ensure the integrity of subsequent processing.

3.3 MalHdb Building

MalHdb is a model we have built that combines different types of neural network branches. Its purpose is to extract complementary features from multimodal or heterogeneous data (such as sequence data and image data) to improve task performance.

In this malware detection and classification task, the first branch of the MalHdb we constructed uses Dense Convolutional Network (DenseNet) [22] to extract features from the input gray-scale maps. DenseNet is an innovative deep convolutional neural network architecture. Its core feature is to concatenate the input of each layer in the channel dimension with the output of all previous layers through a dense connection mechanism, thereby achieving multi-level reuse and efficient propagation of features, as shown in Fig. 4. This feature can also be represented by (5).

$$x_l = H_l([x_0, x_1, x_2, \ldots, x_{l-1}]) \quad (5)$$

In the above equation, where x_l denotes the output of layer, H_l denotes the nonlinear variation of layer ι (including convolution, normalization, and activation function), and $[x_0, x_1, \ldots, x_{\iota-1}]$ denotes the splicing of the output of layers 0 to $\iota - 1$ in the channel dimension.

In detail, this network consists of an initial convolutional layer, 4 Dense Blocks and 3 Transition Layers. The initial layer consists of a 7×7 convolution and 3×3 maximum pooling for extracting low-level features; each dense block consists of multiple 1×1 and 3×3 convolutional layers, with 1×1 convolution used to reduce the number of channels, 3×3 convolution used to extract features, and the outputs of all the previous layers are used as inputs to the current layer through dense connections; the transition layer reduces the feature map size and the number of channels through 1×1 convolution and 2×2 average pooling. Finally, the global average pooling layer and the fully connected layer output the classification results.

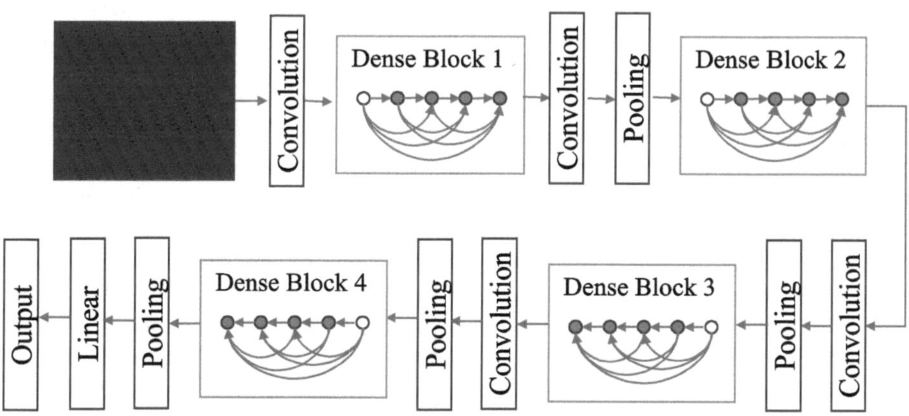

Fig. 4. Dense Connection

The second branch is processes input API sequences, using LSTM model that incorporates an attention mechanism. This model processes the sequence data through an LSTM layer to capture sequence features, introduces an attention mechanism to dynamically assign importance weights each time step in the sequence, and generates context-aware feature representations through weighted summation.

The model enhances the ability to model long sequences and solves the problem that traditional LSTM may lose important information when processing long sequences. The flexibility of feature extraction has been enhanced, allowing the model to focus on the more significant part of the sequence for the task. At the same time, the incorporation of the attention mechanism has improved the model's interpretability, making it easier to analyze where the model is focusing its attention and enabling it to effectively capture the key information within the input sequence.

Once the processing of each branch is completed, the output feature vectors are concatenated, and subsequently processed by the full connectivity layer to finally output the classification results.

4 Experiments

In this section, we will describe in detail how our related experiments were conducted, including an introduction to the dataset, what was done to the dataset, ablation experiments done to verify that our model outperforms the single-branch model, and experiments comparing it to a number of models with decent malware categorization capabilities on multiple datasets.

4.1 Datasets and Preprocessing

Our proposed approach has been experimentally validated on three datasets: BIG2015 [23], BDCI-21 [24], and a private dataset. BIG2015 is a malware classification competition hosted by Microsoft, aimed at automating malware classification through machine learning and deep learning methodologies. The competition offers a publicly accessible dataset with a vast collection of malware samples, along with their corresponding disassembled code (.asm files) and binary files (.bytes files). This dataset comprises 9 malicious code families and the size of the training set is a total of 184G. We then randomly divide it into new training and test sets at a ratio of 8:2. BDCI-21 is a competition dataset in the CCF Big Data and Computational Intelligence Competition (BDCI 2021), dedicated to the malware classification task. This dataset comprises feature data of numerous malware samples, each labeled with malware types or family. We also used its training set, and the size of the training set is a total of 11.5G. The private dataset is composed by combining benign software from the MC-dataset-binary and recent malware downloaded on virusShare, aiming to validate the detection rate of our proposed model against of newer malware. Since we failed to extract valid API call function names from the disassembly files of some of the malware samples provided in the dataset, we removed these samples. Consequently, the final number of malware samples remaining in each category is shown in the Table 2. Finally, we use accuracy and F1 scores to evaluate the performance of our model.

Table 2. Family distribution in three datasets

THE TRAIN DATASET OF MICROSOFT BIG 2015		
Family label	Raw samples	Preprocessed samples
Ramnit	1541	1526
Lollipop	2478	2473
Kelihos_ver3	2942	2597
Vundo	475	241
Simda	42	37
Tracur	751	709
Kelihos_ver1	398	62
Obfuscator.ACY	1228	1186
Gatak	1013	1011
total num	10868	9842
THE TRAIN DATASET OF BDCI-21		
Family label	Raw samples	Preprocessed samples
1	428	427
2	746	746
3	20	20
4	261	258
5	321	315
6	181	181
7	776	776
8	1350	1347
9	594	594
Total num	4677	4664
THE DATA OF PRIVATE		
Family label	Raw samples	Preprocessed samples
Benign	2144	2144
Malware	1935	1935
total num	4079	4079

4.2 Parameter Setting

We implemented MalHdb model using PyTorch, which combines both DenseNet121 and LSTM architectures, for the malware classification task. During the training process, the following hyperparameters were experimentally optimized to enhance the model performance: The DenseNet121 branch used a pretrained model and output image features after removing the original classification header. The input dimension of the LSTM branch was set to 100 to capture the embedded features of API sequences, and the dimension of the hidden layer was set to 128 to enhance the time-dependent modeling capability by stacking two layers of LSTM. The features of the two branches were combined into fusion features via the fully connected layer. These were then further processed by another fully connected layer, which include a 512 dimensional hidden layer. Overfitting is prevented by using BatchNorm, the ReLU activation function, and Dropout with a rate of 0.5. The Adam optimizer was used in the experiments with an initial learning rate of 1e-4, and the ReduceLROnPlateau learning rate scheduler was employed to automatically reduce the learning rate when the validation loss plateaus. We set the training period to 20, the random number seed to 39, the batch size to 32, and the image input size to 224x224 to ensure that the model operated with high-resolution features. The training process was further accelerated by mixed precision training (AMP) and distributed data parallelism (e.g., multi-GPU support). The F1 score of the final model on the validation set is used as an evaluation criterion for model preservation and early stopping strategies.

The environment used is: Python 3.12 (ubuntu22.04), PyTorch 2.3.0, Cuda 12.1, Nvidia GeForce RTX3090 24G.

4.3 Ablation Experiment and Grid Experiment

To demonstrate the effectiveness of the MalHdb, we constructed and to further determine the optimal ratio for the output dimensions of each branch in the MalHdb, we next conducted ablation experiments using Our_1024_64 case study. This involved performing experiments with only one branch at a same time. The notation Our_1024_64 means that the output dimension of the DenseNet branch is 1024, and the output dimension of the LSTM is 64. We used the BIG2015 dataset for these experiments, and its results are shown in Table 3.

Table 3. Ablation Experiments

	BIG2015	
model	Accuracy	F1
DenseNet	0.9836	0.9837
LSTM	0.9385	0.9356
Our_1024_64	**0.9938**	**0.9939**

After conducting the ablation experiments, we tested the output dimensions of each branch via grid search and Table 4 shows the accuracy of some different combinations of experiments using the BIG2015 dataset as an example.

Through the ablation experiments and grid search experiments, we demonstrate that our proposed MalHdb model outperforms the single-branch neural network model, and determine the optimal ratio of the output dimensions of the DenseNet branch to those of the LSTM branch to be 1024:64. Upon further analysis of the experimental results, the core reason is that this ratio perfectly reflects the differences and complementarity

Table 4. Grid search experiments

	BIG2015	
model	Accuracy	F1
Our_512_128	0.9851	0.9852
Our_512_64	0.9826	0.9821
Our_512_32	0.9846	0.9848
Our_1024_128	0.9847	0.9824
Our_1024_64	**0.9938**	**0.9939**
Our_1024_32	0.9863	0.9841

between the two feature sources in terms of information content and expression ability. Firstly, the image branch uses DenseNet to extract structured spatial features from the grayscale images of malicious software. These features contain rich patterns and texture information, and have strong discriminative power against the malicious software family. Therefore, higher dimensions (1024 dimensions) are needed to fully express complex image features. The API sequence branch uses LSTM to extract behavioral features of static API calls. Although these features are important for revealing malicious behavior, they are sparser and more abstract than image features. A 64 dimensional expression is sufficient to capture their main patterns. If the branch dimension is set too high, it is easy to introduce redundant information or even noise, which will affect the overall fusion performance of the model. By implementing asymmetric design in the output dimension, the model achieves clear feature integration in the fusion stage, ensuring the dominant position of image features. At the same time, it effectively utilizes the supplementary behavioral information provided by API sequences, thereby significantly improving classification accuracy and robustness.

4.4 Baseline

To evaluate the performance of our model, we compared and analyzed three datasets using MF-WS [25]: Microsoft BIG-15, CCF BDCI-21, and Private dataset. MF-WS is a malware family classification method based on multimodal fusion and weight self-learning, published by Li et al. in 2023. It applies various feature engineering techniques on three key modalities (byte, format, semantic/statistical), including static byte feature extraction based on Ember [26], text statistical feature modeling based on TF-IDF, and assembly semantic embedding method based on Asm2Vec [27]. At the same time, the model layer combines XGBoost [28] gradient boosting algorithm, weighted soft voting mechanism, and multi model ensemble strategy to achieve robust classification of imbalanced data.

Bytes class: The core idea of this approach is to treat malicious executable files (PE files) as binary sequences and extract key features from them for machine learning model training and evaluation. Specifically, PE files are first processed as raw binary byte sequences, and then multiple features such as byte histograms, byte entropy histograms, and string information are extracted from them.

Section: This technique identifies potential malicious behavior by collecting structured statistical information about PE files to deeply analyze their internal characteristics. Specifically, it first counts the overall number of sections and segments in the file and detects the presence of anomalies in them, such as unnamed sections or sections of size zero. In addition, it analyzes the attributes of each section and counts the number of sections marked as readable, executable, or writable to assess whether the file's permission settings are suspicious. The method also detects the presence of specific functional sections in the file, such as debugging information sections, relocation sections, resource sections, and thread-local storage (TLS) sections.

Byte + Section + Statistics: This approach combines multiple types of features through feature fusion techniques in order to describe and analyze PE files more comprehensively. In addition to the previously mentioned structured statistical features and binary sequence features, it introduces two important types of statistical features: readable strings and assembly code sequences. For readable strings, the method first extracts readable strings from a large number of PE file samples, then counts the frequency of these strings and selects the top 1000 words with the highest frequency of occurrence as a vocabulary list, which is sorted by frequency and then used for feature representation. For assembly code sequences, the method extracts the opcode, the first operand, and the related comment information from the code segments generated by disassembly, and connects these components in order to construct coherent assembly code sequences. These sequences can reflect the execution logic and underlying behavioral patterns of the program.

Weighted soft voting: In this method, the trained model is first used to predict the training set and the predictions are organized in series. For each sample, the model outputs its probability distribution of belonging to each category. Next, a log-loss is calculated for each series based on the true label, which is a value that reflects the model's prediction accuracy on that series. In order to quantify the importance of each series feature, the log loss was taken as a negative logarithm and used as a weighting factor for that series feature, with a larger weight indicating a more significant contribution of that series feature to the final prediction. After obtaining the weights, the final prediction probability is calculated by multiplying and summing the prediction probability of each model by its corresponding weight. For each sample, the category with the highest aggregation probability is selected as the series label for that sample.

4.5 Results

Table 5 illustrates the specific performance of each model on the three datasets. Our model uses DenseNet and LSTM to output a dimension ratio of 1024:64 architecture, as detailed below.

Figure 5 shows the confusion matrix for our model's categorization of each dataset. And from the ablation experiment, confusion matrix of three datasets, and comparison with other models in terms of accuracy and F1 score, it is evident that our proposed MalHdb multimodal method improves accuracy compared to single feature detection

Table 5. Experimental results on three datasets

BIG2015		
Model	**Accuracy**	**F1**
Bytes class	0.9886	0.9766
Section	0.9814	0.9642
Byte + section + statistics	0.9906	0.9875
Weighted soft voting	0.9924	0.9911
Ours	**0.9938**	**0.9939**
BDCI-21		
Model	**Accuracy**	**F1**
Bytes class	0.9796	0.9466
Section	0.9897	0.9536
Byte + section + statistics	0.9912	0.9784
Weighted soft voting	0.9931	0.9879
Ours	**0.9956**	**0.9957**
Private		
Model	**Accuracy**	**F1**
Bytes class	0.9149	0.9011
Section	0.9364	0.9153
Byte + section + statistics	0.9573	0.9498
Weighted soft voting	0.9597	0.9421
Ours	**0.9661**	**0.9661**

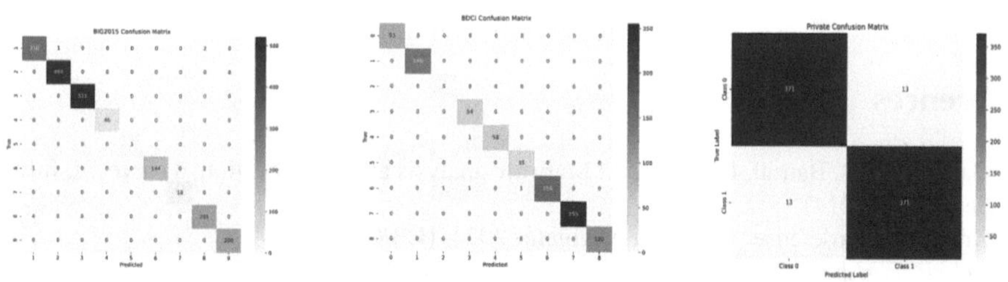

Fig. 5. Confusion matrix of our model on three datasets

methods. Compared with other models, it has shown superior results in malware classification and detection, demonstrating the rationality and advantages of the multimodal fusion method of image and static API sequence that we have chosen. In addition, it is worth noting that our model's F1 score is close to or even higher than accuracy, which is particularly important in multi classification problems, indicating that our model has better overall performance. In contrast, the F1 value combines precision and recall, better reflecting the model's recognition ability for each category, especially for small categories. Therefore, a higher F1 value can ensure that the model not only has higher overall accuracy in practical applications, but also can effectively identify all categories.

5 Conclusion

In this paper, we propose a model architecture based on MalHdb, specifically including a DenseNet branch and an LSTM branch. The malware bytes file is transformed into a grayscale graph and input into the DenseNet branch for feature extraction. The static API function names in the.asm file are converted into sequences using a dictionary and input into the LSTM branch for feature extraction. Subsequently, the features from the two branches are fused to classify and predict the malware by considering both global features and sequence features together at the same time. We also examine the effect of different output dimensions between the two branches on the model's performance and determine the optimal ratio of the two output dimensions. The model we built yields outstanding results on all three datasets.

Although the MalHdb model has achieved good detection performance in fusing grayscale images with static API sequences, its structure still has potential for expansion. For example, system call trajectories can be introduced, or cross modal attention mechanisms and graph neural network structures can be used. In addition, in the face of the real challenge of continuous evolution of malicious software samples, research on the sustainable updating and migration capabilities of transferable models, and exploration of malicious code detection strategies based on incremental learning or domain adaptation, in order to improve the adaptability and robustness of models in practical attack and defense scenarios.

Disclosure of Interests The authors have no competing interests to declare that are relevant to the content of this article.

References

1. Gandotra, E., , Bansal, D., Sofat, S.: Malware analysis and classification: a survey. J. Inform. Sec. **2014** (2014)
2. Orange Cyberdefense. 《Security Navigator 2024》[EB/OL]
3. Deldar, F., Abadi, M.: Deep learning for zero-day malware detection and classification: a survey. ACM Comput. Surv. (2023)
4. Nataraj, L., et al.: Malware images: visualization and automatic classification. In: Proceedings of the 8th International Symposium on Visualization for Cyber Security (2011)
5. Mira, F.: A review paper of malware detection using API call sequences. In: 2019 2nd International Conference on Computer Applications & Information Security (ICCAIS). IEEE (2019)
6. Kalash, M., et al.: Malware classification with deep convolutional neural networks. In: 2018 9th IFIP international conference on new technologies, mobility and security (NTMS). IEEE (2018)

7. Chu, Q., Liu, G., Zhu, X.: Visualization feature and CNN based homology classification of malicious code. Chin. J. Electron. **29**(1), 154–160 (2020)
8. Chen, X., Wei, S., Qin, Z.: Malware family classifi-cation based on deep learning visualization[J]. Comput. Eng. Appl. **57**(22), 131–138 (2021)
9. Qiao, Y., Jiang, Q., Gu, L., et al.: Research on malicious code classification method based on assembly instruction word vector and convolutional neural network. Inform. Netw. Sec. (4), 20–28 (2019)
10. Han, X.: Research on key technology of malicious code detection. University of Science and Technology Beijing, Beijing (2015)
11. Zhao, Y., et al.: Maldeep: A deep learning classification frame-work against malware variants based on texture visualization Sec. Commun. Netw. **2019**(1), 4895984 (2019)
12. Iwamoto, K., Wasaki, K.: Malware classification based on extracted API sequences using static analysis. In: Proceedings of the 8th Asian Internet Engineering Conference (2012)
13. Salehi, Z., Ghiasi, M., Sami, A.: A miner for malware detection based on API function calls and their arguments. In: The 16th CSI International Symposium on Artificial Intelligence and Signal Processing (AISP 2012). IEEE (2012)
14. Amer, E., Zelinka, I.: A dynamic Windows malware detection and prediction method based on contextual understanding of API call sequence. Comput. Secur. **92**, 101760 (2020)
15. Amer, E., Zelinka, I., El-Sappagh, S.: A multi-perspective malware detec- tion approach through behavioral fusion of api call sequence. Comput. Secur. **110**, 102449 (2021)
16. Mathew, J., Ajay Kumara, M.A.: API call based malware detection approach using recurrent neural network—LSTM. In: Intelligent Systems Design and Applications: 18th International Conference on Intelligent Systems Design and Applications (ISDA 2018) held in Vellore, India, December 6–8, 2018, Volume 1. Springer International Publishing (2020)
17. Li, C., et al.: A novel deep framework for dynamic malware detection based on API sequence intrinsic features. Comput. Sec. **116**, 102686 (2022)
18. Li, C., Zheng, J.: API call-based malware classification using recurrent neural networks. J. Cyber Sec. Mobility, 617–640 (2021)
19. Xuan, B., Li, J., Song, Y.: BiTCN-TAEfficientNet malware classification approach based on sequence and RGB fusion. Comput. Secur. **139**, 103734 (2024)
20. Xue, D., et al.: Malware classification using probability scoring and machine learning. IEEE Access **7**, 91641–91656 (2019)
21. Alsumaidaee, Y.A.M., Yahya, M.M., Yaseen, A.H.: Optimizing malware detection and classification in real-time using hybrid deep learning approaches. Inter. J. Safety Security Eng. **15**(1) (2025)
22. Huang, G., et al.: Densely connected convolutional networks. In: Proceedings of the IEEE Conference on Computer Vision and Pattern Recognition (2017)
23. W. C. W. B. C. C. Alessandro Panconesi, M.: Microsoft malware classification challenge (big 2015) (2015.) https://kaggle.com/competitions/malware-classification
24. CCF-BDCI, "Malware family classification based on artificial intelligence (2021). https://www.datafountain.cn/competitions/507/datasets
25. Li, S., Li, Y., Wu, X., Al Otaibi, S., Tian, Z.: Imbalanced malware family classification using multimodal fusion and weight self-learning. IEEE Trans. Intell. Trans. Syst. (2022)
26. Anderson, H.S., Roth, P.: EMBER: an open dataset for training static PE malware machine learning models, CoRR (2018.) http://arxiv.org/abs/1804.04637
27. Ding, S.H., Fung, B.C., Charland, P.: Asm2vec: boosting static representation robustness for binary clone search against code obfuscation and compiler optimization. In: 2019 IEEE Symposium on Security and Privacy (SP), pp. 472–489. IEEE (2019)
28. Chen, T., Guestrin, C: .Xgboost: a scalable tree boosting system. In: Proceedings of the 22nd ACM SIGKDD International Conference on Knowledge Discovery and Data Mining, pp. 785–794 (2016)

MS Faster-RCNN: A Novel Multi-scale Feature Fusion Based Object Detection Scheme

Yibo Sun(✉), Chenlei Liu, Bixiao Xu, Jing Gong, Zhe Sun, and Weitong Chen

The University of Adelaide, Adelaide, Australia
yibo.sun@adelaide.edu.au

Abstract. Object detection is one of the crucial branches of computer vision areas, which is generally utilized to identify and localize the spatial orientation of individual objects within a series of provided images or video streams. In this paper, we propose a novel object detection architecture named MS-Faster R-CNN based on the mechanism of multi-scale feature fusion and the backbone of Faster R-CNN. The fusion strategy mainly uses the structure of the Feature Pyramid Network (FPN), incorporating two links to combine the feature fusion, which enriches the semantics of the fused features and is suitable for various scales of objects. The cascaded Region Proposal Network(RPN) along with the optimized Non-Maximum Suppression (NMS) algorithm are utilized in the candidate box recommendation stage to overcome the problem of over-suppression of small-scale objects in region candidate boxes. The recommendation efficiency of the candidate box can be greatly improved. Finally, the Region of Interest (ROI) Align pooling technology based on the bilinear interpolation method is considered to avoid the loss of accuracy caused by quantization.

Keywords: Object Detection · Multi-scale Features · MS-Faster R-CNN · Feature Pyramid Network

1 Introduction

Traditional object detection algorithms generally consist of three major steps [1, 2]: 1) using sliding windows to scan the input images; 2) extracting the semantic features by choosing fixed feature extraction method; 3) using selected features to do classifications. However, handcrafted features are generally used in traditional object detection models, and it is hard for handcrafted features to be robust for the diverse of objects [3]. Moreover, as manually designed features are not universal, the accuracy of detection results cannot be ensured [4]. Therefore, using deep learning techniques rather than traditional methods can enrich the semantics of features and improve the effectiveness of object detection [5].

Deep learning-based object detection models are generally divided into candidate region-based and regression method-based object detection models. The

M. Yoshikawa et al. (Eds.): ADMA 2025, LNAI 16197, pp. 368–376, 2026.
https://doi.org/10.1007/978-981-95-3453-1_25

candidate region-based object detection models consist of two major steps. It first generates the Region of Interest (ROI), i.e., the candidate regions for detecting object location, and then estimates the object category and the regression of the border position for each of the generated candidate regions, respectively. Most of these deep learning-based object detection models just extract features of the last neural network layer for analysis, resulting in a partial loss on detailed underlying features. Commonly the detection accuracy of small target objects will drop sharply in these methods.

Therefore, using multi-scale feature fusion methods is a general idea to solve this problem. Current multi-scale fusion object detection methods still have limitations. These schemes do not fully exploit potential information from the feature graph, like detailed locations and stronger semantic features. Moreover, many schemes come with a roughly stacked fusion phase which may lead to an increase in redundant information and degrade the detection performance. To solve these problems, we explore a novel multi-scale feature fusion method to improve the performance of small-scale object recognition in real application scenarios. The main contributions of this paper are summarized as follows:

(1) We propose a novel multi-scale object detection model named MS-Faster R-CNN based on the backbone architecture of Faster R-CNN and the structure of FPN. It uses optimized cascaded RPN [6] and an improved NMS algorithm to filter and position refinement of candidate boxes.
(2) The FPN network is optimized for a multi-scale feature fusion strategy. The optimized network is based on two links of the FPN network and uses the first layer to copy the FPN link. It fully makes use of the FPN structure to understand the lower-level and higher-level information of features.
(3) We redesign the NMS algorithm for optimizing the RPN candidate box recommendation module. The algorithm solves the problem that many candidate boxes with different objects though overlapping are mistakenly deleted.

2 Related Work

Multi-scale feature fusion strategy solves the problem of the degradation of small-scale objects. It fully makes use of global information on given images and multi-layer feature extraction techniques to integrate global and local information from different levels.

Bell et al. [7] early proposed a convolutional layer features fusion model, named Inside-Outside Net(ION). Their model is designed with two sub-networks, the outside network for contextual feature extraction, and the inside network for the extraction of three scale features corresponding to the ROI region with convolutional layers and finally fused with contextual features. Similarly, Kong et al. [8] proposed an excellent Faster R-CNN [9] variant(HyperNet), using a layer-hopping approach to extract features for fusion with different layers. [10] proposed a MOD method based on the Faster R-CNN framework, where a feature fusion module is introduced to complement the fine-grained knowledge of small-scale objects in the final features. These models use multi-scale feature fusion methods to combine the features of the last layer and hidden layers based on candidate regions. However, they do not fully exploit the relationship between feature scales.

These multi-scale feature fusion methods have strong adaptability, and the scale information of the final feature is comprehensive. However, it still has the problem of time complexity. Too much scale fusion calculation makes the model timeliness worse. Mask R-CNN models, considered in much literature, all use the top-down pyramid feature fusion method, which has achieved better results on standard datasets but does not consider the directionality of feature fusion. Based on the main architecture of FPN, this paper proposes an innovative multi-scale object detection model named MS-Faster R-CNN, which adds a bottom-up feature fusion link to provide more resolution information.

3 Proposed Method

3.1 MS-Faster R-CNN General Structure

Similar to the detection process of Faster R-CNN, our model first selects the residual network to extract image features. Then uses RPN to complete the candidate box recommendation and filter out a collection of candidate boxes based on the extracted features. Finally, the results of classification and border regression are transformed into the classification and regression layers. Our model has been optimized in many aspects, and the detailed components of our model are described below in Fig. 1.

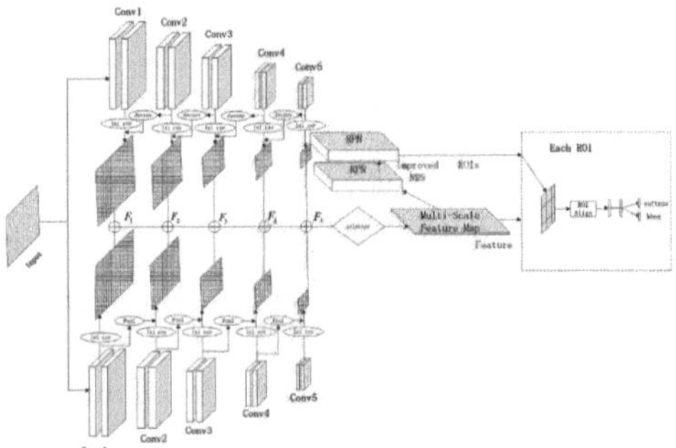

Fig. 1. MS-Faster R-CNN General Structure.

Our proposed method MS-Faster R-CNN can be described as four major modules as below:

(1) ResNet-101, the backbone network for extracting features from input images. The FPN network, based on the ResNet network, contains both top-down semantic down-transfer links and resolution information up-transfer links.

(2) RPN, the module recommends candidate boxes. Unlike Faster R-CNN, the RPN module is a cascaded structure, anchors of different scales take the Feature Map of corresponding levels through the region selectors.
(3) ROI Align pooling layer, for each selected candidate region selected by the RPN network for each corresponding Feature Map fragment is intercepted and downscaled by using the ROI Align pooling layer to form a final feature size of, which is input to the fully connected layer.
(4) The fully connected layer, the last module classifies the category of objects and finds the border position based on the final features. The module is designed to finalize the classification and border regression is employed to find the location more accurately.

3.2 Multi-scale Feature Fusion Strategy

The semantic information encoded in the high-level features of CNNs is profoundly abstract. The foundational features encapsulate intricate resolution details. Thus, the organic integration of features across diverse levels and scales facilitates the absorption of nuanced information by the high-level features from the underlying layers. The underlying features assimilate the semantic information on high-level. However, different fusion strategies have different impacts on the detection results. Even more complex fusion strategies only increase the computational complexity of the model but have a subtle impact on results. Currently, within candidate region-based object detection models, the feature pyramid scheme stands out as a multi-scale feature fusion strategy that consistently yields superior results.

The multi-scale feature fusion strategy adopted by MS-Faster R-CNN builds upon the optimization of FPN, with the FPN links copied twice. One link shows the downward transfer of high-level semantic information, and the other link enables the upward transfer of bottom-level detail information. The detailed structure is shown in Fig. 2 below.

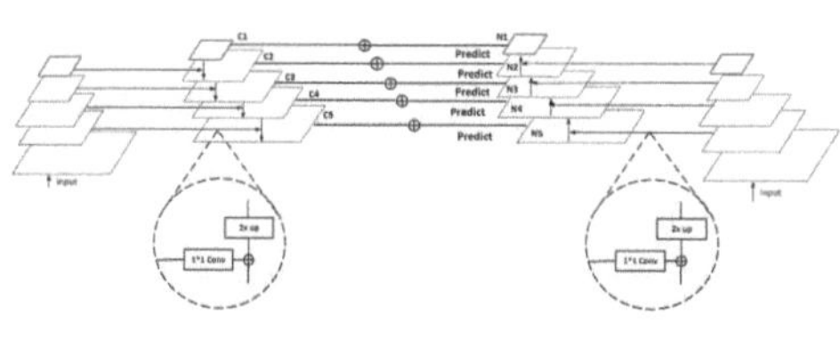

Fig. 2. Feature Fusion Strategy of Our Model.

From Fig. 2, it is obvious that five fusion links in this model. The first component involves the feedforward calculation of a convolutional neural network, which only uses the convolutional calculation to complete the feature extraction of the input image and save the features of each layer. Furthermore, there exist two information transfer links and lateral connections on both sides. The left side of the figure shows the top-down information transfer link and the right lateral connection link. These facilitate the downward transfer of high-level semantic information from the fifth layer to the first layer. The feature fusion between two adjacent layers is achieved by upsampling features from the upper layer. Since the output scales of features from two ResNet layers differ by a sector of two, only two upsampling operations with deconvolution are necessary to match the scale of the upper layer features the same as the lower layer features. In comparison, the features from the lower layer need calculation via 1×1 convolution. Then, the two layers of features are element-wise added to yield the features $\{C_1, C_2, C_3, C_4, C_5\}$.

The primary core module of RPN entails utilizing pre-defined anchors to represent different shapes of objects, resulting in a total of 15 border scale types. The design of the feature selector is shown in Fig. 3, where features of varying scales are fed into the RPN to generate candidate box recommendations and a pooling layer is introduced to conduct feature pooling.

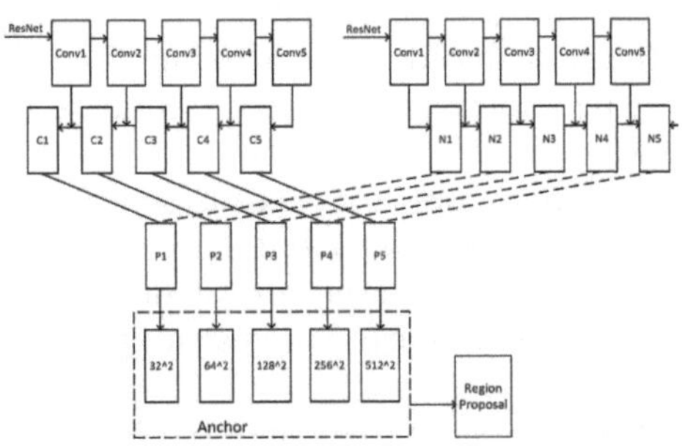

Fig. 3. RPN Selection Features.

3.3 Optimized RPN Candidate Box Recommendation Module

The ultimate objective of object detection is to determine the object category and predict its spatial extent within an image. As described in the previous section, the multi-scale fusion strategy equips the Region Proposal Network (RPN) with multi-scale fusion features, enhancing its ability to recommend a higher-quality collection of candidate boxes. RPN leverages these features to more effectively recommend regions likely to contain foreground objects. However, among the 20,000 candidate boxes generated, many may contain background images leading to the generation of many negative samples. In the prediction stage, this model also does some optimization operations to improve the recommendation efficiency of RPN. Firstly, another RPN module is appended to refine the borders of candidate boxes further. Additionally, an optimized NMS algorithm is employed to suppress the screening of the candidate boxes generated by the initial RPN layer.

Traditional NMS algorithms may mistakenly discard many candidate boxes containing different objects due to the lack of a deterministic positive relationship between the confidence of the classification result and the confidence of the box position. In this paper, an optimized NMS algorithm is devised to solve this issue.

The algorithm contains two major aspects. Firstly, it incorporates the soft-NMS score suppression, where the score equation is recalculated by using Eq. 1, with variable D representing the selected candidate boxes. Secondly, it contains another branch of weight adjustment in the softer-NMS method. The position coordinates of the best candidate boxes are appropriately adjusted based on the score weights, which are computed and described in Eq. 2. However, other candidate boxes that suggest adjustments are not removed from the candidate box set. Instead, they prefer selecting reference candidates for the next round.

$$s_i = s_i \cdot e^{-\frac{iou(M,b_i)^2}{\sigma}}, \quad \forall b_i \notin D \tag{1}$$

$$M'_x = \sum_i^m \frac{b_{i,\text{score}}}{\sum_j^m b_{j,\text{score}}} b_{ix} \tag{2}$$

4 Experiments and Analysis

To validate the effectiveness of the MS-Faster R-CNN model in improving the detection accuracy of small-scale objects, we will train the model through different datasets. Subsequently, we employ the trained model to test on the corresponding test set, counting the detection results based on uniform criteria. By comparing the advantages and disadvantages of the detection results of different models to prove the efficiency of MS-Faster R-CNN in this regard.

4.1 Results Analysis and Comparison

Pascal VOC 2012. Similar to the experiment described above on the VOC2007 test set, we employ the jointly trained models on the VOC2012 test set. The IOU threshold remains at 0.5 and computes the AP values for 20 classes of objects, along with calculating the mAP values. We compare the results obtained using Faster R-CNN, Hyper-Net, RON, SSD and R-FCN and Mask R-CNN as a comparison model. The following Table 1 shows the results of the experiments on the VOC2012 test set.

The analysis presented in Table 1, reveals that MS-Faster R-CNN achieves higher detection accuracy than other models. For example, the mAP value is approximately 11% higher than the basic Faster R-CNN, and around 5%-10% higher than other models that also employ multi-scale feature fusion. Although Mask R-CNN, based on FPN, is currently considered a better model for multi-scale feature fusion detection, its mAP value is 1% lower than MS-Faster R-CNN. From the analysis of small-scale object special cases, such as Bottles and Plants, the MS-Faster R-CNN outperforms Mask R-CNN by 1% and 3.5%, respectively.

Table 1. mAP value comparison on VOC2012.

Class	Faster R-CNN	Hyper-Net	RON	SSD	R-FCN	Mask R-CNN	Ours
Aero	84.6	84.5	86.5	87.9	79.8	81.3	82.1
Bike	78.6	78.2	82.4	82.4	87.2	88.2	89.3
Bird	70.3	73.5	77.6	73.6	81.5	80.9	81.8
Boat	53.7	55.2	60.6	61.3	72.4	73.9	75.2
Bottle	48.9	53.7	55.3	44.5	66.2	69.2	70.3
Bus	77.6	78.5	81.7	82.5	86.2	87.3	88.3
Car	75.9	79.5	80.3	74.6	88.3	89.4	91.4
Cat	86.5	87.8	91.2	92.8	89.2	90.3	91.6
Chair	43.6	49.7	57.3	60.1	57.1	65.5	66.4
Cow	78.5	74.8	81.3	81.3	78.1	87.2	87.3
Table	54.8	52.2	60.9	72.8	66.5	72.1	73.1
Dog	86.8	86.1	87.5	86.7	89.8	91.3	88.4
Horse	81.6	81.8	84.9	87.2	85.2	89.5	90.8
Mbike	80.5	85.3	83.0	77.4	85.3	83.2	86.5
Person	78.5	78.8	81.4	70.1	81.4	80.1	81.3
Plant	40.3	48.8	53.8	50.9	50.3	62.3	65.8
Sheep	72.3	73.8	79.2	78.6	77.2	78.5	79.2
Sofa	60.9	59.3	64.8	82.6	75.7	77.6	79.9
Train	81.2	80.1	84.5	80.2	85.9	86.2	86.5
Tv	63.5	67.5	71.1	76.5	72.3	77.5	78.7
mAP	70.0	71.5	75.3	75.6	77.8	80.6	81.7

It demonstrates that the Ms-Faster R-CNN model in this paper is more efficient and accurate in detecting small-scale objects compared to existing models.

MS COCO. Then, we employ the complex COCO dataset for model training. The COCO dataset contains over 120k available images, of which 80k are designated for the training set and 40k as the test set. Given the complex background of COCO images especially small-scale objects, the test results of the COCO dataset are significantly influenced by the IOU threshold. Therefore the IOU threshold is generally considered when counting the mAP value after the test. Specifically, the following measures are included AP, AP50, AP75, and APS, APm, APl. AP refers to the average value of mAP after traversing the IOU threshold from 0.5 to 0.95 in steps of 0.05. AP50 and AP75 refer to the mAP values under different IOU thresholds (0.5 and 0.75, respectively).

Table 2. Detection accuracy results for 20 common object categories in COCO dataset .

Object	aeroplane	bicycle	bird	boat	bus	cow	motorbike	sofa
AP	55.8	49.2	45.2	56.4	88.6	55.5	68.9	56.9
Object	car	dining table	person	train	cat	dog	potted plant	TV
AP	53.2	72.8	84.8	83.2	63.6	62.5	26.8	56.6
Object	chair	horse	sheep	bottle				
AP	74.9	87.9	57.2	27.7				

From Table 2, it can be observed that the detection accuracy of the model proposed in this paper exceeds 40% for most of the objects, with around 30% for a few small objects. Particularly for objects such as bottles and plants, the detection accuracy reaches 27.7% and 26.8%, respectively, which are 1.5% and 3% higher than the detection results of Mask R-CNN on COCO. To better demonstrate the effectiveness of the model, we will conduct a detailed comparison with several object detection models for achieving better results on the COCO dataset in the following.

5 Conclusion

In this paper, we address the challenge of accurately detecting small-scale objects commonly in object detection models. We propose the optimization of the MS-Faster R-CNN detection model based on the framework of Faster RCNN. Our approach involves utilizing bi-directional FPN to enable multi-scale feature extraction. Furthermore, we introduce a cascaded RPN and redesign the NMS method to refine the candidate box recommendation stage. Furthermore, we employ the ROI Align technique for bilinear interpolation to mitigate the mismatch issue caused by quantization. Our experimental results showcase significant improvements in the accuracy of small-scale object detection, validating the effectiveness of the proposed model.

References

1. Zhao, Z.Q., Zheng, P., Xu, S.T., Wu, X.: Object detection with deep learning: a review. IEEE Trans. Neural Netw. Learn. Syst. **30**(11), 3212–3232 (2019)
2. Jiao, L., et al.: A survey of deep learning-based object detection. IEEE Access **7**, 128837–128868 (2019)
3. Zitnick, C.L., Dollár, P.: Edge boxes: locating object proposals from edges. In: Fleet, D., Pajdla, T., Schiele, B., Tuytelaars, T. (eds.) ECCV 2014. LNCS, vol. 8693, pp. 391–405. Springer, Cham (2014). https://doi.org/10.1007/978-3-319-10602-1_26
4. Dehghani A, Moloney D, Griffin I. Object recognition speed improvement using BITMAP-HoG. In 2016 IEEE International Conference on Image Processing (ICIP), pp. 659-663. IEEE (2016)
5. Szegedy, C., Toshev, A., Erhan, D.: Deep neural networks for object detection. Adv. Neural Inform. Process. Syst. **26**, (2013)
6. Vu, T., Jang, H., Pham, T. X., Yoo, C.: Cascade RPN: delving into high-quality region proposal network with adaptive convolution. Adv. Neural Inform. Processing Syst. **32**, (2019)
7. Bell, S., Zitnick, C. L., Bala, K., Girshick, R.: Inside-outside net: detecting objects in context with skip pooling and recurrent neural networks. In Proceedings of the IEEE Conference on Computer Vision and Pattern Recognition, pp. 2874-2883 (2016)
8. Kong, T., Yao, A., Chen, Y., Sun, F. Hypernet: towards accurate region proposal generation and joint object detection. In Proceedings of the IEEE Conference on Computer Vision and Pattern Recognition, pp. 845-853 (2016)

9. Ren, S., He, K., Girshick, R., Sun, J. Faster r-cnn: towards real-time object detection with region proposal networks. Adv. Neural Inform. Processing Syst. **28** (2015)
10. Guan, W., Zou, Y., Zhou, X.: Multi-scale object detection with feature fusion and region objectness network. In 2018 IEEE International Conference on Acoustics, Speech and Signal Processing (ICASSP), pp. 2596-2600. IEEE (2018)
11. Lin, T. Y., Dollár, P., Girshick, R., He, K., Hariharan, B., Belongie, S. Feature pyramid networks for object detection. In Proceedings of the IEEE Conference on Computer Vision and Pattern Recognition, pp. 2117-2125 (2017)

A Knowledge-Enhanced Network for Multimodal Aspect-Based Sentiment Classification

Qinlong Hu, Guozhe Jin(✉), Yahui Zhao, Rongyi Cui, and Zhenghao Huang

Intelligent Information Processing Laboratory, Department of Computer and Technology, Yanbian University, Yanji 133002, China
jinguozhe@ybu.edu.cn

Abstract. Multimodal Aspect-Based Sentiment Classification (MASC) aims to identify the sentiment polarity of aspect terms in text by leveraging both textual and visual modalities. Although recent methods integrate detailed visual features, they often overlook the challenge that short and sparse multimodal tweets provide limited context for aspect semantics. Caption generation has been explored, yet such descriptions are usually generic and lack task-specific relevance. To address this, we propose a Knowledge-Enhanced Network (AKENM) that enriches missing context with external knowledge. AKENM comprises unimodal feature extraction, knowledge feature extraction, knowledge-guided interaction, and cross-modal fusion modules. By incorporating knowledge-enhanced textual and visual features, our approach better captures aspect-specific semantics and improves sentiment classification. Experiments on Twitter-2015 and Twitter-2017 show that AKENM outperforms competitive models in both accuracy and F1 score.

Keywords: Multimodal Aspect-Based Sentiment Classification · Knowledge Enhancement · Knowledge-guided Interaction

1 Introduction

Multimodal Aspect-Based Sentiment Analysis (MABSA) task includes two subtasks: Multimodal Aspect Term Extraction (MATE) and Multimodal Aspect Sentiment Classification (MASC). This paper focuses on the MASC task.Existing MASC research can generally be divided into three categories. **(i)Image-to-text conversion approaches:** transforming image content into auxiliary text to bridge the gap between text and visual modalities.**(ii)Fine-grained alignment methods:** mapping text tokens to corresponding image regions, reducing irrelevant visual noise.External **(iii)knowledge-enhanced methods:** enriching sparse tweet context with external knowledge to improve sentiment prediction.

M. Yoshikawa et al. (Eds.): ADMA 2025, LNAI 16197, pp. 377–384, 2026.
https://doi.org/10.1007/978-981-95-3453-1_26

However, previous methods face several challenges: **(i)** Twitter datasets consist of concise and informal texts lacking contextual information about aspect terms. While existing approaches attempt to compensate by incorporating context through syntactic analysis or external knowledge, they often fall short in fully capturing the context surrounding aspect terms. **(ii)** Methods that integrate external knowledge often directly append it to the original text, potentially introducing noise. To overcome these challenges, we introduce the Knowledge-Enhanced Multi-modal Aspect-Based Sentiment Analysis Network (AKENM), comprising four key modules: unimodal feature extraction, knowledge feature extraction, knowledge-guided interaction, and fusion and prediction. The primary contributions of this study are as follows:
(1). We propose a knowledge-enhanced network leveraging retrieval-augmented generation to augment background information pertinent to tweets, thereby enhancing sentiment prediction accuracy.
(2). We devise a knowledge-guided interaction module to filter out noise from background knowledge, facilitating the extraction of knowledge-enhanced text and visual features.
(3). We conducted extensive experiments on the Twitter-2015 and Twitter-2017 datasets to validate the effectiveness of the AKENM model.

2 Related Work

In recent years, the growth of social media and mobile technologies has enabled users to express emotions through multiple modalities, such as text, images, and voice, generating vast multimodal data. Early aspect-based sentiment analysis (ABSA) focused solely on text, overlooking multimodal information. To address this, Xu et al. [1] proposed the Multimodal Aspect-Based Sentiment Classification (MASC) task, combining text and images to improve sentiment classification accuracy for aspect terms. Mining sentiment from both modalities remains challenging. Recent deep learning advances have driven progress in this area. Yu et al. introduced TomBERT [2], a BERT-based model with aspect-aware attention for visual features. Yang et al. developed a Multi-View Interaction and Fusion Network AMIFN [3] with image gating to reduce noise. Huang et al. proposed SeqCSG [4], leveraging image descriptions, scene graphs, and cross-modal semantic graphs to capture fine-grained semantic relations. Despite these improvements, short and context-poor tweet texts hinder accurate sentiment analysis. To overcome this, we propose the Knowledge-Enhanced Multimodal Aspect-Based Sentiment Analysis Network (AKENM), integrating external knowledge related to aspect terms to enrich tweet content and enhance classification.

3 Method

In this section, we first introduce the formulation of the MASC task. Then, we provide a detailed description of our proposed Knowledge-Enhanced Multimodal Aspect-Based Sentiment Classification Network (AKENM). The overall architecture of AKENM is shown in Fig. 1.

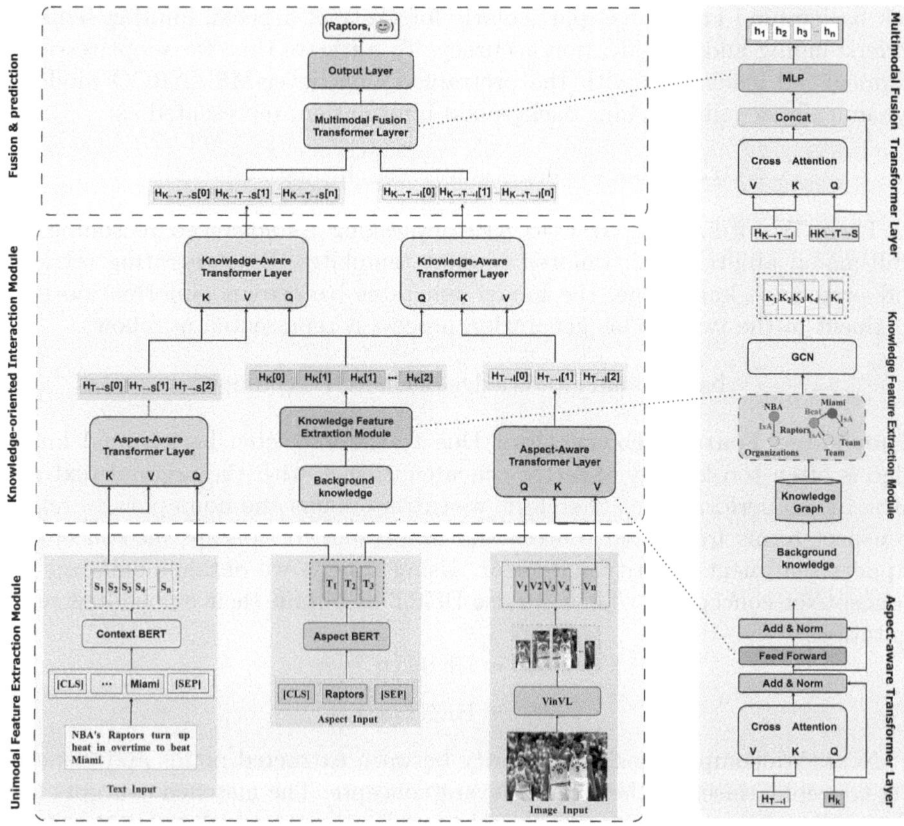

Fig. 1. Overall framework of knowledge-enhanced Multimodal Aspect-Based Sentiment Classification network(AKENM).

3.1 Unimodal Feature Extraction Module

Textual Feature Extraction: we use BERTweet [5] as the encoder for both the text and the aspect term, which can be respectively represented as:

$$H_S = \text{BERTweet}(S) \tag{1}$$

$$H_T = \text{BERTweet}(T) \tag{2}$$

Visual Feature Extraction: Inspired by previous research, we use VinVL [6] model to extract the top k visual objects with the highest confidence, which can be represented as:

$$H_V = \text{VinVL}(I) = [v_1, v_2, \ldots, v_k] \tag{3}$$

3.2 Knowledge Feature Extraction Module

Background Knowledge Generation: The MASC task combines tweets and images to predict aspect term sentiment, yet social media data often lack background knowledge, particularly for abstract aspects, limiting semantic understanding and classification accuracy. To mitigate this, we adopt retrieval-augment -ed generation with the pretrained Contriever-MSMARCO model to dynamically acquire relevant background information, represented as:

$$\text{retrival\_knowledge} = \text{contriever\_msmarco}(\text{query}) \tag{4}$$

Then, We utilize the GPT-4o API, leveraging its advanced reasoning and multimodal abilities with tailored prompt templates. By integrating retrieved and pretrained knowledge, the model generates background information more pertinent to the tweet. This generation process is represented as follows:

$$\text{background\_knowledge} = \text{ChatGPT}(\text{prompt}) \tag{5}$$

Knowledge Feature Extraction: Due to the extracted background knowledge is often too lengthy, directly concatenating it with the original text may harm model performance. Therefore, we extract nouns and noun phrases related to aspect terms from both sources and map them to ConceptNet concepts to support sentiment polarity judgment. Using spaCy, we obtain nouns n_i, and ConceptNet concepts c_j, Then we use BERT to obtain their embedding representations:

$$H_{n_i} = \text{BERT}(n_i) \tag{6}$$

$$H_{c_j} = \text{BERT}(c_j) \tag{7}$$

Next, We compute cosine similarity between extracted nouns and ConceptNet concepts, selecting the top-k relevant concepts. The matched concepts form the set $C = \{c_1, c_2, \ldots, c_k\}$,each linked to multiple triples in ConceptNet, represented as:

$$F = \{f_1, f_2, \ldots, f_m\} \tag{8}$$

$$f_i = (c_i, r_i, c_j),\ (i \neq j) \tag{9}$$

Then, we use C as the set of nodes and the relationships in F as the set of edges to construct the knowledge subgraph G, and employ a GCN to update the node features.

$$G = (V, E) \tag{10}$$

3.3 Knowledge-Oriented Interaction Module

To capture aspect-related semantics, we introduce an aspect-aware Transformer layer. A multi-head cross-attention mechanism uses aspect term features H_T as queries and text features H_S as keys and values, followed by normalization and a feed-forward network, yielding aspect-aware text features $H_{T \to S}$:

$$Z = \text{LN}(H_T + \text{Cross-Attention}(H_T, H_S, H_S)) \tag{11}$$

$$H_{T \to S} = \text{LN}\left(\text{FNN}(Z) + Z\right) \tag{12}$$

Similarly, aspect term features H_T serve as queries with visual features H_V as keys and values to obtain aspect-aware visual features $H_{T \to I}$.To model interactions among knowledge features, aspect-aware text, and visual features, a knowledge-aware Transformer layer uses knowledge features H_K as queries and aspect-aware text features $H_{T \to S}$ as keys and values, followed by normalization and a feed-forward network, producing knowledge-enhanced text features $H_{K \to T \to S}$:

$$K = \mathrm{LN}(H_K + \text{Cross-Attention}(H_K, H_{T \to S}, H_{T \to S})) \tag{13}$$

$$H_{K \to T \to S} = \mathrm{LN}(\mathrm{FNN}(K) + K) \tag{14}$$

Then, we treat the knowledge features H_T as queries and the aspect-aware visual features $H_{T \to I}$ as keys and values to obtain the knowledge-enhanced visual features $H_{K \to T \to I}$.

3.4 Fusion and Prediction Module

To fuse multimodal features, we employ a multimodal fusion Transformer layer with cross-modal multi-head attention. We use Knowledge-enhanced text features $H_{K \to T \to S}$ serve as queries,knowledge-enhanced visual features $H_{K \to T \to I}$ as keys and values, producing context-aware visual features:

$$H_I' = \text{Cross-Attention}(H_{K \to T \to S}, H_{K \to T \to I}, H_{K \to T \to I}) \tag{15}$$

Next, we concatenate $H_{K \to T \to S}$ with H_I' and input them into a MLP layer to generate the final fused feature representation:

$$H_O = \mathrm{MLP}([H_{K \to T \to S}; H_I']) \tag{16}$$

Finally, we perform aspect sentiment prediction based on the fused feature H_O using the Softmax function. The formula is as follows:

$$y = \mathrm{Softmax}(W_a H_o + b) \tag{17}$$

4 Experiments

4.1 Comparative Experiment

We evaluated AKENM against representative multimodal methods(Table 1). While TomBERT suffers from coarse-grained textâĂŞimage interaction, EF-CapT rBERT-DE [7] reduces modality gaps via image-to-text conversion, and FITE [8] further improves fine-grained fusion with facial features. Knowledge-based models, including HIMT [9], AMIFN, and SeqCSG, gain from external knowledge, and ITOAOF [10] performs well through noise filtering. AKENM surpasses all baselines by integrating fine-grained cross-modal interaction with tweet-related background knowledge, outperforming ITOAOF by 1.14% and 1.21% in accuracy and by 1.47% and 1.25% in F1 on Twitter-2015 and Twitter-2017, respectively, demonstrating the efficacy of knowledge supplementation (Table 2).

Table 1. Main results on the two Twitter datasets (unit: %).

Methods	Twitter-2015		Twitter-2017	
	Acc	F1	Acc	F1
TomBERT [2]	77.15	71.75	70.34	68.03
EF-CapTrBERT [7]	78.01	73.25	69.77	68.42
EF-CapTrBERT-DE [7]	77.92	73.90	72.30	70.02
FITE [8]	78.49	73.90	70.90	68.70
FITE-DE [1]	78.64	74.30	72.98	71.97
FITE-DE-Large [8]	78.76	74.79	73.87	73.03
HIMT [9]	78.14	73.68	71.14	69.16
AMIFN [3]	78.69	75.50	72.29	70.21
ITM [11]	78.27	74.19	72.61	71.97
SeqCSG [4]	79.30	75.00	74.60	73.20
ITOAOF [10]	79.45	75.11	74.47	73.05
Ours(AKENM)	**80.59**	**76.58**	**75.68**	**74.30**

4.2 Ablation Experiment

Table 2. Ablation experimental results.

Methods	Twitter-2015		Twitter-2017	
	Acc	F1	Acc.	F1
w/o retrieval	80.35	76.32	75.42	73.98
w/o prompt	79.74	75.64	74.92	73.46
w/o GCN	80.21	76.16	75.27	73.82
w/o Interaction module	79.92	75.86	75.03	73.48
AKENM	**80.59**	**76.58**	**75.68**	**74.30**

(1)w/o Retrieval: After removing the retrieval module, accuracy decreases by 0.24% and 0.26% on the Twitter-2015 and Twitter-2017 datasets, respectively, while the F1 score decreases by 0.26% and 0.32%. This indicates that the external knowledge retrieved contributes to generating more accurate background knowledge. **(2) w/o prompt:** Without the generation module, accuracy drops by 0.85% and 0.76% on Twitter-2015 and Twitter-2017, respectively, and the F1 score drops by 0.94% and 0.84%. This demonstrates the crucial role of the generation module in knowledge integration and refinement. **(3) w/o GCN:** Replacing the GCN with BERTweet to directly extract features from knowledge triples results in accuracy decreases of 0.38% and 0.41% on Twitter-2015 and Twitter-2017, respectively, and F1 score decreases of 0.43% and 0.48%. This shows that

Fig. 2. Three case studies of how background knowledge can help model predictions.

the GCN is beneficial for modeling relational information and improving the quality of knowledge features. **(4) w/o Interaction module:** Removing the knowledge-guided interaction module reduces accuracy by 0.67% and 0.65% on Twitter-2015 and Twitter-2017, respectively, and F1 score by 0.72% and 0.82%. This confirms its importance in enhancing both textual and visual features.

4.3 Case Study

To validate the model, we conducted a case study comparing AKENM with AMIFN, SeqCSG, and ITOAOF(Fig. 2). Results show AKENM improves sentiment accuracy by integrating external knowledge. In Fig. 2(a), it correctly predicts positive sentiment for two basketball players, avoiding neutral misclassification. In Fig. 2(c), it identifies negative sentiment for "Aleppo" by leveraging image context and civil war knowledge, unlike other models.

5 Conclusion

We propose a Knowledge-Enhanced Fusion Network (AKENM) for multimodal Aspect-based sentiment analysis. This network extracts multimodal features through a pretrained model, retrieves background knowledge through retrieval-enhanced generation, and integrates them using a knowledge-guided interaction module. Ultimately, it improves the accuracy of the model.

References

1. Xu, N., Mao, W., Chen, G.: Multiinteractive memory network for aspect based multimodal sentiment analysis. In: Proceedings of the AAAI Conference on Artificial Intelligence (2019)
2. Yu, J., Jiang, J.: Adapting BERT for target-oriented multimodal sentiment classification. IJCAI(2019)
3. Yang, J., Xu, M., Xiao, Y., et al.: AMIFN: aspect-guided multi-view interactions and fusion network for multimodal aspect-based sentiment analysis. Neurocomputing (2024)
4. Huang, Y., Chen, Z., et al.: Target-oriented sentiment classification with sequential cross-modal semantic graph. In: International Conference on Artificial Neural Networks (2023)
5. Dat, N., Thanh, V., Anh, N.: BERTweet: a pre-trained language model for English Tweets. In Proceedings of the 2020 Conference on Empirical Methods in Natural Language Processing: System Demonstrations, pages 9–14. Association for Computational Linguistics (2020)
6. Zhang, P., Li, X., Hu, X., et al.: Vinvl: revisiting visual representations in vision-language models. In: Proceedings of the IEEE/CVF Conference on Computer Vision and Pattern Recognition (2021)
7. Khan, Z., Fu, Y.: Exploiting BERT for multimodal target sentiment classification through input space translation. In: Proceedings of the 29th ACM International Conference on Multi-media, pp. 3034-3042 (2021)
8. Yang, H., Zhao, Y.,Qin, B.: Face-sensitive image-to-emotional-text cross-modal translation for multimodal aspect-based sentiment analysis. In: Proceedings of the 2022 Conference on Empirical Methods in Natural Language Processing, pp. 3324-3335(2022)
9. Yu, J., Chen, K., Xia, R.: Hierarchical interactive multimodal transformer for aspect-based multimodal sentiment analysis. IEEE Trans. Affective Comput. (2022)
10. Wang, Q., Xu, H., Wen, Z., et al.: Image-to-text conversion and aspect-oriented filtration for multimodal aspect-based sentiment analysis. IEEE Trans. Affective Comput. (2023)
11. Yu, J., Wang, J., Xia, R. et al.: Targeted multimodal sentiment classification based on coarse-to-fine grained image-target matching. In: IJCAI (2022)

Enhanced Discriminant Sparse Feature Extraction for Image Classification

Hongyu Cheng, Zhuojie Huang, Lin Jiang, and Jigang Wu(✉)

School of Computer Science and Technology, Guangdong University of Technology, Guangzhou 510006, Guangdong, China
asjgwucn@outlook.com

Abstract. Linear discriminant analysis (LDA) is a widely used method for supervised feature extraction and dimension reduction in pattern recognition and data analysis. However, the existing data is characterized by high dimensionality and complexity, often containing numerous redundant features. These interference features may misguide the discriminative capabilities of traditional LDA. Therefore, how to extract effective discriminative features from plenty of information is the challenge to be addressed. To solve above problems, we propose a novel method called Enhanced Discriminant Sparse Feature Extraction (EDSFE). Specifically, EDSFE differs from traditional LDA methods through the incorporation of novel components that enhance its discriminative feature extraction capabilities. While both EDSFE and traditional LDA aim to maximize the between-class difference and minimize the within-class difference, EDSFE introduces unique innovations to improve its performance. It promotes sparsity and reconstruction capability of the projection matrix through sparse regularization and reconstruction error terms. Extensive experiments conducted on three databases shows that the proposed method performs competitively when compared to other state-of-the-art feature extraction methods.

Keywords: Linear Discriminant Analysis · Dimension Reduction · Feature Extraction · Sparse Representation

1 Introduction

In the era of massive data, the exponential growth of high-dimensional data exacerbates the "Curse of Dimensionality" [1]. Linear Discriminant Analysis (LDA) [2], a supervised dimensionality reduction technique, aims to maximize inter-class differences and minimize intra-class variations, finding wide applications in classification tasks such as human action recognition [3], face recognition [4,5], and personal identification [6]. However, traditional LDA has critical limitations: 1) It fails to address the curse of dimensionality in high-dimensional data, leading to computational inefficiencies and overfitting when computing scatter matrices; 2) It lacks automatic feature selection, retaining redundant and noisy features

M. Yoshikawa et al. (Eds.): ADMA 2025, LNAI 16197, pp. 385–392, 2026.
https://doi.org/10.1007/978-981-95-3453-1_27

in the low-dimensional space, which degrades classifier performance and generalization; 3) It focuses solely on inter-class discrimination while neglecting data reconstruction, limiting its utility in tasks like data analysis and visualization.

To overcome these drawbacks, this paper proposes a novel dimensionality reduction method: Enhanced Discriminant Sparse Feature Extraction (EDSFE), with the following contributions: It handles high-dimensional data via sparse regularization and orthogonal constraints, extracting key features to reduce computational complexity while preserving classification performance. It incorporates a sparse feature selection mechanism to retain discriminative features and eliminate noise, enhancing classification accuracy, robustness, and generalization. It introduces a reconstruction error term to balance discriminative power and data reconstructability, enabling the learned projection to both distinguish classes and accurately reconstruct original data—valuable for comprehensive data analysis and visualization.

The remainder of this paper is structured as: Sect. 2 details the EDSFE method and its solution; Sect. 3 provides experimental results on public facial image datasets; Sect. 4 concludes.

2 The Proposed Method

In this section, we introduce the method proposed in this paper, named Enhanced Discriminant Sparse Feature Extraction (EDSFE) and the alternatively iterative algorithm is designed to solve the optimization problem.

2.1 Learning Model of EDSFE

LDA is a widely used linear dimensionality reduction technique that projects high-dimensional data into a lower-dimensional space, but it suffers from limitations such as sensitivity to noise and dependence on the number of projection directions. To address these issues, we introduce the $l_{2,1}$-norm as a regularization term into LDA for high-dimensional data processing. Specifically, the $l_{2,1}$-norm constrains the projection matrix Q, encouraging the selection of key features (with non-zero weights) while suppressing irrelevant ones (with zero weights). This achieves sparse feature representation, eliminates redundancy while retaining critical information, and enhances model accuracy, interpretability, and generalization. Our initial objective function for improving the robustness of the discriminative subspace is:

$$\min_{Q} \operatorname{Tr}\left(Q^T(S_w - aS_b)Q\right) + \lambda_1\|Q\|_{2,1} \tag{7}$$

where $Q \in \mathbb{R}^{m\times d}$ ($d \leq m$) is the projection matrix, S_b and S_w denote the between-class and within-class scatter matrices, respectively, λ_1 is a weighting parameter, and a is a constant.

To leverage label information and enhance feature discriminability, we redefine the objective function by introducing a projection matrix $P \in \mathbb{R}^{c \times d}$ (c is the number of classes) that aligns extracted features with the label matrix $Y \in \mathbb{R}^{c \times n}$, with an orthogonal constraint $P^T P = I$:

$$\begin{aligned} &\min_{Q,P} \operatorname{Tr}\left(Q^T(S_w - aS_b)Q\right) + \lambda_1\|Q\|_{2,1} + \lambda_2\|PQ^TX - Y\|_F^2 \\ &\text{s.t.} \quad P^TP = I \end{aligned} \tag{8}$$

Here, λ_2 controls the alignment strength, guiding Q to focus on class-specific features and improving classification performance.

To mitigate overfitting, we add a Frobenius norm constraint on PQ^T to regulate the complexity of P and Q, resulting in the final EDSFE objective function:

$$\begin{aligned} &\min_{Q,P} \operatorname{Tr}\left(Q^T(S_w - aS_b)Q\right) + \lambda_1\|Q\|_{2,1} + \lambda_2\|PQ^TX - Y\|_F^2 + \lambda_3\|PQ^T\|_F^2 \\ &\text{s.t.} \quad P^TP = I \end{aligned} \tag{9}$$

where λ_3 balances the regularization effect, restricting model complexity to enhance generalization. Parameters $\lambda_1, \lambda_2, \lambda_3$ collectively balance the contributions of discriminative, sparsity, alignment, and regularization terms during optimization.

2.2 Optimization

The constrained minimization problem (9) is efficiently solved using the Augmented Lagrangian Method (ALM) , expressed as:

$$\begin{aligned} L =& Tr(Q^T(S_w - aS_b)Q) + \lambda_1||Q||_{2,1} + \lambda_2||PQ^TX - Y||_F^2 \\ &+ \lambda_3||PQ^T||_F^2 + \operatorname{Tr}(P^TP - I)\Gamma \end{aligned} \tag{10}$$

where Γ is the Lagrange multiplier penalizing constraint violations. The solution involves iterative optimization of P and Q:

P-step: Fix Q and update P by solving the constrained minimization:

$$\min_{P^TP=I} \lambda_2||PQ^TX - Y||_F^2 + \lambda_3||PQ^T||_F^2 + \operatorname{Tr}((P^TP - I)\Gamma) \tag{11}$$

This reduces to an Orthogonal Procrustes problem. With $SVD(YX^TQ) = USV^T$, the solution is $P = UV^T$ [7].

Q-step: Fix P and update Q by setting the derivative of L with respect to Q to zero:

$$\frac{\partial L}{\partial Q} = 2(S_w - aS_b)Q + \lambda_1 DQ + \lambda_2 XX^TQ - \lambda_2 XY^TP + 2\lambda_3 IQ = 0 \tag{13}$$

where D is a diagonal matrix with $D_{ii} = 1/\|q_i\|_2$ (q_i is the i-th row of Q). Solving for Q gives:

$$Q = \left(2(S_w - aS_b) + \lambda_1 D + \lambda_2 XX^T + 2\lambda_3 I\right)^{-1} \left(\lambda_2 XY^T P\right) \tag{14}$$

The full optimization process is summarized in Algorithm 1.

Algorithm 1. Optimization of problem (9)

Input: Data matrix X and label matrix Y, parameters λ_1, λ_2, λ_3, the maximum number of iteration : *maxiter*.
Initialization: $Q = \mathbf{0}_{m\times d}$; $a = 10^{-4}$;
$P = \text{argmin}Tr(P^T(S_w - aS_b)P)\ s.t. P^TP = I$; Set $iter = 1$, $converged = false$.
Output: Q, P
while not converged && $iter <= maxiter$
1. Update P by using (12).
2. Update Q by solving (14).

end

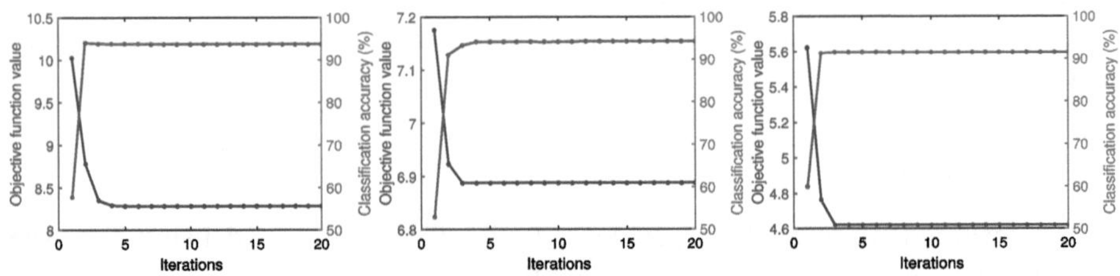

(a) CMU PIE (b) AR (c) Extended Yale B

Fig. 1. Convergence and classification accuracy versus the iterations of the proposed method on (a) the CMU PIE database, (b) the AR database, (c) the Extended Yale B database.

2.3 Computational Complexity and Convergence Analysis

This section analyzes the computational complexity and convergence of the proposed EDSFE algorithm. For training data X with d features, the algorithm involves two main steps: Step 1 primarily incurs cost from singular value decomposition (SVD) of a $c \times d$ matrix (where c is the number of classes), with complexity $O(cd^2 + \max(d^3, d^2c))$. Step 2 involves matrix inversion of an $m \times m$ matrix (where m is the number of samples), with complexity $O(m^3)$. Overall, the computational complexity of EDSFE is $O(t \cdot (m^3 + cd^2 + \max(d^3, d^2c)))$, where t is the number of iterations. In practice, since $m \gg c, d$ in most cases, the complexity simplifies to $O(tm^3)$.

Convergence analysis (Fig. 1) shows that as iterations increase, the objective value decreases and stabilizes, while classification accuracy rises rapidly to a steady state. The method converges within 5 iterations, demonstrating fast and effective convergence.

3 Experiment

In this section, To evaluate the effectiveness of the proposed method, we conducted comparative experiments with several other methods on the Extended Yale B database [8], AR database [9], and CMU PIE database [10].

3.1 Baselines and Experimental Setting

The methods include the dimensionality reduction methods, i.e. LDA [11], OLDA [12], SLDA [13], CRP [14], MPDA [4], LRPE [15], LRPP_GRR [16], FSP [17], AAML [18], DSL_AGR [19].

In the proposed method, feature selection techniques are employed to improve performance. To evaluate the effectiveness of these methods, the Classification accuracy on the testing dataset was measured. Before feature selection, PCA is used for dimensionality reduction. For each database, training sets are created by randomly selecting samples from each class, and this process was repeated 20 times. The mean classification accuracy (%) is calculated and compared across different methods. The classification is performed using the Nearest Neighbor (NN) classifier. Training data comprised randomly selected images of each individual, while the remaining images are used for testing. The experiment used the specific dimensions indicated on the horizontal axis of the respective databases. All images underwent automatic cropping and alignment. The performance of the proposed method is affected by the initial values, so fixed variables in EDSFE are initialized during the experiment.

3.2 Experiments on Face Databases

We evaluated our method on three public face databases. The AR database includes diverse expressions, lighting, and occlusions, we used 3,120 images. The CMU PIE database contains 41,368 images of 68 subjects with variations in poses, expressions, and lighting. The Extended Yale B database comprises 38 individuals with 64 images each under varying illumination, we randomly selected subsets of 10, 15, 20, and 25 images per individual for training, while using the remaining images for testing. Table 1 show our method consistently achieves competitive or superior performance across all databases, validating its effectiveness and advantages over alternatives in practical face recognition scenarios.

Table 1. Mean classification accuracies (%) of different methods on three databases.

					AR Database					
Samples	LDA	OLDA	SLDA	CRP	MPDA	LRPE	LRPP	FSP	AAML	DSL	**Ours**
4	87.33	**90.11**	89.83	75.48	87.94	74.96	78.56	84.11	80.64	80.67	88.7
6	93.60	94.35	94.00	86.02	92.68	86.67	85.08	90.02	83.93	87.39	**94.43**
8	95.56	96.08	95.83	89.31	94.47	90.15	88.21	92.83	88.90	90.82	**96.27**
12	97.47	97.37	97.38	93.39	97.41	94.52	90.29	95.93	91.33	96.11	**98.07**
					CMU PIE Database						
Samples	LDA	OLDA	SLDA	CRP	MPDA	LRPE	LRPP	FSP	AAML	DSL	**Ours**
10	81.09	87.38	85.74	82.97	76.32	75.65	84.59	76.94	83.45	78.39	**91.25**
15	87.53	91.32	90.41	88.11	82.74	82.72	89.75	83.74	89.38	82.59	**94.10**
20	90.57	93.20	92.60	90.72	86.81	85.99	92.49	87.27	92.81	87.00	**95.35**
25	92.41	94.33	93.92	92.50	89.44	88.94	95.82	89.23	95.88	90.38	**96.07**
					Extended Yale B Database						
SamplesA	LDA	OLDA	SLDA	CRP	MPDA	LRPE	LRPP	FSP	AAML	DSL	**Ours**
10	82.01	86.18	83.77	82.95	83.67	83.65	84.82	84.31	**87.93**	85.17	86.25
15	87.57	90.38	88.97	88.72	86.82	87.70	89.07	87.84	**91.62**	89.39	91.39
20	90.24	92.56	91.74	89.35	90.81	90.40	91.41	89.81	93.11	91.30	**93.51**
25	91.94	93.78	93.31	90.68	91.79	91.24	92.38	90.88	94.53	92.66	**94.69**

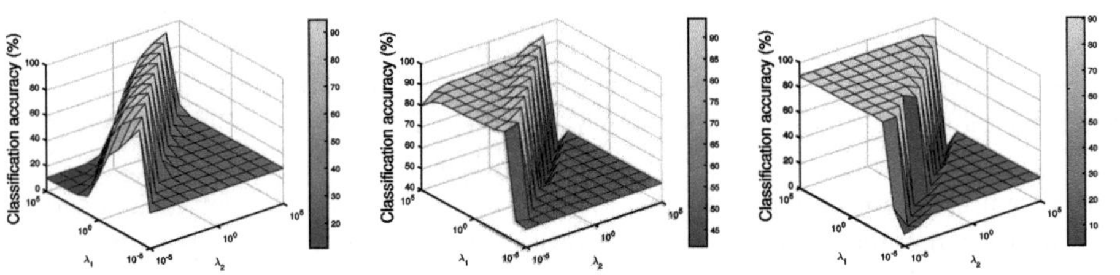

(a) CMU PIE (b) AR (c) Extended Yale B

Fig. 2. The classification accuracies (%) vs. parameters λ_1 and λ_2 with fixed parameter λ_3 = 0.00001 of our on (a) the CMU PIE database, (b) the AR database, (c) the Extended Yale B database.

3.3 Parameter Sensitivity Analysis

The proposed EDSFE requires pre-setting three parameters: λ_1, λ_2, and λ_3. Parameter sensitivity analysis (testing values from 10^{-5} to 10^5 across three databases) shows optimal performance when $\lambda_1 \approx \lambda_2$ (lower values work better), while λ_3 has minimal impact unless near 10^4 (which should be avoided for CMU PIE and AR). Tuning λ_1 and λ_2 is critical (Figs. 2 and 3).

Figure 4 illustrates EDSFE's classification performance across feature dimensions on the three databases. It shows consistent performance over a wide range, with CMU PIE performing well with 5 to 20 subspace dimensions. More training samples enhance flexibility in selecting optimal subspace dimensions, improving adaptability.

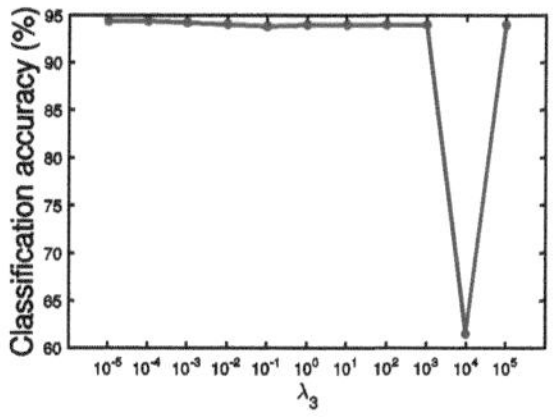

(a) CMU PIE

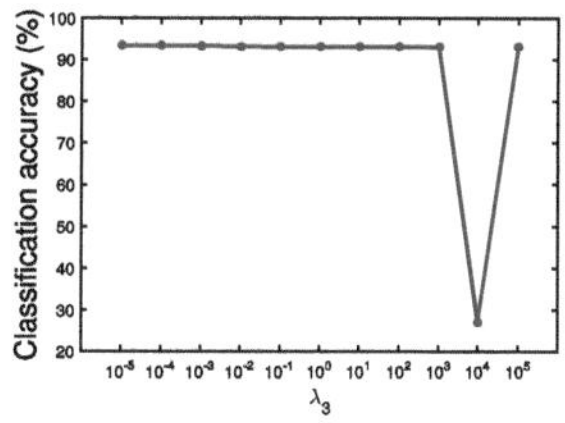

(b) AR

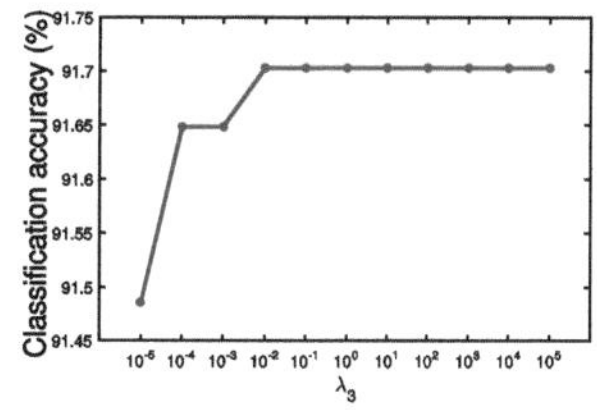

(c) Extended Yale B

Fig. 3. The classification accuracies (%) vs. The parameter λ_3 with fixed parameters $\lambda_1 = 0.001$ and $\lambda_2 = 0.01$ of our on (a) the CMU PIE database, (b) the AR database, (c) the Extended Yale B database.

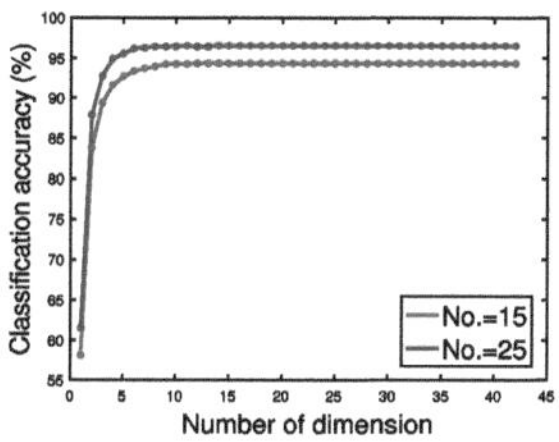

(a) CMU PIE

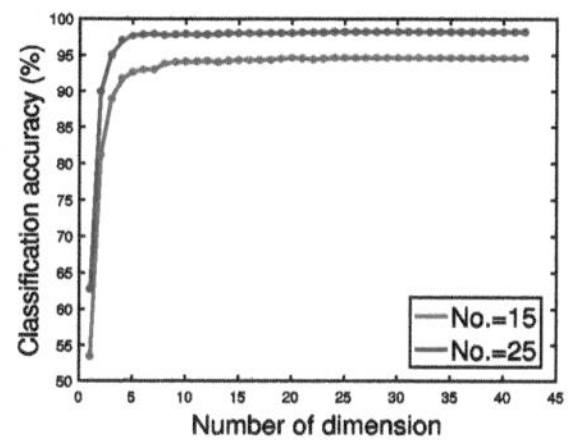

(b) AR

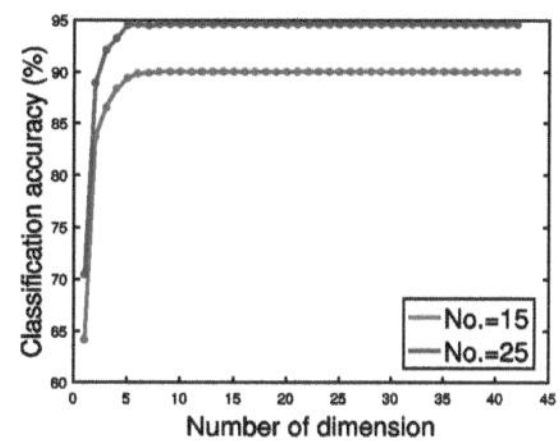

(c) Extended Yale B

Fig. 4. Dimension and classification accuracy versus the dimensions of the proposed method on (a) the CMU PIE database, (b) the AR database, (c) the Extended Yale B database.

4 Conclusion

We have introduced a novel approach EDSFE to address key limitations in the traditional LDA method. EDSFE designs for feature extraction, effectively extracts discriminative and sparse features from data. EDSFE enhances the separability of different classes. This leads to a reduced-dimensional representation, which not only enhances classification accuracy and facilitates feature selection but also aids in data visualization—thereby improving the overall effectiveness of feature extraction tasks. Experimental results on three databases prove that compared with other competitive methods, the proposed method obtains the best performance. Future work for EDSFE includes parameter selection, computational complexity considerations, and exploring efficient algorithms to enhance its practical application and adoption across various domains.

References

1. Jiang, L., Fang, X., Sun, W., Han, N., Teng, S.: Low-rank constraint based dual projections learning for dimensionality reduction. Signal Process. **204**, 108817 (2023)
2. Zhao, S., Zhang, B., Yang, J., Zhou, J., Xu, Y.: Linear discriminant analysis. Nat. Rev. Methods Primers **4**(1), 70 (2024)

3. Iosifidis, A., Tefas, A., Nikolaidis, N., Pitas, I.: Multi-view human movement recognition based on fuzzy distances and linear discriminant analysis. Comput. Vis. Image Underst. **116**(3), 347–360 (2012)
4. Zhou, Y., Sun, S.: Manifold partition discriminant analysis. IEEE Trans. Cybernet. **47**(4), 830–840 (2016)
5. Zhang, X., Chu, D., Tan, R.C.: Sparse uncorrelated linear discriminant analysis for under sampled problems. IEEE Trans. Neural Netw. Learn. Syst. **27**(7), 1469–1485 (2015)
6. Wan, H., Wang, H., Guo, G., Wei, X.: Separability-oriented subclass discriminant analysis. IEEE Trans. Pattern Anal. Mach. Intell. **40**(2), 409–422 (2017)
7. Zou, H., Hastie, T., Tibshirani, R.: Sparse principal component analysis. J. Comput. Graph. Stat. **15**(2), 265–286 (2006)
8. Georghiades, A.S., Belhumeur, P.N., Kriegman, D.J.: From few to many: illumination cone models for face recognition under variable lighting and pose. IEEE Trans. Pattern Anal. Mach. Intell. **23**(6), 643–660 (2001)
9. Martinez, A., Benavente, R.: The ar face database: Cvc technical report, 24 (1998)
10. Sim, T., Baker, S., Bsat, M.: The cmu pose, illumination, and expression (pie) database. In: Proceedings of Fifth IEEE International Conference on Automatic Face Gesture Recognition, pp. 53–58. IEEE (2002)
11. Martinez, A.M., Kak, A.C.: Pca versus lda. IEEE Trans. Pattern Anal. Mach. Intell. **23**(2), 228–233 (2001)
12. Ye, J., Xiong, T.: Null space versus orthogonal linear discriminant analysis. In: Proceedings of the 23rd International Conference on Machine Learning, pp. 1073–1080 (2006)
13. Qiao, Z., Zhou, L., Huang, J.Z.: Sparse linear discriminant analysis with applications to high dimensional low sample size data. IAENG Intern. J. Appli. Math. **39**(1) (2009)
14. Yang, W., Wang, Z., Sun, C.: A collaborative representation based projections method for feature extraction. Pattern Recogn. **48**(1), 20–27 (2015)
15. Zhang, Y., Xiang, M., Yang, B.: Low-rank preserving embedding. Pattern Recogn. **70**, 112–125 (2017)
16. Wen, J., Han, N., Fang, X., Fei, L., Yan, K., Zhan, S.: Low-rank preserving projection via graph regularized reconstruction. IEEE Trans. Cybernet. **49**(4), 1279–1291 (2018)
17. Tang, C.: Feature selective projection with low-rank embedding and dual laplacian regularization. IEEE Trans. Knowl. Data Eng. **32**(9), 1747–1760 (2019)
18. He, J., et al.: Adaptive affinity matrix learning for dimensionality reduction. Inter. J. Mach. Learn. Cybernet. **1–15** (2023)
19. Huang, Z., Zhao, S., Liang, Z., Wu, J.: Discriminative subspace learning with adaptive graph regularization. Comput. J. **67**(9), 2823–2837 (2024)

Meta-CoT-A*-MCTS: Search for Stronger User Preference Alignment in Agent4Rec

Ruilong Huang[1], Bohan Li[1(✉)], Haofen Wang[2], Mengfei Xu[1], Chen Chen[1], and Xinzhe Zhao[1]

[1] Nanjing University of Aeronautics and Astronautics, Nanjing, China
{hrl114514,bhli,reuschen32,xinzhe_zhao,jensonxu}@nuaa.edu.cn
[2] Tongji University, Shanghai, China

Abstract. Recommender systems aim to personalize user experiences by suggesting relevant items based on historical interactions and preferences. However, traditional CoT reasoning struggles to align generative agents' outputs with real-time user preferences, especially when faced with sparse or changing data. In this paper, we introduced the Meta-CoT-A*-MCTS framework by utilizing A* for deterministic pathfinding and MCTS for stochastic exploration, which combines the strengths of A* search and MCTS to improve personalized recommendation systems. The hybrid approach allows for dynamic exploration of reasoning paths, optimizing the balance between computational efficiency and recommendation quality. Our experiments on datasets including MovieLens-1M, MovieLens-10M, Amazon-Book, and Steam demonstrate that that Meta-CoT-A*-MCTS consistently outperforms other methods, achieving higher F1 scores across various benchmarks and model sizes. This framework effectively balances high-quality recommendation generation with computational efficiency, reducing time complexity while maintaining high-quality recommendations.

Keywords: Recommendation System · Generative Agents · Search Algorithm · Large Language Model · User Preference Alignment

1 Introduction

Recommender systems are pivotal in personalizing user experiences by suggesting relevant items based on user preferences [13,21] and historical interactions. The emergence of Large Language Models(LLMs) [3] has significantly boosted the performance of generative agents [19] in recommendation tasks, as LLM-driven agents can simulate user behavior and generate personalized content, fostering highly tailored recommendations.

However, the traditional Chain-of-Thought (CoT) approach faces two main challenges. First, preference modeling suffers from drift due to sparse, noisy, and non-stationary behavioral logs, leading to cold-start errors and temporal staleness [9]. Despite zero-shot rankers, empirical studies reveal a growing mismatch

M. Yoshikawa et al. (Eds.): ADMA 2025, LNAI 16197, pp. 393–401, 2026.
https://doi.org/10.1007/978-981-95-3453-1_28

between generated rationales and evolving user preferences [5]. Second, hallucinations occur when LLMs generate irrelevant or non-existent items [1], reducing trust in system. While CoT can address transparency issues, it lacks mechanisms for backtracking and refining incorrect reasoning steps, leading to misalignments in personalized recommendations [5].

Meta-CoT resolves these issues by incorporating a search over latent reasoning traces, enabling the model to explore multiple reasoning paths and iteratively refine its predictions [20]. This approach improves adaptability and alignment with dynamic user preferences. However, combining Meta-CoT with search strategies like A* and MCTS can lead to high computational complexity, particularly due to the large search space.

To address the challenges of traditional Meta-CoT search strategies, we propose a hybrid **Meta-CoT-A*-MCTS** framework that integrates A* and MCTS. A* optimizes the search by balancing cost and heuristic estimates, while MCTS simulates reasoning paths through multiple rollouts. This combination enhances recommendation quality and reduces time complexity. Our approach reduces computational complexity by leveraging **MCTS as a heuristic for A***, balancing efficiency and recommendation quality, overcoming the limitations of standalone Meta-CoT methods.

2 Related Work

Recent advancements in LLMs have enabled the rise of LLM-empowered generative agents, offering adaptability across diverse applications, from social simulations to recommendation systems [6,19]. Unlike traditional, domain-specific agents [7], these agents leverage memory, planning, and reflection to enhance their performance in tasks like user behavior simulation within recommendation systems, as seen in models like RecAgent [11] and Agent4Rec [19].

Traditional CoT prompting has proven effective for guiding LLMs to break down complex problems into explicit intermediate reasoning steps, improving performance in multi-step benchmarks [12,14,17]. However, CoT's linear reasoning approach can become trapped in local dead-ends, limiting its robustness. Meta-CoT extends CoT by treating the reasoning process itself as a search problem, allowing for multiple potential reasoning chains [15]. Classical search algorithms like A* and MCTS enhance this process by guiding reasoning towards high-reward paths. He et al. apply A*-style search with a learned Q-value heuristic to direct chain expansion without fine-tuning [2], while Xie et al. integrate MCTS to balance exploration and exploitation by simulating reasoning paths [16]. These methods, when combined, provide efficient and structured searches over latent reasoning spaces, significantly improving the generation of coherent, high-quality recommendations.

3 Meta-Agent4Rec

In this section, we propose the Meta-CoT-A*-MCTS framework, which integrates Meta-CoT reasoning with complementary search strategies, specifically

A* and MCTS, to improve personalized recommendations. By embedding Meta-CoT within search processes like A* and MCTS [15,18], we enhance LLM's inference capabilities and allow dynamic exploration of reasoning paths. A* optimizes the search using a heuristic-driven approach, while MCTS explores decision spaces through randomized rollouts, enabling more robust recommendation reasoning in data-sparse environments. These strategies are combined to balance exploration and exploitation, improving both recommendation accuracy and computational efficiency.

3.1 What Is Meta-CoT?

Traditional CoT prompting enhances LLM reasoning by generating intermediate steps towards a final answer [12]. However, CoT is static and linear, making it prone to failure if early assumptions are wrong, as it lacks the ability to revise them [5]. Meta-CoT extends CoT by exploring multiple reasoning paths, allowing for iterative refinement [15]. This enables "System 2" deliberation to complement LLMs' fast "System 1" generation [20], improving decision-making, particularly in complex recommendation tasks.

In Traditional CoT, the model generates a reasoning chain $s_1, s_2, \ldots, s_n$ leading to a final recommendation a, expressed as:

$$p_{\text{model}}(a \mid q) \propto \int_{s_1}^{s_n} p_{\text{model}}(a \mid s_{1:n}, q) \prod_{t=1}^{n} p_{\text{model}}(s_t \mid s_{1:t-1}, q)\, dS, \tag{1}$$

However, CoT's fixed reasoning path can fail when early steps are incorrect. Meta-CoT addresses this by introducing latent variables $z_1, z_2, \ldots, z_K$ to explore multiple reasoning paths in parallel. This allows the model to backtrack and refine hypotheses, enhancing preference alignment in recommendation systems [5,15].

$$p_{\text{model}}(a, s_{1:n} \mid q) \propto \int_{z_1}^{z_K} p_{\text{model}}(a, s_{1:n} \mid z_{1:K}, q) \prod_{t=1}^{K} p_{\text{model}}(z_t \mid z_{1:t-1}, q)\, dZ, \tag{2}$$

3.2 Markov Decision Process for Personalized Recommendation

Recent studies have demonstrated similar Markov Decision Process (MDP) applications for multi-step reasoning in LLM-based models, where the agent learns to optimize the reasoning chain leading to the best recommendation. Meta-CoT leverages MDPs to align both reasoning and actions, iteratively refining the reasoning process to maximize user satisfaction. We model the recommendation task as an MDP with the tuple $(\mathcal{S}, \mathcal{A}, T, R, \gamma)$, where $\mathcal{S}$ represents states (user profile, preferences, partial reasoning chain), and $\mathcal{A}$ denotes actions such as recommending an item or refining reasoning [2].

The reward function $R : \mathcal{S} \times \mathcal{A} \rightarrow R$ assigns a numerical value based on user satisfaction, such as the rating given to an item. The agent's objective is to maximize expected cumulative reward, expressed as:

$$J(\pi) = E_{\pi}\left[\sum_{t=0}^{H} \gamma^{t} r_{t} \mid s_{0}\right], \tag{3}$$

where $r_t = R(s_t, a_t)$ is the reward at time t, and H is the horizon, representing the number of reasoning steps or recommendations made. By framing the recommendation task as an MDP, we can apply planning algorithms to search for the optimal sequence of actions (reasoning steps and recommendations) that maximize user satisfaction.

Algorithm 1. Meta-CoT Search: A* with MCTS-based Heuristic

Input: initial state s_0, goal test $Goal(\cdot)$, transition T, reward R, MCTS budget N, exploration constant c
Output: reasoning path maximizing user preference alignment
Initialize open list Open $\leftarrow \{s_0\}$ with priority $f(s) = g(s) + h(s)$, where $g(s)$ is cumulative reward and $h(s)$ is the heuristic
Initialize $g(s_0) \leftarrow 0$, $h(s_0) \leftarrow$ MCTSHeuristic(s_0, N, c)
while Open $\neq \emptyset$ **do**
  $s \leftarrow$ **pop** state with highest priority from Open
  if $Goal(s)$ **then**
    return solution path (backtrack from s_0)
  end if
  for each action $a \in \mathcal{A}(s)$ **do**
    $s' \leftarrow T(s, a)$, $r \leftarrow R(s, a)$, $g(s') \leftarrow g(s) + r$
    $h(s') \leftarrow$ MCTSHeuristic(s', N, c), $f(s') \leftarrow g(s') + h(s')$
    if $s' \notin$ Closed or $g(s') > g(s')_{\text{previous}}$ **then**
      push s' into Open with priority $f(s')$
    end if
  end for
end while
return failure // *no solution found*
Function: MCTSHeuristic(state s, MCTS budget N, exploration constant c)
Initialize $Q(s, a) \leftarrow 0$, $N(s, a) \leftarrow 0$ for all $a \in \mathcal{A}(s)$
for $i = 1$ to N **do**
  $v \leftarrow s$
  for depth $d = 1$ to H **do**
    $a \leftarrow$ SelectAction(v, c) // *Select using UCB or similar strategy*
    $v \leftarrow T(v, a)$
  end for
  $G \leftarrow$ Rollout(v) // *simulate until terminal state*
  Backpropagate(G, v) // *update Q-values and visit counts*
end for
return $\max_a Q(s, a)$

3.3 Meta-CoT Inference with A* Search and MCTS

Meta-CoT-A*-MCTS integrates A* search and MCTS to enhance the inference process in personalized recommendation systems. A* is a best-first search algorithm that guarantees an optimal path when the heuristic is admissible. In our framework, we adapt A* to maximize expected reward, treating negative rewards as "costs." Each node in the search tree represents a state s, including a partial reasoning chain, user profile, or recommendation. A heuristic function $h(s)$ estimates the additional reward the agent can achieve from state s until the goal [15,18]. This guides the search by predicting the success of different reasoning paths. For example, if the agent hypothesizes user preferences but has yet to make a recommendation, $h(s)$ estimates the likelihood of success based on the reasoning chain. The A* algorithm explores states in order of decreasing $f(s) = g(s) + h(s)$, where $f(s)$ is the total estimated score.

We incorporate an MCTS-based heuristic to simulate reasoning paths and estimate future rewards, improving the heuristic $h(s)$. This multi-step exploration helps anticipate user preferences more accurately than static methods. Combining A* with MCTS creates a balanced approach: A* prioritizes high-potential paths, while MCTS explores alternative reasoning chains. The algorithm is shown in Algorithm 1. The agent initializes $g(s_0) = 0$ and computes $h(s_0)$ via MCTS-based rollouts. The total score is $f(s_0) = g(s_0) + h(s_0)$, and the search begins from s_0. The agent applies A* to prioritize states with the highest $f(s)$, simulating possible actions $a \in \mathcal{A}$. The transition function $T(s, a)$ models how actions update the state, and the reward function $R(s, a)$ represents user feedback. The heuristic for each successor state s' is computed as $h(s') \approx \text{MCTS}(s', N)$, where N is the number of rollouts. MCTS simulates several paths from s', averaging rewards to estimate future success. The agent updates the priority queue, prioritizing states with higher $f(s') = g(s') + h(s')$, ensuring a more informed exploration and exploitation process. This hybrid Meta-CoT-A*-MCTS approach provides an efficient, dynamic method for personalized recommendations, improving both time complexity and accuracy.

3.4 Computational Efficiency and Analysis

This section analyzes the time complexities of four LLM-driven recommendation strategies, with a focus on Meta-CoT-A*-MCTS. This hybrid method combines MCTS as a heuristic for A*, optimizing the search by reducing expanded nodes and improving pruning. As shown in Table 1, Meta-CoT-A*-MCTS outperforms A*, retaining its deterministic pathfinding while adding stochastic exploration through MCTS. In the best case, the complexity is $O(d)$, similar to A*, but with better pruning. Even in the worst case, Meta-CoT-A*-MCTS is more efficient than standalone A*, thanks to the MCTS heuristic guiding the search and minimizing unnecessary exploration.

4 Experiments

The experiment evaluates whether LLM-driven generative agents can maintain long-term preference coherence, crucial for personalized recommendations. We instantiate 1,000 independent agents on four benchmarks: MovieLens-1M [4], MovieLens-10M [4], Amazon-Book [8], and Steam [10]. Each agent is assigned 20 candidate items and asked to accept or reject them under interaction ratios of 1:1, 1:2, 1:3, and 1:9. Experiments use LLaMA 2 models (1.3B, 3B, 7B, and 13B parameters) repeated 10 times with deterministic seeds $\{10, 20, \ldots, 100\}$. Results report the arithmetic mean of the **F1 Score**, with standard deviations. From Fig. 2, we observe that Meta-CoT-A*-MCTS consistently outperforms Meta-CoT-A* and Meta-CoT-MCTS across almost all datasets. For models of the same architecture, larger models are more stable, while smaller models show more fluctuations. The performance improvement from 7B to 13B is marginal,

Table 1. Complexity Summary under Different Scenarios (Structural vs. Execution).

Method	Structural Complexity T			Execution Complexity T_{exec}		
	Best	Normal	Worst	Best	Normal	Worst
CoT	$O(nd)$	$O(nd)$	$O(nd)$	$O(nd\tau)$	$O(nd\tau)$	$O(nd\tau)$
Meta-CoT-A*	$O(d)$	$O(n^{\alpha}d)$	$O(n^{\alpha d})$	$O(d\tau)$	$O(n^{\alpha}d\tau + n^{\alpha}d\log n)$	$O(n^{\alpha d}d\log n\tau)$
Meta-CoT-MCTS	$O(d)$	$O(Td)$	$O(n^{d}d)$	$O(d\tau)$	$O(Td\tau)$	$O(n^{d}d\tau)$
Meta-CoT-A*-MCTS	$O(d)$	$O(n^{\beta}d)$	$O(n^{\beta d})$	$O(d\tau)$	$O(n^{\beta}d\tau)$	$O(n^{\beta d}d\tau)$

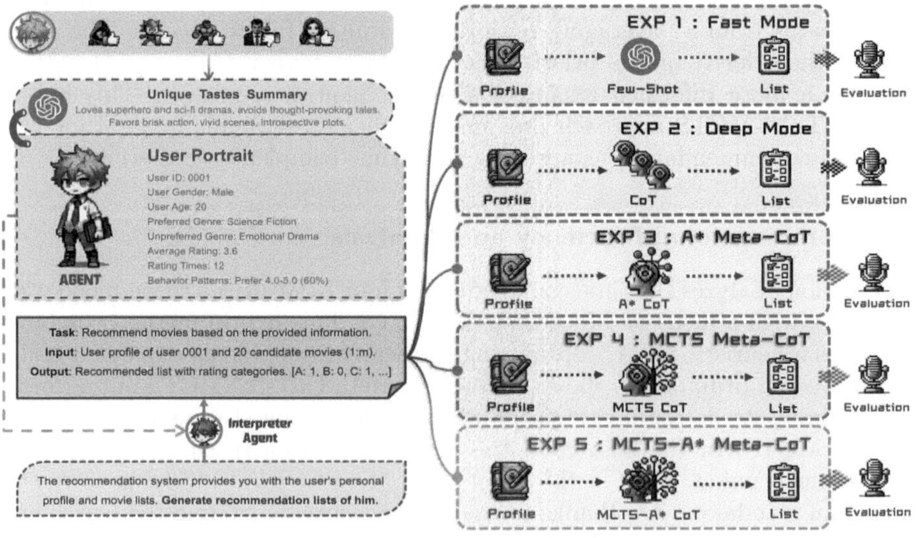

Fig. 1. The integration of A* and MCTS in the Meta-CoT-A*-MCTS approach is particularly noteworthy, as it offers both high-quality recommendations and improved computational efficiency.

suggesting diminishing returns. As interaction ratios increase from 1:1 to 1:9, F1 scores decrease, indicating the growing challenge of maintaining preference coherence with higher imbalances in acceptance and rejection.

5 Conclusion and Acknowledgments

In this paper, we proposed the Meta-CoT-A*-MCTS framework, which integrates A* search with MCTS to enhance personalized recommendation systems by improving both recommendation quality and computational efficiency. Our experiments demonstrated the effectiveness of the hybrid approach. We observed that increasing model size led to improved F1 scores, while Meta-CoT-A*-MCTS consistently outperformed other methods. We would like to express our sincere gratitude to Professor Li, Professor Wang and the members of the DBK Lab for their valuable support. We also thank the developers and contributors to the LLaMA open-source models and Agent4Rec [19] (Fig. 1).

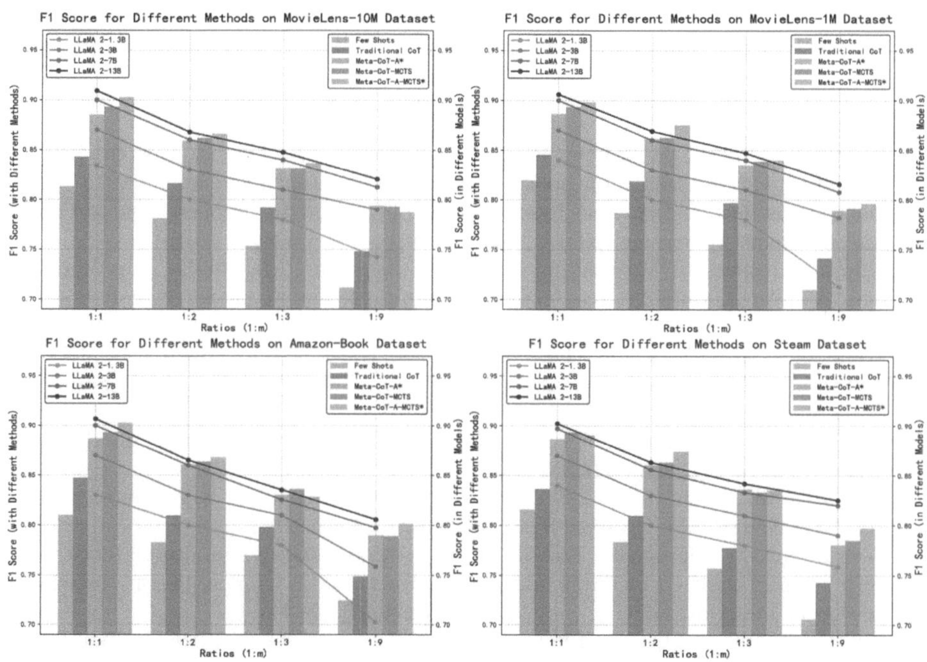

Fig. 2. F1 Score for Different Methods on MovieLens-10M, MovieLens-1M, Amazon-Book, and Steam Datasets. The bar chart compares performance across five different prompting methods or reasoning approaches, while the line chart evaluates the Meta-CoT-A*-MCTS method using models of different sizes (LLaMA 2-1.3B, 3B, 7B, 13B).

References

1. Abbasiantaeb, Z., Yuan, Y., Kanoulas, E., Aliannejadi, M.: Let the llms talk: simulating human-to-human conversational qa via zero-shot llm-to-llm interactions. In: Proceedings of the 17th ACM International Conference on Web Search and Data Mining, pp. 8–17 (2024)
2. Chakraborty, S., Ghosal, S.S., Yin, M., Manocha, D., Wang, M., Bedi, A.S., Huang, F.: Transfer q-star: principled decoding for llm alignment. Adv. Neural. Inf. Process. Syst. **37**, 101725–101761 (2024)
3. Chang, Y., Wang, X., Wang, J., Wu, Y., Yang, L., Zhu, K., Chen, H., Yi, X., Wang, C., Wang, Y., et al.: A survey on evaluation of large language models. ACM Trans. Intell. Syst. Technol. **15**(3), 1–45 (2024)
4. Harper, F.M., Konstan, J.A.: The movielens datasets: history and context. Acm Trans. Interact. Intell. Syst. (tiis) **5**(4), 1–19 (2015)
5. Lyu, Q., et al.: Faithful chain-of-thought reasoning. In: The 13th International Joint Conference on Natural Language Processing and the 3rd Conference of the Asia-Pacific Chapter of the Association for Computational Linguistics (IJCNLP-AACL 2023) (2023)
6. Park, J.S., O'Brien, J., Cai, C.J., Morris, M.R., Liang, P., Bernstein, M.S.: Generative agents: Interactive simulacra of human behavior. In: Proceedings of the 36th Annual acm Symposium on User Interface Software and Technology, pp. 1–22 (2023)
7. Putta, P., et al.: Agent q: advanced reasoning and learning for autonomous ai agents. arXiv preprint arXiv:2408.07199 (2024)
8. Srujan, K., Nikhil, S., Raghav Rao, H., Karthik, K., Harish, B., Keerthi Kumar, H.: Classification of amazon book reviews based on sentiment analysis. In: Information Systems Design and Intelligent Applications. In: Proceedings of Fourth International Conference INDIA 2017, pp. 401–411. Springer (2018)
9. Subramaniyaswamy, V., Logesh, R., Chandrashekhar, M., Challa, A., Vijayakumar, V.: A personalised movie recommendation system based on collaborative filtering. Int. J. High Perform. Comput. Networking **10**(1–2), 54–63 (2017)
10. Wang, D., Moh, M., Moh, T.S.: Using deep learning and steam user data for better video game recommendations. In: Proceedings of the 2020 ACM Southeast Conference, pp. 154–159 (2020)
11. Wang, L., et al.: User behavior simulation with large language model based agents. arXiv preprint arXiv:2306.02552 (2023)
12. Wei, J., et al.: Chain-of-thought prompting elicits reasoning in large language models. Adv. Neural. Inf. Process. Syst. **35**, 24824–24837 (2022)
13. Wu, J., Xu, Y., Zhang, B., Xu, Z., Li, B.: Graph-based dynamic preference modeling for personalized recommendation. In: Pacific-Asia Conference on Knowledge Discovery and Data Mining, pp. 356–368. Springer (2024)
14. Wu, W., Wang, H., Li, B., Huang, P., Zhao, X., Liang, L.: Multirag: a knowledge-guided framework for mitigating hallucination in multi-source retrieval augmented generation. arXiv preprint arXiv:2508.03553 (2025)
15. Xiang, V., et al.: Towards system 2 reasoning in llms: learning how to think with meta chain-of-though. arXiv preprint arXiv:2501.04682 (2025)
16. Xie, Y., et al.: Monte carlo tree search boosts reasoning via iterative preference learning. arXiv preprint arXiv:2405.00451 (2024)
17. Xuan, H., Li, B., Wu, W., Liu, Y., Yin, H.: Knowledge enhancement and temporal aware for multi-behavior contrastive recommendation. ACM Trans. Intell. Syst. Technol. (2025)

18. Yao, H., et al.: Mulberry: empowering mllm with o1-like reasoning and reflection via collective monte carlo tree search. arXiv preprint arXiv:2412.18319 (2024)
19. Zhang, A., Chen, Y., Sheng, L., Wang, X., Chua, T.S.: On generative agents in recommendation. In: Proceedings of the 47th International ACM SIGIR Conference on Research and Development in Information Retrieval, pp. 1807–1817 (2024)
20. Zhang, Z., et al.: Igniting language intelligence: the hitchhiker's guide from chain-of-thought reasoning to language agents. ACM Comput. Surv. (2023)
21. Zhuo, J., et al.: Tidgrec: dual-graph modeling with target-intention filtering for session-based recommendation (2025)

An Entity-Relation Extraction Framework via Symmetry-Aware Augmentation and Priority-Constrained Optimization

Xiaojun Sheng[1,2], Yiyan Li[1,2], Minmin Li[1,3](✉), Shunli Wang[1,2], Yafei Wang[1,2], and Renzhong Guo[2]

[1] Guangdong Laboratory of Artificial Intelligence and Digital Economy (SZ), Shenzhen 518123, China

[2] Shenzhen University, Shenzhen 518060, China

[3] Key Laboratory of Urban Land Resources Monitoring and Simulation, Ministry of Natural Resources, Shenzhen 518034, China

liminmin@gml.ac.cn

Abstract. Entity-relation extraction aims to jointly identify entities and their semantic relations from unstructured text, yet it remains hindered by relation sparsity, long-tail distributions, and optimization conflicts between entity and relation subtasks. We propose SPADE, a unified framework that addresses these challenges through two core components: (1) a Symmetry-aware and Self-referential Relation Augmentation strategy that injects structurally valid training instances to improve relational diversity and robustness; and (2) a Joint Optimization with Priority Constraints method that enforces hierarchical learning dynamics via Lagrangian multipliers, prioritizing entity boundary detection to stabilize multi-task training. Extensive experiments on five benchmark datasets across multiple domains show that SPADE consistently outperforms strong baselines, achieving up to +3.1% absolute improvement in triplet-level F1 on low-resource settings. Our code and models will be released to support future research.

1 Introduction

Entity-relation extraction (ERE) aims to identify structured triples—(subject, relation, object)—from unstructured text, supporting applications such as knowledge graph construction [21], semantic retrieval [11], and question answering [15]. While early work focused on sentence-level ERE, recent research emphasizes the document-level setting, where relational reasoning must span multiple, often distant, sentences. Despite progress in pipeline architectures [14], joint models [23], and LLMs [6], document-level ERE remains challenging due to two factors: (1) relation sparsity and long-tail distribution [13], and (2) optimization conflicts caused by parallel training of entity and relation subtasks [7], ignoring their inherent hierarchy. Prior work generally falls into two categories: encoder-sharing models such as TDEER [5] and TPLinker [18], which suffer from weak

M. Yoshikawa et al. (Eds.): ADMA 2025, LNAI 16197, pp. 402–409, 2026.
https://doi.org/10.1007/978-981-95-3453-1_29

task interaction, and unified decoders like OneRel [10] and UNIRE [17], which improve consistency but neglect the asymmetry between entity and relation tasks. More recent methods incorporate richer interactions via hypergraphs [20], bidirectional refinement [8], and text-to-graph generation [23], but still lack explicit modeling of task hierarchy or relational structure.

To address these issues, we propose **SPADE** (**S**ymmetric and **P**riority-**A**ware **D**ocument **E**xtractor), a unified framework comprising: (i) **SSRA**, a structure-aware augmentation scheme that injects symmetric, bidirectional, and self-referential triplets; and (ii) **JOPC**, a dynamic training strategy enforcing task hierarchy via Lagrangian optimization. SPADE aligns training dynamics with task dependencies and achieves state-of-the-art performance across five benchmarks, especially in long-context or low-resource settings.

In summary, this work makes the following contributions:

- We propose SSRA to enrich training with structurally valid triplets.
- We design JOPC to prioritize entity recognition during joint optimization.
- We validate SPADE on five benchmarks, demonstrating consistent improvements over strong baselines.

2 Methodology

2.1 Task Formulation

Given a document D, the goal of ERE is to extract a set of structured triplets (e_s, r, e_o), where e_s and e_o are entities, and r is the relation type.

2.2 Overview of SPADE

SPADE is a unified framework for document-level ERE, built upon a query-based set prediction architecture [12]. It consists of four stages: (i) structure-aware augmentation, (ii) contextual encoding via a PLM (e.g., DeBERTa), (iii) parallel triplet decoding, and (iv) joint optimization with task prioritization. The architectural novelty lies in two key components: **SSRA** and **JOPC**.

2.3 SSRA: Symmetry-Aware Augmentation

To address the long-tail distribution of relations, we propose a principled data augmentation strategy leveraging relational structures.

(1) Relational Property-Aware Augmentation. We categorize relations as:

- **Symmetric** (e.g., *married to*): augment (e_s, r, e_o) with (e_o, r, e_s).
- **Directionless** (e.g., *member of*): inject reversed pairs for bidirectional robustness.
- **Asymmetric** (e.g., *founded by*): generate inverse relations (e_o, r', e_s).

(2) Entity Self-Referential Augmentation. We add synthetic triplets (e_i, NR, e_i) (NR: no-relation), serving as soft negatives and reinforcing entity boundary detection.

This augmentation enriches training diversity and improves generalization, especially in sparse or low-resource settings.

2.4 JOPC: Joint Optimization with Priority Constraints

Standard joint training overlooks task hierarchy: relation prediction is only meaningful when entity spans are correct. To address this, we formulate a constrained optimization objective:

$$\min_{\theta} L_r \quad \text{s.t.} \quad L_s \leq \hat{L}_s, \quad L_o \leq \hat{L}_o \tag{1}$$

We introduce Lagrangian multipliers λ_s, λ_o to enforce this constraint during training:

$$L = \frac{L_r + \lambda_s(L_s - \hat{L}_s) + \lambda_o(L_o - \hat{L}_o)}{1 + \lambda_s + \lambda_o} \tag{2}$$

Here, L_s, L_o are entity span losses; L_r is relation classification loss. Multipliers are updated adaptively to prioritize entity learning first. This "first-things-first" dynamic ensures stable multi-task training and better alignment with the underlying task structure.

3 Experiments

3.1 Experimental Setup

We evaluate SPADE on five benchmarks from Wikipedia (WebNLG [9], news (NYT [9]), DocRED [22]), and biomedical domains (CDR [4], GDA [19]), covering both sentence-level and document-level extraction. Dataset statistics are in Table 1.

We implement SPADE using Huggingface Transformers with DeBERTa [3] as encoder. Following SPN4RE [12], we use AdamW with learning rates 1e−5 (encoder) and 2e−5 (decoder), training for 30 epochs and decoding up to 100 triplets per document. Micro-averaged F1 is reported. All experiments are run on a single NVIDIA A100 80 GB GPU.

Table 1. Statistics of datasets.

Dataset	Domain	Train	Val.	Test
NYT	News	56k	5k	5k
WebNLG	Wiki	35k	1.7k	1.7k
DocRED	Wiki	3k	300	700
CDR	Bio	500	500	500
GDA	Bio	19k	4.7k	4.7k

Table 2. SPADE achieves strong performance on sentence level benchmarks (NYT, WebNLG). Our method matches or exceeds previous SOTA systems (e.g., MFSF, ERGM), demonstrating that structural augmentation and priority-aware optimization also benefit high-resource, saturated benchmarks. † indicates results reproduced from [24].

Model	NYT			WebNLG		
	P	R	F1	P	R	F1
ChatGPT-3.5†	52.1	53.2	52.5	–	–	–
ChatGPT-4†	62.6	63.4	62.9	–	–	–
SPN4RE [12]	92.5	92.2	92.3	93.1	93.6	93.4
ERFD-RTE [1]	**94.0**	91.4	92.7	91.2	87.4	89.3
ERGM [2]	93.3	91.5	92.4	94.2	91.2	92.7
MFSF [16]	93.6	91.7	92.6	**94.9**	92.3	93.5
SPADE	92.6$_{\pm 0.08}$	**92.4**$_{\pm 0.07}$	**92.9**$_{\pm 0.05}$	93.3$_{\pm 0.06}$	**94.0**$_{\pm 0.09}$	**93.9**$_{\pm 0.07}$

3.2 Benchmark Results

Table 2 shows that SPADE achieves F1 scores of 92.9% on NYT and 93.9% on WebNLG, matching or surpassing prior state-of-the-art models such as MFSF and ERGM. Although performance on these high-resource benchmarks is nearing saturation, SPADE still offers consistent gains in precision and recall, largely attributable to SSRA's ability to inject semantically valid augmented examples that enhance relational coverage.

As shown in Table 3, SPADE significantly improves triplet-level F1 across DocRED (+2.38%), CDR (+3.09%), and GDA (+2.13%) compared to SPN4RE. These improvements are particularly notable in entity F1 (up to +3.46%), validating our key design: accurate entity recognition is essential for stable relation classification. By enforcing task prioritization via JOPC, SPADE mitigates gradient conflicts in joint training. The combined effect of SSRA and JOPC enables robust generalization in long-context and low-resource scenarios.

Table 3. SPADE significantly improves document-level extraction across all metrics. Compared to SPN4RE, SPADE achieves consistent gains in entity, relation, and triplet-level F1, especially on low-resource datasets like CDR and long-context ones like DocRED.

Component	DocRED	CDR	GDA
SPN4RE [12]			
Entity	57.28	52.86	62.31
Relation	68.32	96.56	99.12
Triplet	50.12	48.32	60.31
SPADE			
Entity	**58.57**$_{\pm 0.07}$	**55.45**$_{\pm 0.09}$	**65.77**$_{\pm 0.10}$
Relation	**67.68**$_{\pm 0.08}$	**95.80**$_{\pm 0.05}$	**97.38**$_{\pm 0.07}$
Triplet	**52.50**$_{\pm 0.03}$	**51.41**$_{\pm 0.08}$	**62.44**$_{\pm 0.05}$

3.3 Ablation Study

To investigate the contribution of each core component, we conduct ablation experiments on the DocRED dataset by selectively disabling the SSRA and JOPC modules. Results are reported in Table 4.

Table 4. Ablation study on DocRED confirms the complementary effects of SSRA and JOPC. Each component individually improves both entity and relation metrics, and their combination (SPADE) yields the strongest triplet-level F1.

Model	P	R	F1	R-P	R-R	R-F1	E-P	E-R	E-F1
SPN4RE (baseline)	51.8	49.5	50.6	68.3	66.4	67.3	56.8	57.7	57.2
+ SSRA	53.6	53.2	53.4	70.2	67.8	69.0	58.1	59.6	58.8
+ JOPC	55.2	50.1	52.5	72.5	66.0	69.1	60.3	60.9	60.6
SPADE (full)	**57.8**	**55.3**	**56.5**	**74.6**	**71.8**	**73.2**	**63.5**	**65.7**	**64.6**

Effect of SSRA. SSRA improves overall F1 from 50.6% to 53.4%, notably boosting recall by enriching long-tail relation types through symmetry-aware and self-referential triplets. This structural augmentation increases diversity without compromising semantic correctness, benefiting rare relations (Fig. 1). It serves both as a data amplifier and a regularizer, promoting better generalization under label imbalance.

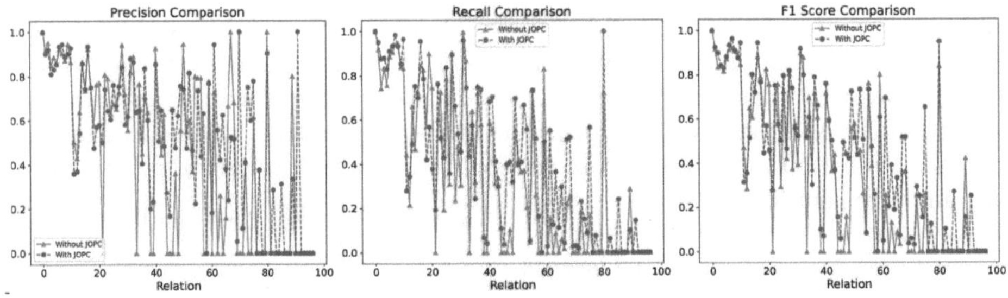

Fig. 1. SSRA improves recall on rare relation types while preserving precision on common ones. This validates SSRA's dual role in enhancing diversity and acting as a regularizer.

Effect of JOPC. JOPC raises F1 to 52.5%, primarily by enhancing entity span consistency. Unlike SSRA, which expands data coverage, JOPC dynamically adjusts loss weighting to prioritize entity recognition until it stabilizes. This "first-things-first" approach mitigates optimization interference, leading to smoother convergence and more accurate triplet extraction (Figs. 2, 3).

Together, SSRA and JOPC address distinct but complementary challenges that data sparsity and training instability–forming the core of SPADE's performance gains.

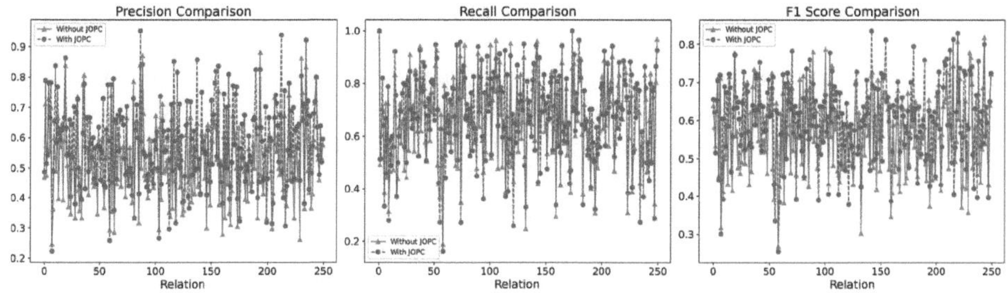

Fig. 2. JOPC improves entity consistency and stabilizes multi-task learning. Entity F1 across test documents shows that JOPC enhances boundary detection.

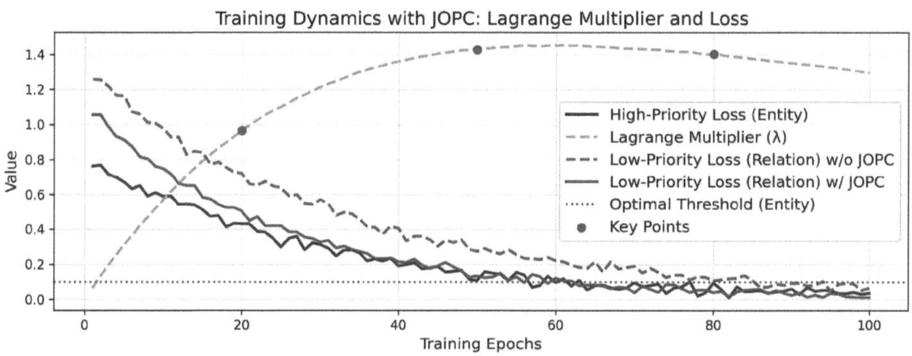

Fig. 3. Training curves show entity losses stabilize before relation convergence, reflecting the intended task hierarchy.

Type	DocRE	Re-DocRED	CDR	GDA	Examples	Predict	Target
C	49.75	41.46	47.21	62.92	Eiffel Tower is in Paaris.	(Eiffel Tower, Located In, Paris)	(Eiffel Tower, Located In, Paris)
WR	1.51	1.30	1.15	0.00	Google acquired YouTube.	(Google, Partner Of, YouTube)	(Google, Acquired, YouTube)
WE	37.71	44.45	40.01	31.94	Obama was born in Honolulu.	(Obama, Born In, Honolulu)	(Barack Obama, Born In, Honolulu)
W	11.03	12.68	11.65	5.14	Apple Park is in Cupertino.	(Google, Owns, Mountain View)	(Apple, Located In, Cupertino)

Fig. 4. Error distribution shows entity misclassification is the dominant failure mode. Wrong-Entity (WE) accounts for the majority of errors across datasets, reinforcing the need for entity-first optimization strategies like JOPC.

3.4 Error Analysis

To better understand model limitations, we categorize prediction errors into four types: **Correct (C)**: Triplet is fully correct. **Wrong-Relation (WR)**: Correct entities, incorrect relation. **Wrong-Entity (WE)**: Correct relation, incorrect entity (subject/object). **Wrong (W)**: Entire triplet is incorrect.

As shown in Fig. 4, **Wrong-Entity (WE)** errors dominate across datasets, reflecting the challenges of boundary detection under co-reference and ambiguity. In contrast, **Wrong-Relation (WR)** errors are rare (<2%), confirming that relation classification is reliable when entity spans are correct.

This asymmetry validates SPADE's strategy: strengthening entity learning (via JOPC) and diversifying training data (via SSRA) effectively addresses the main failure source. Robust entity modeling is thus crucial for improving overall ERE accuracy.

4 Conclusion

We propose **SPADE**, a document-level ERE framework addressing data sparsity, long-tail relations, and task interference. It integrates two key components: (1) **SSRA**, a structure-aware augmentation method leveraging relational symmetry and self-reference; and (2) **JOPC**, a Lagrangian-based optimization strategy that prioritizes entity recognition for stable multi-task learning. Combined with a non-autoregressive, query-based decoder, SPADE consistently improves performance across diverse benchmarks, demonstrating the value of structure-aligned augmentation and hierarchy-aware optimization in document-level ERE.

Acknowledgments. This work was supported by National Key R&D Program of China (2022YFB3903705) and the Open Fund of Key Laboratory of Urban Land Resources Monitoring and Simulation, Ministry of Natural Resources (KF-2023-08-17).

References

1. Chen, J., Hu, J., Li, T., Teng, F., Du, S.: An effective relation-first detection model for relational triple extraction. Expert Syst. Appl. **238**(Part B), 122007 (2024)
2. Gao, C., et al.: ERGM: a multi-stage joint entity and relation extraction with global entity match. Knowl. Based Syst. **271**, 110550 (2023)
3. He, P., Gao, J., Chen, W.: Debertav3: improving deberta using electra-style pre-training with gradient-disentangled embedding sharing. In: ICLR. OpenReview.net (2023)
4. Li, J., et al.: BioCreative V CDR task corpus: a resource for chemical disease relation extraction. Database **2016** (2016)
5. Li, X., Luo, X., Dong, C., Yang, D., Luan, B., He, Z.: TDEER: an efficient translating decoding schema for joint extraction of entities and relations. In: EMNLP, no. 1, pp. 8055–8064. ACL (2021)
6. Li, X., Chen, K., Long, Y., Zhang, M.: LLM with relation classifier for document-level relation extraction. CoRR **abs/2408.13889** (2024)

7. Liu, B., et al.: Integration of relation filtering and multi-task learning in globalpointer for entity and relation extraction. Appl. Sci. **14**(15), 6832 (2024)
8. Qian, Y., Ren, E., Xu, H.: Joint entity and relation extraction based on bidirectional update and long-term memory gate mechanism. In: CCL. Lecture Notes in Computer Science, vol. 14761, pp. 174–190. Springer (2024)
9. Riedel, S., Yao, L., McCallum, A.: Modeling relations and their mentions without labeled text. In: Proceedings of ECML PKDD, pp. 148–163 (2010)
10. Shang, Y., Huang, H., Mao, X.: Onerel: joint entity and relation extraction with one module in one step. In: AAAI, pp. 11285–11293. AAAI Press (2022)
11. Shehata, D.: Information retrieval with entity linking. CoRR **abs/2404.08678** (2024)
12. Sui, D., Zeng, X., Chen, Y., Liu, K., Zhao, J.: Joint entity and relation extraction with set prediction networks. IEEE Trans. Neural Netw. Learn. Syst. (2023)
13. Tan, Q., He, R., Bing, L., Ng, H.T.: Document-level relation extraction with adaptive focal loss and knowledge distillation. arXiv preprint arXiv:2203.10900 (2022)
14. Tuo, M., Yang, W.: Review of entity relation extraction. J. Intell. Fuzzy Syst. **44**(5), 7391–7405 (2023)
15. Wang, L., et al.: Genre: Generative multi-turn question answering with contrastive learning for entity–relation extraction. Complex Intell. Syst. 1–15 (2024)
16. Wang, T., et al.: Joint entity and relation extraction with fusion of multi-feature semantics. J. Intell. Inf. Syst. **63**(1), 21–42 (2025)
17. Wang, Y., Sun, C., Wu, Y., Zhou, H., Li, L., Yan, J.: Unire: a unified label space for entity relation extraction. In: ACL/IJCNLP, no. 1, pp. 220–231. ACL (2021)
18. Wang, Y., Yu, B., Zhang, Y., Liu, T., Zhu, H., Sun, L.: Tplinker: single-stage joint extraction of entities and relations through token pair linking. In: COLING, pp. 1572–1582. International Committee on Computational Linguistics (2020)
19. Wu, Y., Luo, R., Leung, H.C., Ting, H.F., Lam, T.W.: Renet: a deep learning approach for extracting gene-disease associations from literature. In: Proceedings of RECOMB, pp. 272–284 (2019)
20. Yan, Z., Yang, S., Liu, W., Tu, K.: Joint entity and relation extraction with span pruning and hypergraph neural networks. In: EMNLP, pp. 7512–7526. ACL (2023)
21. Yang, Z., Huang, Y., Feng, J.: Learning to leverage high-order medical knowledge graph for joint entity and relation extraction. In: ACL (Findings), pp. 9023–9035. ACL (2023)
22. Yao, Y., et al.: Docred: a large-scale document-level relation extraction dataset. arXiv preprint arXiv:1906.06127 (2019)
23. Zaratiana, U., Tomeh, N., Holat, P., Charnois, T.: An autoregressive text-to-graph framework for joint entity and relation extraction. In: AAAI, pp. 19477–19487. AAAI Press (2024)
24. Zhang, Y., Sadler, T., Taesiri, M.R., Xu, W., Reformat, M.: Fine-tuning language models for triple extraction with data augmentation. In: Proceedings of the 1st Workshop on Knowledge Graphs and Large Language Models (KaLLM 2024), pp. 116–124. ACL, Bangkok, Thailand, August 2024. https://doi.org/10.18653/v1/2024.kallm-1.12

Harnessing Deep LLM Participation for Robust Entity Linking

Jiajun Hou, Chenyu Zhang, and Rui Meng(✉)

Guangdong Provincial/Zhuhai Key Laboratory of IRADS, Beijing Normal-Hong Kong Baptist University, Zhuhai, China
ruimeng@uic.edu.cn

Abstract. Entity Linking (EL), the task of mapping textual entity mentions to their corresponding entries in knowledge bases, constitutes a fundamental component of natural language understanding. Recent advancements in Large Language Models (LLMs) have demonstrated remarkable potential for enhancing EL performance. Prior research has leveraged LLMs to improve entity disambiguation and input representation, yielding significant gains in accuracy and robustness. However, these approaches typically apply LLMs to isolated stages of the EL task, failing to fully integrate their capabilities throughout the entire process.

In this work, we introduce DeepEL, a comprehensive framework that incorporates LLMs into every stage of the entity linking task. Furthermore, we identify that disambiguating entities in isolation is insufficient for optimal performance. To address this limitation, we propose a novel self-validation mechanism that utilizes global contextual information, enabling LLMs to rectify their own predictions and better recognize cohesive relationships among entities within the same sentence. Extensive empirical evaluation across ten benchmark datasets demonstrates that DeepEL substantially outperforms existing state-of-the-art methods, achieving an average improvement of 2.6% in overall F1 score and a remarkable 4.6% gain on out-of-domain datasets. These results underscore the efficacy of deep LLM integration in advancing entity linking performance.

1 Introduction

With the proliferation of Internet technology, an unprecedented volume of natural language data is being generated across social media platforms. This growth has significantly accelerated research in natural language processing (NLP). Entity Linking (EL), the process of mapping entity mentions in text to corresponding entries in knowledge bases, has become crucial across various applications, including Question Answering [13], Information Retrieval [7], and Machine Translation [2].

J. Hou and C. Zhang—Equal Contribution.

M. Yoshikawa et al. (Eds.): ADMA 2025, LNAI 16197, pp. 410–417, 2026.
https://doi.org/10.1007/978-981-95-3453-1_30

The complexity of EL stems from linguistic phenomena such as synonymy (multiple surface forms referring to the same entity, e.g., "IBM" and "International Business Machines") and homonymy (identical terms representing different entities, e.g., "Apple" as fruit or company). These challenges necessitate entity linking models with robust contextual understanding. Research has evolved from dictionary-based approaches to current methods leveraging deep learning [1,10,16]. However, the performance of deep learning models remains inconsistent due to limitations in training data and parameter capacity, deteriorating significantly when encountering unseen entities.

Large Language Models (LLMs), built upon multi-head attention mechanisms, offer a promising solution. Recent investigations include ChatEL [6], which employs LLMs for entity disambiguation as a multiple-choice task, and LLMAEL [17], which focuses on generating entity descriptions with LLMs. While these methods have achieved competitive results, their use of large language models remains limited, typically applied to only one component of the entity linking process rather than the entire process. Moreover, their LLM integration is simplistic, failing to leverage advanced capabilities such as self-validation.

Given LLMs' remarkable generalization capabilities, we propose **DeepEL**, a framework for comprehensive LLM integration throughout the EL task. First, DeepEL leverages LLMs to describe entities requiring linkage, providing richer contextual information for generating more accurate candidate lists. Subsequently, it formulates entity disambiguation as a multiple-choice task using the candidate list and knowledge base descriptions. Finally, DeepEL incorporates a self-validation mechanism that cross-checks each entity against other entities' linking results in the same sentence, significantly enhancing accuracy through global consistency checking. The main contributions are summarized as follows:

- We present DeepEL, an entity linking framework that leverages LLMs at all stages–candidate generation, entity disambiguation, and self-validation, to fully exploit their capabilities.
- We introduce a novel self-validation mechanism based on global contextual information, improving disambiguation accuracy by providing LLMs with information about other entities within the same context.
- We conduct comprehensive comparisons with five state-of-the-art models across ten benchmark datasets, demonstrating DeepEL's superior performance. We have made our code publicly available[1] to guarantee reproducibility.

2 Problem Definition

Entity Linking (EL) is the task of mapping mentions in text to their corresponding entities in a knowledge base. We formally define the problem as follows: Let KB denote a knowledge base containing a set of entities $\{e_1, e_2, \ldots, e_n\}$, where

[1] https://github.com/SStan1/DeepEL.

each e_i represents a unique real-world object. Given an input text T containing multiple entity mentions $M = \{m_1, m_2, \ldots, m_k\}$, the text can be represented as $c = \ldots \,||\, t_1 \,||\, m_1 \,||\, t_2 \,||\, m_2 \,||\, \ldots$, where t_i denotes text spans and m_i denotes entity mentions. The objective of entity linking is to determine the correct mapping from each mention m_i to its corresponding entity $e_i \in KB$, resulting in a set of mention-entity pairs $\{(m_i, e_i)\}_{i=1}^{k}$.

3 Methodology

Our proposed framework, DeepEL, achieves deep LLM integration across all stages of the entity linking process via three primary modules (Fig. 1):

1. **Entity Candidate Generation:** We first let the LLM describe the entities, then use a pre-trained entity linking model to generate a high-quality candidate list based on both the LLM's descriptions and the original sentences.
2. **Entity Disambiguation:** We organize the entity disambiguation task as a multiple-choice question for LLMs to perform entity selection.
3. **Self-Validation:** To better consider contextual information, we input linking information about other entities in the same sentence into the LLM and ask it to self-validate its previous judgments.

3.1 Entity Candidate Generation

To address the limitation that LLMs lack direct access to structured knowledge bases, we develop a hybrid approach combining LLM interpretation with traditional entity linking models.

Given an input text T containing entity mentions $M = \{m_1, m_2, \ldots, m_k\}$, we implement a two-stage process:

1. **LLM Interpretation**: We prompt the LLM to generate a contextual interpretation I_{m_i} for each mention $m_i \in M$.
2. **Dual-Source Candidate Retrieval**: We use BLINK to generate two complementary candidate lists:
   - $C_{original}$: candidates retrieved using the original text T;
   - C_{LLM}: candidates retrieved using the LLM's interpretations $\{I_{m_1}, \ldots, I_{m_k}\}$.

Each candidate list is constrained to the top ten entities. For each candidate entity e, we retrieve its canonical description from the knowledge base:

$$C_{original} = \{(e_1, d_1), \ldots, (e_n, d_n)\} \text{ where } d_i = KB.description(e_i) \quad (1)$$

We merge these lists using a priority-based algorithm:

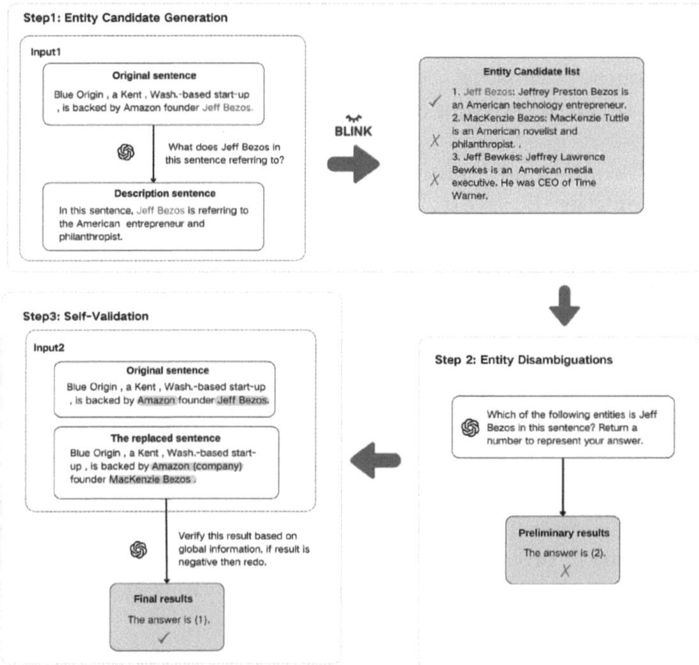

Fig. 1. DeepEL's workflow. Text marked in red indicates the target entity mention currently being linked. Highlighted text represents already-linked entity mentions within the same sentence, providing contextual linking information for the target entity.

$$C_{final} = f_{merge}(C_{original}, C_{LLM}) \text{ such that } |C_{final}| \leq 10 \tag{2}$$

This dual-source approach enhances candidate quality by combining traditional context-based retrieval with LLM interpretations, capturing both explicit textual cues and implicit semantic connections.

3.2 Entity Disambiguation

For each mention m_i, we formulate the entity disambiguation task as a multiple-choice problem:

$$f_{select}(m_i, I_{m_i}, C_{final}) \rightarrow (e_j, d_j), \text{ where } (e_j, d_j) \in C_{final} \cup \{\emptyset\}, \tag{3}$$

where $\emptyset$ indicates that none of the candidates is correct. We require the LLM to output the index j of its chosen entity, with 0 representing $\emptyset$.

3.3 Self-validation

To enhance accuracy through contextual coherence, we implement a self-validation mechanism leveraging global information from the entire sentence context.

Given text T containing k entity mentions $M = \{m_1, ..., m_k\}$ and their initially predicted entities $E = \{e_1, ..., e_k\}$, we construct:

$$T_g = T[m_1 \rightarrow e_1, m_2 \rightarrow e_2, ..., m_k \rightarrow e_k] \tag{4}$$

The global context G integrates three components:

$$G = (T, T_g, D), \tag{5}$$

where $D = \{d_1, ..., d_k\}$ are the entity descriptions. We prompt the LLM to perform self-validation based on this global context. For entities that fail validation, we re-prompt the model to perform entity re-selection from the original candidate list.

This self-validation process enhances linking accuracy by ensuring global consistency across all entity mentions within the same context. For efficiency, this validation is performed only once.

4 Experiment

In this section, we comprehensively compare DeepEL with state-of-the-art baselines across ten benchmark datasets. We further analyze the quality of entity candidate lists generated in Step 1 and evaluate the effectiveness of the self-validation method in Step 3. Due to space limitations, please refer to the source code for specific implementation details.

4.1 Experimental Setup

Following ChatEL [6], our experiments utilize eight standard datasets from their work, along with two additional commonly used datasets. By convention, we categorize these datasets into in-domain and out-of-domain groups for evaluation:

- **Out-of-Domain**: OKE15 [12], OKE16 [12], DER [5], KORE [8], REU, and RSS [15]. These datasets assess a model's ability to generalize to new domains and adapt to unfamiliar contexts.
- **In-Domain**: ACE04 [14], AIDA [9], MSN [4], and AQU [11]. These datasets evaluate model performance within familiar contexts.

We compare with the following baseline methods: REL [10], BLINK [16], GENRE [3], ReFinED [1] and ChatEL [6]. To ensure consistency and comparability across both in-domain and out-of-domain datasets, we adopt the evaluation protocol from ChatEL [6]. Specifically, we report the in-KB micro-F1 score, which excludes entities not present in the knowledge base from evaluation.

Table 1. Shows the F1 score of DeepEL on ten standard datasets, where avg-OOD represents the average of the model's performance on the out-domain dataset. The results of the best performing model for each of these datasets are bolded and the second best are underlined.

	Out-of-domain						In-domain					
	KORE	OKE15	REU	RSS	DER	OKE16	AQU	ACE04	MSN	AIDA	Average	avg-OOD
REL	0.618	0.705	0.662	0.680	0.411	0.749	**0.881**	**0.897**	**0.930**	0.805	0.7338	0.6375
GENRE	0.542	0.640	0.697	0.708	0.541	0.708	0.849	0.848	0.780	<u>0.837</u>	0.7150	0.6393
ReFinED	0.567	<u>0.781</u>	0.680	0.708	0.507	<u>0.794</u>	0.861	0.864	0.891	**0.840**	0.7493	0.6728
BLINK	0.618	0.763	0.712	0.767	0.709	**0.805**	<u>0.863</u>	0.793	0.865	0.807	0.7702	0.7290
ChatEL	<u>0.787</u>	0.758	<u>0.789</u>	<u>0.822</u>	<u>0.717</u>	0.752	0.767	<u>0.893</u>	0.881	0.821	<u>0.7987</u>	<u>0.7708</u>
DeepEL	**0.873**	**0.833**	**0.812**	**0.857**	**0.768**	0.762	0.753	0.888	<u>0.905</u>	0.799	**0.8250**	**0.8175**

4.2 Main Results

Our experimental evaluation compared DeepEL with five state-of-the-art entity linking models across multiple datasets. As shown in Table 1, DeepEL demonstrates exceptional performance, surpassing ChatEL by 2.6% in average F1 score, validating our methodological framework's effectiveness.

Most significantly, DeepEL exhibits remarkable generalization capability. On out-of-domain datasets, it achieves a substantial 4.6% improvement in average F1 score compared to ChatEL. This provides compelling evidence that our approach successfully leverages LLMs' advanced inference and contextual understanding capabilities, enabling robust performance on previously unseen data.

Table 2. Performance of DeepEL using different LLM models.

	Out-of-domain						In-domain				
	KORE	OKE15	REU	RSS	DER	OKE16	AQU	ACE04	MSN	AIDA	Average
Llama-2–70b	0.814	0.708	0.650	0.789	0.678	0.601	0.563	0.753	0.788	0.678	0.7022
GPT 3.5	0.769	0.708	0.712	0.803	0.718	0.619	0.634	0.773	0.847	0.634	0.7302
GPT 4	**0.873**	**0.833**	**0.812**	**0.857**	**0.768**	**0.762**	**0.753**	**0.888**	**0.905**	**0.799**	**0.8250**

Different LLMs in DeepEL. We evaluate DeepEL using various LLM backbones, including GPT-3.5, GPT-4, and Llama-2–70b. Table 2 reveals that scores exhibit a clear gradient corresponding to model capabilities, with GPT-4 achieving the highest scores. This demonstrates that DeepEL functions as an adaptable framework, with performance scaling alongside underlying LLM capabilities, providing significant potential for future applications. Additionally, the relatively close scores between GPT-3.5 and Llama-2 highlight DeepEL's generalizability across diverse model families, suggesting effective deployment with various LLMs while consistently enhancing entity linking performance.

4.3 Ablation Study

We conduct ablation studies to assess the contribution of individual model components, focusing on: (1) The effect of the Entity Description strategy, and (2) The impact of the Global Self-validation module. We first evaluate candidate set generation without LLM rewrites (denoted as "W/O De"), and then test the model without the self-validation step ("W/O Val").

Across ten benchmarks, removing entity descriptions (i.e., Use $C_{original}$ generated by BLINK directly) consistently harms performance, as shown in Table 3. This highlights that LLM-generated descriptions enhance the link between mentions and target entities, improving candidate quality. Eliminating self-validation also leads to decreased performance, confirming that global self-checking helps reduce errors by validating answer correctness and adjusting choices accordingly.

Interestingly, AIDA and MSN datasets show slightly better results without Step 1. Since they are in-domain and heavily trained, BLINK performs more accurately, making it easier for LLMs to reach correct answers even without entity descriptions.

Table 3. Ablation study of Entity descriptions and Global Self-validation.

	Out-of-domain						In-domain				
	KORE	OKE15	REU	RSS	DER	OKE16	AQU	ACE04	MSN	AIDA	Average
DeepEL	**0.873**	**0.833**	**0.812**	**0.857**	**0.768**	**0.762**	**0.753**	**0.888**	0.905	0.799	**0.8250**
DeepEL W/O De	0.806	0.790	0.796	0.841	0.751	0.734	0.681	0.887	0.898	**0.806**	0.7990
DeepEL W/O Val	0.870	0.820	0.794	0.850	0.762	0.742	0.744	0.869	**0.906**	0.790	0.8147

5 Conclusion

In this study, we present the DeepEL framework, a comprehensive approach to entity linking that comprises three sequential stages. Initially, a pre-trained model generates a list of candidate entities based on descriptive inputs derived from LLMs. In the second stage, a multiple-choice disambiguation mechanism selects from the candidate set, effectively narrowing down potential entities. Finally, a self-validation phase systematically reassesses and refines preliminary selections, ensuring more accurate and contextually appropriate final outputs. DeepEL deeply integrates LLMs' capabilities at each stage, fully leveraging their strengths in the entity linking task. Compared to existing approaches, DeepEL achieves superior performance without requiring fine-tuning, delivering consistent accuracy gains across diverse benchmarks.

Acknowledgements. This work was supported by the Guangdong Basic and Applied Basic Research Foundation (2023A1515110624) and the Guangdong Provincial Key Laboratory of IRADS (2022B1212010006).

References

1. Ayoola, T., Tyagi, S., Fisher, J., Christodoulopoulos, C., Pierleoni, A.: Refined: an efficient zero-shot-capable approach to end-to-end entity linking. In: NAACL-HLT (Industry Papers), pp. 209–220. ACL (2022)
2. Babych, B., Hartley, A.: Improving machine translation quality with automatic named entity recognition. In: Proceedings of the 7th International EAMT Workshop on MT and Other Language Technology Tools, Improving MT through Other Language Technology Tools: Resources and Tools for Building MT, EAMT 2003, Pp. 1–8. ACL, USA (2003)
3. Cao, N.D., Izacard, G., Riedel, S., Petroni, F.: Autoregressive entity retrieval. In: ICLR, OpenReview.net (2021)
4. Cucerzan, S.: Large-scale named entity disambiguation based on wikipedia data. In: EMNLP-CoNLL, pp. 708–716. ACL (2007)
5. Derczynski, L., et al.: Analysis of named entity recognition and linking for tweets. Inf. Process. Manag. **51**(2), 32–49 (2015)
6. Ding, Y., Zeng, Q., Weninger, T.: Chatel: entity linking with chatbots. In: LREC/COLING, pp. 3086–3097. ELRA and ICCL (2024)
7. Hambarde, K.A., Proença, H.: Information retrieval: recent advances and beyond. IEEE Access **11**, 76581–76604 (2023)
8. Hoffart, J., Seufert, S., Nguyen, D.B., Theobald, M., Weikum, G.: KORE: keyphrase overlap relatedness for entity disambiguation. In: CIKM, pp. 545–554. ACM (2012)
9. Hoffart, J., et al.: Robust disambiguation of named entities in text. In: EMNLP, pp. 782–792. ACL (2011)
10. van Hulst, J.M., Hasibi, F., Dercksen, K., Balog, K., de Vries, A.P.: REL: an entity linker standing on the shoulders of giants. In: SIGIR, pp. 2197–2200. ACM (2020)
11. Milne, D.N., Witten, I.H.: Learning to link with wikipedia. In: CIKM, pp. 509–518. ACM (2008)
12. Nuzzolese, A.G., Gentile, A.L., Presutti, V., Gangemi, A., Garigliotti, D., Navigli, R.: Open knowledge extraction challenge. In: SemWebEval@ESWC. Communications in Computer and Information Science, vol. 548, pp. 3–15. Springer (2015)
13. Pan, X., Sun, K., Yu, D., Chen, J., Ji, H., Cardie, C., Yu, D.: Improving question answering with external knowledge. In: MRQA@EMNLP, pp. 27–37. ACL (2019)
14. Ratinov, L., Roth, D., Downey, D., Anderson, M.: Local and global algorithms for disambiguation to wikipedia. In: ACL, pp. 1375–1384. ACL (2011)
15. Röder, M., Usbeck, R., Hellmann, S., Gerber, D., Both, A.: N^3 - A collection of datasets for named entity recognition and disambiguation in the NLP interchange format. In: LREC, pp. 3529–3533. European Language Resources Association (ELRA) (2014)
16. Wu, L., Petroni, F., Josifoski, M., Riedel, S., Zettlemoyer, L.: Scalable zero-shot entity linking with dense entity retrieval. In: EMNLP, no. 1, pp. 6397–6407. ACL (2020)
17. Xin, A., et al.: LLMAEL: large language models are good context augmenters for entity linking. CoRR **abs/2407.04020** (2024)

Alternating Aggregation Low-Rank Adaptation Approach for Federated Large Models

Tao Zhang, Chao Zhang, Feiyang Yuan, Lele Zheng(✉), and Yiyun Guo

School of Computer Science and Technology, Xidian University, Xi'an, China
taozhang@xidian.edu.cn,
{zhangchaoxidian,yuanfeiyang,llzheng,yyyguo}@stu.xidian.edu.cn

Abstract. Low-rank adaptation (LoRA) is a parameter-efficient fine-tuning (PEFT) method for pre-trained large language models tailored to specific downstream tasks. Due to its flexibility and computational efficiency, LoRA has become the preferred approach within the PEFT framework. However, when applied to federated learning environments, LoRA often experiences instability. This instability primarily arises from two factors: (1) the direct integration of the traditional federated averaging algorithm with the LoRA adapter, which can cause errors in parameter updates during model aggregation, and (2) the amplification of noise introduced to satisfy differential privacy requirements. To address these challenges, this paper introduces an innovative Alternating Federated Low-Rank Adaptation (AF-LoRA) method. AF-LoRA enhances model training stability by implementing an alternating upload and aggregation mechanism for matrices **A** and **B**, while maintaining only 50% of the communication cost of the standard LoRA method. Specifically, in each communication round, only one matrix is uploaded for global aggregation, while the other matrix is locally optimized for the next update. Extensive experimental results demonstrate that, compared to traditional federated LoRA methods, AF-LoRA significantly outperforms in both standard and privacy-preserving federated learning scenarios.

Keywords: LoRA · PEFT · LLMs · alternating upload and aggregation

1 Introduction

In recent years, large language models (LLMs) have rapidly advanced and found widespread applications across various domains, including chatbots and multimodal systems [13]. However, LLMs typically contain billions, or even tens of billions, of parameters, making full-parameter training highly resource-intensive and accessible only to a few industrial laboratories. As a result, most researchers opt to fine-tune existing pre-trained LLMs to adapt them to downstream tasks. Fine-tuning also requires vast amounts of data, which a single institution can

M. Yoshikawa et al. (Eds.): ADMA 2025, LNAI 16197, pp. 418–425, 2026.
https://doi.org/10.1007/978-981-95-3453-1_31

rarely provide, necessitating collaboration among multiple institutions for joint fine-tuning.

Fine-tuning pre-trained LLMs under the federated learning framework [2] addresses the issue of insufficient data in single institutions. For resource-limited clients, researchers have developed parameter-efficient fine-tuning (PEFT) techniques [3], such as prompt fine-tuning [5], adapter methods (training lightweight inserted layers with frozen pre-trained weights) [7], and low-rank adaptation (LoRA) [6] (modifying pre-trained weights via two trainable low-rank matrices **A** and **B**, with frozen pre-trained weights). LoRA has gained wide attention in federated fine-tuning of LLMs, as over-parameterized LLMs lie in a lower intrinsic dimensional space [1,6]. Its advantages include enabling the creation of independent LoRA modules for different tasks, thereby improving usability without incurring inference delays.

LoRA reduces LLMs training parameters by decomposing pre-trained weights into two low-rank matrix products, mitigating excessive communication overhead in federated fine-tuning. However, direct combination with FedAvg causes aggregation errors, degrading performance. To solve this problem, researchers have explored some new solutions. For example, Sun et al. [12] proposed a federated frozen **A** LoRA (FFA-LoRA) method, which freezes matrix **A**, only updates and aggregates matrix **B**, and adds differential privacy noise to matrix **B**. Also, Guo et al. [4] proposed a federated shared low-rank adaptive (FedSA-LoRA) method, in which matrix **A** is trained locally on the client and only matrix **A** is uploaded for update. However, the FFA-LoRA method's fixation of matrix **A** weakens LoRA's learning ability and leads to poor performance. FedSA-LoRA implements a personalized federation scheme, where each client has a different matrix **B**, and ultimately there is no unified global model. Therefore, this paper proposes an alternating federated low-rank adaptive (AF-LoRA) method. Compared with FFA-LoRA and FedSA-LoRA, AF-LoRA is able to train matrices **A** and **B** without increasing communication overhead and without introducing additional aggregation errors. In addition, this paper studies how to achieve privacy-preserving LoRA fine-tuning under the AF-LoRA framework.

In summary, this paper contributes the following:

1. We analyze the aggregation error problem caused by directly combining LoRA with FedAvg, and the reason why the noise impact is amplified when adding Gaussian noise for privacy protection under Federated LoRA.
2. We propose a novel method called AF-LoRA, which alternately uploads the low-rank matrices **A** and **B** in each communication round. This design enables accurate aggregation in each round while reducing the communication cost to 50% of that of standard LoRA. We further extend AF-LoRA to privacy-preserving federated learning scenarios.

2 Our Method: AF-LoRA

2.1 Alternating Federated Low-Rank Adaptation

Inspired by FFA-LoRA and FedSA-LoRA, we propose Alternating Federated LoRA (AF-LoRA) for federated settings, which alternately uploads matrices **A** and **B** in each client-server communication round. For the LoRA module's weight matrix $\mathbf{W} \in \mathbb{R}^{d \times k}$, model updates are decomposed as:

$$\mathbf{W} = \mathbf{W_0} + \mathbf{\Delta W} = \mathbf{W_0} + \mathbf{AB}, \text{where } \mathbf{B} \in \mathbb{R}^{d \times r}, \mathbf{A} \in \mathbb{R}^{r \times k} \tag{1}$$

where $\mathbf{W_0}$ is initialized as pre-trained weights, **A** follows a random Gaussian distribution and **B** is zero-initialized (consistent with standard LoRA). In each round, $\mathbf{W_0}$ is frozen, while both **A** and **B** are trainable.

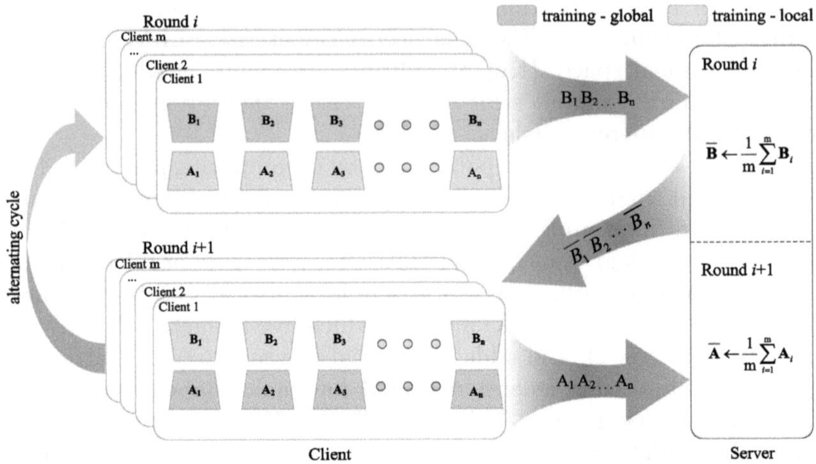

Fig. 1. Training and aggregation procedure of AF-LoRA: In each round, both matrices **A** and **B** are trainable. In the first round, matrix **B** is uploaded to the server for aggregation. In the subsequent round, matrix **A** is uploaded for aggregation, and this alternating process continues in the following rounds. This alternating aggregation reduces the communication cost to 50% of that of standard LoRA.

Once initialization is complete, the number of clients is set to m. After the first round of local training at each client, the clients upload their respective matrices $\mathbf{B}_i$ to the server for aggregation. Once the aggregation is complete, the aggregated matrix **B** is distributed to each client. At this point, the local update content of the weight matrix for client i is as follows:

$$\mathbf{\Delta W}_i = \frac{1}{m}(\mathbf{B_1} + \mathbf{B_2} + \cdots + \mathbf{B_m})\mathbf{A}_i \tag{2}$$

Once the client's first round of updates is complete, the second round of local training commences. Upon finishing the second round of local training, clients upload respective matrices $\mathbf{A}_i$ to the server for aggregation. When the server completes the aggregation, it distributes the aggregated matrix **A** to each client. At this stage, the local update content of client i's weight matrix is as follows:

$$\mathbf{\Delta W}_i = \frac{1}{m}\mathbf{B}_i(\mathbf{A_1} + \mathbf{A_2} + \cdots + \mathbf{A_m}) \tag{3}$$

Algorithm 1. AF-LoRA

Input: the number of global epochs T, LoRA rank set r, the number of clients m, learning rate η, the number of client training rounds N, the pre-trained model weights W_0, the initialized LoRA matrices $\mathbf{A}_0$ and $\mathbf{B}_0$

Output: the global model parameters w^T

The parameter server initializes $\mathbf{A}$ and $\mathbf{B}$, send W_0, $\mathbf{A}_0$, $\mathbf{B}_0$ to all clients
for $t = 1, 2, \ldots, T$ **do**
  for $i = 1, 2, \ldots, N$ **do**
    $\nabla_A, \nabla_B \leftarrow \nabla L_i(W_0; \mathbf{A}_i, \mathbf{B}_i)$
    $\mathbf{A}_i \leftarrow \mathbf{A}_i - \eta \nabla_A$
    $\mathbf{B}_i \leftarrow \mathbf{B}_i - \eta \nabla_B$
  end for
  if $t \bmod 2 = 1$ **then**
    Send the updated LoRA matrices $\mathbf{B}$ to the parameter server
  else
    Send the updated LoRA matrices $\mathbf{A}$ to the parameter server
  end if
  if $t \bmod 2 = 1$ **then**
    $\overline{\mathbf{B}} \leftarrow \frac{1}{m} \sum_{i=1}^{m} \mathbf{B}_i$
  else
    $\overline{\mathbf{A}} \leftarrow \frac{1}{m} \sum_{i=1}^{m} \mathbf{A}_i$
  end if
end for
Clients upload both matrices $\mathbf{A}$ and $\mathbf{B}$ after the entire training is completed
$\overline{\mathbf{A}} \leftarrow \frac{1}{m} \sum_{i=1}^{m} \mathbf{A}_i$
$\overline{\mathbf{B}} \leftarrow \frac{1}{m} \sum_{i=1}^{m} \mathbf{B}_i$
Send $\overline{\mathbf{A}}$ and $\overline{\mathbf{B}}$ to all clients

Subsequent training follows this pattern: clients alternate uploading matrices **A** and **B** during client-server communication rounds. In each round, the uploaded matrix is aggregated on the server, while the unuploaded one continues updating locally on the clients. The full AF-LoRA training process is shown in Fig. 1.

AF-LoRA avoids aggregation errors from traditional LoRA-FedAvg combinations and overcomes issues in FFA-LoRA and FedSA-LoRA. It uses alternating upload and aggregation of low-rank matrices **A** and **B**, enabling more accurate capture of cross-client data features. This enhances performance and generalization in complex distributed scenarios. AF-LoRA training procedure is summarized in Algorithm 1.

2.2 AF-LoRA with Differential Privacy

Guo et al. [4] noted that low-rank matrix **A** learns general knowledge, while **B** captures client-specific knowledge, with high cosine similarity of **A** across clients. Thus, uploading **A** has minimal privacy leakage risk. For AF-LoRA privacy protection, Gaussian noise is only added to **B** during upload, not **A**. Since unuploaded local matrices remain on the client and are not shared, they avoid potential data leakage risks, and AF-LoRA avoids introducing non-Gaussian cross-noise terms (e.g., $\xi_{\mathbf{B}}\xi_{\mathbf{A}}$ in DP-LoRA [8]), resulting in noise terms only in $\xi_{\mathbf{B}}\mathbf{A}$. Without clipping, first-round updates are $\mathbf{W}_0 + (\mathbf{B} + \xi_{\mathbf{B}})\mathbf{A} = \mathbf{W}_0 + \mathbf{BA} + \xi_{\mathbf{B}}\mathbf{A}$, second-round updates are $\mathbf{W}_0 + \mathbf{BA}$, alternating thereafter.

Alternately uploading matrices with Gaussian noise only added to $\mathbf{B}$ achieves privacy protection while avoiding excessive noise, reducing its impact on model performance. This alternating mechanism balances noise addition, allowing the model to maintain good performance under privacy guarantees.

3 Experiments

In this section, we evaluate and compare the performance of the method proposed in this paper with other methods on two types of tasks: natural language understanding and natural language generation. For natural language understanding tasks, we use the RoBERTa-base [9] to conduct evaluations on the GLUE benchmark [14], which includes MNLI, SST2, QNLI, QQP, RTE, and COLA. For natural language generation tasks, we use the GPT-2 [11] to conduct evaluations on the E2E NLG Challenge dataset [10]. All experiments were conducted using dual NVIDIA GeForce RTX 3090 GPUs with half-precision arithmetic employed for computations. Experimental results are reported as the mean of outcomes from three independent runs with distinct random initializations.

3.1 Natural Language Understanding

For natural language understanding tasks, the dataset was evenly partitioned across 3 clients. All experiments utilized the AdamW optimizer with batch size 128, 3 local update steps, learning rate of $1e-3$, 50 communication rounds, and scaling factor of $\alpha = 16$. LoRA was applied exclusively to the query weight matrix $\mathbf{W}_q$ and value weight matrix $\mathbf{W}_v$ within the attention layers, with the pre-trained RoBERTa-base model (125M parameters) serving as the backbone.

Table 1 presents the results of the RoBERTa-base model at $r = 4$. Among them, AF-LoRA outperforms existing federated fine-tuning methods on 6 tasks, and in some cases approaches the performance of centralized LoRA. Specifically, the accuracy rates on MNLI_matched and QNLI reach 86.04% and 92.71% respectively, both higher than the accuracy rates of 85.89% and 92.13% achieved by the current state-of-the-art methods, which validates the effectiveness of AF-LoRA. However, on the QQP dataset, the accuracy of AF-LoRA is slightly lower than that of others. This may be because the QQP dataset exhibits unique characteristics in terms of text semantic features and distribution.

3.2 Natural Language Understanding with Differential Privacy

The experiments with differential privacy are also conducted on the GLUE benchmark for natural language understanding. For the LoRA and FFA-LoRA methods, we adopt the same experimental settings as described in the FFA-LoRA

Table 1: Performance of different methods on the GLUE benchmark. For all tasks, we report mean accuracy evaluated across 3 runs.

Method	MNLI (matched)	MNLI (mismatched)	SST2	QNLI	QQP	RTE	COLA
Centralized LoRA$_{r=4}$	86.72	86.81	94.61	92.75	91.38	75.45	84.62
LoRA$_{r=4}$	85.89	85.26	94.04	92.13	**90.23**	73.64	83.22
FFA-LoRA$_{r=4}$	84.62	84.81	94.27	91.40	89.21	71.84	82.65
Fedex-LoRA$_{r=4}$	85.78	85.90	94.15	91.01	88.83	63.54	82.45
AF-LoRA$_{r=4}$	**86.04**	**85.91**	**94.50**	**92.71**	89.00	**75.09**	**83.41**

paper [12]. The learning rates η are taken from the sets $\{0.01, 0.02, 0.05, 0.1\}$ and $\{0.1, 0.2, 0.5, 1\}$, and for AF-LoRA, the learning rates are $\{0.001, 0.02\}$. For these three algorithms, we fix the rank $r = 8$. Regarding the privacy parameters, the privacy budget ϵ takes values from the set $\{6, 3, 1\}$, and with the fixed $\delta = 1e-5$.

The experimental results are presented in Table 2. Adding differential privacy to gradients significantly degrades the performance of LoRA, FFA-LoRA, and AF-LoRA across all tasks. However, it can be observed that AF-LoRA outperforms both DP-LoRA and FFA-LoRA regardless of whether privacy protection is applied. Specifically, with a privacy budget of $\epsilon = 1$, AF-LoRA achieves an accuracy improvement of 7.86% and 7.38% over FFA-LoRA on MNLI_matched and QNLI, respectively, and an improvement of 49.11% and 9.98% compared to DP-LoRA on the same tasks. These results indicate that the performance advantage of AF-LoRA is more prominent in more complex three-class classification tasks such as MNLI.

Table 2: Performance of different methods on the GLUE benchmark with differential privacy. For all tasks, we report mean accuracy evaluated across 3 runs.

Privacy Budget	Method	MNLI (matched)	MNLI (mismatched)	SST-2	QQP	QNLI
$\epsilon = 6$	DP-LoRA	39.46	39.69	93.70	82.11	84.99
	FFA-LoRA	78.81	80.00	**93.73**	83.31	87.27
	AF-LoRA	**83.87**	**83.65**	93.12	**85.59**	**90.15**
$\epsilon = 3$	DP-LoRA	35.82	35.85	93.32	82.08	83.94
	FFA-LoRA	77.42	78.69	**93.59**	83.03	86.18
	AF-LoRA	**83.55**	**83.16**	92.78	**84.35**	**89.95**
$\epsilon = 1$	DP-LoRA	33.80	33.80	92.14	81.28	78.93
	FFA-LoRA	75.05	76.50	92.42	82.50	81.53
	AF-LoRA	**82.91**	**82.76**	**92.43**	**83.47**	**88.91**

3.3 Natural Language Generation

For natural language generation tasks, we selected the E2E NLG Challenge dataset [10] to fine-tune and evaluate the GPT-2. The LoRA module is only deployed on the $\mathbf{W}_q$ and $\mathbf{W}_v$ matrices of the self-attention layers of the model. During the fine-tuning process, we considered two values of the rank parameter $r = \{4, 1\}$, and set the local training epochs to 3 and 10 respectively. For both experiments, we complete model optimization through 8 rounds of communication and aggregation operations.

The experimental results in Table 3 show that there are significant differences in the performance of the fine-tuned GPT-2 model when the rank $r = \{4, 1\}$, respectively. Under all evaluation metrics and experimental parameter settings, the AF-LoRA method demonstrates remarkable advantages over other mainstream federated fine-tuning techniques. This verifies the stable reliability and performance superiority of AF-LoRA in natural language generation tasks.

Table 3: Performance with GPT-2 on the E2E NLG Challenge, comparing various federated LoRA methods at ranks $r = \{4, 1\}$.

Method	E2E NLG Challenge				
	BLEU ↑	NIST ↑	MET ↑	ROUGE-L ↑	CIDER ↑
Centralized LoRA$_{r=4}$	68.94	8.83	46.80	71.25	2.52
FedIT$_{r=4}$	67.61	8.62	46.45	70.28	2.43
FFA-LoRA$_{r=4}$	67.21	8.57	46.05	69.98	2.41
Fedex-LoRA$_{r=4}$	68.49	8.72	46.56	70.61	2.45
AF-LoRA$_{r=4}$	**68.52**	**8.74**	**46.77**	**70.74**	**2.49**
Centralized LoRA$_{r=1}$	67.41	8.68	46.01	69.51	2.41
FedIT$_{r=1}$	66.16	8.55	45.56	68.32	2.30
FFA-LoRA$_{r=1}$	65.78	8.49	45.01	67.82	2.26
Fedex-LoRA$_{r=1}$	66.34	8.54	45.98	68.89	2.32
AF-LoRA$_{r=1}$	**66.57**	**8.59**	**46.07**	**69.21**	**2.37**

4 Conclusion

In this paper, we identify the limitations of existing state-of-the-art federated fine-tuning approaches and propose Alternating Federated Low-Rank Adaptation (AF-LoRA), a novel method that enables precise aggregation in federated settings. By alternately uploading and aggregating the two trainable low-rank matrices, **A** and **B**, AF-LoRA significantly reduced communication overhead by up to 50% while improving model performance compared to standard LoRA. Extensive experiments on natural language understanding and generation tasks showed that AF-LoRA consistently outperformed existing federated LoRA variants across various rank configurations. Furthermore, we evaluate AF-LoRA under differential privacy constraints. Even with varying privacy budgets, our method achieves superior performance compared to the baselines such as DP-LoRA and FFA-LoRA.

Acknowledgments. This study was funded by the National Natural Science Foundation of China under Grant 62220106004.

References

1. Aghajanyan, A., Gupta, S., Zettlemoyer, L.: Intrinsic dimensionality explains the effectiveness of language model fine-tuning. In: Proceedings of the 59th Annual Meeting of the Association for Computational Linguistics and the 11th International Joint Conference on Natural Language Processing, ACL/IJCNLP 2021, pp. 7319–7328 (2021)
2. Babakniya, S., et al.: Slora: Federated parameter efficient fine-tuning of language models. arXiv preprint arXiv:2308.06522 (2023)
3. Ding, N., Qin, Y., Yang, G., Wei, F., Yang, Z.: Parameter-efficient fine-tuning of large-scale pre-trained language models. Nat. Mac. Intell. **5**(3), 220–235 (2023)
4. Guo, P., Zeng, S., Wang, Y., Fan, H., Wang, F., Qu, L.: Selective aggregation for low-rank adaptation in federated learning. In: The Thirteenth International Conference on Learning Representations, ICLR 2025 (2025)
5. Guo, T., Guo, S., Wang, J., Tang, X., Xu, W.: Promptfl: let federated participants cooperatively learn prompts instead of models - federated learning in age of foundation model. IEEE Trans. Mob. Comput. **23**(5), 5179–5194 (2024)
6. Hu, E.J., et al.: Lora: low-rank adaptation of large language models. In: The Tenth International Conference on Learning Representations, ICLR 2022 (2022)
7. Li, X.L., Liang, P.: Prefix-tuning: optimizing continuous prompts for generation. In: Proceedings of the 59th Annual Meeting of the Association for Computational Linguistics and the 11th International Joint Conference on Natural Language Processing, ACL/IJCNLP 2021, pp. 4582–4597 (2021)
8. Liu, X.Y., et al.: Differentially private low-rank adaptation of large language model using federated learning. ACM Trans. Manag. Inf. Syst. **16**(2), 1–24 (2025)
9. Liu, Y., et al.: Roberta: a robustly optimized BERT pretraining approach. arXiv preprint arXiv:1907.11692 (2019)
10. Novikova, J., Dusek, O., Rieser, V.: The E2E dataset: new challenges for end-to-end generation. In: Proceedings of the 18th Annual SIGdial Meeting on Discourse and Dialogue, pp. 201–206 (2017)
11. Radford, A., et al.: Language models are unsupervised multitask learners. OpenAI blog **1**(8), 9 (2019)
12. Sun, Y., Li, Z., Li, Y., Ding, B.: Improving LORA in privacy-preserving federated learning. In: The Twelfth International Conference on Learning Representations, ICLR 2024, Vienna, Austria, May 7-11, 2024 (2024)
13. Touvron, H., et al.: Llama: open and efficient foundation language models. arXiv preprint arXiv:2302.13971 (2023)
14. Wang, A., Singh, A., Michael, J., Hill, F., Levy, O., Bowman, S.R.: GLUE: a multi-task benchmark and analysis platform for natural language understanding. In: 7th International Conference on Learning Representations, ICLR 2019 (2019)

SemantiHunt: A New Behavioral Semantics-Driven Method for Network Threat Hunting

Haiyan Wang[1], Yuxiang Hu[2], Rui Zong[1], Aiting Yao[1], Juan Zhao[3], Xiangyu Song[3], and Zhaoquan Gu[1,4](✉)

[1] Pengcheng Laboratory, Shenzhen 518000, China
guzhaoquan@hit.edu.cn

[2] University of Electronic Science and Technology of China, Chengdu 610000, China

[3] Chang'an University, Xi'an 710000, China

[4] Harbin Institute of Technology, Shenzhen 518000, China

Abstract. As a proactive defense approach, network threat hunting has become a research hotspot in modern cybersecurity due to its advantages, including strong initiative, high detection accuracy, rapid response, and continuous optimization capabilities. However, traditional subgraph matching methods based on combinatorial optimization suffer from high computational complexity, making them unsuitable for real-time threat detection in large-scale provenance graphs with millions of nodes. Although existing Graph Neural Network (GNN) approaches demonstrate superior efficiency and robustness, they often overlook the semantic information of entire behavior chains during the embedding process. To address these challenges, we propose SemantiHunt, a behavioral semantics-driven method for network threat hunting. Our method takes "behavior" as the basic modeling unit. We first use a pre-trained BERT model to perform initial semantic embedding for each behavior, and then design a behavior-centric message passing mechanism to efficiently capture and integrate the fine-grained semantic features of continuous behavior chains within the graph structure. Experimental results show that SemantiHunt can more effectively distinguish attack paths from normal operation paths in complex attack chain identification tasks. Compared with the current advanced method Provg-Searcher, the overall accuracy rate is increased by an average of about 0.81%.

Keywords: Network Threat Hunting · Subgraph Matching · Graph Neural Network · Behavioral Semantic

1 Introduction

Cyber Threat Hunting [5] is a proactive defense technology that assumes threats exist and uses threat intelligence, behavioral analysis, and automated tools to find malicious activities. It enhances security teams' situational awareness

M. Yoshikawa et al. (Eds.): ADMA 2025, LNAI 16197, pp. 426–434, 2026.
https://doi.org/10.1007/978-981-95-3453-1_32

and helps meet compliance requirements, especially in high-value industries like finance and government. Provenance Graphs [12] are key tools in cyber threat hunting, showing causal relationships between system entities to help trace attack paths. Threat hunting methods [7–9] include anomaly detection, which finds deviations from normal behavior but may have high false positives, and rule matching, which detects known threats but struggles with new attacks. Cyber Threat Intelligence (CTI)-driven [1] hunting is a growing focus, offering comprehensive attack chain analysis but often limited to matching individual IOCs. Milajerdi et al. proposed modeling threat hunting as a Graph Pattern Matching problem [6], which focuses on the structure of entire attack campaigns by aligning attack behavior query graphs with provenance graphs, enabling effective identification of complex attack paths.

In the graph pattern matching framework, attack patterns described by Cyber Threat Intelligence (CTI) can be modeled as a Query Graph, where nodes represent key system entities and edges denote interactions between these entities [10]. The system's actual behavior is modeled through a Provenance Graph, which extracts causal relationships from audit logs. The core task of cyber threat hunting is to search for subgraphs within the provenance graph that match the query graph to uncover the attacker's activity path. Compared to traditional rule-based matching, subgraph matching provides a more comprehensive way to capture the full attack chain and reveal the complete trajectory of the intrusion. For example, the Poirot system constructs CTI-driven query graphs and performs pattern matching against the provenance graph to uncover the entire sequence of attack activities [6]. However, as the size of the provenance graph grows, traditional subgraph matching algorithms face significant efficiency bottlenecks [11]. Furthermore, attackers may deliberately insert redundant activities to interfere with the matching process, potentially leading to false positives or missed detections. Therefore, improving efficiency while maintaining matching accuracy has become a key challenge in this field. To improve the efficiency and robustness of traditional subgraph matching methods, researchers have begun exploring Graph Neural Network (GNN)-based subgraph matching models [2]. However, GNNs face two challenges: (1) Insufficient behavioral semantics capture due to separate processing of node and edge attributes and a node-centric message-passing mechanism. (2) Lack of interpretability, making forensic analysis difficult for security analysts.

The aforementioned methods still face two critical challenges: insufficient capture of the semantic information of attack behaviors and a lack of interpretability in detection results. To address these bottlenecks, we propose SemantiHunt, a behavior semantics-based cyber threat hunting approach. Our main contributions are as follows:

- We propose a behavior-centric modeling approach for cyber threat hunting. Instead of modeling individual nodes or events, our method uses "behaviors" as the fundamental unit of modeling. This enhances the representation of operational semantics and contextual correlations within attack chains.

- We employ a semantic embedding mechanism based on a pre-trained language model. By leveraging a pre-trained BERT model to generate semantic representations of behaviors, our method effectively captures their inherent semantic features, addressing the limitations of traditional graph embeddings in understanding behavioral semantics.
- We design a behavior-centric message-passing mechanism to model fine-grained behavioral chain semantics. By integrating graph structure with a behavior-centered message-passing strategy, our approach more effectively captures semantic dependencies across sequences of behaviors, thereby improving the accuracy of complex attack path identification.

2 SemantiHunt: A Behavioral Semantics-Driven Method for Network Threat Hunting

Our SemantiHunt framework has two modules: graph construction and subgraph matching, as shown in 1. The first extracts events from logs to build an initial provenance graph and constructs process self-graphs and a query graph. The second integrates node and edge attributes, uses BERT for embeddings, encodes features via a message passing network, and calculates subgraph similarity using sequential embedding learning, triggering alerts if the threshold is exceeded.

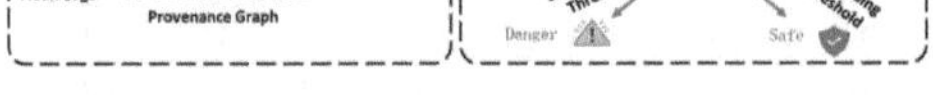

Fig. 1. The SemantiHunt framework.

Graph Construction Module: This module includes the construction of both the query graph and the provenance graph.

(1) *Query Graph Construction*: The query graph is built using multi-source threat intelligence like security blogs, industry reports, dark web forums, and subscription services. It converts known attack patterns into DAGs, where nodes are threat entities and edges show their relationships. This module automates query graph construction for subgraph matching in network threat hunting.

(2) *Provenance Graph Construction*: We build a provenance graph from kernel audit logs, turning system behaviors into a DAG. Nodes are key entities like processes and files, and edges show interactions such as process spawning and

file operations. This graph reveals system behavior structure and aids in tracing attack chains. To improve search efficiency and matching accuracy, we employ the provenance graph preprocessing method proposed in [6] to perform multi-step preprocessing on the provenance graph, as shown in Fig. 2. We simplify the provenance graph by retaining key entities and events, merging threads, and using abstraction. We manage dependency explosion and over-smoothing with node version control and aggregation nodes. Process-centered self-graphs are extracted and simplified via iterative label propagation and hash aggregation, preserving core behavior features for efficient network threat hunting.

Subgraph Matching Module: The subgraph matching module aims to predict the subgraph similarity score between the query graph and the process self-graph. The module takes the query graph G_Q and the process self-graph G_S as input and outputs their subgraph similarity score. If the score exceeds the given threshold, an alert will be triggered. The workflow of the subgraph matching module is shown in Fig. 3.

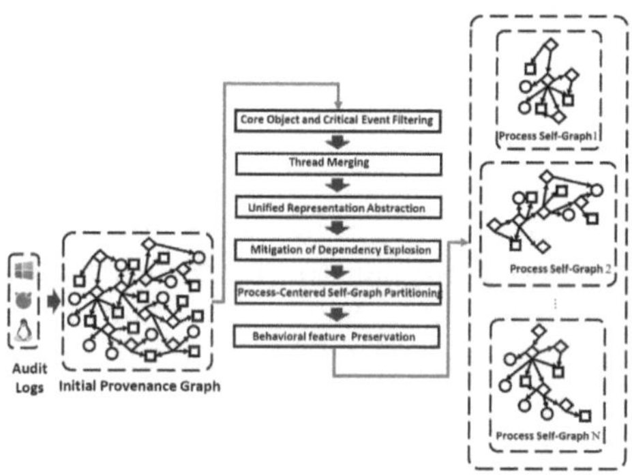

Fig. 2. Provenance Graph Preprocessing.

(1) *Initial Behavioral Embedding:* Both the query graph G_Q and the process self-graph G_S undergo an initial behavioral embedding. A behavior is described collectively by the initiating entity, the target entity, and the interaction between them. Specifically, taking an edge $(u, v) \in E_Q$ in the query graph as an example, in this study, the behavior $B_{(u,v)}$ is defined as a triplet, $B_{(u,v)} = \big(u,\ (u,v),\ v\big)$, where u represents the source node of the behavior, v represents the target node, and (u, v) represents the directed edge connecting these two nodes. For each behavior $B_{(u,v)}$, the attributes of the source node, the edge, and the target node are concatenated in sequence to form a text sequence that expresses the corresponding behavioral semantics (BS). $BS_{(u,v)} = L_Q(u) \oplus$ "" $R_Q(u,v)$ "" $\oplus L_Q(v)$, where $L_Q(u)$ represents the attributes of the source node u, $R_Q(u,v)$ represents the attributes of the edge (u, v), and $L_Q(v)$ represents the attributes of the target node v. The symbol $\oplus$ denotes the operation of string concatenation, and "" represents an empty string space. After obtaining the behavioral

semantics from the graph, we utilize a BERT pre-trained model to perform an initial embedding. Specifically, the behavioral semantics are fed into the BERT pre-trained model and then mapped to a d_0-dimensional vector space through a Multilayer Perceptron (MLP), thereby obtaining the initial embedding representation $h_{(u,v)}^{(0)} \in R^{d_0}$ of the behavior, $h_{(u,v)}^{(0)} = \text{MLP}\left(\text{BERT}\left(\text{BS}_{(u,v)}\right)\right)$. Similarly, for an edge $(x, y) \in E_S$ in the process self-graph, after the initial behavioral embedding, its embedding representation $h_{x,y}^{(0)} \in R^{d_0}$ can be obtained.

(2) *Behavior-Centric Message Passing:* After getting the initial embeddings, T rounds of behavior-centric message passing occur. Each behavior gets messages from neighbors in both forward and backward phases due to the directed graph. The behavior $B_{u,v}$ receives messages from all behaviors that have node u as their target node. It then updates the embedding representation of behavior $B_{u,v}$ during the forward phase, denoted as $h_{u,v}^{t_f} \in R^{d_t}$, $h_{(u,v)}^{t_f} = \sigma\left(W^{t_f} h_{(u,v)}^{t-1} + U^{t_f} h_{(\cdot,u)}^{t-1} + b^{t_f}\right)$, where $W^{t_f}, U^{t_f} \in R^{d_t \times d_{t-1}}$ are learnable weight matrices, $b^{t_f} \in R^{d_t}$ is a learnable bias vector, and σ is the activation function, with the LeakyReLU function being used in the experiments. Additionally, $h_{(\cdot,u)}^{t-1} \in R^{d_{t-1}}$ is the aggregated message from behaviors that target node u, $h_{\cdot,u}^{t-1} = \text{AGGR}\left(\left\{h_{(i,u)}^{t-1} \mid (i,u) \in E_Q\right\}\right)$, where AGGR represents an aggregation function, with common choices including mean, sum, and MLP, with the mean being used in the experiments. In the backward message passing phase, behavior $B_{u,v}$ receives messages from all behaviors that have node v as their target node. Similar to the forward message passing phase, the embedding representation of behavior $B_{u,v}$ during the backward phase, denoted as $h_{(u,v)}^{t_b} \in R^{d_t}$, is updated based on the previous round's embedding representation and the messages received during the backward phase, and $h_{(u,v)}^{t_b} = \sigma\left(W^{t_b} h_{(u,v)}^{t-1} + U^{t_b} h_{(\cdot,v)}^{t-1} + b^{t_b}\right)$, where $W^{t_b}, U^{t_b} \in R^{d_t \times d^{t-1}}$ are learnable weight matrices, $b^{t_b} R^{d_t}$ is a learnable bias vector, and $h_{(\cdot,v)}^{t-1} \in R^{d_t}$ is the aggregated message from behaviors targeting node v. Ultimately, based on the updated embedding representations of behavior $B_{u,v}$ from both the forward and backward message passing phases, the embedding representation for the next

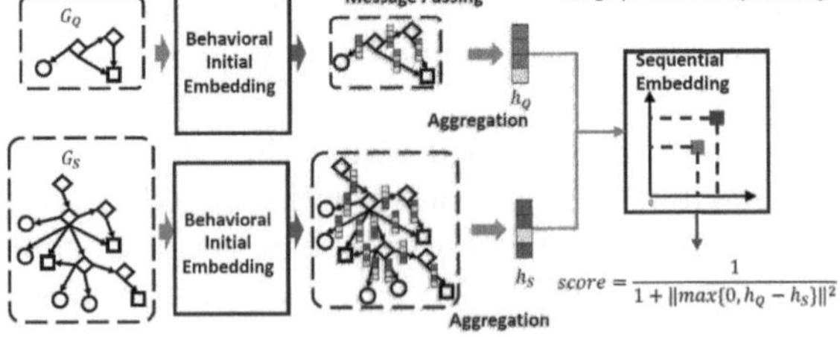

Fig. 3. The workflow of the subgraph matching module.

round is obtained, denoted as $h^t_{u,v} \in R^{d_t}$, $h^t_{(u,v)} = \sigma\left(W^t\left(h^{t_f}_{(u,v)} \parallel h^{t_b}_{(u,v)}\right) + b^t\right)$, where $W^t \in R^{d_t \times 2d_t}$ is a learnable weight matrix, $b^t \in R^{d_t}$ is a learnable bias vector, and $\parallel$ denotes vector concatenation.

After T rounds of message passing, the final embedding representation of behavior $B_{u,v}$ is recorded as: $z_{(u,v)} = h^T_{(u,v)} \in R^{d_T}$, where $h^T_{(u,v)}$ represents the embedding obtained after T rounds of message passing. Similarly, after T rounds of message passing, the final embedding representation of behavior $B_{x,y}$ in the process self-graph can be obtained as $z_{(x,y)} = h^T_{(x,y)} \in R^{d_T}$. All behaviors in both the query graph and the process self-graph undergo initial behavioral embedding and T rounds of message passing to obtain their final embedding representations. Since subgraph matching is a graph-level task, it is necessary to aggregate the embedding vectors of all behaviors in the graph to obtain the embedding vector representation of the entire graph. The process self-graph's embedding vector is $h_Q = \sigma(Q \cdot AGGR(h^T_{<u,v>} | <u,v> \in E_Q)), h_S = \sigma(S \cdot AGGR(h^T_{<x,y>} | <x,y> \in E_S))$, where $Q \in R^{d_T \times d_T}$ is a learnable weight matrix shared by all query graphs, and $S \in R^{d_T \times d_T}$ is a learnable weight matrix shared by all process self-graphs. AGGR represents the graph aggregation operation, with summation being used in the experiments.

(3) *Subgraph Matching Relationship Learning*: The subgraph relationship is essentially a partial order relation. Given two graphs $G_1$and G_2, if the vertex set and edge set of G_1 are subsets of G_2, then G_1 is called a subgraph of G_2, denoted as $G_1 \subseteq G_2$. The subgraph relationship satisfies the three properties of a partial order relation. Firstly, reflexivity requires that any graph is a subgraph of itself, i.e., $G \subseteq G$; secondly, antisymmetry indicates that if $G_1 \subseteq G_2$ and $G_2 \subseteq G_1$ both hold, then it must be the case that $G_1 = G_2$; finally, transitivity implies that if $G_1 \subseteq G_2$ and $G_2 \subseteq G_3$, then $G_1 \subseteq G_3$ must also hold. In the subgraph matching problem, an isomorphic relationship between the query graph and target graph implies a containment relationship. Order Embedding captures this partial order by mapping graphs to a low-dimensional space, preserving the containment through vector coordinate comparison. This method provides a theoretical basis for isomorphism determination in subgraph matching.

3 Experimental Evaluation and Analysis

The experiments used a server with 12 vCPUs of Intel Xeon Platinum 8352V at 2.10GHz, an RTX 3080 × 2 GPU with 20GB memory, 48GB RAM, and 80GB disk space. The software environment includes Ubuntu 20.04, Python 3.8, and PyTorch 1.11.0. We use the DARPA TC dataset, including Theia, Trace, Cadets, and FiveDirections. These datasets record OS-level events like system calls and network flows in provenance graphs showing causal relationships between entities. Query graphs are built from attack reports. After constructing provenance graphs, we divide them into process self-graphs and sequentially perform steps like core object filtering, thread merging, unified representation abstraction, dependency explosion mitigation, and behavioral feature preservation.

Table 1. The experimental results for Acc and AUC

Method	Theia		Trace		Cadets		FiveDirections	
	Acc	AUC	Acc	AUC	Acc	AUC	Acc	AUC
IsoRankN [4]	61.32	61.26	53.47	52.04	60.45	54.93	58.76	55.63
SimGNN [3]	81.36	87.46	73.86	82.24	82.48	88.49	80.52	87.62
DeepHunter [13]	81.55	88.83	72.81	81.42	82.11	88.38	79.34	86.51
Poirot [6]	95.33	95.42	95.93	96.41	96.29	97.56	95.17	95.45
Provg-Searcher [2]	97.73	97.82	97.44	97.46	95.74	97.81	97.77	97.92
SemantiHunt	98.81	98.85	98.38	98.49	97.13	98.56	97.58	97.61

Table 2. The experimental results for the Prec and Recall

Method	Theia		Trace		Cadets		FiveDirections	
	Prec	Recall	Prec	Recall	Prec	Recall	Prec	Recall
IsoRankN [4]	62.37	60.19	66.48	22.94	82.27	27.89	64.51	21.79
SimGNN [3]	76.86	90.23	70.13	87.98	78.24	90.15	72.68	87.88
DeepHunter [13]	77.71	86.39	69.47	84.35	77.58	90.07	73.89	87.79
Poirot [6]	93.09	97.95	95.21	97.08	97.59	96.01	95.76	97.23
Provg-Searcher [2]	97.34	97.11	97.49	96.37	97.87	97.73	97.98	97.74
SemantiHunt	98.18	97.95	98.59	98.21	98.66	98.91	97.73	97.32

The experimental results for the Acc and AUC metrics are shown in Table 1, the Prec and Recall evaluation metrics are presented in Table 2. From the perspectives of Acc and AUC, SemantiHunt achieved the highest or nearly the highest scores across various datasets. Particularly on the Theia, Trace, and Cadets datasets, its accuracy and AUC were higher than those of the Provg-Searcher method. This indicates that the strategy based on constructing complete behaviors, using the BERT pre-trained model, and behavior-centric message passing effectively enhances the model's overall understanding and detection capability of network threat behaviors. Comparing the Prec and Recall results, SemantiHunt also demonstrated excellent performance. For instance, on the Trace dataset, SemantiHunt achieved high scores of 98.59 and 98.21, respectively, and maintained the highest or nearly the highest performance on other datasets. This fully validates that the method in this chapter can ensure high-accuracy detection while effectively capturing more positive samples, overcoming the high false negative or false positive issues that previous methods might have. In summary, SemantiHunt integrates node and edge attributes to construct complete behaviors, utilizes the BERT pre-trained model for initial embedding, and is designed with a behavior-centric message passing mechanism. It surpasses existing mainstream methods in global performance metrics such as Acc and AUC, and performs well in fine-grained detection metrics such as Prec and Recall. The experimental results fully demonstrate that SemantiHunt has strong generalization and stability in network threat detection tasks.

4 Conclusion

This paper proposed A Behavioral Semantics-Driven Method for Network Threat Hunting, named SemantiHunt. This method includes four major processes: construction of query graphs and provenance graphs, initial behavioral embedding, behavior-centric message passing, and subgraph matching. By taking behavioral sequences as the core modeling unit and capturing and integrating the fine-grained semantic associations of continuous behavior chains within the graph structure, SemantiHunt significantly enhances the subgraph matching model's ability to identify attack paths. This effectively addresses the shortcomings of existing methods in semantic understanding and detection accuracy, providing a precise solution to the problem of network threat hunting. Future work could explore unifying data into a multimodal provenance graph and designing cross-modal subgraph matching algorithms for complex attack patterns. Combining meta-learning or self-supervised learning could create dynamic models that quickly adapt to new threats. Integrating the model with an automated response system could close the loop from detection to blocking.

Acknowledgments. This work is supported by the Major Key Project of PCL (Grant No. PCL20224A05).

References

1. Alaeifar, P., Pal, S., Jadidi, Z., Hussain, M., Foo, E.: Current approaches and future directions for cyber threat intelligence sharing: A survey. J. Inf. Secur. Appl. **83**, 103786 (2024)
2. Altinisik, E., Deniz, F., Sencar, H. T.: Provg-searcher: a graph representation learning approach for efficient provenance graph search. In: the 2023 ACM SIGSAC Conference on Computer and Communications Security, pp. 2247–2261, 2023
3. Bai, Y., Ding, H., Bian, S., Chen, T., Sun, Y., Wang, W.: Simgnn: a neural network approach to fast graph similarity computation. In: Proceedings of the Twelfth ACM International Conference on Web Search and Data Mining, pp. 384–392, 2019
4. CS, L., K, L., B.M., S. R., B. B.: Isorankn: spectral methods for global alignment of multiple protein networks. Bioinformatics, 25(12), 2009
5. Kulkarni, M.S., Ashit, D.H., Chetan, C.N.: A proactive approach to advanced cyber threat hunting. In: 2023 7th International Conference on Computation System and Information Technology for Sustainable Solutions (CSITSS), pp. 1–6, 2023
6. Milajerdi, S.M., Eshete, B., Gjomemo, R., Venkatakrishnan, V.: Poirot: aligning attack behavior with kernel audit records for cyber threat hunting. In: CCS '19: Proceedings of the 2019 ACM SIGSAC Conference on Computer and Communications Security,pp. 1795–1812, 2019

7. Regragui, Y., Mazighi, A., Ballihi, L., Orhanou, G.: Impact evaluation of feature selection algorithms on machine learning-based intrusion detection. In: 2024 11th International Conference on Wireless Networks and Mobile Communications (WINCOM), pp. 1–6, 2024
8. Sato, M., Yamaki, H., Takakura, H.: Unknown attacks detection using feature extraction from anomaly-based ids alerts. In: 2012 IEEE/IPSJ 12th International Symposium on Applications and the Internet, pp. 273–277, 2012
9. Settanni, G., et al.: A collaborative cyber incident management system for european interconnected critical infrastructures. J. Inf. Secur. Appl. **34**, 166–182 (2017)
10. Sunuwar, D., Singh, M.: Comparative analysis of relational and graph databases for data provenance: Performance, queries, and security considerations. In. 2023 World Conference on Communication and Computing (WCONF), pp. 1–7, 2023
11. Vasylkivskyi, O., Tkachenko, S., A'Ggel Al-Zabi, B.R.: Selection isomorphic subgraphs from graph. In: 2011 11th International Conference The Experience of Designing and Application of CAD Systems in Microelectronics (CADSM), pp. 243–245, 2011
12. Wang, S., et al.: Threatrace: detecting and tracing host-based threats in node level through provenance graph learning. IEEE Trans. Inf. Forensics Secur. **17**, 3972–3987 (2022)
13. Wei, R., Cai, L., Zhao, L., Yu, A., Meng, D.: Deephunter: a graph neural network based approach for robust cyber threat hunting. In: Garcia-Alfaro, J., Li, S., Poovendran, R., Debar, H., Yung, M. (eds.) Security and Privacy in Communication Networks, pp. 3–24. Springer International Publishing, Cham (2021)

Mining Temporal Structures for Emotion Recognition in Conversation via a Temporal-Aware Attention Network

Juntao Wang(✉) and Tsunenori Mine

Kyushu University, Fukuoka, Japan
juntao.wang26@gmail.com , mine@ait.kyushu-u.ac.jp

Abstract. Emotion recognition in conversation (ERC) aims to identify the emotions expressed in utterances by speakers during a conversation. Effective ERC requires modeling both contextual and temporal dependencies. While prior work emphasizes speaker and discourse cues, temporal structure remains underexplored. We propose a Temporal-Aware Attention Network (TAA-Net), which integrates graph-inspired relational encoding into the self-attention mechanism. By incorporating relative temporal positions and time intervals between utterances, TAA-Net enables more precise modeling of temporal dependencies. Experiments on three ERC benchmarks demonstrate the effectiveness of our approach.

Keywords: Emotion recognition in conversation · Temporal modeling

1 Introduction

Emotion recognition in conversation (ERC) aims to identify the emotion expressed in each utterance within a conversation. In recent years, this task has received growing attention due to its potential to enhance empathetic responses in human-computer interaction systems. Prior studies have demonstrated that contextual information plays a vital role in emotion understanding.Although various architectures have been employed to model contextual cues (e.g., speaker identity and discourse history), most methods rely on the inherent temporal modeling capabilities of their backbones.Recurrent Neural Networks process input sequentially and thus accumulate temporal influence implicitly, but they lack mechanisms explicitly designed to capture temporal dependencies.Graph Neural Networks excel at capturing relational structures, yet often struggle to preserve the sequential continuity crucial for dialogue understanding.Recently, large language models (LLMs) have been introduced into ERC [6,11], leveraging their powerful pre-trained knowledge to boost performance. While LLMs provide strong general language understanding, they do not explicitly model task-specific structures such as temporal progression and speaker turn dynamics–both essential for effective ERC. Moreover, the substantial computational cost

M. Yoshikawa et al. (Eds.): ADMA 2025, LNAI 16197, pp. 435–442, 2026.
https://doi.org/10.1007/978-981-95-3453-1_33

of LLMs makes adaptation to new dialogue domains particularly challenging. To address the need for explicit temporal structure modeling in ERC, we propose a Temporal-Aware Attention Network (TAA-Net). This framework captures temporal features through an attention mechanism that focuses on relative temporal positions and time intervals between utterances. Specifically, we employ Long Short-Term Memory (LSTM) networks to track the emotional progression across the conversation timeline. In parallel, the Temporal-Aware Attention module introduces graph-inspired relational encoding into the self-attention mechanism to model positional dependencies between utterances, and incorporates a time-aware decay function to reflect the diminishing influence of distant contextual information. By explicitly modeling temporal information, our method demonstrates competitive performance on three benchmark ERC datasets compared to strong baseline models. Further ablation studies confirm the effectiveness of the proposed temporal module in enhancing emotion recognition performance.

2 Methodology

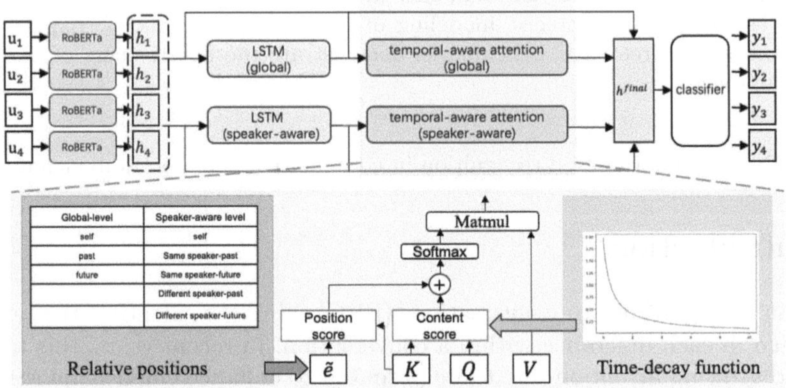

Fig. 1. The overall architecture of the model.

In the ERC task, a conversation is represented as a sequence of N utterances, denoted by $D = \{u_1, u_2, \ldots, u_N\}$. Each utterance u_i is spoken by a speaker $s_{\phi(u_i)} \in \{s_1, s_2, \ldots, s_M\}$, where $M \geq 2$ is the number of distinct speakers in the conversation, and ϕ is a mapping from the utterance index to its corresponding speaker index. For each $\lambda \in [1, M]$, let U_λ be the subset of U consisting of all utterances spoken by speaker s_λ. The goal of the ERC task is to predict the emotion label y_i for each utterance u_i in the conversation.

The overall architecture of our proposed model is illustrated in Fig. 1. In the following, we describe each component of the framework in detail.

2.1 Feature and Context Modeling

Following prior work [3], we adopt the RoBERTa-large model fine-tuned on the ERC dataset for utterance-level emotion classification. We freeze the model parameters during training and use it as a feature extractor. Specifically, for each input utterance u_i at time step i in the conversation, we obtain its context-independent representation h_i from the `[CLS]` token of the last hidden layer of the pre-trained model. The utterance representation h_i has a dimensionality of d.

To enhance the feature's expressiveness, we apply a linear transformation followed by a ReLU activation before feeding them into the subsequent LSTM-based context modeling module:

$$\mathbf{H}_i = \mathrm{ReLU}(W_1 \cdot \mathbf{h}_i + \mathbf{b}_1), \tag{1}$$

where $\mathbf{h}_i \in \mathbb{R}^d$ is the context-independent utterance representation at timestep i, $\mathbf{H}_i \in \mathbb{R}^{d_H}$ is the transformed input feature, and $W_1 \in \mathbb{R}^{d_H \times d}$, $\mathbf{b}_1 \in \mathbb{R}^{d_H}$ are learnable parameters of the linear transformation, where d_H denotes the hidden dimension size.

To more effectively incorporate speaker information, following prior work on conversational context modeling [5], we employ two separate bidirectional LSTM networks to model global and speaker-level contextual dependencies.

For global context modeling, we employ a bidirectional LSTM to capture sequential dependencies among adjacent utterances across the entire conversation:

$$\mathbf{l}_i^g = \overleftrightarrow{\mathrm{LSTM}}_g(\mathbf{H}_i), \tag{2}$$

where $\mathbf{l}_i^g \in \mathbb{R}^{2d_H}$ represents the global-level contextual representation of the i-th utterance, and $\overleftrightarrow{\mathrm{LSTM}}_g$ denotes the bidirectional LSTM dedicated to modeling global context.

For speaker-level context modeling, we use a separate bidirectional LSTM to capture intra-speaker dependencies, i.e., the flow of information among utterances spoken by the same speaker:

$$\mathbf{l}_i^s = \overleftrightarrow{\mathrm{LSTM}}_s(\mathbf{H}_i(s_{\phi(i)})), \tag{3}$$

where $\mathbf{l}_i^s \in \mathbb{R}^{2d_H}$ is the speaker-level contextual representation of utterance u_i, $\lambda = \phi(u_i)$ is the index of the speaker who utters u_i, and U_λ denotes the ordered set of utterances spoken by speaker p_λ. The input $\mathbf{H}_i(s_{\phi(i)})$ refers to the local context around u_i within U_λ, preserving the speaker-specific utterance order.

2.2 Temporal-Aware Modeling

To enhance the temporal sensitivity in ERC, we propose a temporal-aware attention module designed to capture the relative temporal structure of utterances in a dialogue.

Inspired by DialogueGCN's [4] modeling of temporal dependencies in graph networks, we propose a corresponding approach within the attention mechanism [10].

We define relative positions between utterances (u_i, u_j) at two levels for modeling temporal dependencies:

$$\mathcal{R}_{\text{global}} = \{\text{past}, \ \text{future}, \ \text{self}\},$$

$$\mathcal{R}_{\text{speaker}} = \{\text{same-past}, \ \text{different-past}, \ \text{same-future}, \ \text{different-future}, \ \text{self}\},$$

where *past* indicates $i < j$, *future* indicates $i > j$, and *self* indicates $i = j$. *same* and *different* indicate whether the speaker function $\phi(u_i) = \phi(u_j)$ or not.

Each relation type $r^g_{ij} \in \mathcal{R}_{\text{global}}$ and $r^s_{ij} \in \mathcal{R}_{\text{speaker}}$ is first mapped to a trainable embedding:

$$\mathbf{e}^g_{ij} = \text{Embed}_g(r^g_{ij}) \in \mathbb{R}^{d_r}, \quad \mathbf{e}^s_{ij} = \text{Embed}_s(r^s_{ij}) \in \mathbb{R}^{d_r},$$

where d_r denotes the dimensionality of the relation embeddings.

Then, each embedding is projected into the feature space through separate linear transformations:

$$\tilde{\mathbf{e}}^g_{ij} = \mathbf{W}^g_r \mathbf{e}^g_{ij} \in \mathbb{R}^{d_H}, \quad \tilde{\mathbf{e}}^s_{ij} = \mathbf{W}^s_r \mathbf{e}^s_{ij} \in \mathbb{R}^{d_H},$$

where $\mathbf{W}^g_r, \mathbf{W}^s_r \in \mathbb{R}^{d_H \times d_r}$ are learnable projection matrices for global relations and speaker relations, respectively.

To capture these two levels of temporal relations, we design a two-branch Temporal-Aware Attention mechanism, where each branch independently models either global or speaker-aware relations.

The input contextual representation $\mathbf{l}^* \in \mathbb{R}^{2d_H}$, where $* \in \{g, s\}$, is first projected into query, key, and value spaces:

$$Q^{(*)}, \ K^{(*)}, \ V^{(*)} = \mathbf{W}^{(*)}_q \mathbf{l}^{(*)}, \ \mathbf{W}^{(*)}_k \mathbf{l}^{(*)}, \ \mathbf{W}^{(*)}_v \mathbf{l}^{(*)},$$

where all projection matrices $\mathbf{W}^{(*)}_q, \mathbf{W}^{(*)}_k, \mathbf{W}^{(*)}_v \in \mathbb{R}^{d_H \times 2d_H}$.

Our attention score consists of two components: content attention and position attention. Content attention follows the standard self-attention mechanism, computing the extent to which u_i attends to u_j:

$$\text{content-score}^{(*)}_{i,j} = \frac{Q^{(*)}_i \cdot (K^{(*)}_j)^\top}{\sqrt{d_H}}. \tag{4}$$

We introduce a position attention term, which captures temporal structure by reflecting the relative positional relationship between i and j. Specifically, we use the relation feature $\tilde{\mathbf{e}}^{(*)}_{ij}$ corresponding to the pair (i, j), and treat it as a structural query that interacts with the key vector $K^{(*)}_j$:

$$\text{position-score}^{(*)}_{i,j} = \tilde{\mathbf{e}}^{(*)}_{ij} \cdot K^{(*)}_j. \tag{5}$$

To further enhance the model's sensitivity to temporal distance, we introduce a reciprocal decay function that modulates the contribution of contextual utterances based on their relative distance to the target, reflecting the intuition that nearby utterances tend to be more emotionally informative. Unlike exponential decay, which often leads to excessively sharp attenuation of distant context, the reciprocal form enables a smoother decay, balancing the emphasis on nearby utterances while preserving useful long-range dependencies. We also conducted preliminary experiments, which confirmed that the reciprocal decay performs better than exponential decay. The reciprocal decay function is defined as follows:

$$f_{\text{decay}}(d_{ij}) = \frac{1}{d_{ij} + c}, \tag{6}$$

where $d_{ij} = |i-j|$ denotes the absolute positional distance between utterances u_i and u_j, and $c > 0$ is a smoothing constant to avoid division by zero. To preserve the self-attention behavior, we set $f_{\text{decay}}(0) = 1$.

The final attention score is then computed by applying a time-decay function to the content score and adding the position score:

$$\text{score}_{i,j}^{(*)} = f_{\text{decay}}(d_{ij}) \cdot \text{content-score}_{i,j}^{(*)} + \text{position-score}_{i,j}^{(*)}. \tag{7}$$

The time-decay function is applied only to the content-score, not to the position-score, in order to maintain the integrity of temporal dependency modeling while avoiding the introduction of noise.

The temporal-aware context representation is computed as a weighted sum over all value vectors, with attention weights obtained via softmax:

$$\mathbf{o}_i^{(*)} = \sum_{j=1}^{N} \frac{\exp\left(\text{score}_{i,j}^{(*)}\right)}{\sum_{j=1}^{N} \exp\left(\text{score}_{i,j}^{(*)}\right)} \cdot \mathbf{V}_j^{(*)}. \tag{8}$$

We concatenate the utterance representation $\mathbf{H}_i$, the two contextual representations $\mathbf{l}_i^g$, $\mathbf{l}_i^s$, and the two temporal-aware representations $\mathbf{o}_i^g$, $\mathbf{o}_i^s$ to form the final feature representation:

$$\mathbf{h}_i^{\text{final}} = [\mathbf{H}_i \parallel \mathbf{l}_i^g \parallel \mathbf{l}_i^s \parallel \mathbf{o}_i^g \parallel \mathbf{o}_i^s], \tag{9}$$

where $\parallel$ denotes concatenation. Finally, a classification layer is applied to predict the emotion label for each utterance. Specifically, the prediction is computed as:

$$\mathbf{y}_i = \text{softmax}(\mathbf{W}_c \mathbf{h}_i^{\text{final}} + \mathbf{b}_c), \tag{10}$$

where $\mathbf{W}_c \in \mathbb{R}^{|\mathcal{Y}| \times 7d_H}$ and $\mathbf{b}_c \in \mathbb{R}^{|\mathcal{Y}|}$ are trainable parameters, and $|\mathcal{Y}|$ is the number of emotion classes.

3 Experiments

3.1 Datasets and Implementation Details

We conducted experiments on three widely-used conversational emotion datasets: IEMOCAP [2], MELD [9], and EmoryNLP [13]. Summary statistics are provided in Table 1.

Table 1. The statistics of three ERC datasets.

Dataset	Conversations			Emotion labels
	Train	Val	Test	
IEMOCAP	120		31	happy, sad, neutral, angry, excited, frustrated
MELD	1038	114	280	joy, anger, fear, disgust, sadness, surprise, neutral
EmoryNLP	659	89	79	sad, mad, scared, powerful, peaceful, joyful, neutral

Hyperparameters were selected through manual tuning based on performance on each validation set. For the IEMOCAP, MELD, and EmoryNLP datasets, the batch sizes are set to 8, 32, and 32, respectively; the learning rates are 0.00005, 0.0002, and 0.0001; and the hidden layer size d_H is fixed at 300 across all experiments. We use cross-entropy loss for optimization and adopt the Adam optimizer. All experiments are conducted on a single NVIDIA RTX 4090 GPU with 24 GB of memory. All reported results are obtained from a single run under the same fixed random seed across all datasets.

3.2 Comparison Methods

Following previous studies, we adopt the weighted-average F1 score and accuracy as evaluation metrics to assess model performance.

We compare our approach with several representative baseline models from prior work: **Recurrent-based models:** DialogueRNN [8], DialogueCRN [5], and DualRAN [7]; **Graph-based models:** DialogueGCN [4], SGED+DAG-ERC [1], and DCGCN [12].

3.3 Results

Table 2. Performance comparison on IEMOCAP, MELD, and EmoryNLP using Weighted F1 and Accuracy. **Bold** indicates the best result, and <u>underline</u> indicates the second-best.

Model	IEMOCAP		MELD		EmoryNLP	
	w-F1	Acc	w-F1	Acc	w-F1	Acc
DialogueRNN-RoBERTa	64.76	-	63.61	-	37.44	-
DialogueCRN	67.53	67.39	65.77	<u>66.93</u>	38.79	<u>41.04</u>
DualRAN	**69.73**	**69.62**	66.24	**67.70**	39.22	-
DialogueGCN-RoBERTa	64.91	-	63.02	-	38.10	-
SGED+DAG-ERC	68.53	-	65.46	-	**40.24**	-
DCGCN	68.31	-	<u>66.25</u>	-	<u>40.23</u>	-
TAA-Net	<u>68.65</u>	<u>68.58</u>	**66.56**	66.78	39.66	**41.77**

Table 2 presents the results on the IEMOCAP, MELD, and EmoryNLP datasets. On the EmoryNLP dataset, our method achieves the highest accuracy (41.77) and a competitive weighted F1 score of 39.66, closely approaching the top results (40.24 and 40.23). On MELD, TAA-Net delivers the best weighted F1 score (66.56), while its accuracy (66.78) is competitive with the second-best result (66.93), highlighting its strong discriminative ability. Although DualRAN slightly outperforms all models on IEMOCAP, TAA-Net secures a solid second place with a weighted F1 of 68.65 and an accuracy of 68.58, indicating strong generalization across speaker dynamics. Despite not ranking first on all metrics, TAA-Net consistently ranks among the top-performing models across all benchmarks, demonstrating robust and reliable performance. These results underscore the model's cross-domain adaptability, and its effective modeling of temporal and contextual dependencies in conversations. We also conducted ablation experiments to assess the effectiveness of our Temporal-Aware Modeling. In particular, we removed both the position attention score and the reciprocal decay function, which resulted in a significant drop in performance across all three datasets, as shown in Table 3. These results underscore the critical role of the Temporal-Aware Modeling in enhancing the overall performance of the model and its ability to capture temporal dynamics in conversations.

Table 3. Experimental results of ablation study.

Model	IEMOCAP		MELD		EmoryNLP	
	w-F1	Acc	w-F1	Acc	w-F1	Acc
TAA-Net	68.65	68.58	66.56	66.78	39.66	41.77
w/o Temporal-Aware Modeling	67.41	67.34	65.18	64.87	38.27	40.04
Drop	-1.24	-1.24	-1.38	-1.91	-1.39	-1.73

4 Conclusion

In this study, we proposed a Temporal-Aware Attention Network (TAA-Net) for emotion recognition in conversation, which explicitly models temporal dependencies through a time-aware attention mechanism with relational encoding and decay control. Results on three ERC benchmarks suggest that TAA-Net effectively captures temporal structures and consistently delivers strong performance across diverse conversational settings.

References

1. Bao, Y., Ma, Q., Wei, L., Zhou, W., Hu, S.: Speaker-guided encoder-decoder framework for emotion recognition in conversation. arXiv preprint arXiv:2206.03173 (2022)

2. Busso, C., et al.: Iemocap: interactive emotional dyadic motion capture database. Lang. Resour. Eval. **42**(4), 335–359 (2008)
3. Ghosal, D., Majumder, N., Gelbukh, A., Mihalcea, R., Poria, S.: COSMIC: cOmmonSense knowledge for eMotion identification in conversations. In: Cohn, T., He, Y., Liu, Y. (eds.) Findings of the Association for Computational Linguistics: EMNLP 2020. pp. 2470–2481. Association for Computational Linguistics, Online (2020). https://doi.org/10.18653/v1/2020.findings-emnlp.224, https://aclanthology.org/2020.findings-emnlp.224/
4. Ghosal, D., Majumder, N., Poria, S., Chhaya, N., Gelbukh, A.: DialogueGCN: a graph convolutional neural network for emotion recognition in conversation. In: Inui, K., Jiang, J., Ng, V., Wan, X. (eds.) Proceedings of the 2019 Conference on Empirical Methods in Natural Language Processing and the 9th International Joint Conference on Natural Language Processing (EMNLP-IJCNLP). pp. 154–164. Association for Computational Linguistics, Hong Kong, China (2019)https://doi.org/10.18653/v1/D19-1015, https://aclanthology.org/D19-1015/
5. Hu, D., Wei, L., Huai, X.: DialogueCRN: contextual reasoning networks for emotion recognition in conversations. In: Zong, C., Xia, F., Li, W., Navigli, R. (eds.) Proceedings of the 59th Annual Meeting of the Association for Computational Linguistics and the 11th International Joint Conference on Natural Language Processing (Volume 1: Long Papers). pp. 7042–7052. Association for Computational Linguistics, Online (2021).https://doi.org/10.18653/v1/2021.acl-long.547, https://aclanthology.org/2021.acl-long.547/
6. Lei, S., Dong, G., Wang, X., Wang, K., Wang, S.: Instructerc: reforming emotion recognition in conversation with a retrieval multi-task LLMS framework. CoRR (2023)
7. Li, J., Wang, X., Zeng, Z.: A dual-stream recurrence-attention network with global-local awareness for emotion recognition in textual dialog. Eng. Appl. Artif. Intell. **128**, 107530 (2024)
8. Majumder, N., Poria, S., Hazarika, D., Mihalcea, R., Gelbukh, A., Cambria, E.: Dialoguernn: an attentive RNN for emotion detection in conversations. In: Proceedings of the AAAI conference on artificial intelligence. vol. 33, pp. 6818–6825 (2019)
9. Poria, S., Hazarika, D., Majumder, N., Naik, G., Cambria, E., Mihalcea, R.: MELD: a multimodal multi-party dataset for emotion recognition in conversations. In: Korhonen, A., Traum, D., Màrquez, L. (eds.) Proceedings of the 57th Annual Meeting of the Association for Computational Linguistics. pp. 527–536. Association for Computational Linguistics, Florence, Italy (2019). https://doi.org/10.18653/v1/P19-1050, https://aclanthology.org/P19-1050/
10. Vaswani, A., et al.: Attention is all you need. Adv. Neural Inf. Proc. Syst. **30** (2017)
11. Xue, J., Nguyen, M.P., Matheny, B., Nguyen, L.M.: Bioserc: integrating biography speakers supported by LLMS for ERC tasks. In: International Conference on Artificial Neural Networks, pp. 277–292. Springer (2024)
12. Yang, Z., Li, X., Cheng, Y., Zhang, T., Wang, X.: Emotion recognition in conversation based on a dynamic complementary graph convolutional network. IEEE Trans. Affect. Comput. **15**(3), 1567–1579 (2024). https://doi.org/10.1109/TAFFC.2024.3360979
13. Zahiri, S.M., Choi, J.D.: Emotion detection on tv show transcripts with sequence-based convolutional neural networks. In: AAAI Workshops. vol. 18, pp. 44–52 (2018)

Detecting and Mitigating Positional Bias in Zero-Shot Anomaly Detection

Ayano Ito[1(✉)], Takeaki Sakabe[1], Yuko Sakurai[1], and Satoshi Oyama[2]

[1] Nagoya Institute of Technology, Nagoya, Japan
{a.ito.588,t.sakabe.858}@stn.nitech.ac.jp, sakurai@nitech.ac.jp
[2] RIKEN AIP, Nagoya City University, Nagoya, Japan
oyama@ds.nagoya-cu.ac.jp

Abstract. In recent years, zero-shot anomaly detection algorithms have attracted attention as effective methods in practical settings where collecting anomalous samples is difficult. However, similar to the well-known order bias observed in large language models (LLMs), there is a possibility that these algorithms may also exhibit positional bias depending on the location of input anomalies in images. In this study, we investigate positional bias in the representative zero-shot anomaly segmentation model, Segment Any Anomaly+ (SAA+), through empirical experiments. The results reveal a clear tendency for the model to detect anomalies more frequently in certain regions, especially the upper center of the image. To mitigate this bias, we propose two novel calibration methods and evaluate their effectiveness in terms of anomaly detection performance and bias mitigation. Experimental results demonstrate that the proposed methods improve both precision and recall while mitigating positional bias.

Keywords: Zero-Shot Anomaly Detection · Positional Bias · Confidence Score Calibration · Bias Mitigation

1 Introduction

Anomaly detection is widely used in various fields such as manufacturing, healthcare, and infrastructure monitoring. However, because anomalous samples are inherently rare, it is often difficult to collect sufficient labeled data for supervised learning. To address this challenge, zero-shot anomaly detection segmentation has gained attention as an approach that enables anomaly detection without requiring images of anomalous instances of the target object.

In the field of computer vision, zero-shot anomaly segmentation, which detects anomalies without using labeled data for the anomaly class, has emerged as a promising research topic. The development of powerful foundation models such as CLIP (Contrastive Language-Image Pretraining) [8] and SAM (Segment Anything Model) [5] has greatly accelerated progress in this area by enabling

M. Yoshikawa et al. (Eds.): ADMA 2025, LNAI 16197, pp. 443–450, 2026.
https://doi.org/10.1007/978-981-95-3453-1_34

task-agnostic feature extraction and segmentation. One representative method is WinCLIP [3], which divides an image into windows and detects anomalous regions by computing text-image similarity using CLIP for each window. Building on this idea, ClipSAM [6] further improves performance by combining CLIP with SAM. On the MVTec AD dataset [1], ClipSAM has demonstrated superior performance compared to existing zero-shot anomaly detection methods, including WinCLIP. In this study, we focus on a representative method called Segment Any Anomaly+ (SAA+) [2]. SAA+ generates segmentation masks and confidence scores from an input image and prompt, and then applies three types of filters-area, saliency, and quantity-to improve anomaly detection accuracy.

In the field of large language models (LLMs), it has been widely reported that the order in which choices are presented can significantly influence model outputs, a phenomenon known as *order bias* [9]. While this type of bias has been actively studied in LLMs, spatial biases in visual tasks have received comparatively little attention, especially in models that rely on prompting mechanisms, such as SAA+. In this study, we explore the possibility that similar biases may emerge in zero-shot visual anomaly detection. Our preliminary investigation reveals that SAA+ tends to assign higher confidence scores to anomalies located in the upper center of the image, suggesting the presence of positional bias in the model's scoring mechanism.

Therefore, this study aims to quantitatively verify the existence and impact of positional bias in SAA+, and proposes two calibration methods to mitigate it. The first method adjusts confidence scores based on position-wise precision obtained from preliminary experiments, while the second employs multinomial logistic regression to estimate the probability of anomaly for each position. We demonstrate through experiments that both methods contribute to improving detection accuracy and mitigating positional bias.

2 Segment Any Anomaly+ (SAA+)

SAA+ is a type of zero-shot anomaly segmentation model. SAA+ uses the GroundingDINO [7] and SAM models to obtain anomaly candidate regions R and confidence scores S from a input image I and input prompts T. Additionally, R and S are filtered using three refined such as *Area Filtering*, *Saliency Filtering*, and *Quantity Filtering* to improve accuracy.

Area Filtering: In general, the size of anomalies in objects is smaller than the size of the objects themselves. Therefore, SAA+ can exclude candidate anomaly regions that do not meet the conditions by specifying the maximum percentage of the area occupied by the size of anomalies in objects. Furthermore, anomalies are often located inside objects. Therefore, SAA+ uses an indicator called IoU, which represents the degree of overlap between two areas, to filter out candidate anomalies.

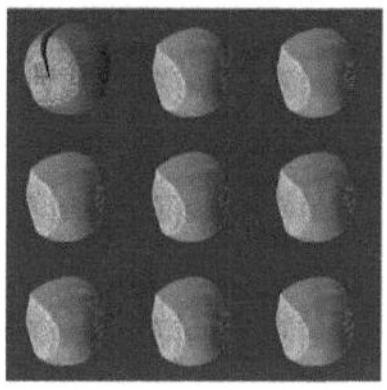

Fig. 1. Example of hazelnut dataset.

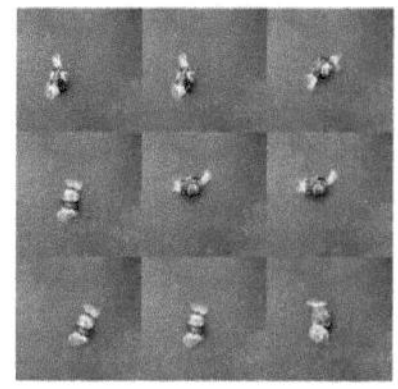

Fig. 2. Example of chocolate dataset.

Saliency Filtering: SAA+ calculates saliency map, or how different an object is from other areas, and uses that value for filtering. This is the same process humans use when searching for anomalous objects, which is to find differences from surrounding objects.

Quantity Filtering: Generally, the number of anomalies in the test object is limited. Therefore, by specifying that there are at most K anomalies, SAA+ selects the top K anomaly region candidates with high confidence scores.

3 Preliminary Experiment: Verification of Positional Bias in SAA+

We conduct a preliminary experiment to examine the possibility that SAA+ assigns high confidence scores to specific anomaly areas.

3.1 Datasets Generation

We use the publicly available ADFI datasets, *Hazelnut* and *Package_sorting* [4], to investigate positional bias. The *Hazelnut* dataset contains images for inspecting defective hazelnuts, consisting of two categories: normal hazelnuts and hazelnuts with cracks. The *Package_sorting* dataset includes six types of chocolate packages used for package sorting tasks. In this study, we construct new datasets based on these two sources to evaluate positional bias. The data generation procedures are described below, and example images are shown in Figs. 1 and 2.

Hazelnut Dataset Generation

1. Randomly select defective and non-defective products from the Hazelnut dataset.
2. Place one defective product and eight non-defective products in a 3×3 grid.
3. Change the position of the defective product to generate nine image patterns per set.

We generated a total of 1,800 images, which were organized into 200 sets.

Fig. 3. Location of positional bias (orange area). (Color figure online)

Chocolate Dataset Generation

1. Randomly select two types of chocolate, A and B, from the Package_sorting dataset.
2. Place A in one location and B in eight locations in a 3×3 grid.
3. Change the position of A to generate nine patterns per set of images.

We also generated a total of 200 sets, corresponding to 1,800 images in total.

3.2 Preliminary Experimental Setup

We verified the positional bias using the generated data sets. To assess positional bias, we ran SAA+ on each dataset and counted the number of times each position was detected as anomalous. If there were no positional bias, each position should be detected as anomalous 200 times. In addition, we counted the number of successful anomaly detections for each position and calculated the corresponding precision. The precision is expressed as the number of successful anomaly detections divided by the number of times detected as anomalous, as shown in equation (1).

$$\text{Precision} = \frac{\text{Number of successful anomaly detections}}{\text{Number of times detected as anomalous}} \tag{1}$$

The input prompt for the hazelnut dataset was ['anomaly. defect. crack.', 'food']. The input prompt for the chocolate dataset was ['anomaly. defect. color defect.', 'food']. For both datasets, SAA+ was executed with the maximum percentage of the area of the anomaly set to 100% of the object and the maximum number of anomalies set to 1.

3.3 Preliminary Experimental Results

As a result of running SAA+ 1,800 times, it was found that the upper center area highlighted in orange in Fig 3 was detected as anomalous more often than other areas. It was also found that the middle center area was detected as anomalous more frequently, although not as frequently as the upper center area. The numerical results are shown in Tables 1 and 2. In the hazelnut dataset, SAA+ detected 824 times anomalies in the upper center. In the chocolate dataset, it detected 436 times anomalies in the upper center. Furthermore, the lower section

Table 1. Results for hazelnut dataset

Position	Number of times detected as anomalous	Number of successful anomaly detections	Precision
Upper left	233	60	0.27
Upper center	**824**	**101**	**0.123**
Upper right	97	60	0.619
Middle left	62	47	0.758
Middle center	356	82	0.23
Middle right	60	47	0.783
Lower left	46	43	0.935
Lower center	76	49	0.645
Lower right	46	42	0.913

Table 2. Results for chocolate dataset

Position	Number of times detected as anomalous	Number of successful anomaly detections	Precision
Upper left	281	91	0.332
Upper center	**436**	**98**	**0.225**
Upper right	189	89	0.471
Middle left	152	74	0.487
Middle center	294	89	0.303
Middle right	150	70	0.467
Lower left	85	56	0.659
Lower center	142	72	0.507
Lower right	71	56	0.789

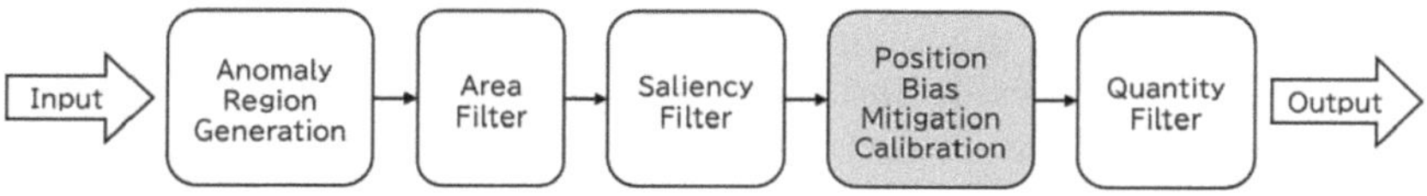

Fig. 4. SAA+ with Confidence Score Calibration to Mitigate Position.

was detected as anomalous less often than the upper section. Looking at the precision, we can see that the upper center had the lowest rate in both data sets, while the lower left and lower right had higher rates than the others. From these results, it can be said that SAA+ has a bias that tends to detect the upper center as anomalous.

4 Positional Bias Mitigation Method

In this study, as shown in Fig. 4, we propose two new calibration methods to mitigate positional bias between the saliency filter and the quantity filter. By incorporating these calibration methods, we aim to correct the confidence scores of anomaly regions and improve their ranking, thereby mitigating positional bias.

4.1 Proposed Method 1: Precision-Based Confidence Score Adjustment

The first calibration method adjusts the confidence score using a weight based on the deviation of positional precision from 0.5. Let denote the precision at position i be Precision_i. The calibration weight is defined as:

$$\text{Weight}_i = 1 + \beta \times (\text{Precision}_i - 0.5) \tag{2}$$

Here, β is a sensitivity parameter that determines the degree of upscaling or downscaling. A precision above 0.5 increases the weight, while a precision below 0.5 decreases it. This calibration is expected to help mitigate positional bias by promoting fairer prioritization.

4.2 Proposed Method 2: Positional Probability Estimation via Multinomial Logistic Regression

The second method is to use multinomial logistic regression to calculate the probability that an object at each location is anomalous. In this case, multinomial logistic regression is performed with the confidence scores for each location as input (9 dimensions) and the probability that an object at each location is anomalous as output (9 dimensions).

SAA+ may have multiple confidence scores for a single object. Therefore, in order to match the input format of the multinomial logistic regression model, only the maximum confidence score for each position is retained. Additionally, positions for which no confidence score was calculated are set to 0. By performing this processing, the maximum anomalous value for each position in each image is obtained. This value and the objects at each position that are detected as anomalous are used as correct labels for training using multinomial logistic regression.

By incorporating the trained model into SAA+, the probability that objects at each position are anomalous is calculated from the confidence scores. The corresponding anomalous candidate areas are detected as anomalous in order of the obtained probabilities from highest to lowest.

5 Experimental Results

5.1 Experimental Setup

The same image dataset as in the experiment that verified the existence of bias is used. We compared the number of anomaly detections and the number of successful detections by SAA+ before and after applying the calibration methods. In this experiment, β in equation (2) was set to 0.1 based on the results of preliminary experiments. In the method using multinomial logistic regression, the dataset used in the preliminary experiment is divided into 10 parts for cross-validation, thereby separating the training data from the test data. In this study, the results from the test data were used for comparison.

The precision and the recall are used as indicators to evaluate the overall accuracy. As described in Sect. 3.2, the precision is the number of successful anomaly detection relative to the number of times detected as anomalous. Recall is the number of successful anomaly detections relative to the number of true anomalies (1800 in this case), as shown in equation (3). The input prompt is the same as in the preliminary experiment.

$$\text{Recall} = \frac{\text{Number of successful anomaly detections}}{\text{Number of true anomalies}} \tag{3}$$

5.2 Experimental Results

The results are shown in Fig. 5. Both proposed methods reduced the number of instances detected as anomalous at places where bias existed. In particular, on

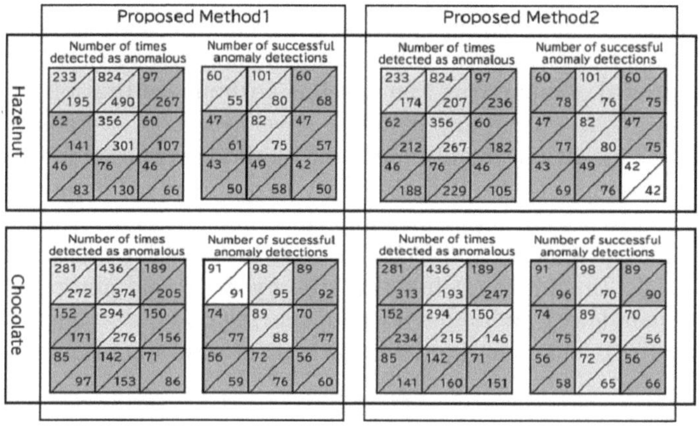

Fig. 5. Visualization of detection counts before and after applying the calibration methods. The top-left figure shows results before calibration, and the bottom-right shows results after calibration. Red regions indicate an increase in detection counts, while blue regions indicate a decrease. (Color figure online)

Table 3. Accuracy of the hazelnut dataset

	Precision	Recall
Before Calibration	0.297	0.294
Proposed Method 1	0.311	0.308
Proposed Method 2	0.360	0.360

Table 4. Accuracy of the chocolate dataset

	Precision	Recall
Before Calibration	0.388	0.386
Proposed Method 1	0.399	0.397
Proposed Method 2	0.364	0.364

the hazelnut dataset, the upper center position was detected as anomalous more than 800 times before applying any calibration. This number was reduced to 490 by the precision-based method and further to 207 by the multinomial logistic regression method. These results suggest that the proposed calibration methods are effective in mitigating positional bias.

As shown in Tables 3 and 4, the precision and recall have increased except for the method using multinomial logistic regression on the chocolate dataset. From this, we can say that the overall accuracy has improved.

However, in some cases, the number of successful anomaly detections decreased after calibration. Moreover, even with the proposed methods, a certain degree of positional bias remained in the detection counts. These findings suggest that the current calibration strategies do not fully eliminate bias and that further refinement of the scoring and ranking process may be necessary.

6 Conclusion

In this study, we tested the existing zero-shot anomaly detection model, SAA+, for the presence of a positional bias. Preliminary experiments revealed the presence of positional bias. Therefore, we introduced two calibration methods to mitigate positional bias. Experimental results demonstrated that both precision and recall improved compared to the baseline without calibration.

As future work, it will be important to develop a more systematic approach for determining calibration weights, since the methods proposed in this study rely on empirical rules derived from preliminary experiments. Furthermore, it may be necessary to investigate input prompts that tend to result in high confidence scores for anomalous objects and to devise learning data for multinomial logistic regression models. In addition, while this study verified and mitigated bias using a grid, SAA+ can detect anomalies without placing objects on a grid. Therefore, it is necessary to consider methods for mitigating positional bias without using a grid.

Acknowledgement. This work was partially supported by JSPS KAKENHI Grant Number JP24K01112 and by JST CREST Grant Number JPMJCR21D1.

References

1. Bergmann, P., Fauser, M., Sattlegger, D., Steger, C.: MVTec AD – A comprehensive real-world dataset for unsupervised anomaly detection. In: Proceedings of the IEEE/CVF conference on computer vision and pattern recognition (CVPR-2019), pp. 9592–9600 (2019)
2. Cao, Y., et al.: Segment any anomaly without training via hybrid prompt regularization. arXiv (2023)
3. Jeong, J., et al.: Winclip: Zero-/few-shot anomaly classification and segmentation. In: Proceedings of the IEEE/CVF Conference on Computer Vision and Pattern Recognition (CVPR-2023), pp. 19606–19616 (2023)
4. Karakurai Inc.: ADFI, Real-World Dataset for Anomaly Detection (2025). https://adfi.jp
5. Kirillov, A., et al.: Segment anything. In: Proceedings of the IEEE/CVF International Conference on Computer Vision (ICCV-2023), pp. 4015–4026 (2023)
6. Li, S.: ClipSAM: CLIP and SAM collaboration for zero-shot anomaly segmentation. Neurocomputing **618**, 129122 (2025)
7. Liu, S., et al.: Grounding DINO: marrying DINO with grounded pre-training for open-set object detection. In: The 18th European Conference on Computer Vision (ECCV-2024), pp. 38–55 (2024)
8. Radford, A., et al.: Learning transferable visual models from natural language supervision. In: The 38th International conference on machine learning (ICML-2021), pp. 8748–8763 (2021)
9. Zhao, T.Z., Wallace, E., Feng, S., Klein, D., Singh, S.: Calibrate before use: improving few-shot performance of language models. In: Proceedings of the 38th International Conference on Machine Learning (ICML-2021), vol. 139, pp. 12697–12706 (2021)

A Path-Aware Framework for Multi-hop Question Answering via Structured Reasoning

Shihao Hu, Jiantong Zhang, and Tao Luo(✉)

Beijing Laboratory of Advanced Information Network, Beijing University of Posts and Telecommunications, Beijing 100876, China
{hushihao,zjt,tluo}@bupt.edu.cn

Abstract. We introduce Scope-then-Traverse, a novel framework for multi-hop question answering that resolves the fundamental tension between evidence breadth and reasoning depth. The framework decouples the QA process into two synergistic stages. The first stage, Knowledge Scoping, generates a complete logical blueprint by decomposing the complex query into independent sub-questions and dependent templates. Subsequently, it performs parallel retrieval using only the high-certainty, low-ambiguity independent sub-questions to build a high-recall evidence pool. This strategy ensures comprehensive evidence coverage while avoiding premature commitment to a single, potentially flawed reasoning path. The second stage, Path-Aware Traversal, constructs a coherent reasoning path within this pre-scoped pool, guided by the blueprint from Stage 1. At each step, a fused scoring mechanism selects the next evidence node by jointly optimizing for local relevance and global path cohesion. Experiments on HotpotQA, 2WikiMultiHopQA, and MuSiQue demonstrate that our approach achieves strong performance by grounding answers in explicit and verifiable reasoning paths.

Keywords: Multi-hop Question Answering · Retrieval-Augmented Generation · Large Language Models

1 Introduction

Multi-hop Question Answering (QA) is a core challenge in natural language understanding, requiring models to synthesize information spread across multiple documents to answer complex queries. For example, answering "Who is the mother of the director of the movie 'The War of Poland and Russia'?" involves identifying the director and then retrieving facts about their mother—two facts rarely co-located in a single source. This necessitates precise multi-step reasoning and integration of cross-document evidence.

Large Language Models (LLMs) have demonstrated strong performance in question answering by leveraging their internal knowledge and reasoning capabilities. However, their reliance on static, pre-trained parameters inherently limits their ability to handle dynamic or specialized knowledge, often resulting in

M. Yoshikawa et al. (Eds.): ADMA 2025, LNAI 16197, pp. 451–458, 2026.
https://doi.org/10.1007/978-981-95-3453-1_35

factual inaccuracies or hallucinations—particularly for knowledge-intensive or time-sensitive queries. To overcome these limitations, Retrieval-Augmented Generation (RAG) [1] has emerged as a promising paradigm, where relevant passages are retrieved from an external corpus and used to condition the generation process. By grounding answers in explicit evidence, RAG improves both the factual accuracy and trustworthiness of LLM outputs.

However, applying RAG to multi-hop QA reveals a fundamental tension between evidence breadth and reasoning depth. Ensuring sufficient breadth by retrieving all potentially relevant facts can lead to a noisy and redundant evidence set that degrades the quality of the model's generated answers [2]. On the other hand, emphasizing depth through strict step-by-step reasoning may result in a narrow search that is vulnerable to early errors and may overlook crucial but non-sequential evidence [3].

To mitigate these challenges, prior work explores three primary strategies. Iterative retrieval [4,5] builds reasoning chains step-by-step, but its myopic, greedy nature makes it highly susceptible to error propagation; a single early misstep can derail the entire reasoning path. Query decomposition [6,7] effectively improves recall by generating sub-questions, but in doing so, it flattens the problem's logical structure, burdening the final LLM with a noisy and unstructured evidence set. Meanwhile, others incorporate structured knowledge graphs [8], yet these pre-compiled structures are often query-agnostic and lack the flexibility to handle the specific relational nuances of novel, open-domain questions.

To address the fundamental tension between evidence breadth and reasoning depth in multi-hop QA, we propose Scope-then-Traverse, a novel two-stage framework. Our approach systematically tackles this challenge through a series of coordinated technical contributions. The main contributions of this work are summarized as follows:

- A Decoupled Two-Stage Architecture: We propose a new framework that decouples the QA process into two specialized stages: Knowledge Scoping for achieving evidence breadth, and Path-Aware Traversal for ensuring reasoning depth. This architecture provides a principled solution to the inherent trade-off.
- Blueprint-Guided Scoping Mechanism: For the Scoping stage, we introduce a mechanism that first generates a complete logical blueprint of the query. It then builds a high-recall, low-noise evidence pool by retrieving only with the high-certainty, entity-centric independent sub-questions, thus mitigating the risk of early commitment errors.
- Fused Scoring for Path-Aware Traversal: Central to our Traversal stage is a novel fused scoring mechanism. This function guides the construction of a coherent reasoning path by being the first to jointly optimize for both forward-looking local query relevance and backward-looking global path cohesion.

We conduct extensive experiments on three challenging benchmarks, demonstrating that our approach achieves strong performance, particularly on tasks requiring deep compositional reasoning.

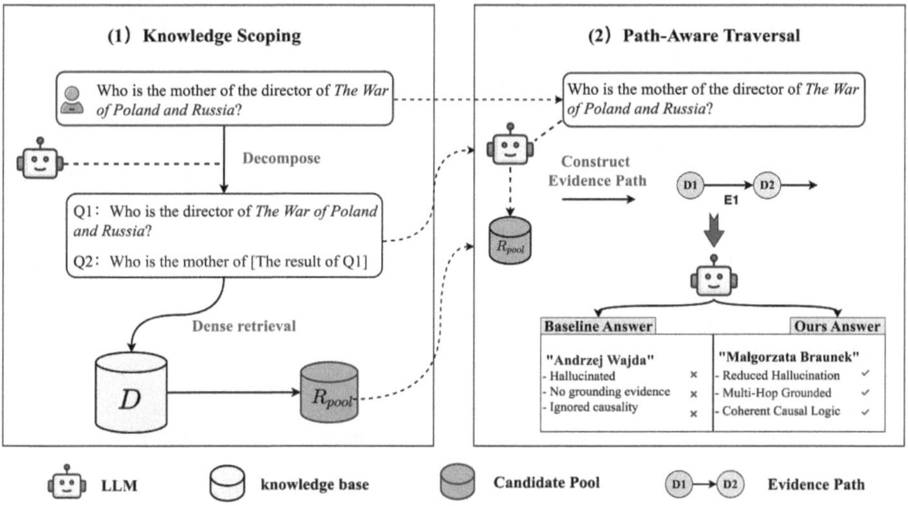

Fig. 1. An overview of our proposed two-stage framework. (1) Knowledge Scoping: The complex question is decomposed into independent and dependent sub-questions. Independent ones are executed in parallel to build a high-recall candidate pool. (2) Path-Aware Traversal: An evidence path is iteratively constructed within the pool by generating anticipatory queries and selecting the next node via a fused scoring mechanism.

2 Method

Our framework decouples the multi-hop QA process into two distinct stages, Knowledge Scoping and Path-Aware Traversal, each designed to address one half of the breadth-depth dilemma. The Scoping stage is optimized for breadth, building a comprehensive evidence pool. The Traversal stage is optimized for depth, constructing a coherent reasoning path within that pool, as illustrated in Fig. 1. The overall process is detailed in Algorithm 1.

2.1 Knowledge Scoping

1. Hierarchical Query Decomposition. A core challenge in multi-hop QA is that a single query vector is often insufficient. To address this, we leverage an LLM to act as a query analyst. It first decomposes the original complex query q into a logical sequence of sub-questions. Crucially, the LLM then categorizes these into two types:

- **Independent Sub-questions (e.g., Q1):** Answerable without prior context (e.g., "Who directed'The War of Poland and Russia'?").
- **Dependent Templates (e.g., Q2):** Containing slots that rely on earlier answers (e.g., "Who is the mother of [the result of Q1]?").

Algorithm 1. Scope-then-Traverse: A Two-Stage Multi-Hop QA Framework

Input: Complex question q
Output: Final answer A
— Stage 1: Knowledge Scoping —
$B \leftarrow$ LLM.Decompose(q) ▷ Decompose into a structured blueprint B
$R_{\text{pool}} \leftarrow \bigcup_{i \in B.\text{indep\_queries}}$ Rerank(DenseRetrieve(i)) ▷ Retrieve using only independent queries from the blueprint
— Stage 2: Path-Aware Traversal —
$v_0 \leftarrow$ SelectAnchorNode(B, R_{pool}) ▷ Select node most relevant to entities from independent queries
$P \leftarrow [(v_0, \text{"START"})]$
$V_{\text{visited}} \leftarrow \{v_0.\text{id}\}$
$t \leftarrow 0$
while $t < T_{\max}$ **and** $t <$ len(B.dep_templates) **do**
  $template_t \leftarrow B$.dep_templates[t] ▷ Get the next dependent template from the ordered blueprint
  $d_{t \to t+1} \leftarrow$ LLM.GenerateQuery($q, P, template_t$) ▷ Instantiate template; may return [STOP]
  if $d_{t \to t+1} =$ "[STOP]" **then**
    break
  end if
  $C_t \leftarrow$ RetrieveCandidates($d_{t \to t+1}, R_{\text{pool}}, V_{\text{visited}}$)
  if C_t is empty **then**
    break
  end if
  $v_{t+1} \leftarrow \arg\max_{d \in C_t} ((1 - \beta) \cdot \text{Score}_{\text{local}}(d) + \beta \cdot \text{Score}_{\text{cohesion}}(d))$
  Append $(v_{t+1}, d_{t \to t+1})$ to P
  Add v_{t+1}.id to V_{visited}
  $t \leftarrow t + 1$
end while
# — Final Answer Generation —
$A \leftarrow \text{LLM}_{\text{gen}}(q, P)$ ▷ Generate answer from the structured evidence path
return A

Together, these sub-questions form a structured blueprint that guides the subsequent path-aware traversal.

2. Targeted Parallel Retrieval. The design of our scoping strategy is centered on maximizing evidence breadth while minimizing retrieval noise. During this stage, only the independent sub-questions are executed in parallel for retrieval. The rationale for this targeted approach is two-fold: 1) Focusing on High-Certainty Anchors: Independent sub-questions typically target the stable, named entities central to the query (e.g., a specific person or movie). Retrieving based on these focused, low-ambiguity queries is more precise and effective than using the original, often multifaceted query. This ensures that documents concerning the core reasoning anchors are effectively retrieved. 2) Postponing Commitment to Mitigate Risk: Unlike iterative methods that must commit to a single document at the first step, our approach gathers a superset of potentially relevant evidence. This postpones the path selection decision to the Traversal stage, creating a safe search space and mitigating the risk of irrecoverable early-stage errors. The retrieved documents are then aggregated and deduplicated to form a unified candidate pool, R_{pool}. This query-specific "micro-knowledge-base" provides a high-recall foundation for the traversal stage, free from the noise of ambiguous dependent queries.

2.2 Path-Aware Traversal

1. Iterative Path Construction via Fused Scoring. To achieve reasoning depth, our traversal process is guided by a fused scoring mechanism designed to balance two critical objectives. This mechanism addresses both the "forward-looking" and "backward-looking" needs of reasoning: 1) Forward-Looking Informativeness: the next node must be locally relevant to the immediate query $d_{t\to t+1}$ to ensure progress. This is measured by our Local Relevance Score. 2) Backward-Looking Coherence: the node must also be globally coherent with the existing path to prevent semantic drift. This is measured by our Path Cohesion Score.

By jointly optimizing these two aspects, our framework constructs a path that is not only a sequence of locally relevant documents but a globally coherent narrative. The specific scores are defined as follows:

- **Local Relevance Score ($\text{Score}_{\text{local}}$):** This score measures how well a candidate document answers the current anticipatory query $d_{t-1\to t}$. It is obtained through a two-stage process—first using a dense retriever to retrieve candidates from R_{pool}, followed by reranking with a cross-encoder to reflect fine-grained semantic alignment.
- **Path Cohesion Score ($\text{Score}_{\text{cohesion}}$):** This score evaluates semantic coherence between a candidate document d and the accumulated path (v_0, $\ldots, v_{t-1}$) using an exponentially decayed similarity:

$$\text{Score}_{\text{cohesion}}(d, P_{t-1}) = \mathcal{N} \sum_{j=0}^{t-1} \gamma^{(t-1)-j} \cdot \text{sim}(\text{Enc}(d), \text{Enc}(v_j)), \tag{1}$$

  where $\gamma \in (0, 1]$ controls the decay rate, $\mathcal{N}$ is a normalization constant, and $\text{sim}(\cdot,\cdot)$ denotes a similarity function measuring the semantic relatedness between two encoded documents.

The next node is selected by a weighted fusion of the two scores:

$$v_t = \arg\max_{d\in C_t} \left((1-\beta)\cdot \text{Score}_{\text{local}}(d) + \beta \cdot \text{Score}_{\text{cohesion}}(d)\right), \tag{2}$$

2. Self-adaptive Termination. Our framework employs dynamic termination. After each step, an LLM evaluator assesses if the path P_t is sufficient to answer q and issues a [STOP] signal if so. This prompt-driven, self-adaptive termination ensures the evidence path is both logically minimal and semantically sufficient.

Table 1. Main results (%) on the MuSiQue, 2WikiMultiHopQA, and HotpotQA test sets. We compare our framework against several RAG baselines. The best score in each column is highlighted in **bold**.

Method	MuSiQue		2Wiki		HotpotQA	
	F1	EM	F1	EM	F1	EM
Naive RAG (BM25)	22.82	15.01	50.61	42.50	56.92	42.00
Naive RAG (BGE)	24.15	16.64	52.34	44.18	58.05	43.51
QD-RAG [6]	38.72	35.56	54.15	51.89	61.44	48.88
ITER-RETGEN [4]	40.59	36.11	57.88	49.60	61.17	47.23
Self-RAG [5]	44.12	36.50	63.47	55.20	**69.71**	55.53
RQ-RAG [7]	46.88	41.43	68.37	60.01	69.38	**55.85**
Ours (Scope-then-Traverse)	**49.48**	**42.70**	**70.04**	**62.76**	69.43	55.70

3. Answer Generation from the Evidence Path. The constructed evidence path is not a flat set of documents, but a structured sequence that captures both evidential content and the inferential steps linking them. Once the traversal is complete, the final answer is synthesized by an LLM conditioned on this full structured path. The context provided to the generator includes the original question q, the sequence of retrieved evidence nodes $\{v_0, \ldots, v_T\}$, and the intermediate anticipatory query $\{d_{i \to i+1}\}$ that encode the logical transitions:

$$\text{answer} = \text{LLM}_{\text{gen}}(q, v_0, d_{0 \to 1}, v_1, \ldots, v_T) \quad (3)$$

Grounding the generation process in this explicit evidence path enables the model to produce accurate and interpretable answers, closely aligned with the multi-hop logic of the question.

3 Experiments

3.1 Setup

We evaluate our method on three multi-hop QA benchmarks: HotpotQA [9], MuSiQue [10], and 2WikiMultiHopQA [11], using 5,000 randomly sampled questions from each dataset. Standard EM and F1 metrics are used. For a fair comparison, all methods use Llama-3.3-70B-Instruct as the LLM, with BGE-large-en-v1.5 and BGE-reranker-large for retrieval and reranking. Baselines include Naive RAG (BM25, BGE), iterative methods (ITER-RETGEN [4], Self-RAG [5]), and decomposition-based approaches (QD-RAG [6], RQ-RAG [7]).

3.2 Results and Analysis

The main results (Table 1) show notable performance differences across benchmarks. On MuSiQue and 2WikiMultiHopQA, our framework achieves the highest F1 and EM scores. These datasets, which require complex compositional

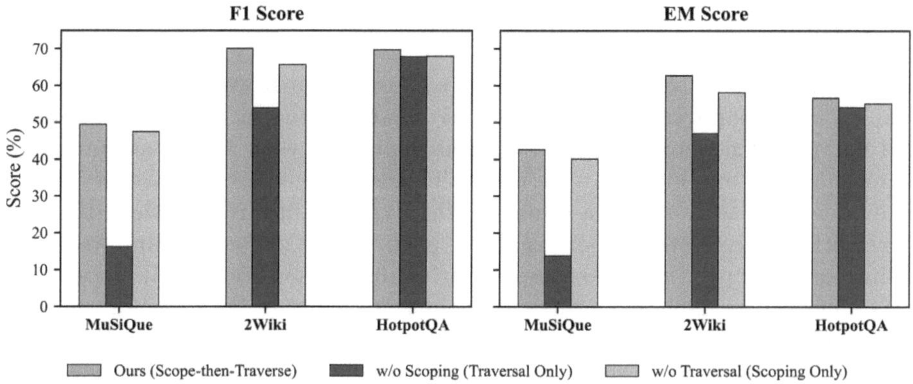

Fig. 2. Ablation study results (EM and F1 scores) across three datasets, demonstrating the necessity of both the Scoping and Traversal stages.

reasoning across disjoint sources, highlight the strength of our two-stage design: Knowledge Scoping ensures comprehensive evidence coverage, while Path-Aware Traversal organizes it into a coherent reasoning chain to ground generation. On HotpotQA, our model performs competitively with state-of-the-art baselines (Self-RAG, RQ-RAG), with tightly clustered scores. This dataset contains fewer complex compositional questions and more direct ones; this reduces the relative advantage of our two-stage structure, as optimized iterative/refinement methods already handle such cases effectively. These findings underscore a key insight: the complexity and structure of a dataset's reasoning requirements are critical for determining the most suitable retrieval-augmented generation architecture.

3.3 Ablation Studies

To validate the contribution of each stage, we conducted two ablations (Fig. 2).

- **w/o Scoping (Stage 1):** Removing the knowledge scoping module significantly impairs model performance, especially on tasks requiring compositional reasoning. This indicates that comprehensive evidence breadth is crucial for multi-hop QA, laying the foundation for subsequent reasoning.
- **w/o Traversal (Stage 2):** Removing the path traversal stage and directly feeding the entire candidate pool into the generation model also leads to consistent performance drops across all datasets. Although powerful language models can perform a certain degree of implicit reasoning on "flattened" evidence, the lack of a structured reasoning path (reasoning depth) undermines reasoning accuracy—this impact is particularly pronounced in tasks requiring complex logical integration.

4 Conclusion and Future Work

In this paper, we introduced a two-stage "scope-then-traverse" framework for complex multi-hop question answering. By first building a high-recall evidence pool via query decomposition and then navigating it with a path-aware traversal mechanism, our method systematically addresses the challenge of balancing evidence breadth with reasoning depth. Experimental results show that our framework achieves strong performance, particularly on benchmarks demanding intricate compositional reasoning. Our analysis also reveals that the optimal RAG architecture may be contingent on the complexity of the task. Future work could explore dynamic, adaptive frameworks that can select between an explicit traversal strategy and a simpler evidence-aggregation approach based on the predicted difficulty of a given question.

Acknowledgements. This work was supported by the National Natural Science Foundation of China under Grant No. 62171047.

References

1. Lewis, P., Perez, E., Piktus, A., et al.: Retrieval-augmented generation for knowledge-intensive NLP tasks[J]. Adv. Neural. Inf. Process. Syst. **33**, 9459–9474 (2020)
2. Cuconasu, F., Trappolini, G., Siciliano, F., et al.: The power of noise: redefining retrieval for rag systems [C]. In: Proceedings of the 47th International ACM SIGIR Conference on Research and Development in Information Retrieval, pp. 719-729 (2024)
3. Mavi, V., Jangra, A., Jatowt, A.: Multi-hop question answering [J]. Found. Trends® Inf. Retrieval **17**(5), 457-586 (2024)
4. Shao, Z., Gong, Y., Shen, Y., et al.: Enhancing retrieval-augmented large language models with iterative retrieval-generation synergy [C]. Find. Assoc. Comput. Linguist. EMNLP 2023, 9248-9274 (2023)
5. Asai, A., Wu, Z., Wang, Y., et al.: Self-rag: Learning to retrieve, generate, and critique through self-reflection[C]. The Twelfth International Conference on Learning Representations (2023)
6. Ammann, P.J.L., Golde, J., Akbik, A.: Question decomposition for retrieval-augmented generation [J]. arXiv preprint arXiv:2507.00355 (2025)
7. Chan, C.M., Xu, C., Yuan, R., et al.: RQ-RAG: learning to refine queries for retrieval augmented generation [J]. arXiv preprint arXiv:2404.00610 (2024)
8. Edge, D., Trinh, H., Cheng, N., et al.: From local to global: a graph RAG approach to query-focused summarization [J]. arXiv preprint arXiv:2404.16130 (2024)
9. Yang, Z., Qi, P., Zhang, S., et al.: HotpotQA: a dataset for diverse, explainable multi-hop question answering [J]. arXiv preprint arXiv:1809.09600 (2018)
10. Trivedi, H., Balasubramanian, N., Khot, T., et al.: MuSiQue: Multihop Questions via Single-hop Question Composition[J]. Trans. Assoc. Comput. Linguist. **10**, 539–554 (2022)
11. Ho, X., Nguyen, A.K.D., Sugawara, S., et al.: Constructing a multi-hop QA dataset for comprehensive evaluation of reasoning steps [J]. arXiv preprint arXiv:2011.01060 (2020)

A Multi-descriptor Stacking-Based Framework for Parkinson's Disease Detection from Handwriting

Sana Trigui[1](✉), Hala Bezine[1], and Basant Agarwal[2]

[1] REGIM-Lab: Research Group on Intelligent Machines Laboratory, National School of Engineers, University of Sfax, BP 1173, 3038 Sfax, Tunisia
sana.trigui@enis.tn, hala.bezine@ieee.org

[2] Department of Computer Science and Engineering, Central University of Rajasthan, Bandar Seendri, India
basant@curaj.ac.in

Abstract. Neurodegenerative disorders such as Parkinson's Disease (PD) are progressive neurological conditions that severely affect both motor and cognitive abilities, often manifesting through distinctive handwriting abnormalities. Early and accurate detection of PD is essential to enable timely therapeutic intervention and to improve patient outcomes. In this study, we propose a hybrid feature extraction framework for PD detection based on offline handwriting analysis. Our approach integrates Fourier Transform (FT), Histogram of Oriented Gradients (HOG), and Discrete Wavelet Transform (DWT) to capture both global structures and fine-grained variations in handwriting. To boost classification accuracy, a stacking ensemble model is adopted, combining multiple machine learning classifiers to exploit their complementary predictive capabilities. Experimental results on a PD-specific handwriting dataset validate the proposed method's effectiveness, achieving an average classification accuracy of 98.49% across multiple tasks. These findings underscore the promise of offline handwriting analysis as a non-invasive, cost-efficient, and scalable diagnostic aid for neurodegenerative diseases.

Keywords: Parkinson's disease · Handwriting analysis · Feature fusion · Stacking ensemble · Machine learning

1 Introduction

Neurodegenerative diseases, particularly Parkinson's Disease (PD), are becoming increasingly prevalent due to global population aging. PD is a chronic and progressive neurological disorder caused by the degeneration of dopaminergic neurons in the substantia nigra, leading to both motor and non-motor impairments. It is the second most common neurodegenerative disease after Alzheimer's, affecting over 10 million people worldwide [1], with prevalence rising significantly in individuals over 60 [2].

Clinically, PD manifests through hallmark motor symptoms such as resting tremors, bradykinesia, rigidity, and postural instability [3], as well as non-motor symptoms

M. Yoshikawa et al. (Eds.): ADMA 2025, LNAI 16197, pp. 459–467, 2026.
https://doi.org/10.1007/978-981-95-3453-1_36

including cognitive decline, depression, sleep disturbances, and autonomic dysfunctions [4]. Early diagnosis remains challenging, particularly because non-motor symptoms often precede visible motor impairments. Current diagnostic approaches rely on clinical assessment and neuroimaging modalities such as MRI. Additionally, recent research has explored non-invasive alternatives like speech analysis. However, these methods require specialized equipment, limiting their use in low-resource settings. In contrast, handwriting analysis has emerged as a promising tool, capturing both motor and cognitive alterations through the study of writing patterns. Offline handwriting analysis, which involves scanned or photographed samples, is particularly appealing due to its low cost, non-invasiveness, and accessibility.

Recent advances in pattern recognition have enabled the development of automatic PD detection systems using either handcrafted or deep learning-based features. Among handcrafted descriptors, Histogram of Oriented Gradients (HOG), Fourier Transform (FT), and Wavelet Transform (WT) have shown effectiveness in extracting discriminative handwriting traits [5]. However, relying on a single descriptor may overlook important information. To address this, ensemble methods such as stacking have been proposed to combine multiple feature sets and classifiers for improved accuracy. The remainder of this paper is organized as follows: Sect. 2 presents a review of related works on handwriting-based PD detection. Section 3 details the proposed methodology. Section 4 describes the experimental setup and reports the evaluation results. Finally, Sect. 5 concludes the paper and outlines future research directions.

2 Related Works

Handwriting analysis is a promising, non-invasive tool for PD detection. Early studies primarily used traditional machine learning pipelines that extract handcrafted features from handwriting samples, followed by classification with standard algorithms. Common descriptors include Histogram of Oriented Gradients (HOG), which captures edge orientations and structural details, as well as Fourier Transform (FT) and Wavelet Transform (WT), which extract frequency-domain and multiresolution features, respectively. These features are typically classified using Support Vector Machines (SVM) or Random Forests (RF), yielding encouraging results. For example, Das et al. [6] combined Discrete Wavelet Transform (DWT) with HOG features, achieving 97.8% accuracy. Gupta and Chanda [7] applied FT-based features on spiral drawings, reaching 81.66% accuracy with an SVM classifier. Ranjan et al. [8] also confirmed the effectiveness of HOG with classical classifiers, with accuracies above 85%. Comparative evaluations by Sree Sai et al. [9] highlighted Random Forest as a top-performing classifier in this domain. Moetesum et al. [10] utilized a pre-trained AlexNet, achieving 83% accuracy on the PaHaW dataset. Gazda et al. [11] fine-tuned several CNN architectures, reaching 92.7% accuracy on NewHandPD. Kamran et al. [12] merged multiple datasets and applied transfer learning, achieving an impressive 99.22% accuracy. Pereira et al. [15] proposed a deep learning approach focusing on micrographia by analyzing spiral and meander drawings with Convolutional Neural Networks (CNNs). Haddadi et al. introduced a deep learning model optimized with the Harris Hawks Optimization (HHO) algorithm, outperforming several pre-trained models (e.g., AlexNet, ResNet) with 94.12% accuracy, and achieving 100% accuracy when averaging across circle, spiral, and meander tasks. Despite their superior performance, deep learning methods face challenges such as limited large-scale annotated handwriting datasets, high computational demands, and limited interpretability, which may hinder clinical adoption. These limitations emphasize the continued relevance of handcrafted feature-based approaches, especially when combined with ensemble learning techniques like stacking, which offer a balanced trade-off between interpretability, robustness, and accuracy.

3 Proposed Method

The proposed handwriting analysis pipeline is specifically developed to detect the presence or absence of PD by integrating complementary feature extraction techniques with an ensemble machine learning approach. As illustrated in Fig. 1, the framework follows a structured sequence of stages: preprocessing, feature extraction, feature concatenation and normalization, feature selection, and classification. Each component is carefully designed to enhance the discriminative power of handwriting features and improve classification accuracy. The subsequent subsections provide a detailed explanation of each stage.

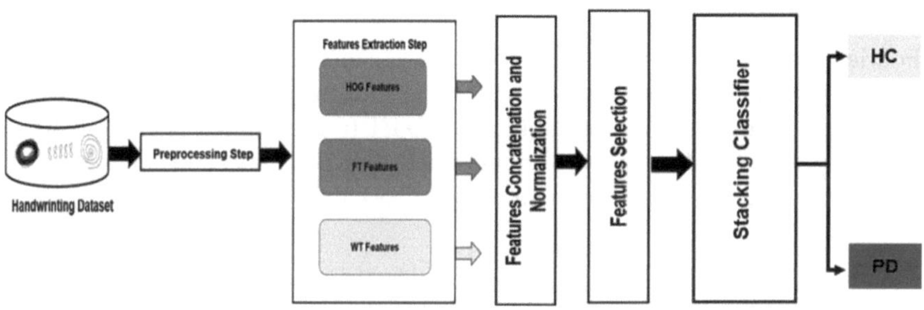

Fig. 1. The architecture of the proposed model

3.1 Processing Step

This study uses the Parkinson's Corpus, containing offline and online handwriting samples from 30 PD patients and 30 healthy controls, each performing five standardized tasks [13]. We focus on offline data due to its accessibility and suitability for low-resource settings, as it requires only scanned images without special hardware. Offline handwriting captures key PD motor impairments like tremors and irregularities, detectable via image analysis. Preprocessing involves resizing images to 128 $\times$ 128 pixels to standardize and reduce computation, converting to grayscale to retain relevant intensity information, and normalizing pixel values to [0, 1] for uniformity and improved model stability. These steps ensure consistent, ready-to-use data for feature extraction and classification.

3.2 Features Extraction

The feature extraction process is a cornerstone of our methodology, enabling the identification of distinctive and relevant information from handwriting images. We employ a combination of advanced techniques carefully selected to capture both global structures and fine-grained details (Fig. 2).

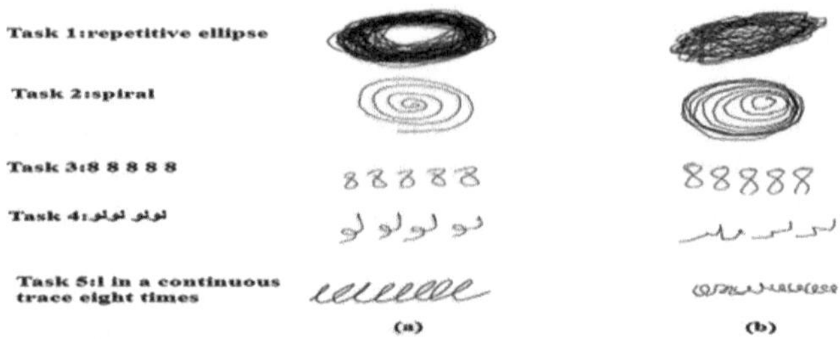

Fig. 2. Examples of tasks completed by: (a) HC participants (b) PD patients.

Fourier Transform (FT) Features

The Fourier Transform (FT), introduced by Fourier in 1822, converts spatial data into the frequency domain, decomposing handwriting into frequency components to detect tremors, irregular strokes, and instability. The 2D FT further enhances analysis by identifying both global trends (e.g., stroke curvature, directionality) and fine details (e.g., micro-tremors, jagged edges). Given an image f(x,y), the 2D FT is:

$$F(u, v) = \sum_{x=0}^{M-1} \sum_{y=0}^{N-1} f(x, y).e^{-2\pi i(\frac{ux}{M} + \frac{vy}{N})} \tag{1}$$

The magnitude spectrum |F(u,v)|, computed from the real and imaginary parts of F(u,v), reflects the energy distribution across frequencies. From this spectrum, we extract statistical descriptors including mean, standard deviation, entropy, energy, and extrema values (max/min magnitudes), which capture handwriting complexity. For implementation, we used the 2D Discrete Fourier Transform (DFT) via the Fast Fourier Transform (FFT).

The Histogram of Oriented Gradients (HOG) Features

The Histogram of Oriented Gradients (HOG) is a feature extraction technique widely used in handwriting and object analysis. It captures local gradient patterns that reveal details like tremors or stroke irregularities, useful for detecting motor disorders such as Parkinson's disease. Gradients are computed, orientation histograms are built for localized regions, and normalized across blocks for invariance to illumination and contrast. In our implementation, we used the following configuration: Cell size: 8 × 8 pixels, Block size: 2 × 2 cells (16 × 16 pixels), Block stride: 8 pixels, Number of orientation bins: 9. The resulting descriptors were concatenated into a fixed-length vector for classification.

Discrete Wavelet Transform (DWT) Features
The DWT extracts multi-scale features from handwriting by decomposing images into approximation (LL) and details (LH, HL, HH) sub-bands. This captures fine and coarse patterns related to PD, such as tremors and stroke irregularities. In practice, grayscale images are decomposed using a multi-level 2D wavelet, and all coefficient matrices are flattened and concatenated into a single feature vector. Mathematically, the DWT of an image I(x,y) can be expressed as:

$$W(a,b) = \frac{1}{\sqrt{a}}\sum\nolimits_m\sum\nolimits_n I(x_m, y_n).\psi(\frac{x-b_x}{a}, \frac{y-b_y}{a}) \tag{2}$$

where: I(x,y) is the image, (x_m, y_n) are the coordinates of the pixel in the image, ψ is the 2D wavelet function, are the position parameters. In practice, we employed the Haar wavelet family with three decomposition levels. Approximation (LL) and detail coefficients (LH, HL, HH) at each level were flattened and concatenated. The final feature vector combines frequency and spatial details across scale

3.3 Features Concatenation and Normalization

To build a robust handwriting representation, features extracted from different methods (e.g., Fourier Transform, HOG, Wavelet) are first converted into one-dimensional vectors, then concatenated into a single feature vector. This fusion captures complementary information. To ensure all features have equal influence during classification, Min-Max normalization is applied, scaling values to the [0, 1] range.

3.4 Features Selection

To improve classification accuracy and reduce redundancy, Recursive Feature Elimination with Cross-Validation (RFECV) is used. This method iteratively removes the least important features based on their contribution to model performance, using cross-validation to ensure generalizability. Unlike filter methods, RFECV is model-aware and accounts for feature interactions, making it well-suited for high-dimensional fused features. It results in a compact and effective feature subset, enhancing both efficiency and robustness of the PD detection system.

3.5 Classification Step

In the final stage, a stacking ensemble combines multiple classifiers SVM, Random Forest, and Logistic Regression to improve PD detection. Each is trained separately, and their predictions are passed to a Gradient Boosting meta-classifier, which learns how to best combine them for optimal final classification.

4 Experimentation and Results

To thoroughly evaluate the effectiveness and robustness of the proposed method, we conducted a series of experiments in three stages. First, we assess the performance of our approach on the Parkinson's Corpus previously described. Then, we extend the evaluation to two additional public datasets HandPD [17] and NewHandPD [18] to validate the generalizability of our framework. Finally, we compare our results with several state-of-the-art approaches that have also used the same public dataset, enabling a consistent and fair performance comparison.

4.1 Evaluation on the Parkinson's Corpus

The evaluation of different feature extraction methods on the Parkinson's dataset reveals varying performance levels depending on the descriptors used. As reported in Table 1, the Fourier Transform (FT) consistently achieves high accuracy, reaching 96.67% in Task 1 and maintaining strong performance across the board. In contrast, the Histogram of Oriented Gradients (HOG) yields comparatively lower results, with accuracy dropping to 84.44% in Task 5, suggesting that HOG alone may be less effective in capturing Parkinson-related handwriting impairments. The Wavelet Transform (WT) offers intermediate performance, often approaching that of FT, with scores such as 95.56% in Task 3. Importantly, the combination of feature types systematically enhances the classification performance. For instance, combining FT and WT yields 97.30% in Task 3, outperforming either method used individually. A further improvement is observed when all three features are fused. The TF + HOG + WT combination reaches a perfect 100% accuracy in Task 3, and exceeds 96% across all tasks. This confirms the advantage of a hybrid feature extraction approach.

Table 1. Evaluation of learning features using PD dataset

Features	Task 1	Task 2	Task 3	Task 4	Task 5
TF	96.67	95.56	90.12	93.54	92.78
HOG	91.11	87.22	95.17	90.56	84.44
WT	91.08	95.31	95.56	92.00	92.00
TF + HOG	97.00	96.00	96.50	95.50	94.80
TF + WT	97.20	96.50	97.30	96.80	95.40
HOG + WT	95.50	95.00	98.20	95.30	95.10
TF + HOG + WT	98.00	96.67	100.00	99.44	98.33

4.2 Evaluation on HandPD and NewHandPD Datasets

To test our framework's robustness, we evaluated it on two public datasets: HandPD (35 subjects) [17] and NewHandPD (66 subjects) [17], covering tasks like spirals, meanders, and circles. Table 2 compares our results with those of Haddadi et al. [14] and Pereira et al. [15]. Our method consistently achieves higher accuracy, especially in subtle motor impairment tasks. For example, in the meander task, we reach 97.93% (HandPD) and 98.62% (NewHandPD), outperforming Haddadi (90.3%) and Pereira (83.11%). In the spiral task, we achieve 96.06% and 98.36%, exceeding their 91% and 90.38%. For the circle task, our accuracy is 94.53%, compared to Haddadi's 100%. While their deep learning approach with Harris Hawks Optimization performs well on simple shapes, it struggles with complex patterns, and Pereira's CNN shows moderate results without advanced optimization. Moreover, Pereira did not evaluate some specific tasks, focusing on overall classification. This highlights our method's strength in providing detailed, task-specific, and consistently high performance across handwriting tasks.

Table 2. Comparison with related works

Datasets	Task	Our Method	Haddadi et Arabani [18]	Pereira et al. [19]
HandPD	meander	**97.93**	90.3	83.11
	spiral	**96.06**	91	90.38
NewHandPD	Circle	**94.53**	100	–
	spiral	**98.36**	98.07	–
	meander	**98.62**	96.15	–

4.3 Comparison with Prior Work Using the Same Dataset

In the final evaluation, we compare our proposed offline handwriting method with previous work on the same Parkinson's Corpus dataset [13] by Allebawi et al. [16] Their online handwriting approach, combining a Beta-elliptical model, fuzzy perceptual detector, and Bi-LSTM classifier, achieved accuracies of 93.33% (Task 1: Repetitive Ellipse), 89.99% (Task 2: Spiral), and 86.66% (Task 3: "88888") [16]. In contrast, our stacking ensemble method using handcrafted features (FT, HOG, and DWT) on offline data outperforms these results across all tasks. This demonstrates that a carefully designed offline system, without relying on dynamic signals like pen pressure or velocity, can match or exceed complex online models for Parkinson's disease detection (Fig. 3).

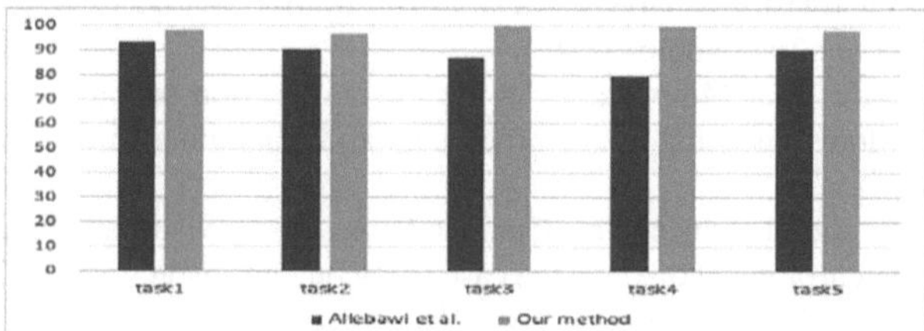

Fig. 3. Performance comparison between Allebawi et al.'s method and our method across five tasks

5 Conclusion

This study demonstrates the effectiveness of a hybrid feature extraction framework combining Fourier FT, HOG, and DWT for PD detection from offline handwriting samples. Using a stacking ensemble classifier, the proposed method achieves higher accuracy and robustness in identifying handwriting anomalies compared to existing techniques. However, the reliance on offline data excludes dynamic features such as pen pressure and stroke velocity, which could further improve results. Future work should integrate online handwriting data, explore deep learning feature extraction methods, and expand datasets to include diverse populations for better generalization. Overall, this research highlights the promise of automated handwriting analysis as a scalable and non-invasive tool for early diagnosis, bridging machine learning and clinical practice.

References

1. Poewe, W., Seppi, K., Tanner, C., et al.: Parkinson disease. Nat. Rev. Dis. Primers. **3**, 17013 (2017). https://doi.org/10.1038/nrdp.2017.13
2. Pringsheim, T., Jette, N., Frolkis, A., Steeves, T.D.L.: The prevalence of Parkinson's disease: a systematic review and meta-analysis. Mov. Disord. **29**(13), 1583–1590 (2014). https://doi.org/10.1002/mds.25945
3. Jankovic, J.: Parkinson's disease: clinical features and diagnosis. J. Neurol. Neurosurg. Psychiatry **79**(4), 368–376 (2008). https://doi.org/10.1136/jnnp.2007.131045
4. Aarsland, D., Andersen, K., Larsen, J.P., Lolk, A., Kragh-Sørensen, P.: Prevalence and characteristics of dementia in Parkinson disease: an 8-year prospective study. Arch. Neurol. **60**(3), 387–392 (2003). https://doi.org/10.1001/archneur.60.3.387
5. Akter, L.: Early identification of Parkinson's disease from hand-drawn images using histogram of oriented gradients and machine learning techniques. In: 2020 Emerging Technology in Computing, Communication and Electronics (ETCCE), Bangladesh, pp. 1–6 (2020). https://doi.org/10.1109/ETCCE51779.2020.9350870
6. Das, H.S., Das, A., Neog, A., Mallik, S., Bora, K., Zhao, Z.: Early detection of Parkinson's disease using fusion of discrete wavelet transformation and histograms of oriented gradients. Mathematics **10**(22) (2022). https://doi.org/10.3390/math10224218
7. Das Gupta, J., Chanda, B.: Novel features for diagnosis of Parkinson's disease from off-line archimedean spiral images. In: 2019 IEEE 10th International Conference on Awareness Science and Technology (iCAST), pp. 1–6. IEEE (2019). https://doi.org/10.1109/ICAwST.2019.8923159
8. Ranjan, N.M., Mate, G., Bembde, M.: Detection of Parkinson's disease using machine learning algorithms and handwriting analysis. J. Data Min. Manag. **8**(1), 21–29 (2023). https://doi.org/10.46610/jodmm.2023.v08i01.004
9. Sai, K.S., Sumanth, T.R., Mounika, T., et al.: Early detection of Parkinson's disease from spiral and wave drawings using image processing and machine learning. Int. J. Res. Trends Innov. **8**(4) (2023)
10. Moetesum, M., Siddiqi, I., Vincent, N., Cloppet, F.: Assessing visual attributes of handwriting for prediction of neurological disorders: a case study on Parkinson's disease. Pattern Recogn. Lett. **121**, 19–27 (2019). https://doi.org/10.1016/j.patrec.2018.04.00
11. Gazda, M., Hires, M., Drotar, P.: Multiple-fine-tuned convolutional neural networks for Parkinson's disease diagnosis from offline handwriting. IEEE Trans. Syst. Man, Cybern. Syst. **52**(1), 78–89 (2022). https://doi.org/10.1109/TSMC.2020.3048892

12. Kamran, I., Naz, S., Razzak, I., Imran, M.: Handwriting dynamics assessment using deep neural network for early identification of Parkinson's disease. Fut. Gener. Comput. Syst. **117**, 234–244 (2021). https://doi.org/10.1016/j.future.2020.11.020
13. Allebawi, M.F., et al.: A new online Arabic handwriting dataset for analyzing Parkinson's disease. In: Proceedings of the International Conference on Cyberworlds (CW). IEEE (2023)
14. Hadadi, S., Arabani, S.P.: A novel approach for Parkinson's disease diagnosis using deep learning and Harris Hawks optimization algorithm with handwritten samples. Multimed. Tools Appl. (2024). https://doi.org/10.1007/s11042-024-18584-3
15. Pereira, C.R., Pereira, D.R., Papa, J.P., Rosa, G.H., Yang, X.-S.: Convolutional neural networks applied for Parkinson's disease identification. In: Holzinger, A., Jurisica, I. (eds.) Machine Learning for Health Informatics, pp. 377–390. Springer, Cham (2016)
16. Allebawi, M.F., Dhieb, T., Jarraya, I., et al.: Parkinson's disease detection from online handwriting based on beta-elliptical approach and fuzzy perceptual detector. IEEE Access. **12**, 56936–56950 (2024). https://doi.org/10.1109/ACCESS.2024.3387342
17. Pereira, C.R., et al.: A step towards the automated diagnosis of Parkinson's disease: analyzing handwriting movements. In: IEEE 28th International Symposium on Computer-Based Medical Systems 171–176 (2015)
18. Pereira, C.R., Weber, S.A.T., Hook, C., Rosa, G.H., Papa, J.P.: Deep learning-aided Parkinson's disease diagnosis from handwritten dynamics. In: 2016 29th SIBGRAPI Conference on Graphics, Patterns and Images (SIBGRAPI 2016), pp. 340–346 (2017). https://doi.org/10.1109/SIBGRAPI.2016.054

Correction to: Domain Graph-Structured Multi-source Domain Adaptation with Dual Integration

Jiayi Wang, Xin Zheng, Yi Li, and Yanqing Guo

Correction to: Chapter 10 in: M. Yoshikawa et al. (Eds.): *Advanced Data Mining and Applications*, LNAI 16197, https://doi.org/10.1007/978-981-95-3453-1_10

In the original version of this chapter, the name of the third and fourth authors were wrong. This has been corrected. Correctly it should read as author name: "Yi Li" & "Yanqing Guo".

The updated version of this chapter can be found at
https://doi.org/10.1007/978-981-95-3453-1_10

M. Yoshikawa et al. (Eds.): ADMA 2025, LNAI 16197, p. C1, 2026.
https://doi.org/10.1007/978-981-95-3453-1_37

Author Index

M. Yoshikawa et al. (Eds.): ADMA 2025, LNAI 16197, pp. 469–471, 2026.
https://doi.org/10.1007/978-981-95-3453-1

MIX
Papier aus verantwortungsvollen Quellen
Paper from responsible sources
FSC® C105338

If you have any concerns about our products,
you can contact us on
ProductSafety@springernature.com

In case Publisher is established outside the EU,
the EU authorized representative is:
Springer Nature Customer Service Center GmbH
Europaplatz 3, 69115 Heidelberg, Germany

Printed by Libri Plureos GmbH
in Hamburg, Germany